Domain Decomposition Methods in Sciences and Engineering

Domain Decomposition Methods in Sciences and Engineering

8th International Conference, Bejing, P. R. China

Edited by

R. Glowinski
University of Houston, Texas, USA

J. Périaux
Dassault Aviation, Saint-Cloud, France

Z-C. Shi
State Key Lab of Scientific & Engineering Computing,
Academia Sinica, Bejing, P. R. China

O. Widlund
Courant Institute of Mathematical Sciences, New York, USA

John Wiley & Sons

Chichester • New York • Weinheim • Brisbane • Singapore • Toronto

Other Wiley Editorial Offices

John Wiley & Sons, Inc., 605 Third Avenue,
New York, NY 10158-0012, USA

VCH Verlagsgesellschaft mbH, Pappelallee 3,
D–69469 Weinheim, Germany

Jacaranda Wiley Ltd, 33 Park Road, Milton,
Queensland 4064, Australia

John Wiley & Sons (Asia) Pte Ltd, 2 Clementi Loop #02-01,
Jin Xing Distripark, Singapore 129809

John Wiley & Sons (Canada) Ltd, 22 Worcester Road,
Rexdale, Ontario M9W 1L1, Canada

British Library Cataloguing in Publication Data

A catalogue record for this book is available from the British Library

ISBN 0 471 96560 X

Produced from camera-ready copy supplied by the authors.
Printed and bound in Great Britain by Bookcraft (Bath) Ltd.
This book is printed on acid-free paper responsibly manufactured from sustainable forestation,
for which at least two trees are planted for each one used for paper production.

Contents

Preface

The Eighth International Conference on Domain Decomposition Methods for Partial Differential Equations took place in Beijing, P. R. China, from May 16 to May 20, 1995. It was organized by the State Key Laboratory of Scientific and Engineering Computing and the Institute of Computational Mathematics, both of the Chinese Academy of Sciences (CAS).

One of the main reasons for organizing this conference in Beijing was the growing popularity of Domain Decomposition Methods in China over the last few years, mostly due to the need for mathematical tools and associated numerical algorithms for efficiently solving Partial Differential Equations of various types on distributed computers and networked workstations. Domain Decomposition Methods for Partial Differential Equations have become topics of intense interest in Scientific and Engineering Computing over recent years because they not only provide computing strategies suitable for high performance computing systems but also can be very effective for the numerical solution of broad classes of large scale problems from Sciences and Engineering. From that point of view, the Domain Decomposition Conferences have become very important components of the Computational Science landscape.

The main goal of this conference was to bring together researchers with different backgrounds, all of whom are working on Domain Decomposition topics, in order to discuss and review the most recent results and also to promote interaction between numerical analysts, applied mathematicians, and computer scientists. The conference featured sixteen invited lectures, 54 contributed and poster presentations, and was attended by about 100 participants.

The topics from the conference ranged from basic theoretical research to industrial applications, including the numerical analysis of domain decomposition methods with or without overlappings, of multigrid and multilevel methods, of domain decompositions with non-matching grids, etc., as well as applications to problems from Fluid and Structural Mechanics, from Electro-Magnetics, and from Petroleum Engineering. Several lectures were devoted to the hardware and software aspects of the implementation of domain decomposition methods on parallel platforms.

These proceedings contain the text of 51 presentations. They have been divided into five parts, namely:

    I.   Basic Domain Decomposition Algorithms
    II.   Domain Decomposition and Multilevel Methods
    III.   Domain Decomposition and Parallel Computing
    IV.   Domain Decomposition Methods for Advection-Diffusion, Transport and Wave Problems
    V.   Domain Decomposition Methods for Flow Problems

We are most grateful to members of the Organizing Committee: Junzhi Cui (CAS), Wenzhuang Cui (CAS) Xiaomel Li (Hunan), Xiancheng Hu (Tsinghua University), Hongci Huang (Hong Kong), Tsi-Min Shi (Hong Kong), Jiachang Sun (CAS), Yaxiang Yuan (CAS), Tianxiao Zhou (Cl, Xi'an). Special thanks go to the following individuals: Dr. Zhenghui Xie, Ms. Shizhen Zhang, and Mr. Guizhang Chen, whose contributions have been essential to the success of the 8th International Conference of Domain Decomposition Methods.

Finally, we would like to acknowledge the support of various institutions and organizations, particularly the Chinese Academy of Sciences, the Science and Technology Commission of China, the Natural National Science Foundation of China, the French Embassy in Beijing, GAMNI/SMAI, Paris, and SGI Computer Systems. This support allowed, in addition, to provide travel support for several participants from Asia, Europe, and the USA.

Roland Glowinski
University of Houston
Houston, Texas, USA

Jacques Périaux
Dassault Aviation
Saint -Cloud, France

Zhong-ci Shi
State Key Lab of Scientific &
Engineering Computing,
Academia Sinica, Beijing, PRC

O. Widlund
Courant Institute of
Mathematical Sciences
New York, USA

June 1996

Part I Basic Domain Decomposition Algorithms

Efficient Preconditioners for Boundary Element Methods and their Use in Domain Decomposition Methods

O. STEINBACH AND W. L. WENDLAND [1]

1 Introduction

Boundary element methods are well suited for partial differential equations with piecewise constant coefficients and for problems in infinite regions. Moreover, domain decomposition methods are a powerful technique for parallelization and coupling with other discretization methods like finite elements. The formulation of boundary element methods for mixed boundary value problems and their use in domain decomposition methods depends on the boundary integral equations used and their discretizations as e.g. collocation or Galerkin methods. We discuss a general approach to solve mixed boundary value problems by preconditioned iterative methods and we give a preconditioning technique resulting from boundary integral equations, too. Common examples are problems in potential theory or in linear elasticity.

Let us consider a selfadjoint and elliptic boundary value problem of 2nd order in a bounded domain $\Omega \subset I\!\!R^n$ $(n = 2, 3)$ with a Lipschitz continuous boundary Γ, which is decomposed into two distinct parts Γ_D and Γ_N, where boundary conditions of Dirichlet and Neumann type, respectively, are given:

$$
\begin{aligned}
L(x)u(x) &= f(x) && \text{for } x \in \Omega, \\
\gamma_0 u(x) &= g(x) && \text{for } x \in \Gamma_D, \\
\gamma_1 u(x) &= h(x) && \text{for } x \in \Gamma_N.
\end{aligned}
\tag{1}
$$

[1] Universität Stuttgart, Mathematisches Institut A, Pfaffenwaldring 57, 70569 Stuttgart, Germany. E–Mail: wendland@mathematik.uni-stuttgart.de.
This work was partially supported by the Priority Research Programme "Boundary Element Methods" of the German Research Foundation DFG under Grant Nb. We 659/8–5.

Domain Decomposition Methods in Sciences and Engineering, edited by R. Glowinski *et al.*
© 1997 John Wiley & Sons, Ltd.

For a non–overlapping domain decomposition

$$\overline{\Omega} = \bigcup_{i=1}^{p} \overline{\Omega}_i \ , \quad \Omega_i \cap \Omega_j = \emptyset \ , \quad \Gamma_{ij} = \Gamma_i \cap \Gamma_j$$

with Lipschitz continuous subdomain boundaries $\Gamma_i = \partial\Omega_i$, we suppose that we are given fundamental solutions of the differential operator in (1) locally, i.e. for all $i = 1, \ldots, p$ there exist functions $E^i(x, y)$ satisfying

$$L(x)\, E^i(x, y) \ = \ \delta(x - y) \cdot \mathbf{I}(x) \quad \text{for } x, y \in \Omega_i \ . \tag{2}$$

Setting

$$u_i(x) \ = \ \gamma_0^i u(x) \ , \quad t_i(x) \ = \ \gamma_1^i u(x) \quad \text{for } x \in \Gamma_i \ ,$$

the solution $u(\cdot)$ of (1) is given by the local representation formulae

$$u(x) \ = \ \int_{\Gamma_i} t_i(y) \cdot \gamma_0^i E^i(x, y)\, ds_y \ - \ \int_{\Gamma_i} u_i(y) \cdot \gamma_1^i E^i(x, y)\, ds_y \ + \ \int_{\Omega_i} f(y) \cdot E^i(x, y)\, dy \tag{3}$$

for $x \in \Omega_i$, where the densities t_i and u_i have to satisfy transmission conditions

$$u_i(x) - u_j(x) \ = \ 0 \ , \quad t_i(x) + t_j(x) \ = \ 0 \quad \text{for } x \in \Gamma_{ij} \tag{4}$$

in addition to the boundary conditions

$$t_i(x) = h(x) \quad \text{for } x \in \Gamma_i \cap \Gamma_N \ , \quad u_i(x) = g(x) \quad \text{for } x \in \Gamma_i \cap \Gamma_D \ .$$

Associated with the fundamental solutions (2), we define common boundary integral operators by

$$(V_i t_i)(x) \ = \ \int_{\Gamma_i} t_i(y) \cdot \gamma_0^i E^i(x, y)\, ds_y \ ,$$

$$(K_i u_i)(x) \ = \ \int_{\Gamma_i} u_i(y) \cdot \gamma_1^i E^i(x, y)\, ds_y \ ,$$

$$(K_i' t_i)(x) \ = \ \int_{\Gamma_i} t_i(y) \cdot \gamma_1^i(x) E^i(x, y)\, ds_y \ ,$$

$$(D_i u_i)(x) \ = \ -\gamma_1^i \int_{\Gamma_i} u_i(y) \cdot \gamma_1^i E^i(x, y)\, ds_y$$

for $x \in \Gamma_i$ and the boundary traces of the volume or Newton potentials by

$$N_j^i(x) \ = \ \gamma_j^i \int_{\Omega_i} f(y) \cdot E^i(x, y)\, dy \tag{5}$$

for $x \in \Gamma_i$ and $j = 0, 1$. By using the Calderon projector, we can write the boundary integral equations resulting from (3) as the overdetermined system

$$\begin{pmatrix} u_i(x) \\ t_i(x) \end{pmatrix} = \begin{pmatrix} \tfrac{1}{2}I - K_i & V_i \\ D_i & \tfrac{1}{2}I + K_i' \end{pmatrix} \begin{pmatrix} u_i(x) \\ t_i(x) \end{pmatrix} - \begin{pmatrix} N_0^i \\ N_1^i \end{pmatrix} f(x) \tag{6}$$

for $x \in \Gamma_i$ and $i = 1, \ldots, p$. The boundary integral operators V_i, K_i, D_i are pseudodifferential operators of orders $-1, 0, 1$, respectively, and their mapping properties on Lipschitz domains are well known [Cos88]. The right hand side includes still volume integrals which require a discretization of Ω, too. In this case, the advantage of dimension reduction by boundary element methods is partly lost. To overcome this drawback, one can homogenize problem (1). In general, the volume integrals (5) are to be replaced by boundary integrals, too. We will discuss such methods in connection with fictitious domain methods coupled with boundary element methods in [SW].

From (6) we find two explicit representations of a Dirichlet–Neumann map, including the so–called Steklov–Poincaré operator S_i:

$$
\begin{aligned}
t_i \;&=\; S_i u_i + \tilde{N}^i \\[2mm]
&=\; V_i^{-1}\left(\tfrac{1}{2}I + K_i\right) u_i \;+\; V_i^{-1} N_0^i \\[2mm]
&=\; \left[\left(\tfrac{1}{2}I + K_i'\right) V_i^{-1}\left(\tfrac{1}{2}I + K_i\right) + D_i\right] u_i - N_1^i + \left(\tfrac{1}{2}I + K_i'\right) V_i^{-1} N_0^i,
\end{aligned}
\tag{7}
$$

which is a selfadjoint, strongly elliptic pseudodifferential operator of order -1, corresponding to the solution of the local Dirichlet problem, where $u_i = g$ on Γ_i is given. For the inverse operation, the Neumann–Dirichlet map is described by

$$
\begin{aligned}
u_i - r_i \;&=\; U_i t_i + \hat{N}^i \\[2mm]
&=\; \left(\tfrac{1}{2}I \overset{\cdot}{+} K_i\right)^{-1}\left(V_i t_i - N_0^i\right) \\[2mm]
&=\; \left(\tfrac{1}{2}I - K\right) \dot{D}_i^{-1}\left[\left(\tfrac{1}{2}I - K_i'\right) t_i + N_1^i\right] + V_i t_i - N_0^i,
\end{aligned}
\tag{8}
$$

where $\dot{D}_i^{-1}$ denotes the pseudoinverse of D_i with respect to the rigid motions r_i associated with the differential operator in (1). This map corresponds to the solution of a Neumann problem, where solvability conditions must be satisfied.

If we introduce the skeleton Γ_S by

$$
\Gamma_S = \bigcup_{i=1}^{p} \Gamma_i ,
$$

we are able to give variational formulations of the transmission problem (4) in related Sobolev spaces $H^s(\Gamma_S)$.

Variational formulation 1. *Find $u \in H^{1/2}(\Gamma_S)$ with $u_{|\Gamma_D} = g$ such that*

$$
\sum_{i=1}^{p} \int_{\Gamma_i} S_i u_{|\Gamma_i}(x) \cdot v_{|\Gamma_i}(x)\, ds_x \;=\; \int_{\Gamma_N} h(x) \cdot v(x)\, ds_x \;-\; \sum_{i=1}^{p} \int_{\Gamma_i} \tilde{N}^i(x) \cdot v(x)\, ds_x
$$

holds for all $v \in H^{1/2}(\Gamma_S)$ with $v_{|\Gamma_D} = 0$.

With $u_i = u_{|\Gamma_i}$ the first transmission condition in (4) is satisfied strongly, while the second condition has to be satisfied in a weak sense with $t_i = S_i u_{|\Gamma_i}$. This formulation is equivalent to the variational formulation used in domain decomposition methods for finite element methods and is therefore well suited for coupling of finite and boundary element discretization methods [Cos87, Hsi90, Lan94, Wen88]. However, we can change the meaning of the Cauchy data to get a second formulation corresponding to nonconforming finite elements [HSW] :

Variational formulation 2. *Find* $t \in \prod_{i=1}^{p} H^{-1/2}(\Gamma_i)$ *with* $t(x) = h(x)$ *for* $x \in \Gamma_N$ *such that*

$$\sum_{i=1}^{p} \int_{\Gamma_i} U_i t_{|\Gamma_i}(x) \cdot \tau_{|\Gamma_i}(x)\, ds_x = \int_{\Gamma_D} g(x) \cdot \tau(x)\, ds_x - \sum_{i=1}^{p} \int_{\Gamma_i} \hat{N}^i(x) \cdot \tau(x)\, ds_x$$

holds for all $\tau \in \prod_{i=1}^{p} H^{-1/2}(\Gamma_i)$.

Setting $t_i = t_{|\Gamma_{ij}}, t_j = -t_{|\Gamma_{ij}}$ for $i < j$, the second transmission condition is satisfied strongly, while the first one with $u_i = T_i t_i$ has to be fulfilled in a weak sense.

The existence and uniqueness of solutions of the variational formulations above follow directly from the coerciveness of the bilinear forms involved; for the Neumann problem this is true only with respect to appopriate quotient spaces generated by the rigid motions r_i. Moreover, these variational formulations give us a large variety for using boundary element methods for the discretization. The second one is well suited for the macro–element technique using Neumann series for realizing the Poincaré–Steklov map U_i by adaptive strategies as to keep symmetry and $H^{-1/2}$–ellipticity also in the discrete form [HSW, Tür95].

To discretize the local Steklov–Poincaré operators S_i included in the variational formulation 1, we have to introduce approximate operators

$$S_i^h u_{h|\Gamma_i} := t_h^i \in H_h^{-1/2}(\Gamma_i)$$

by the solution of the local finite–dimensional variational problems

$$\langle V_i t_h^i, \tau_h^i \rangle_{\Gamma_i} = \langle (\tfrac{1}{2}I + K_i) u_{h|\Gamma_i}, \tau_h^i \rangle_{\Gamma_i} \quad \forall \tau_h^i \in H_h^{-1/2}(\Gamma_i) . \tag{9}$$

Due to the properties of the local single layer potential operators V_i we have the quasi–optimal error estimates [Wen83]

$$\|S_i u_{h|\Gamma_i} - S_i^h u_{h|\Gamma_i}\|_{H^{-1/2}(\Gamma_i)} \le c \inf_{\tau_h^i \in H_h^{-1/2}(\Gamma_i)} \|S_i u_{h|\Gamma_i} - \tau_h^i\|_{H^{-1/2}(\Gamma_i)} .$$

Then there exists an unique solution $\tilde{u}_h$ of the perturbed variational problem

$$\sum_{i=1}^{p} \int_{\Gamma_i} S_i^h \tilde{u}_h(x) \cdot v_h(x)\, ds_x = f(v_h) \tag{10}$$

and we get the error estimate

$$\|u - \tilde{u}_h\|_{H^{1/2}(\Gamma_S)} \le C(h_0) \left\{ \inf_{v_h \in H_h^{1/2}(\Gamma_S)} \|u - v_h\|_{H^{1/2}(\Gamma_S)} \right.$$
$$\left. + \sum_{i=1}^{p} \inf_{\tau_h^i \in H_h^{-1/2}(\Gamma_i)} \|S_i u_{h|\Gamma_i} - \tau_h^i\|_{H^{-1/2}(\Gamma_i)} \right\}$$

for all $h < h_0$, which ensures the convergence of this approach based on the non–symmetric formulation due to the approximation property of the trial spaces used.

The discretization of the variational problem (10) leads now to the discrete system of linear equations

$$\tilde{M}_h^\top V_h^{-1} \left(\frac{1}{2} M_h + K_h \right) \underline{u} = \underline{f} \,, \tag{11}$$

where

$$V_h = \operatorname{diag} \left(V_i^h \right)_{i=1}^{p}$$

is block diagonal and therefore well suited for the parallel computation of V_h^{-1}. To invert the locally stored blocks V_i^h one can e.g. use direct methods like LU–decomposition. This can be done in a previous step of the iteration process. Furtheron, preconditioned iterative methods itself seems to be an efficient tool to realize the action of $V_i^{h,-1}$; in this case local preconditioners C_V^i are necessary. For the global iteration, due to the non–symmetry we have to use a generalized method of orthogonal directions like GMRES or BiCGStab, and a preconditioner for the discrete Steklov–Poincaré operator is needed, too. However, in any global iteration step we have to solve a local Dirichlet problem in Ω_i; involving the whole boundary Γ_i, which may be inefficient. Setting

$$\underline{t} = V_h^{-1} \left(\frac{1}{2} M_h + K_h \right) \underline{u} \,,$$

we get the coupled system

$$\begin{pmatrix} V_h & -\frac{1}{2} \tilde{M}_h - K_h \\ M_h^\top & 0 \end{pmatrix} \begin{pmatrix} \underline{t} \\ \underline{u} \end{pmatrix} = \begin{pmatrix} \underline{f} \\ \underline{0} \end{pmatrix} \,. \tag{12}$$

The first row of blocks corresponds to the local approximations in (9), which can be handled by Galerkin or collocation methods, while the coupling conditions have to be discretized by Galerkin concepts in any case. For collocation methods we get almost $\tilde{M}_h \simeq I$, whereas for Galerkin methods we have $\tilde{M}_h = M_h$. This system can be solved again by preconditioned iterative methods, where we have only three matrix multiplications per iteration step. As block preconditioners we can use the same components as above.

Before discretizing the variational problem 1 we can replace $t_i = S_i u_{|\Gamma_i}$ by the corresponding expression in the Calderon projector (6) and we arrive at the symmetric formulation [Cos87, Sir79]:

$$\sum_{i=1}^{p} \left\{ \langle D_i u_{|\Gamma_i}, v_{|\Gamma_i} \rangle_{\Gamma_i} + \tfrac{1}{2} \langle t_i, v_{|\Gamma_i} \rangle_{\Gamma_i} + \langle t_i, K_i v_{|\Gamma_i} \rangle_{\Gamma_i} \right\} = f_1(v)$$

$$\langle V_i t_i, \tau_i \rangle_{\Gamma_i} - \tfrac{1}{2} \langle u_{|\Gamma_i}, \tau_i \rangle_{\Gamma_i} - \langle K_i u_{|\Gamma_i}, \tau_i \rangle_{\Gamma_i} = f_2(\tau_i) \,,$$

whose Galerkin discretization leads to the skew–symmetric positive definite system

$$\begin{pmatrix} V_h & -K_h & -\frac{1}{2} M_h - K_h \\ K_h^\top & D_{h,NN} & D_{h,CN} \\ \frac{1}{2} M_h^\top + K_h^\top & D_{h,NC} & D_{h,CC} \end{pmatrix} \begin{pmatrix} \underline{t} \\ \underline{u}_N \\ \underline{u}_C \end{pmatrix} = \begin{pmatrix} \underline{f}_D \\ \underline{f}_N \\ \underline{f}_C \end{pmatrix} \,. \tag{13}$$

The vector $\underline{u}_C$ corresponds to all unknowns along the coupling interfaces, whereas $\underline{u}_N$ denotes only the unknowns at the original Neumann boundary.

2   Iterative solution methods

In this section we want to describe different possibilities to solve system (13) by iterative methods, which are well suited for parallel computations, too. We restrict ourselves to the symmetric formulation; however, to solve (12), which can be transformed into a saddle point problem, we can apply related techniques. Instead of (13) we solve the transformed system

$$
\begin{pmatrix} C_V^{-1/2} & 0 \\ 0 & C_S^{-1/2} \end{pmatrix}
\begin{pmatrix} V_h & -\tilde{K}_h \\ \tilde{K}_h^\top & D_h \end{pmatrix}
\begin{pmatrix} C_V^{-1/2} & 0 \\ 0 & C_S^{-1/2} \end{pmatrix}
\begin{pmatrix} \underline{\tilde{t}} \\ \underline{\tilde{u}} \end{pmatrix}
= \begin{pmatrix} \underline{\tilde{f}} \\ \underline{0} \end{pmatrix}
\tag{14}
$$

with $\tilde{K}_h = (K_h | \frac{1}{2} M_h + K_h)$; where C_V and C_S are symmetric and positive definite preconditioners for V_h and $S_h = D_h + K_h^\top V_h^{-1} K_h$, respectively, i.e. we suppose the inequalities

$$
\left.
\begin{aligned}
c_1^V \, (C_V \underline{t}, \underline{t}) &\leq (V_h \underline{t}, \underline{t}) \leq c_2^V \, (C_V \underline{t}, \underline{t}) \\
c_1^S \, (C_S \underline{u}, \underline{u}) &\leq ((D_h + K_h^\top V_h^{-1} K_h) \underline{u}, \underline{u}) \leq c_2^S \, (C_S \underline{u}, \underline{u})
\end{aligned}
\right\}
\tag{15}
$$

for all $\underline{t}$ and $\underline{u}$. We remark that the new system matrix in (14),

$$
H = \begin{pmatrix} C_V^{-1/2} V_h C_V^{-1/2} & -C_V^{-1/2} \tilde{K}_h C_D^{-1/2} \\ C_D^{-1/2} \tilde{K}_h^\top C_V^{-1/2} & C_D^{-1/2} D_h C_D^{-1/2} \end{pmatrix},
$$

is of the same structure as the original one in (13).

To solve the skew–symmetric and positive definite system

$$
\begin{pmatrix} A & -B \\ B^\top & D \end{pmatrix}
\begin{pmatrix} \underline{x}_1 \\ \underline{x}_2 \end{pmatrix}
= \begin{pmatrix} \underline{f}_1 \\ \underline{f}_2 \end{pmatrix},
\tag{16}
$$

which corresponds to (13) as well as to (14), we introduce a regular transformation matrix T and solve the transformed system

$$
T \cdot H \underline{x} = T \cdot \underline{f} \, .
$$

Moreover, we suppose the spectral equivalence inequalities

$$
\left.
\begin{aligned}
c_1^A \, (\underline{x}_1, \underline{x}_1) &\leq (A \underline{x}_1, \underline{x}_1) \leq c_2^A \, (\underline{x}_1, \underline{x}_1) \\
c_1^S \, (\underline{x}_2, \underline{x}_2) &\leq ((D + B^\top A^{-1} B) \underline{x}_2, \underline{x}_2) \leq c_2^S \, (\underline{x}_2, \underline{x}_2)
\end{aligned}
\right\}
\tag{17}
$$

corresponding to (15), where we require $c_1^A > 1$ by scaling the preconditioner C_V.

For $T = I$ we still have the original preconditioned system, which can be solved by generalized methods of conjugate directions, such as GMRES or methods of biorthogonal directions [Bea92, Stec] . Taking $T = H^\top$ we arrive at the system of normal equations with a symmetric and positive definite matrix $H^\top H$, for which we can use the standard conjugate gradient iteration; but here the spectral condition number is squared, and therefore this variant may be inefficient.

Following [BP88], the transformed matrix

$$\tilde{M} = \begin{pmatrix} I & 0 \\ -B^\top & I \end{pmatrix} \begin{pmatrix} A & -B \\ B^\top & D \end{pmatrix} = \begin{pmatrix} A & -B \\ B^\top(I-A) & D + B^\top B \end{pmatrix}$$

is selfadjoint with respect to the specially chosen inner product

$$[\underline{x}, \underline{x}] := ((A - I)\underline{x}_1, \underline{x}_1) + (\underline{x}_2, \underline{x}_2) \tag{18}$$

due to $c_1^A > 1$ in (17). Moreover, there hold spectral equivalence inequalities

$$c_1^R \cdot \min\{1, c_1^S\} \cdot [\underline{x}, \underline{x}] \le [\tilde{M}\underline{x}, \underline{x}] \le c_2^R \cdot \max\{1, c_2^S\} \cdot [\underline{x}, \underline{x}]$$

with the constants [BP88]

$$c_1^R = \left(1 + \frac{\alpha}{2} + \sqrt{\frac{\alpha^2}{4} + \alpha}\right)^{-1}, \quad c_2^R = \frac{1 + \sqrt{\alpha}}{1 - \alpha}, \quad \alpha = 1 - \frac{1}{c_2^A}.$$

Therefore, a scaling of the preconditioner C_S is necessary, too.

Obviously, this method requires the symmetry of A to define the inner product (18). We are interested in an approach, which allows a generalization to unsymmetric perturbations of symmetric matrices as in (12). Therefore we consider the second transformed matrix

$$\hat{M} = \begin{pmatrix} A - I & 0 \\ -B^\top & I \end{pmatrix} \begin{pmatrix} A & -B \\ B^\top & D \end{pmatrix} = \begin{pmatrix} A^2 - A & (I - A)B \\ B^\top(I - A) & D + B^\top B \end{pmatrix},$$

which is symmetric and positive definite. However, the constants in the related spectral equivalence inequalities are no longer asymptotically optimal for $A \to I$ as in the approach proposed by [BP88]. So we have to find an appropriate preconditioning technique for $\tilde{M}$. To this end we follow [HLM92] and consider the symmetric and positive definite Matrix

$$\bar{M} = \begin{pmatrix} \bar{A} & \bar{B} \\ \bar{B}^\top & \bar{D} \end{pmatrix}$$

with the factorization

$$\bar{M} = \begin{pmatrix} I & 0 \\ \bar{B}^\top \bar{A}^{-1} & I \end{pmatrix} \begin{pmatrix} \bar{A} & 0 \\ 0 & \bar{D} - \bar{B}^\top \bar{A}^{-1} \bar{B} \end{pmatrix} \begin{pmatrix} I & \bar{A}^{-1} \bar{B} \\ 0 & I \end{pmatrix}.$$

If the spectral equivalence inequalities

$$c_1^{\bar{A}} \cdot (\underline{x}_1, \underline{x}_1) \le (\bar{A}\underline{x}_1, \underline{x}_1) \le c_2^{\bar{A}} \cdot (\underline{x}_1, \underline{x}_1)$$

and

$$c_1^{\bar{S}} \cdot (\underline{x}_2, \underline{x}_2) \le ((\bar{D} - \bar{B}^\top \bar{A}^{-1} \bar{B})\underline{x}_2, \underline{x}_2) \le c_2^{\bar{S}} \cdot (\underline{x}_2, \underline{x}_2)$$

both are satisfied, then the matrix

$$C_{\bar{M}} = \begin{pmatrix} I & 0 \\ \bar{B}^\top & I \end{pmatrix} \begin{pmatrix} I & \bar{B} \\ 0 & I \end{pmatrix}$$

is spectrally equivalent to $\bar{M}$, i.e. we have the inequalities

$$\gamma_1 \left(C_{\bar{M}}\underline{x}, \underline{x}\right) \leq \left(\bar{M}\underline{x}, \underline{x}\right) \leq \gamma_2 \left(C_{\bar{M}}\underline{x}, \underline{x}\right)$$

with the positive constants [HLM92]

$$\gamma_1 = \min\{c_1^{\bar{A}}, c_1^{\bar{S}}\} \left[1 + \frac{1}{2}\left(\mu - \sqrt{\mu^2 + 4\mu}\right)\right],$$

$$\gamma_2 = \max\{c_2^{\bar{A}}, c_2^{\bar{S}}\} \left[1 + \frac{1}{2}\left(\mu + \sqrt{\mu^2 + 4\mu}\right)\right]$$

and the spectral condition number

$$\mu = \rho \left(\left[\bar{D} - \bar{B}^\top \bar{A}^{-1} \bar{B}\right]^{-1} \left[\bar{B}^\top (\bar{A}^{-1} - I)\bar{A}(\bar{A}^{-1} - I)\bar{B}\right] \right) .$$

With $\bar{A} = A^2 - A$, from (17) there follows directly

$$(c_1^A - 1)c_1^A \left(\underline{x}_1, \underline{x}_1\right)\left((A^2 - A)\underline{x}_1, \underline{x}_1\right) \leq (c_2^A - 1)c_2^A \left(\underline{x}_1, \underline{x}_2\right),$$

and a straightforward computation gives

$$\bar{S} = \bar{D} - \bar{B}^\top \bar{A}^{-1} \bar{B} = D + B^\top A^{-1} B \; ;$$

and therefore $c_1^{\bar{S}} = c_1^S, c_2^{\bar{S}} = c_2^S$. From

$$\begin{aligned}
T &= \bar{B}^\top (\bar{A}^{-1} - I)\bar{A}(\bar{A}^{-1} - I)\bar{B} \\
&= B^\top (I - A)((A - I)^{-1}A^{-1} - I)(A^2 - A)(A^{-1}(A - I)^{-1} - I)(I - A)B \\
&= B^\top (A^{-1} - A + I)(A^2 - A)(A^{-1} - A + I)B
\end{aligned}$$

we conclude, that T tends to zero for $A \to I$. Because of the stability of the discrete Steklov–Poincaré operator S we have then asymptotically $\mu \to 0$ for $A \to I$. We will discuss this approach in more details in [Steb]. Finally we remark, that we can generalize this method to the non–symmetric case (12) by choosing the transformation

$$\hat{M} = \begin{pmatrix} A^\top - I & 0 \\ -C & I \end{pmatrix} \begin{pmatrix} A & -B \\ C & 0 \end{pmatrix} = \begin{pmatrix} A^\top A - A & (I - A)B \\ C(I - A) & CB \end{pmatrix} \; ;$$

the related system of linear equations can be solved by any preconditioned iterative method discussed above. Due to the (non–symmetric) discretization of self–adjoint operators involved, the block matrices $A \sim A^\top$ and $B^\top \sim C$, respectively, are perturbations of symmetric matrices, such that the BiCGStab algorithm behaves asymptotically like the classical conjugate gradient method.

3   The construction of optimal preconditioners

The iterative methods described above are based on preconditioners with respect to the discrete single layer potential V_h, the hypersingular integral operator D_h

and the discrete Steklov–Poincaré operator S_h, respectively. Here these operators are pseudodifferential operators of order ± 1. Efficient algebraic preconditioners are multigrid methods, which in a generalized form can also used for operators of negative order [Bra93]. In general, if preconditioners are constructed from equivalence relations of bilinear forms and their matrix representations, then one has to invert spectral equivalent matrices, where the work for doing this job should be in the same order as a matrix multiplication. In two–dimensional boundary element methods one can use periodic Sobolev spaces to construct efficient preconditioners [HKW80, Rja90, Wen80]. For uniform meshes, this approach leads to circulant matrices, which can be handled by the Fast Fourier Transformation. There are generalizations to adaptive meshes and to three–dimensional problems [KW92]; in the case of axially–symmetric surfaces, the discretization leads to block circulant matrices [MR90]. However, generalizations to three–dimensional problems are restricted to structered surface meshes in a corresponding ordering. Here we want to give a general concept to construct efficient preconditioners independent of geometrical properties.

Let us consider strongly elliptic and self–adjoint pseudodifferential operators of orders $\pm 2\alpha$ in the Sobolev spaces on Γ,

$$A \ : \ H^s(\Gamma) \to H^{s-2\alpha}(\Gamma) \,, \quad B \ : \ H^{s-2\alpha}(\Gamma) \to H^s(\Gamma)$$

satisfying the coerciveness inequalities

$$c_1^A \|t\|_{H^s(\Gamma)}^2 \ \leq \ \langle At, t\rangle_{H^{s-\alpha}} \ \leq \ c_2^A \|t\|_{H^s(\Gamma)}^2 \qquad\qquad \forall\, t \in V = H^s(\Gamma)\big/\ker A$$

$$c_1^B \|u\|_{H^{s-2\alpha}(\Gamma)}^2 \ \leq \ \langle Bu, u\rangle_{H^{s-\alpha}} \ \leq \ c_2^B \|u\|_{H^{s-2\alpha}(\Gamma)}^2 \qquad \forall\, u \in W = H^{s-2\alpha}(\Gamma)\big/\ker B \,. \tag{19}$$

For the finite–dimensional $\ker B$ we use an orthonormal basis representation

$$\ker B \ = \ \big\{ v \in H^{s-2\alpha}(\Gamma) \ : \ Bv = 0 \big\} \ = \ \mathrm{span}\,\{v_k\}_{k=1}^m$$

with

$$\langle v_k, v_\ell\rangle_{H^{s-2\alpha}(\Gamma)} \ = \ \delta_{k\ell} \quad \text{for all } k, \ell = 1, \ldots, m \,.$$

If we introduce the factor space $V^0(B)$ by

$$V^0(B) \ = \ \big\{ t \in H^s(\Gamma) \ : \ \langle t, v_k\rangle_{H^{s-\alpha}(\Gamma)} = 0 \ \text{ for all } v_k \in \ker B \big\}$$

we can define the pseudoinverse operator to B,

$$\dot{B}^{-1} \ : \ V^0(B) \to W \,,$$

which is a selfadjoint operator satisfying the inequalities

$$\frac{1}{c_2^B}\,\|t\|_{H^s(\Gamma)}^2 \ \leq \ \langle \dot{B}^{-1}t, t\rangle_{H^{s-\alpha}(\Gamma)} \ \leq \ \frac{1}{c_1^B}\,\|t\|_{H^s(\Gamma)}^2$$

for all $t \in V^0(B)$. Then from the assumptions there follows immediately

Lemma 1. *For all $t \in V \cap V^0(B)$ there hold the inequalities*

$$\tilde{\gamma}_1 \cdot \langle \dot{B}^{-1}t, t\rangle_{H^{s-\alpha}(\Gamma)} \ \leq \ \langle At, t\rangle_{H^{s-\alpha}(\Gamma)} \ \leq \ \tilde{\gamma}_2 \cdot \langle \dot{B}^{-1}t, t\rangle_{H^{s-\alpha}(\Gamma)} \,.$$

with positive constants $\tilde{\gamma}_1 = c_1^A \cdot c_1^B$ and $\tilde{\gamma}_2 = c_2^A \cdot c_2^B$.

To construct a preconditioner for all $u \in V$ we have to split the space V into an orthogonal sum

$$V \,=\, V^0(B) \oplus \left[V^0(B)\right]^{\perp} \,.$$

If we define the Bessel potential operator

$$J \,:\, H^{s-2\alpha}(\Gamma) \rightarrow H^s(\Gamma)$$

such that the relation

$$\langle Ju, v\rangle_{H^{s-\alpha}(\Gamma)} \,=\, \langle u, v\rangle_{H^{s-2\alpha}(\Gamma)}$$

holds for all $v \in H^{s-2\alpha}(\Gamma)$, we can formulate the basic result [Stea]:

Theorem 2. *The operator defined by the bilinear form*

$$c(t,\tau) \,=\, \langle \dot{B}^{-1}t_0, \tau_0\rangle_{H^{s-\alpha}(\Gamma)} \,+\, \sum_{k=1}^{m}\langle t, v_k\rangle_{H^{s-\alpha}(\Gamma)}\langle \tau, v_k\rangle_{H^{s-\alpha}(\Gamma)} \qquad (20)$$

with

$$t_0(x) \,=\, t(x) - \sum_{j=1}^{m}\langle t, v_j\rangle_{H^{s-\alpha}(\Gamma)}(Jv_j)(x) \in V^0(B)$$

is spectrally equivalent to A, i.e. there hold the inequalities

$$\gamma_1 \cdot c(t,t) \,\leq\, \langle At, t\rangle_{H^{s-\alpha}(\Gamma)} \,\leq\, \gamma_2 \cdot c(t,t) \,.$$

with positive constants $\gamma_1 = c_1^A \cdot \min\{1, c_1^B\}$ *and* $\gamma_2 = c_2^A \cdot \max\{1, c_2^B\}$.

Let a finite approximation space $V_h \subset V$ be given by

$$V_h \,=\, \mathrm{span}\,\{\varphi_k^{\nu}(\cdot)\}_{k=1}^{N}$$

with the trial functions φ_k^{ν}, e.g. smoothest splines of polynomial degree ν, and the discretization parameter N. Then it follows directly from Theorem 2, that the matrices A_h and C_h defined by their elements

$$A_h[\ell, k] \,=\, \langle A\varphi_k^{\nu}, \varphi_\ell^{\nu}\rangle_{H^{s-\alpha}(\Gamma)} \,, \quad C_h[\ell, k] \,=\, c(\varphi_k^{\nu}, \varphi_\ell^{\nu})$$

are spectrally equivalent, i.e. the spectral condition number of the preconditioned system is bounded by

$$\kappa(C_h^{-1}A_h) \,\leq\, \frac{c_2^A \cdot \max\{1, c_2^B\}}{c_1^A \cdot \min\{1, c_1^B\}}$$

and therefore independent of all bad parameters like mesh adaptivity or the degree of trial functions. Moreover, there are no restrictions to the space dimension of the problem considered. However, to use the proposed preconditioner, one has to realize the action of C_h^{-1} in a very efficient way.

For the map Jv_k of the functions $v_k \in \ker B$ we have to suppose a representation in terms of the basis of V_h, i.e.

$$(Jv_k)(x) \,=\, \sum_{j=1}^{N} v_k^j \cdot \varphi_j^{\nu}(x) \,,$$

then we find by partial Gram–Schmidt orthogonalization the transformed basis

$$V_h = \text{span}\{\varphi_1^{\nu}, \ldots, \varphi_N^{\nu}\} = \text{span}\{\varphi_1^{\nu,0}, \ldots, \varphi_{N-m}^{\nu,0}, Jv_1, \ldots, Jv_m\}$$

with

$$\varphi_k^{\nu,0}(x) = \varphi_k^{\nu}(x) - \sum_{j=1}^{m} \langle \varphi_k^{\nu}, v_j \rangle_{H^{s-\alpha}(\Gamma)} \cdot (Jv_j)(x) \in V^0(B) .$$

For a given function $t_h(x) \in V_h$ we find the basis transformation by

$$t_h(x) = \sum_{k=1}^{N} t_k \cdot \varphi_k^{\nu}(x) = \sum_{k=1}^{N-m} \tilde{t}_k \cdot \varphi_k^{\nu,0}(x) + \sum_{k=1}^{m} \tilde{t}_{N-m+k} \cdot (Jv_k)(x)$$

with the relations

$$t_k = \begin{cases} \tilde{t}_k + \sum_{\ell=1}^{m} v_{\ell}^k \left(\tilde{t}_{N-m+\ell} - \sum_{j=1}^{N-m} \langle \varphi_j^{\nu}, v_{\ell} \rangle_{H^{s-\alpha}(\Gamma)} \tilde{t}_j \right), & k = 1, \ldots, N-m, \\[2em] \sum_{\ell=1}^{m} v_{\ell}^k \left(\tilde{t}_{N-m+\ell} - \sum_{j=1}^{N-m} \langle \varphi_j^{\nu}, v_{\ell} \rangle_{H^{s-\alpha}(\Gamma)} \tilde{t}_j \right), & k = N-m+1, \ldots, N \end{cases}$$

or in matrix representation

$$\underline{t} = T^{\top} \underline{\tilde{t}} .$$

If we denote by $\underline{f}$ the vector of the right hand side generated by

$$f_k = f(\varphi_k^{\nu})$$

corresponding to the linear form of the original variational problem, we find for the discretization with respect to the transformed basis, i.e. for

$$\tilde{f}_k = f(\varphi_k^{\nu,0}) \ \text{ for } k = 1, \ldots, N - m$$

and

$$\tilde{f}_{N-m+k} = f(Jv_k) \ \text{ for } k = 1, \ldots, m$$

the matrix representation

$$\underline{\tilde{f}} = T \underline{f} ,$$

which ensures the symmetry of the proposed preconditioner.

Due to the orthonormal basis representation of V_h, discretization leads now to the decoupled stiffness matrix

$$\tilde{C}_h = \begin{pmatrix} C_0 & 0 \\ 0 & I \end{pmatrix} \tag{21}$$

with the block

$$C_0[\ell, k] = \langle \dot{B}^{-1} \varphi_k^{\nu,0}, \varphi_\ell^{\nu,0} \rangle_{H^{s-\alpha}(\Gamma)}$$

for all $k, \ell = 1, \ldots, N - m$. We still have to discretize the pseudoinverse operator $\dot{B}^{-1}$, which is in general not given in explicit form. Hence we have to find an approximation

u_0^h of the exact function $u_0 = \dot{B}^{-1} t_0 \in W$ for any function $t_0 \in V^0(B)$. This can be done by solving the finite–dimensional variational problem

$$\langle Bu_0^h, v_0^h \rangle_{H^{s-\alpha}(\Gamma)} = \langle t_0, v_0^h \rangle_{H^{s-\alpha}(\Gamma)}$$

for all test functions $v_h^0 \in W$. From the assumptions for B there follows, that this problem has an unique solution and that we have the quasioptimal error estimate

$$\|u_0 - u_0^h\|_{H^{s-2\alpha}(\Gamma)} \leq c(B) \inf_{v_0^h \in W} \|u_0 - v_0^h\|_{H^{s-2\alpha}(\Gamma)} .$$

This estimate implies

$$\left| \langle \dot{B}^{-1} t_0, t_0 \rangle_{H^{s-\alpha}(\Gamma)} - \langle u_h^0, t_0 \rangle_{H^{s-\alpha}(\Gamma)} \right| \leq c(B) \inf_{v_0^h \in W} \|\dot{B}^{-1} t_0 - v_0^h\|_{H^{s-2\alpha}(\Gamma)} \|t_0\|_{H^s(\Gamma)}^2$$

and therefore we have the spectral equivalence inequalities

$$c_1(B, h_0) \cdot \langle \dot{B}_h^{-1} t_0, t_0 \rangle_{H^{s-\alpha}(\Gamma)} \leq \langle \dot{B}^{-1} t_0, t_0 \rangle_{H^{s-\alpha}(\Gamma)} \leq c_2(B, h_0) \cdot \langle \dot{B}_h^{-1} t_0, t_0 \rangle_{H^{s-\alpha}(\Gamma)}$$

for all mesh parameters $h \leq h_0$, where h_0 is determined only by constants depending on B. Hence we can replace the block matrix C_0 in (21) by the spectrally equivalent discretization based on the approximation u_0^h. This leads to the discrete representation

$$C_0^h = \tilde{M}_h B_h^{-1} \tilde{M}_h^\top .$$

For this discretization we have to use a subspace

$$W_h = \operatorname{span}\{\varphi_1^{\mu,0}, \ldots, \varphi_{N-m}^{\mu,0}\} \subset W$$

with the transformed trial functions

$$\varphi_k^{\mu,0}(x) = \varphi_k^\mu(x) - \sum_{j=1}^m \langle \varphi_k^\mu, v_j \rangle_{H^s(\Gamma)} \cdot v_j(x) \in W$$

for $k = 1, \ldots, N - m$. Then we get the discrete representation of the operator B by

$$B_h[\ell, k] = \langle B\varphi_k^{\mu,0}, \varphi_\ell^{\mu,0} \rangle_{H^{s-\alpha}(\Gamma)} = \langle B\varphi_k^\mu, \varphi_\ell^\mu \rangle_{H^{s-\alpha}(\Gamma)}$$

using the original trial functions $\varphi_k^\mu(\cdot)$. For the modified mass matrix we find

$$\tilde{M}_h[\ell, k] = \langle \varphi_k^\nu, \varphi_\ell^\mu \rangle_{H^{s-\alpha}(\Gamma)} - \sum_{j=1}^m \langle \varphi_\ell^\mu, v_j \rangle_{H^s(\Gamma)} \cdot \langle \varphi_k^\nu, v_j \rangle_{H^{s-\alpha}(\Gamma)} ,$$

i.e. a rank–m pertubation of the original mass matrix M_h. Since we can choose the trial functions $\varphi_k^\mu(\cdot)$ in an appropriate way, we may assume, that the inverse matrix $\tilde{M}_h^{-1}$ exists. Algorithms to compute this inverse are based on Sherman–Morrison–Woodbury formulae [OR70] using the inverse matrix M_h^{-1}, which is more suitable, because M_h is almost sparse and diagonally dominant.

Now we are in the position to present the proposed preconditioner in a compact form, where the action to realize C_h^{-1} is given by

$$C_h^{-1} = T^\top \begin{pmatrix} \tilde{M}_h^{-1} B_h \tilde{M}_h^{-1} & 0 \\ 0 & I \end{pmatrix} T \,. \tag{22}$$

In the next section we give an example, which is important for applications in potential theory as well as in linear elasticity. Using the symmetric formulation to handle mixed boundary value problems or domain decomposition methods numerically with boundary elements, we will see, that this approach is optimal for an efficient preconditioning, since we can choose the operators A and B in such a way, that the additional work for the realization of the proposed preconditioner is neglicable.

4 Examples

We want to demonstrate the proposed preconditioning technique for the potential problem with the differential operator

$$L(x)u(x) = -\operatorname{div} \alpha(x)\nabla u(x) \,,$$

where we suppose piecewise constant coefficients

$$\alpha(x) = \alpha_i \quad \text{for } x \in \Omega_i \,.$$

Therefore, the local fundamental solution for the Laplacian is given by

$$E^i(x,y) = \frac{1}{2(n-1)\pi} \cdot \begin{cases} -\log|x-y| & \text{for } n = 2 \,, \\ |x-y|^{-1} & \text{for } n = 3 \,. \end{cases}$$

The local single layer potential operators

$$(V_i t_i)(x) = \int_{\Gamma_i} E^i(x,y) \cdot t_i(y)\, ds_y$$

and the hypersingular operators

$$(D_i u_i)(x) = -\frac{\partial}{\partial n_x} \int_{\Gamma_i} \frac{\partial}{\partial n_y} E^i(x,y) \cdot u_i(y)\, ds_y$$

for $x \in \Gamma_i$ satisfy all assumptions of the previous sections with $A = V, B = D$ and $s = -1/2, 2\alpha = -1$. Moreover, the ratios of the constants in the inequalities (19) are independent of the local material parameter α_i.

Further we have $\ker V = \{0\}$ and $\ker D = \{1\}$. From

$$v_1(x) = 1 = \sum_{k=1}^{N} \varphi_k^\nu(x)$$

we find the basis transformation T by

$$u_h(x) = \sum_{k=1}^{N} u_k \varphi_k^\nu(x) = \sum_{k=1}^{N-1} \tilde{u}_k\, \varphi_k^{\nu,0}(x) + \omega \cdot v_1(x)$$

with

$$\varphi_k^{\nu,0}(x) = \varphi_k^\nu(x) - \frac{m_k^\nu}{|\Gamma|} \cdot v_1(x) \ \text{ and } \ m_k^\nu = \int_\Gamma \varphi_k^\nu(x)\, ds_x \ .$$

Therefore we have the representation

$$\tilde{M}_h = M_h - \frac{1}{|\Gamma|}\, \underline{m}^\nu (\underline{m}^\mu)^\top$$

for the modified mass matrix, where the inverse is given by the Shermann–Morrison formula

$$\tilde{M}_h^{-1} = M_h^{-1} - \frac{1}{\alpha - |\Gamma|} M_h^{-1} \underline{m}^\nu (\underline{m}^\mu)^\top M_h^{-1} \ , \ \alpha = (\underline{m}^\mu)^\top M_h^{-1} \underline{m}^\nu \ .$$

Note that there are situations, where M_h^{-1} and therefore $\tilde{M}_h^{-1}$ does not exist; e.g. in three dimensions, if we choose for the flux piecewise constant trial functions, which are defined with respect to elements, and for the potential piecewise linear trial functions defined with respect to nodes, and in general, if the numbers of elements and nodes differ, then the mass matrix will be rectangular. Therefore, one has to be very careful to select the exact trials to discretize B. If we choose piecewise linear trial functions with $\nu = \mu = 1$, M_h will be a strongly diagonally dominant sparse matrix, where the inverse M_h^{-1} can be computed very efficiently. In the two–dimensional case, M_h is tridiagonal, where we can use a Cholesky factorization with $O(N)$ multiplications. In general, this preconditioner needs $O(N^2)$ multiplications to apply B_h to a vector and some operations of lower order to realize the basis transformations T and the computation of $\tilde{M}_h^{-1}$.

Because the hypersingular operator D is positive definite with respect to the $|\cdot|_{H^{1/2}(\Gamma)}$ semi–norm, i.e.

$$\langle Du, u \rangle_{L_2(\Gamma)} \geq c_1^D \, |u|_{H^{1/2}(\Gamma)}^2 \ ,$$

there follows immediately

$$\langle (c_1^D \cdot I + D)u, u \rangle_{L_2(\Gamma)} \geq c_1^D \, ||u||_{H^{1/2}(\Gamma)}^2$$

with the modified pseudodifferential operator $c_1^D \cdot I + D$, where a basis transformation is no longer required. This seems to be quite useful for applications in linear elasticity.

For the multiplication of the discrete operator D_h one should use fast multiplication algorithms like the panel clustering [HN89, Sau92] for developing more efficient algorithms.

References

[Bea92] Barret R. and et. al. (1992) *Templates for the solution of linear systems: Building blocks for iterative methods.* SIAM.

[BP88] Bramble J. H. and Pasciak J. E. (1988) A preconditioning technique for indefinite systems resulting from mixed approximations of ellitpic problems. *Math. Comput.* 50: 1–17.

[Bra93] Bramble J. H. (1993) *Multigrid Methods.* Longman.

[Cos87] Costabel M. (1987) Symmetric methods for the coupling of finite elements and boundary elements. In Brebbia C. A., Kuhn G., and Wendland W. L. (eds) *Boundary Elements IX*, pages 411–420. Springer, Berlin.

[Cos88] Costabel M. (1988) Boundary integral operators on lipschitz domains: Elementary results. *SIAM J. Math. Anal.* 19(6): 613–626.

[HKW] Hsiao G. C., Khoromskij B. N., and Wendland W. L.Boundary integral operators and domain decomposition methods. Advances in Computational Mathematics, submitted.

[HKW80] Hsiao G. C., Kopp P., and Wendland W. L. (1980) A galerkin collocation method for some integral equations of the first kind. *Computing* 25: 89–130.

[HLM92] Haase G., Langer U., and Meyer A. (1992) Domain decomposition with inexact subdomain solvers. *J. of Num. Lin. Alg. with Appl.* 1: 27–42.

[HN89] Hackbusch W. and Nowak Z. P. (1989) On the fast matrix multiplication in the boundary element method by panel clustering. *Numer. Math.* 54: 463–491.

[Hsi90] Hsiao G. C. (1990) The coupling of boundary element and finite element methods. *ZAMM* 70: T493–T503.

[HSW] Hsiao G. C., Schnack E., and Wendland W. L.A hybrid coupled finite–boundary element method. Preprint.

[HSW79] Hsiao G. C., Stephan E. P., and Wendland W. L. (1979) On the integral equation method for the plane mixed boundary value problem with the Laplacian. *Math. Methods Appl. Sci.* 1: 265–321.

[HW91] Hsiao G. C. and Wendland W. L. (1991) Domain decomposition methods in boundary element methods. In Glowinski R. and et. al. (eds) *Fourth International Symposium on Domain Decomposition Methods in Engineering and Applied Sciences*, pages 41–49. SIAM, Philadelphia.

[HW92] Hsiao G. C. and Wendland W. L. (1992) Domain decomposition via boundary element methods. In Alder H. and et. al. (eds) *Numerical Methods in Engineering and Applied Sciences*, pages 198–207. CIMNE, Barcelona.

[KW92] Khoromskij B. N. and Wendland W. L. (1992) Spectrally equivalent preconditioners in substructuring techniques. *East–West J. Numer. Math.* 1: 1–25.

[Lan94] Langer U. (1994) Parallel iterative solution of symmetric coupled fe/be–equations via domain decomposition. *Contemp. Math.* 157: 335–344.

[MR90] Meyer A. and Rjasanow S. (1990) An effective direct solution method for certain boundary element equations in 3d. *Math. Meth. Appl. Sc.* 13: 43–53.

[OR70] Ortega J. M. and Rheinboldt W. C. (1970) *Iterative Solution of Nonlinear Equations in Several Variables.* Academic Press.

[Rja90] Rjasanow S. (1990) Vorkonditionierte iterative Auflösung von Randelementgleichungen für die Dirichlet–Aufgabe. Wissenschaftliche Schriftenreihe 7, TU Chemnitz.

[Sau92] Sauter S. A. (1992) *Über die effiziente Verwendung des Galerkinverfahrens zur Lösung Fredholmscher Integralgleichungen.* PhD dissertation, Universität Kiel.

[Sir79] Sirtori S. (1979) General stress analysis by means of integral equations and boundary elements. *Meccanica* 14: 210–218.

[Stea] Steinbach O.The construction of optimal preconditioners in the boundary element method. In preparation.

[Steb] Steinbach O.Fast solvers for boundary integral equations related to mixed boundary value problems. In preparation.

[Stec] Steinbach O.Iterative projection methods of generalized orthogonal directions in the boundary element method. In preparation.

[SW] Steinbach O. and Wendland W. L.Boundary element methods for inhomogeneous problems by fictitious domain methods. In preparation.

[Tür95] Türke K. (1995) *Eine Zweigitter–Methode zur Kopplung von FEM und BEM.* PhD dissertation, Universität Karlsruhe.

[Wen80] Wendland W. L. (1980) On galerkin collocation methods for integral equations of elliptic boundary value problems. In Albrecht J. and Collatz L. (eds) *Numerical Treatment of Integral Equations*, ISNM 53, pages 244–275. Birkhäuser, Basel.

[Wen83] Wendland W. L. (1983) Boundary element methods and their asymptotic convergence. In Fillipi P. (ed) *Theoretical Acoustics and Numerical Techniques*, CISM Courses and Lectures No. 227, pages 135–216. Springer, Wien–New York.

[Wen88] Wendland W. L. (1988) On asymptotic error estimates for combined BEM and FEM. In Stein E. and Wendland W. L. (eds) *Finite Element and Boundary Element Techniques from Mathematical and Engineering Point of View*, pages 273–333. Springer, Berlin.

Preconditioners for Spectral and Mortar Finite Element Methods

Olof B. Widlund [1]

1 Introduction

Domain decomposition methods have been developed quite systematically for conforming lower order finite element approximations of elliptic problems. Less attention has been paid to the large, often very ill-conditioned, linear algebraic systems of equations that arise in discretizations based on spectral elements, $p-$version finite elements, and mortar finite element methods.

In this paper, a brief review is given of some of our relatively recent work on domain decomposition methods for spectral elements, carried out jointly with Luca Pavarino, and more recent work on mortar elements which is joint with Yves Achdou, Mario Casarin, and Yvon Maday.

The iterative substructuring (Schur complement) algorithms form a main family of domain decomposition methods for elliptic problems. Like other domain decomposition methods, they are preconditioned conjugate gradient methods. The region Ω of the given elliptic problem is subdivided into non-overlapping subregions Ω_i and a discretization is introduced. In the pure form of the iterative substructuring algorithms, the interior variables of all the subregions are first eliminated using a direct method. A preconditioner, for the remaining, Schur complement, system of linear algebraic equations, is then constructed from solvers of certain local problems

---

[1] Courant Institute of Mathematical Sciences, 251 Mercer Street, New York, NY 10012. Email: widlund@cs.nyu.edu. This work was supported in part by the National Science Foundation under Grant NSF-CCR-9503408, in part, by the U. S. Department of Energy under contracts DE-FG02-92ER25127 and in part by CNRS while the author visited Université Pierre et Marie Curie, Paris, France.

and, in addition, a solver of a coarse, global problem similar to that of the coarsest level of a multigrid algorithm. The coarse problem is often the key to good performance, and it can be quite exotic; cf. Widlund [Wid94]. Very large systems of linear algebraic equations, arising when elliptic problems are discretized by finite elements, finite differences, or spectral methods, have been solved successfully by such iterative methods. A number of systematic experimental studies have been carried out with domain decomposition algorithms for both lower and higher order finite elements. Early experimental work for quite large lower order finite element problems is reported in Smith [Smi93]. Very large spectral problems have been solved, using similar preconditioners, by Paul Fischer and Einar Rønquist; see, e.g., [FR94]. For pioneering work on $p-$version finite elements, see Mandel [Man90, Man94]. We note that problems based on higher order discretizations are often exceptionally ill-conditioned. Therefore the challenge to construct good preconditioners is greater than for lower order methods.

A theory for $h-$version finite element methods in three dimensions is summarized in Dryja, Smith, and Widlund [DSW94]; see also Dryja and Widlund [DW95] for the underlying general *Schwarz theory* and an analysis of Neumann-Neumann methods. The author has also completed several papers on spectral elements in joint work with Luca Pavarino; see, in particular, [PW94b, PW94a], and section 4 of this paper. For further work on domain decomposition methods for spectral elements, including Schwarz methods with overlap and iterative substructuring algorithms for Stokes and Navier-Stokes equations, see the recent doctoral dissertation of Mario Casarin [Cas96].

Mortar elements form an interesting class of nonconforming finite elements; see section 5 of this paper for a brief description. Several studies on domain decomposition methods for mortar finite elements have already been completed by Achdou, Kuznetsov, [AK95], and together with Pironneau, [AKP95], Dryja [Dry96], and LeTallec and Sassi [LTS93]. An alternative algorithm, a *hierarchical basis* domain decomposition method has been developed by the author for problems in the plane in joint work with Mario Casarin; see [CW95] and section 6 of this paper. There is also an ongoing joint study, with Yves Achdou and Yvon Maday, [AMW96b, AMW96a], also for plane elliptic problems. Here, in addition, we will describe an algorithm, developed with Yvon Maday, [MW96] for problems in three dimensions.

We recall that the number of iterations required to decrease the energy norm of the error in a conjugate gradient iteration, by a fixed factor, is proportional to $\sqrt{\kappa(B^{-1}A)}$. Here κ is the spectral condition number, A the coefficient matrix of the original system, and B that of the preconditioner. Our central goals are the design of methods with small condition numbers and the development of tools for estimating upper and lower bounds for the spectrum of $B^{-1}A$. All good results in our theory show that the condition number of the iteration operator is independent of the number of subregions and grows only polylogarithmically with the number of degrees of freedom of an individual local problem; the bounds are proportional to $(1+\log(H/h))^q$ or $(1+\log p)^q$, with q a small integer. Here, for the $h-$version, H and h are the diameters of a typical subregion and its elements, respectively. For the spectral case, p is the degree of the polynomials used in each subregion of a spectral approximation.

2   The elliptic problem and spectral finite elements

We begin by formulating a linear, elliptic model problem on a bounded domain Ω in R^3, with a Lipschitz continuous boundary $\partial\Omega$, as a classical calculus of variation problem: Find $u \in V$ such that

$$a(u,v) = \int_\Omega k(x)\nabla u \cdot \nabla v \, dx \;=\; f(v), \; \forall \, v \in V \, . \tag{1}$$

The coefficient $k(x) > 0$ can be discontinuous. For simplicity, let $k(x) = k_i$, a constant, in each subregion Ω_i. We impose a homogeneous Dirichlet condition on a nonempty set $\partial\Omega_D \subset \partial\Omega$; this condition is incorporated into the definition of the space V. A Neumann boundary condition is given on $\partial\Omega_N = \partial\Omega \setminus \partial\Omega_D$; inhomogeneous Neumann boundary data are incorporated into the right hand side of (1).

Here, we only consider one variant of the spectral finite element methods. We assume that Ω is the union of subregions Ω_i which are cubes or images of a reference cube under reasonable smooth mappings. These subregions serve as elements of the spectral element discretization. We use conforming Q_p elements and replace the integrals of the bilinear form of (1), element by element, by Gauss-Lobatto-Legendre (GLL) quadrature. The resulting finite element space is denoted by $V^p \subset V$ and the new bilinear form, $a_Q(\cdot,\cdot)$, is given by,

$$a_Q(u,v) = (k\nabla u, \nabla v)_Q = \sum_i k_i(\nabla u, \nabla v)_{Q,\Omega_i}, \quad \forall u,v \in V^P(\Omega).$$

Here, on the reference cube $\overline{\Omega}_{ref} = [-1,1]^3$, we use the inner product

$$(u,v)_{Q,\Omega_{ref}} = \sum_{i=0}^{p}\sum_{j=0}^{p}\sum_{k=0}^{p} u(\xi_i,\xi_j,\xi_k)v(\xi_i,\xi_j,\xi_k)\rho_i\rho_j\rho_k,$$

where the ξ_i and the ρ_i are the GLL nodes and weights, respectively.

It is known that ellipticity is preserved; cf. [BM92]. A convenient nodal basis on the reference cube is provided by

$$l_i(x)l_j(y)l_k(z), \qquad 0 \le i,j,k \le p.$$

The $l_i(x)$ are the basic Lagrange interpolating polynomials of degree p defined by $l_i(\xi_j) = \delta_{ij}, 0 \le i,j \le p$; see further [BM92], where the special structure of the stiffness matrix is also discussed. We note that this set of basis functions naturally divides into interior, face, edge, and vertex functions.

The finite element problem is thus obtained by replacing, in (1), the bilinear form $a(\cdot,\cdot)$ by $a_Q(\cdot,\cdot)$ and by restricting u and v to the space V^p. Numerical quadrature can also be used to replace the right hand side of (1).

The finite element variational problem is turned into a linear system of algebraic equations, $Kx = b$, where K is the stiffness matrix and b the load vector. As a direct consequence of the ellipticity and symmetry of the bilinear form, we have $K^T = K > 0$.

3   Block-Jacobi and related domain decomposition methods

There are a number of ways of introducing domain decomposition methods; here we select the one we believe to be the simplest. We work in a framework of block-Jacobi/conjugate gradient methods. The stiffness matrix K is then preconditioned by a matrix K_J, which is the direct sum of diagonal blocks of K. Each block corresponds to a set of degrees of freedom with corresponding basis elements, which span a subspace V_i. The space V^p is a direct sum of the subspaces $V_i, i = 0, \cdots, N$. A good choice of the subspaces, after a suitable change of basis in V^p, is the key to success.

Associated with each subspace V_i is an orthogonal projection P_i onto V_i defined by

$$a_Q(P_i u, v) = a_Q(u, v), \quad \forall v \in V_i, \quad u \in V^p,$$

or an operator T_i, defined in terms of an alternative bilinear form $\tilde{a}_i(\cdot, \cdot)$,

$$\tilde{a}_i(T_i u, v) = a_Q(u, v), \quad \forall v \in V_i, \quad u \in V^p.$$

In simple cases, where a subspace V_i corresponds to a group of variables associated with adjacent mesh points, P_i corresponds to the inverse of a diagonal block of K embedded among zero blocks, times K. To obtain T_i, we can replace the inverse of the designated matrix block by a preconditioner of a local problem associated with these variables.

The global subspace can be chosen as the finite element problem of degree one, i.e. as V^1, or, more generally, as another suitable subspace with just one, or a few, degrees of freedom for each element Ω_i of the triangulation of Ω. We have considered the consequences of choosing V^1 as well as more exotic choices. Extra care has to be taken if V^1 is selected; we refer to Pavarino and Widlund [PW95] for a discussion of how it is possible to combine the use of V^1 and fast convergence.

The spectrum relevant to the iterative methods is that of the operator

$$T = \sum_{i=0}^{N} T_i.$$

The eigenvalues of $K^{-1} K_J$, and of T^{-1}, can be shown to be the stationary values of the Rayleigh quotient

$$\frac{\sum_{i=0}^{N} \tilde{a}_i(u_i, u_i)}{a(u, u)}, \quad u = \sum_{i=0}^{N} u_i, \quad u_i \in V_i.$$

Typically, the most challenging part, in work of this kind, is to provide an upper bound for this expression. An upper bound on $a(u_i, u_i)/\tilde{a}_i(u_i, u_i), \forall u_i \in V_i$, for each i, is also required to obtain a lower bound of the eigenvalues of $K_J^{-1} K$. Good bounds can only be obtained by carefully selecting the subspaces V_i, and the bilinear forms $\tilde{a}_i(\cdot, \cdot)$, which together fully specify the iterative method.

Our class of preconditioner can be extended to the case when the finite element space is not a direct sum of subspaces that correspond to blocks of a block-Jacobi splitting. The relevant operators T_i are still well defined for any set of subspaces V_i.

The basic formula, which is easy to prove, is then

$$a(T^{-1}u, u) = inf_{u=\sum u_i} \sum_{i=0}^{N} \tilde{a}_i(u_i, u_i);$$

see Zhang [Zha91]. We note that if the subspaces V_i do not form a direct sum, then there is some freedom of choice in the representation of u at least for some elements of V. This can be turned into a considerable advantage. We can, e.g., start with a direct sum decomposition of the finite element space and then enrich some or all of the subspaces, increasing their dimensions. This is often related to an increase of the overlap of a decomposition of the given region Ω into subdomains. We can also strengthen a preconditioner by adding additional subspaces, e.g., a coarse, global space.

There is an underlying general theory. We note that any block-Jacobi method can be considered as a special *additive Schwarz* method where a direct sum decomposition is used. There are also Gauss-Seidel-like, *multiplicative*, and several *hybrid* Schwarz methods. The derivation of bounds for these families of algorithms, once bounds are proven for the additive case, is by now completely routine; see, e.g., [DSW94, DW95].

4   A choice of subspaces for spectral elements in three dimensions

Our choice of subspaces is directly related to important geometric objects: *interiors, faces, edges,* and *vertices* of the elements. We can merge edges and vertices, creating *wire baskets* and we will describe a *wire basket based* global space V_0; see below. We recall that the subspaces V_i and the energy forms $\tilde{a}_i(\cdot, \cdot)$ define an iterative method completely.

The subspaces and bilinear forms are defined as follows:

We use an interior space for each element: $Q_p \cap H_0^1(\Omega_i)$. Exact solvers are used for these subspaces.

There is also a space for each face Γ_{ij}, where $\overline{\Gamma}_{ij} = \overline{\Omega}_i \cap \overline{\Omega}_j$. The functions of this space vanish on and outside $\partial\Omega_{ij} = \partial(\Omega_i \cup \Gamma_{ij} \cup \Omega_j)$. It is crucial to have a good recipe for the extension of the values on the designated face to the interior of the two relevant elements. The minimal energy (discrete harmonic) extension is used. This corresponds to a change of basis and it greatly improves the performance of the resulting block-Jacobi preconditioner. We note that the use of exact solvers for these subspaces can be relatively expensive but that good alternatives have been developed; see, e.g., Casarin [Cas96].

Our coarse space, of piecewise discrete harmonic functions, is associated with the wire baskets of the elements. Any function in this space is defined fully by its values on the wire basket. In our algorithm, we simply extend the values given on the boundary of a face by assigning the average of the boundary values to all the interior GLL points of the face and complete the extension by using discrete harmonic functions in the interior of the subregions. A simple bilinear form, defined by

$$\tilde{a}_0(u, u) = C(1 + \log p) \sum_i k_i \inf_{c_i} \|u - c_i\|_{L_2(W_i)}^2,$$

is used leading to a special linear system of algebraic equations with one degree of freedom per element, c_i, and a larger subsystem with a diagonal matrix.

Our method is closely modeled on an algorithm developed for lower order elements by Barry Smith; see [Smi91, Smi93]. The following result is established in [PW94a].

Theorem 1

$$\kappa(K_J^{-1}K) \le C(1 + \log p)^2.$$

Here C is a constant, which is independent of p, the values of the k_i, as well as the number of subregions.

Our paper also contains a strict numerical upper bound for the condition number of the method as a function of p. There is also a second, quite different proof due to Mario Casarin, see [Cas95a, Cas95b, Cas96]. Similar bounds hold for a number of other algorithms. We also note that if V^1 is used as the coarse space, and it is combined with carefully chosen local spaces, then the same type of result holds provided that the coefficient function $k(x)$ is quasi-monotone; see Pavarino and Widlund [PW95]. For related work on multigrid methods and a definition of quasi-monotonicity, see Dryja, Sarkis, and Widlund [DSW96].

5   Mortar finite elements

We give a brief description of the mortar elements first introduced by Christine Bernardi, Yvon Maday, and Anthony Patera in [BMP94]. We will focus exclusively on the second generation of these methods; see Ben Belgacem and Maday [BB93, BBM93b]. We confine our study to the case of Poisson's equation, i.e. $k_i = 1$, $\forall i$.

The mortar elements have a number of interesting features. They are nonconforming finite element methods where the partitioning of the region Ω into subregions Ω_i is not necessarily geometrically conforming. Thus, in three dimensions, vertices and edges of one subregion can intersect the interior of edges and/or faces of its neighbors; in two dimensions vertices can intersect edges of neighboring subregions. In each subregion, we are free to choose a standard finite element or spectral element method without much regard to its neighbors. Even if the subregions geometrically conform, the finite element meshes need not. In the spectral case, we can use polynomial spaces of different order in different subregions and we can also mix finite elements and spectral elements.

The mortar element functions are, generally, discontinuous across the interface

$$\Gamma = \overline{\cup_{i=1}^{I} \partial\Omega_i \setminus \partial\Omega}.$$

between the subregions. This interface defines the partition of the region into subregions, and it can sometimes serve as a coarse finite element mesh. The original bilinear form $a(\cdot, \cdot)$ must now be replaced by a bilinear form $a^{\Gamma}(\cdot, \cdot)$, which corresponds to a broken norm. It is defined as the sum of contributions from the individual subregions:

$$a^{\Gamma}(u_h, v_h) = \sum_i a_{\Omega_i}(u_h, v_h). \tag{2}$$

Our analysis only involves arguments about individual subregions and their next neighbors. The subregions are assumed to be shape regular and all neighboring subregions to have diameters of comparable size. However, there is no need to assume that all subregions have comparable diameters.

In the rest of this section, we will focus on piecewise linear finite elements and on two dimensions. To simplify our discussion, we also assume that all of the subregions are triangular (or quadrilateral) and that the triangulation of each subregion is quasi-uniform and that it is obtained by successive refinements of the subregions; cf., e.g., [Yse86]. This, in particular, will be the framework for the preconditioner discussed in the next section. We denote the diameter of the subregion Ω_i by H_i, and the smallest diameter of any of its elements by h_i. Our results depend only on the minimal angle of the overall triangulation, and ℓ, the maximum of the number of refinement levels $\ell(i)$ of the subregions Ω_i.

Thus, each Ω_i is subdivided by a nested family of standard conforming finite element triangulations: $\mathcal{T}_0^i = \{\Omega_i\}, \mathcal{T}_1^i, \mathcal{T}_2^i, \ldots, \mathcal{T}_{\ell(i)}^i$. The quasi-uniform triangulation $\mathcal{T}_{k+1}^i$ is obtained from the next coarser triangulation, $\mathcal{T}_k^i$, by subdividing each of its triangles into four shape-regular, but not necessarily equal, triangles. We assume that the triangles of level $k + 1$ have diameters of an order approximately one half of the diameter of those of level k.

A set of *mortars* $\{\gamma_m\}_{m=1}^M$ is obtained by selecting open edges of the subregions such that

$$\Gamma = \cup_{m=1}^M \overline{\gamma}_m, \quad \gamma_m \cap \gamma_n = \emptyset \;\text{ if } m \neq n.$$

Each edge is viewed as belonging to just one subregion. The other edges are the *non-mortars* and are denoted by δ_n. The restrictions of the triangulations of the different subregions to the mortars and nonmortars will typically not match; they are denoted by γ_m^h and δ_n^h, respectively. Discontinuous mortar finite element functions have two different traces on the interface Γ given by one-sided limits of finite element functions defined on the individual subregions. The continuity across the interface of a conforming finite element method is replaced by weak continuity across the individual nonmortars:

For each n, we define a space of test functions $\mathcal{W}^h(\delta_n)$ given by the restriction, to the nonmortar δ_n, of the finite element space defined on the subregion of which δ_n is an edge. The elements of $\mathcal{W}^h(\delta_n)$ are subject to the further constraints that they are constant in the first and last mesh intervals of δ_n^h.

The *mortar projection* π_n maps all of $L_2(\delta_n)$ onto the finite element space defined on the nonmortar mesh δ_n^h. Given $w \in L_2(\delta_n)$ and values $u^{(n)}(v_{n_1})$ and $u^{(n)}(v_{n_2})$ at the two endpoints v_{n_1} and v_{n_2} of δ_n, we determine the values of $\pi_n(w, u^{(n)}(v_{n_1}), u^{(n)}(v_{n_2}))$ on δ_n^h by

$$\int_{\delta_n} (w - \pi_n(w, u^{(n)}(v_{n_1}), u^{(n)}(v_{n_2})))\psi ds = 0, \quad \forall \psi \in \mathcal{W}^h(\delta_m). \tag{3}$$

We note that only the values at the nodes interior to δ_n are determined by this condition; the values $u^{(n)}(v_{n_1})$ and $u^{(n)}(v_{n_2})$ are genuine degrees of freedom of the finite element model.

We note that in the spectral case, a class of polynomials are used as test functions. They are of a degree two less than the space of traces of the spectral element functions given on the subregion of which δ_n is an edge.

After these preparations, the mortar finite element space V^h can now be fully defined: The restriction of V^h to Ω_i, $V^h(\Omega_i)$, is a regular conforming finite element space as described above, and the jump across each nonmortar δ_n satisfies the set of constraints given by (3).

The discrete problem is then: Find $u \in V^h$ such that

$$a^{\Gamma}(u, v) = f^{\Gamma}(v), \quad \forall v \in V^h, \tag{4}$$

where $a^{\Gamma}(u, v)$ is defined in formula (2) and, similarly, $f^{\Gamma}(v)$ is the sum of contributions from the different subregions.

The rate of convergence of the solution of (4) to the solution of (1) is comparable to that of a conforming discretization; cf. [BBM93a], [BMP94], and references therein for theoretical and experimental results.

6 A preconditioner for geometrically conforming problems

We now consider the case of a geometrically conforming decomposition of a plane region. We note that even though we will use a hierarchical basis in the design of our preconditioner, we can primarily work with a nodal basis. We will use a nodal basis of the mortar finite element space which is associated with all nodes interior to the subregions and on $\partial\Omega_N$, all nodes interior to the mortars, and all nodes at all the vertices of the subregions except those which fall on $\partial\Omega_D$.

Our algorithm, fully described and analyzed in [CW95], is based on earlier work by Smith and Widlund [SW90]. We note that the analysis of the new algorithm is considerably more intricate than that of the earlier paper in which only conforming finite elements were considered.

At the expense of an exact solution of a homogeneous Dirichlet problem for each subregion, we reduce problem (4) to that of finding the piecewise discrete harmonic part of the solution. We recall that a finite element function u is discrete harmonic in the subregion Ω_i if

$$a^{\Gamma}(u, v) = 0 \quad \forall v \in V^h \cap H_0^1(\Omega_i),$$

and that a discrete harmonic function provides the unique minimal energy extension of finite element boundary data given on the boundary $\partial\Omega_i$. This static condensation step reduces the size of the discrete system. We note that it is not necessary to compute a matrix representation of the related Schur complement. We only need matrix-vector products with the Schur complement and these can be computed at the expense of solving a Dirichlet problem for each subregion. After finding sufficiently accurate values on Γ, the solution of (4) is then computed everywhere by solving a finite element problem for each subregion Ω_i with Dirichlet data given on $\partial\Omega_i \setminus \partial\Omega_N$.

The values interior to each subregion correspond to a local subspace subspace of V^h. The other subspaces, which define our iterative substructuring method, are associated with the vertices of the subregions, their edges, and there is also a simple global subspace.

The set of vertices of the subregions associated with degrees of freedom of V^h, i.e. those with values not given by the Dirichlet data on $\partial\Omega_D$, is denoted by $\mathcal{V}$. Each crosspoint of Γ corresponds to several nodes of $\mathcal{V}$ and to one degree of freedom for each of the subregions that meet at that point; these nodes are in the same geometrical position, but are assigned to different subregions. In order to describe and analyze our algorithm, we define a special vertex basis function ϕ_{v_ℓ} for each of these degrees of freedom and derive estimates of their norms; for these estimates, see [CW95].

For each vertex v_ℓ of $\mathcal{V}$, let $\phi_{v_\ell} \in V^h(\Omega)$ be given the value 1 at v_ℓ, while all other degrees of freedom on Γ are set to zero. This completely defines ϕ_{v_ℓ} since the interior nodal values on the nonmortars are given by the mortar projections, and those in the interior of the Ω_i by discrete harmonic extensions.

A one-dimensional vertex space is associated with each $v_\ell \in \mathcal{V}$:

$$V_{v_\ell} = \text{span of } \phi_{v_\ell}.$$

We use the exact bilinear form $a^\Gamma(\cdot, \cdot)$ for these spaces.

Before we can introduce the edge subspaces, and their bilinear forms, of our iterative method, we need to review some aspects of Yserentant's hierarchical basis method; cf. [Yse86]. We denote by $\mathcal{N}_k^i$, $k = 0, 1, \ldots, \ell(i)$, the set of vertices of the triangles of $\mathcal{T}_k^i$, by V_k^i the space of continuous functions on $\overline{\Omega}_i$ that are linear in the triangles of $\mathcal{T}_k^i$, and by V^i the most refined space $V_{\ell(i)}^i$. All elements of all V_k^i vanish on $\partial\Omega_i \cap \partial\Omega_D$. An interpolation operator $I_k^i : V^i \to V_k^i$, is defined by

$$I_k^i u(x) = u(x) \quad \forall x \in \mathcal{N}_k^i.$$

Following Yserentant [Yse86], we define a discrete norm, for any set $\Lambda \subset \overline{\Omega}_i$, by

$$|||u|||_\Lambda^2 = \sum_{k=1}^{\ell(i)} \sum_{x \in \mathcal{N}_k^i \setminus \mathcal{N}_{k-1}^i \cap \overline{\Lambda}} |(I_k^i u - I_{k-1}^i u)(x)|^2. \tag{5}$$

Let W_k^i be the image of $I_k^i - I_{k-1}^i$; this is the subspace of functions of V_k^i that vanish on $\mathcal{N}_{k-1}^i$. A hierarchical basis of V^i can now be defined recursively. The hierarchical basis of V_0^i is the standard finite element nodal basis restricted to the single triangle Ω_i. It is clear that $V_k^i = V_{k-1}^i + W_k^i$, $k \geq 1$. In each step, we augment the hierarchical basis of V_{k-1}^i by the level k nodal basis functions which span $W_k^i \subset V_k^i$. For a function u represented in this basis, the discrete norm $|||u|||_\Lambda^2$ is simply the Euclidean norm of its coefficients and thus very easy to compute. Moreover, the transformation between the standard nodal basis and the hierarchical basis is very fast and easy to implement; see [SW90] and [Yse86]. This is especially true in the present context since the change of basis can be carried out edge by edge, and in parallel.

A subspace $V_{i(m)} = V_{\gamma_m}$ is now associated with each mortar γ_m. The bilinear form for this subspace is given by $\tilde{a}_{i(m)}(u, u) = |||u|||_{\gamma_m}^2$. The elements of this local space vanish on $\Gamma \setminus \gamma_m$ and they are continued into the interior of the subregions as discrete harmonic functions.

Finally, a coarse space, which is conforming, is given by

$$V_0 = \{u \in V^h \mid u \text{ is linear on each } \Omega_i\} \cap V.$$

The bilinear form associated with V_0 is $a^\Gamma(\cdot, \cdot)$ which coincides with $a(\cdot, \cdot)$ on this subspace since the restriction of any element of V_0 to an edge is a linear function and therefore satisfies the mortar jump condition.

The Schwarz framework provides a preconditioned equation $Tu = b$, in terms of these spaces and bilinear forms, which has the same solution as (4). The main result of [CW95] is the following theorem.

Theorem 2 *The condition number of T satisfies*

$$\kappa(T) \leq C(1 + \ell)^2.$$

In our relatively extensive numerical experiments, we have found that the rate of convergence tends to be slightly better than for the algorithm described in [SW90]; see [CW95].

7   Algorithms for geometrically nonconforming problems

We now give a short description of recent joint work with Yves Achdou and Yvon Maday; see [AMW96b, AMW96a]. This work addresses more general classes of problems; the partition of the two-dimensional region into subregions can be geometrically non-conforming and spectral elements, as well as lower order finite elements are considered. These algorithms are also iterative substructuring methods with interior local subspaces and complementary families of subspaces with piecewise discrete harmonic elements.

In these algorithms, the piecewise discrete harmonic part of the mortar element space is partitioned into a direct sum of subspaces. The coarse space is of higher dimension than that of the method described in the previous section, but it is also well defined in the geometrically non-conforming case. One degree of freedom of the coarse space is associated with each vertex. The corresponding basis function takes on the value one at the designated vertex and vanishes at all the others. On each mortar it is a linear function; the values at the nonmortars are, as always, determined by those on the mortars and at the vertices. We note that in a geometrically conforming, lower order finite element case, this coarse space is strictly contained in the sum of the coarse space and the vertex spaces considered in the previous section. A main technical challenge is to provide a sufficiently good bound on the energy norm of these basis functions, and a general element of the coarse space. For the spectral case, several new technical tools were required; see [AMW96a] for full details.

In addition, we use one subspace for each mortar γ_m. Just as in the previous section, an element in the space $V_{i(m)} = V_{\gamma_m}$ is defined by its values in the interior of γ_m and it vanishes on $\Gamma \setminus \gamma_m$. The basic results of [AMW96b, AMW96a] concerns algorithms which use exact solvers for these spaces, but a somewhat weaker result is also obtained for a class of inexact solvers in the lower order finite element case.

The resulting polylogarithmic bounds, see [AMW96b, AMW96a], are quite similar to those of the other sections of this paper.

8   An algorithm for mortar elements in three dimensions

We now turn to a brief discussion of an iterative substructuring method for problems in three dimensions, which has been designed and analyzed jointly with Yvon Maday. It is a *face based* method, a family of methods originating in the 1994 PhD thesis of Marcus Sarkis; see [Sar94]. Our method is described by families of subspaces, which fully define the domain decomposition method; in this algorithm the original bilinear form, $a^\Gamma(\cdot,\cdot)$, and exact solvers are used for all the subspaces. Our results, so far, are only for lower order finite elements, but we hope to extend them to the spectral case.

The coarse space has discrete harmonic basis functions, one for each mortar. A basis function is equal to 1 at all of the interior mesh points of one mortar, γ_m, and vanishes at all other nodes on the interface Γ except those on the boundary of γ_m. On each edge, and at each vertex, of γ_m, they take on positive constant values chosen so that the basis functions form a partition of unity. The values on the nonmortars are determined by insisting that the coarse space is a subspace of the original mortar finite element space.

There is one interior space for each subregion with functions that vanish on, and outside, the boundary of the subregion; this subspace is quite similar to an interior subspace for a conforming finite element space.

For each mortar, there is a local space defined by arbitrary values at the interior nodes of the mortar in question, with all other degrees of freedom on the interface Γ set to zero. This is a natural generalization of the local spaces discussed previously.

For each degree of freedom on the wire basket, there is a one-dimensional subspace defined by one standard mortar basis function.

We note that these subspaces do not form a direct sum decomposition of the mortar finite element space; the local spaces themselves span the entire space.

The following result will be established in a forthcoming joint paper with Maday.

Theorem 3 *For the method described in this section,*

$$\kappa(T) \le C(1 + \log (H/h))^2.$$

Here C is a constant, which is independent of H, h, as well as the number of subregions.

9   Acknowledgements

The author wishes to thank his coworkers Yves Achdou, Mario Casarin, Maksymilian Dryja, Yvon Maday, and Luca Pavarino for their assistance and many interesting discussions.

References

[AK95] Achdou Y. and Kuznetsov Y. A. (1995) Substructuring preconditioners for finite element methods on nonmatching grids. *East-West J. Numer. Math.* 3(1): 1–28.

[AKP95] Achdou Y., Kuznetsov Y. A., and Pironneau O. (1995) Substructuring preconditioners for the Q_1 mortar element method. *Numer. Math.* 71(4): 419–449.

[AMW96a] Achdou Y., Maday Y., and Widlund O. B. (1996) Iterative substructuring preconditioners for the mortar finite element method in two dimensions. Technical report, Department of Computer Science, Courant Institute. In preparation.

[AMW96b] Achdou Y., Maday Y., and Widlund O. B. (1996) Méthode itérative de sous-structuration pour les éléments avec joints. *C.R. Acad. Sci. Paris* 322: 185–190.

[BB93] Ben Belgacem F. (January 1993) *Discretisations 3D Non Conformes pour la Méthode de Decomposition de Domaine des Elément avec Joints: Analyse Mathématique et Mise en Œvre pour le Probleme de Poisson.* PhD thesis, Université Pierre et Marie Curie, Paris, France. Tech. Rep. HI-72/93017, Electricité de France, Paris, France.

[BBM93a] Ben Belgacem F. and Maday Y. (1993) The mortar element method for three dimensional finite elements. Unpublished paper based on Yvon Maday's talk at the Seventh International Conference of Domain Decomposition Methods in Scientific and Engineering Computing, held at Penn State University, October 27-30, 1993.

[BBM93b] Ben Belgacem F. and Maday Y. (1993) Non-conforming spectral element method for second order elliptic problem in 3D. Technical Report R93039, Laboratoire d'Analyse Numérique, Université Pierre et Marie Curie – Centre National de la Recherche Scientifique.

[BM92] Bernardi C. and Maday Y. (1992) *Approximations Spectrales de Problèmes aux Limites Elliptiques*, volume 10 of *Mathèmatiques & Applications*. Springer-Verlag France, Paris.

[BMP94] Bernardi C., Maday Y., and Patera A. T. (1994) A new non conforming approach to domain decomposition: The mortar element method. In Brezis H. and Lions J.-L. (eds) *Collège de France Seminar*. Pitman. This paper appeared as a technical report about five years earlier.

[Cas95a] Casarin M. A. (1995) Quasi-optimal Schwarz methods for the conforming spectral element discretization. In Melson N. D., Manteuffel T. A., and McCormick S. F. (eds) *Proceedings of the 1995 Copper Mountain Conference on Multigrid Methods*. NASA, Hampton VA.

[Cas95b] Casarin M. A. (September 1995) Quasi-optimal Schwarz methods for the conforming spectral element discretization. Technical Report 705, Department of Computer Science, Courant Institute.

[Cas96] Casarin M. A. (March 1996) *Schwarz Preconditioners for Spectral and Mortar Finite Element Methods with Applications to Incompressible Fluids.* PhD thesis, Courant Institute of Mathematical Sciences. Tech. Rep. 717, Department of Computer Science, Courant Institute.

[CW95] Casarin M. A. and Widlund O. B. (December 1995) A hierarchical preconditioner for the mortar finite element method. Technical Report 712, Department of Computer Science, Courant Institute.

[Dry96] Dryja M. (1996) Additive Schwarz methods for elliptic mortar finite element problems. In Malanowski K., Nahorski Z., and Peszynska M. (eds) *Modeling and Optimization of Distributed Parameter Systems with Applications to Engineering.* IFIP, Chapman & Hall, London. To appear.

[DSW94] Dryja M., Smith B. F., and Widlund O. B. (December 1994) Schwarz analysis of iterative substructuring algorithms for elliptic problems in three dimensions. *SIAM J. Numer. Anal.* 31(6): 1662–1694.

[DSW96] Dryja M., Sarkis M. V., and Widlund O. B. (1996) Multilevel Schwarz methods for elliptic problems with discontinuous coefficients in three dimensions. *Numer. Math.* 72(3): 313–348.

[DW95] Dryja M. and Widlund O. B. (February 1995) Schwarz methods of Neumann-Neumann type for three-dimensional elliptic finite element problems. *Comm. Pure Appl. Math.* 48(2): 121–155.

[FR94] Fischer P. F. and Rønquist E. (1994) Spectral element methods for large scale parallel Navier- Stokes calculations. *Comput. Methods Appl. Mech. Engrg* 116: 69–76. Proceedings of ICOSAHOM 92, a conference held in Montpellier, France, June 22-26, 1992.

[LTS93] Le Tallec P. and Sassi T. (1993) Domain decomposition with nonmatching grids: Augmented Lagrangian approach. *Mathematics of Computation* to appear.

[Man90] Mandel J. (1990) Two-level domain decomposition preconditioning for the p-version finite element version in three dimensions. *Int. J. Numer. Meth. Eng.* 29: 1095–1108.

[Man94] Mandel J. (1994) Iterative solvers for p-version finite element method in three dimensions. *Comput. Methods Appl. Mech. Engrg* 116: 175–183. Proceedings of ICOSAHOM 92, a conference held in Montpellier, France, June 1992.

[MW96] Maday Y. and Widlund O. B. (1996) Some iterative substructuring methods for mortar finite elements: The lower order case. Technical report, Courant Institute of Mathematical Sciences. In preparation.

[PW94a] Pavarino L. F. and Widlund O. B. (May 1994) Iterative substructuring methods for spectral elements: Problems in three dimensions based on numerical quadrature. Technical Report 663, Courant Institute of Mathematical Sciences, Department of Computer Science. To appear in Computers Math. Applic.

[PW94b] Pavarino L. F. and Widlund O. B. (March 1994) A polylogarithmic bound for an iterative substructuring method for spectral elements in three dimensions. Technical Report 661, Courant Institute of Mathematical Sciences, Department of Computer Science. To appear in SIAM J. Numer. Anal., 33-4.

[PW95] Pavarino L. F. and Widlund O. B. (1995) Preconditioned conjugate gradient solvers for spectral elements in 3D. In Habashi W. G. (ed) *Solution Techniques for Large-Scale CFD Problems*, pages 249–270. John Wiley & Sons. Proceedings of the International Workshop on Solution Techniques for Large-Scale CFD Problems held at CERCA, Montréal, Canada, September 26–28, 1994.

[Sar94] Sarkis M. V. (September 1994) *Schwarz Preconditioners for Elliptic Problems with Discontinuous Coefficients Using Conforming and Non-Conforming Elements.* PhD thesis, Courant Institute, New York University.

[Smi91] Smith B. F. (1991) A domain decomposition algorithm for elliptic problems in three dimensions. *Numer. Math.* 60(2): 219–234.

[Smi93] Smith B. F. (March 1993) A parallel implementation of an iterative substructuring algorithm for problems in three dimensions. *SIAM J. Sci. Comput.* 14(2): 406–423.

[SW90] Smith B. F. and Widlund O. B. (1990) A domain decomposition algorithm using a hierarchical basis. *SIAM J. Sci. Stat. Comput.* 11(6): 1212–1220.

[Wid94] Widlund O. B. (1994) Exotic coarse spaces for Schwarz methods for lower order and spectral finite elements. In Keyes D. E. and Xu J. (eds) *Seventh In-*

ternational Conference of Domain Decomposition Methods in Scientific and Engineering Computing, volume 180 of *Contemporary Mathematics*, pages 131–136. AMS. Held at Penn State University, October 27-30, 1993.

[Yse86] Yserentant H. (1986) On the multi-level splitting of finite element spaces. *Numer. Math.* 49: 379–412.

[Zha91] Zhang X. (September 1991) *Studies in Domain Decomposition: Multilevel Methods and the Biharmonic Dirichlet Problem.* PhD thesis, Courant Institute, New York University.

Algorithms for the Mortar Element Method

YVES ACHDOU[1] and YURI KUZNETSOV[2]

ABSTRACT

We consider the saddle point type linear systems obtained from the discretization of a second order symmetric elliptic equation by the mortar element method. An iterative method in a subspace for solving these systems is described. This algorithm is based on a special class of preconditioners. Several preconditioners are then proposed.

1 INTRODUCTION

The mortar element method introduced in [7], [6], [16] is a finite element method based on domain decomposition which permits to use meshes non necessarily matching at subdomain interfaces, or different finite element approximations in different subdomains. Conformity is impossible since continuity across the interfaces between subdomains cannot be achieved (the meshes do not match), and one has to impose only some kind of weak continuity: for each interface Γ_{kl}, one has to introduce a suitable space W_{kl} of finite element functions supported on Γ_{kl}, and the continuity constraint is that the $L^2(\Gamma_{kl})$ projection of the jump across Γ_{kl} on the space W_{kl} vanishes.

Such a method has many possible advantages:

- It is genuinely suited for parallel computing.

[1] CMAP,Ecole Polytechnique 91128 Palaiseau cedex: achdou&cmapx.polytechnique.fr

[2] Institute of Numerical Mathematics, Russian Academy of Sciences, 32a Leninskij Prospect, Moscow 117334 : kuznetsov&unitron1.inm.ras.ru

Domain Decomposition Methods in Sciences and Engineering, edited by R. Glowinski *et al.*

- It provides flexibility for the construction of the finite element mesh. This flexibility may be exploited to avoid updating the finite element mesh (sliding meshes [5]) or on the contrary for adapting the mesh.

The mortar method has been used in [4] for designing a solver for the Navier Stokes equations.

The aim of this paper is to present a class of solvers for the linear systems obtained when applying the mortar element method to second order elliptic partial differential equations. Here we consider the saddle point formulation of the discrete problem where the weak continuity across interfaces is treated as a constraint and the related Lagrange multiplier is a discretization of the normal derivative at interfaces, and we choose to eliminate the degrees of freedom (d.o.f.) interior to subdomains. Other attractive algorithms can be designed: see [13] for a 3D algorithm avoiding the elimination of the d.o.f. interior to subdomains, [16],15] for a Neumann-Neumann algorithm, [3], [17] for substructuring algorithms based on a two level block diagonal preconditioners with suitably chosen coarse spaces.

The section 2 will be devoted to a brief review on the mortar element method. In §3, we discuss an iterative method for saddle point problems, namely the so called preconditioned conjugate gradient in a subspace of constraints introduced in [14]. In §4, we apply this method for designing an algorithm for the mortar method. In §5, we discuss possible preconditioners.

2   THE DISCRETE PROBLEM

We consider the symmetric elliptic equation

$$- div\alpha(x)gradu + \beta(x)u = f \quad \text{in } \Omega, \qquad u = 0 \quad \text{on } \partial\Omega, \tag{2.1}$$

where Ω is a domain of $\mathbb{R}^N$ ($N = 2, 3$), and α (resp. β) is a positive (resp. nonnegative) function. For simplicity we suppose that Ω is polygonal and that $N = 2$.

2.1   The geometry.

Let $\{\Omega_k\}$ be a partition of Ω into K non-overlapping open polygonal subdomains:

$$\overline{\Omega} = \cup_{k=1}^{K}\overline{\Omega}_k \qquad \text{and} \qquad \Omega_k \cap \Omega_l = \emptyset \quad \text{if } k \neq l. \tag{2.2}$$

For simplicity, we also suppose that the domain decomposition is geometrically conforming, which means that the intersection of the closures of two subdomains is either empty or a vertex or a whole edge. For any $1 \leq k, l \leq K$, let Γ_{kl} be the closed straight segment, possibly degenerate : $\Gamma_{kl} = \overline{\Omega}_k \cap \overline{\Omega}_l$.

2.2   The discretization.

With each $1 \leq k \leq K$, we associate a family of quasi uniform triangular finite element meshes $\mathcal{T}_{k,h}$ of Ω_k with the classical regularity assumption for F.E.M., and we denote X_{kh} the related space of P_1 finite element functions vanishing on $\partial\Omega$. Let h_k be the maximal diameter of the elements of $\mathcal{T}_{k,h}$. Let X_h denote the product space :

$$X_h = \prod_{1 \leq k \leq K} X_{kh}. \tag{2.3}$$

Note that the meshes do not need to match at interfaces. Therefore, in order to build a finite element space approaching $H^1(\Omega)$ one has to write a weak continuity constraint at the subdomain interfaces. Let us define the space of the Lagrange multipliers for the continuity constraint: we denote by Tr_k the trace on $\partial\Omega_k$. If $|\Gamma_{kl}| \neq 0$, the space $\tilde{W}_{k,l,h} = \left\{ Tr_k v_{|\Gamma_{kl}}, \quad v \in X_{kh} \right\}$ has dimension $N_{kl} + 2$, ($N_{kl} + 2$ is the number of vertices of $\mathcal{T}_{k,h}$ lying on Γ_{kl}).

For each interface Γ_{kl}, one can build the Lagrange multiplier space either from $\tilde{W}_{k,l,h}$ or from $\tilde{W}_{l,k,h}$. One possibility is to choose the space corresponding to the finer mesh: assuming the Lagrange multiplier space is built from $\tilde{W}_{k,l,h}$, let us choose $W_{k,l,h}$ as the subspace of $\tilde{W}_{k,l,h}$ of codimension 2 of functions which are constant near the two ends of Γ_{kl}. Let us call W_h the Lagrange multiplier space:

$$W_h = \prod_{1 \leq k < l \leq K:\, |\Gamma_{kl}| \neq 0} W_{k,l,h}. \tag{2.4}$$

Calling b the bilinear form

$$\begin{aligned} b: \quad & X_h \times W_h \to \mathbb{R}, \\ & b(\mathbf{v}_h, \mu_h) = \sum_{k<l\,:\,|\Gamma_{kl}|\neq 0} \int_{\Gamma_{kl}} \mu_{klh}(v_{kh} - v_{lh}), \end{aligned} \tag{2.5}$$

We are now able to define the subspace Y_h of X_h :

$$Y_h \equiv \{ \mathbf{v}_h \in X_h : \quad \forall \mu_h \in W_h, \quad b(\mu_h, \mathbf{v}_h) = 0 \}. \tag{2.6}$$

Calling a the bilinear form :

$$\begin{aligned} a: \quad & X_h \times X_h \to \mathbb{R}, \\ & a(\mathbf{u}_h, \mathbf{v}_h) \equiv \sum_{k=1}^{K} \int_{\Omega_k} \alpha \nabla u_{kh} \cdot \nabla v_{kh} + \beta u_{kh} v_{kh} \end{aligned} \tag{2.7}$$

the discretization of (2.1) is to find $\mathbf{u}_h \in Y_h$ such that

$$\forall \mathbf{v}_h \in Y_h, \quad a(\mathbf{u}_h, \mathbf{v}_h) = \sum_{k=1}^{K} \int_{\Omega_k} f\, v_{kh}, \tag{2.8}$$

which is clearly a well posed problem. It is easily proved that (2.8) is equivalent to the following well posed saddle point problem: find $(\mathbf{u}_h, \lambda_h) \in X_h \times W_h$ such that

$$\begin{aligned} \forall \mathbf{v}_h \in X_h, \quad & a(\mathbf{u}_h, \mathbf{v}_h) + b(\mathbf{v}_h, \lambda_h) = (f, \mathbf{v}_h), \\ \forall \mu_h \in W_h, \quad & b(\mathbf{u}_h, \mu_h) = 0. \end{aligned} \tag{2.9}$$

For the numerical analysis of the method, we refer to [7], [6]. The choice of the space W_h was made in order to achieve the Babuska-Brezzi inf-sup condition, (see [6]).

Remark 1 In [1], a variant of the method has been studied, where the jump operator is modified by means of mass lumping: indeed, defining the bilinear forms $b^{kl} : X_{kh} \times W_{klh} \to \mathbb{R}$ and $b^{lk} : X_{lh} \times W_{klh} \to \mathbb{R}$ by

$$b(\mathbf{v}_h, \mu_h) = \sum_{|\Gamma_{kl}|>0} b^{kl}(v_{kh}, \mu_{klh}) + b^{lk}(v_{lh}, \mu_{klh}), \qquad (2.10)$$

and assuming that the space W_{klh} is constructed with the mesh of Ω_k, the idea is to replace in (2.10) the bilinear form b^{kl} by $\tilde{b}^{kl}$ obtained by performing mass lumping on the matrix of b^{kl}, and by keeping b^{lk} unchanged. A new jump bilinear form $\tilde{b}$ is thus obtained by assembly, and this leads to a new approximation method. With this new approximation, the same error estimates as in the original method can be obtained provided the meshes on which the Lagrange multiplier space W_h is built are sufficiently close to being uniform. As explained later, this modified jump operator may permit to design easily preconditioners with optimal arithmetical complexity.

3   PRECONDITIONED ITERATIVE METHODS IN A SUBSPACE OF CONSTRAINTS

Consider the saddle point problem:

$$\mathcal{S}\begin{pmatrix} V \\ \Lambda \end{pmatrix} = \begin{pmatrix} S & B^T \\ B & 0 \end{pmatrix} \begin{pmatrix} V \\ \Lambda \end{pmatrix} = \begin{pmatrix} G \\ 0 \end{pmatrix} \qquad (3.1)$$

where S is a block diagonal matrix with K blocks S^k. We assume that $\mathcal{S}$ and BB^T are non singular, and that the blocks S^k are symmetric and positive semi-definite. We define the preconditioner for matrix $\mathcal{S}$ as

$$\mathcal{R} \equiv \begin{pmatrix} R & B^T \\ B & 0 \end{pmatrix} \qquad (3.2)$$

where R has the same block structure as S. We assume that $\mathcal{R}$ is non singular and that the blocks R^k are symmetric and positive semi-definite with $Ker(R^k) \subset Ker(S^k)$.

Following [14], [12], [2], we apply for solving system (3.1) the Preconditioned Conjugate Gradient method in the Subspace

$$V_B = (I - \mathcal{R}^{-1}\mathcal{S})^2 \hat{V}_B \subset \hat{V}_B, \qquad (3.3)$$

with

$$\hat{V}_B = \left\{ \begin{pmatrix} V \\ \Lambda \end{pmatrix} : BV = 0 \right\}. \qquad (3.4)$$

Is is possible to use the preconditioned conjugate gradient algorithm because from the assumptions above, the following results can be proved:

1. the matrix $\mathcal{R}^{-1}\mathcal{S}$ keeps the subspace V_B invariant, *i.e.* $\mathcal{R}^{-1}\mathcal{S}V_B \subset V_B$,
2. the matrix $\mathcal{S}$ defines a scalar product in V_B,
3. the matrix $\mathcal{R}^{-1}\mathcal{S}$ is symmetric and positive definite in V_B with respect to the energy scalar product generated by the matrix $\mathcal{S}$, *i.e.* $\mathcal{S}\mathcal{R}^{-1}\mathcal{S}$ is symmetric and positive definite in V_B.

Remark 2 Under the condition: $R - S$ positive (or negative) definite, it is possible to choose $V_B = (I - \mathcal{R}^{-1}S)\hat{V}_B \subset \hat{V}_B$ instead of (3.3), see [8], [11].

4   ELIMINATION OF THE D.O.F. INTERIOR TO SUBDOMAINS

We supply X_h and W_h with their natural basis of nodal functions. Then the matrix form of system (2.9) is

$$\begin{pmatrix} A & \mathbf{B}^T \\ \mathbf{B} & 0 \end{pmatrix} \begin{pmatrix} U \\ \Lambda \end{pmatrix} = \begin{pmatrix} F \\ 0 \end{pmatrix} \tag{4.1}$$

The matrix A is a block diagonal matrix (one block per subdomain), each block corresponds to a discrete Neumann problem in subdomain Ω_k, except if $\overline{\Omega_k} \cap \partial\Omega \neq \emptyset$. It is possible to eliminate the d.o.f. located in the interior of subdomains by solving discrete Dirichlet problems. This leads to the system

$$S\begin{pmatrix} V \\ \Lambda \end{pmatrix} \equiv \begin{pmatrix} S & B^T \\ B & 0 \end{pmatrix} \begin{pmatrix} V \\ \Lambda \end{pmatrix} = \begin{pmatrix} G \\ 0 \end{pmatrix} \tag{4.2}$$

where B denotes the nonzero block of $\mathbf{B}$, and S is the block diagonal matrix whose k^{th} block S^k corresponds to a discretized version of the Steklov-Poincaré operator of subdomain Ω_k: $H^{\frac{1}{2}}(\Omega \cap \partial\Omega_k) \to H^{-\frac{1}{2}}(\Omega \cap \partial\Omega_k)$, $v \to \frac{\partial u_k}{\partial n}$, where u_k is defined by :

$$-div\,\alpha\,grad\,u_k + \beta\,u_k = 0 \quad \text{in } \Omega_k, \quad u_k = v \quad \text{on } \Omega \cap \partial\Omega_k, \quad u_k = 0 \quad \text{on } \partial\Omega \cap \partial\Omega_k.$$

Clearly, the matrix S satisfies all the assumptions of §3 and it is thus possible to use the algorithm described above. We then have to choose properly the matrix R and therefore the preconditioner $\mathcal{R}$. This will be the topic of the next section. Note finally that the elimination procedure is not compulsory (see [13] for an algorithm avoiding elimination).

5   PRECONDITIONERS

We first introduce two inexpensive preconditioners, which are not optimal as regards the condition number estimates, but which lead to linear systems which can be solved at an arithmetical cost proportional to their number of unknowns. Then we discuss inner iterative methods for better preconditioners in terms of condition number estimates.

5.1   Two inexpensive preconditioners

The first preconditioner will be introduced for the following symmetric elliptic p.d.e. (2.1) where Ω is a domain of $\mathbb{R}^N$ ($N = 2, 3$), and α (resp. β) are positive (resp. nonnegative) functions, for simplicity constant in each subdomain. The values of α and β in Ω_k are denoted α_k and β_k, and the jumps of α and β across the interfaces can be arbitrarily large. Then following [9], we choose R as the block diagonal matrix whose k^{th} block is

$$R^k \equiv h_k^{N-2}\alpha_k(I^k - P^k) + d_k h_k^{N-1}\beta_k P^k. \tag{5.1}$$

Here d_k is the diameter of subdomain Ω_k (we assume that the aspect ratio of the subdomains is bounded by a constant), I^k is the identity, and P^k is the matrix of the operator which maps a function defined on $\partial\Omega_k$ to its mean value. We have the following result

Proposition 5.1 *The condition number* $\kappa(\mathcal{R}^{-1}\mathcal{S})$ *with respect to subspace* V_B *satisfies*

$$\kappa(\mathcal{R}^{-1}\mathcal{S}) \leq C \max_k \frac{d_k}{h_k}, \tag{5.2}$$

where the positive constant C *does not depend on* α, β, h_k *and* d_k.

Efficients algorithms can be designed for solving the preconditioning problem: indeed, it is easily observed that R is a low rank perturbation of a diagonal matrix D ($R = D + L$, $rank(L) = K$). Therefore we split the matrix $\mathcal{R}$ into

$$\mathcal{R} = \mathcal{D} + \mathcal{L} \equiv \begin{pmatrix} D & B^T \\ B & 0 \end{pmatrix} + \begin{pmatrix} L & 0 \\ 0 & 0 \end{pmatrix}. \tag{5.3}$$

Since the rank of $\mathcal{L}$ is exactly K, the preconditioning system will consist essentially of solving twice the linear systems with $\mathcal{D}$, and once a *coarse problem* of size K. In two or three dimensions, the system with $\mathcal{D}$ can be solved by eliminating first the unknown U. This leads to a linear system with matrix $B^T D^{-1} B$ which can be solved by a preconditioned iterative method where the preconditioner would be obtained by performing mass lumping on $B^T D^{-1} B$. For a desired precision, the total cost of solving the preconditioner problem is proportional to the number of unknowns.

Alternatively, in two dimensions a direct solver can also be proposed: we first reorder the unknowns into two groups: the second group is made of the d.o.f. of U located at the crosspoints of the domain decomposition and the first group contains the remaining d.o.f. of U and the d.o.f. of Λ. With this ordering, the matrix $\mathcal{D}$ becomes

$$\begin{pmatrix} D_e & B_e^T & 0 \\ B_e & 0 & B_c \\ 0 & B_c^T & D_c \end{pmatrix} \tag{5.4}$$

where e stands for *edges* and c for *crosspoints*. The idea is to eliminate first the unknowns of U_e and Λ, which yields a sparse system whose dimension is proportional to the number of crosspoints and which can be solved by means of a direct Choleski method. To eliminate the d.o.f. of U_e and Λ, we group together the unknowns of U_e and Λ corresponding to same interfaces, and the submatrix

$$\begin{pmatrix} D_e & B_e^T \\ B_e & 0 \end{pmatrix}$$

becomes a block diagonal matrix (one block per interface) and the block related to interface Γ_{kl} is denoted

$$\begin{pmatrix} D_e^{kl} & 0 & B_e^{kl\,T} \\ 0 & D_e^{lk} & B_e^{lk\,T} \\ B_e^{kl} & B_e^{lk} & 0 \end{pmatrix}. \tag{5.5}$$

For solving the systems with such a matrix, the unknowns of U_e^{kl}, U_e^{lk} are eliminated. This yields a system with the band matrix $B_e^{kl} D^{kl\,-1} B_e^{kl\,T} + B_e^{lk} D^{lk\,-1} B_e^{lk\,T}$, which can be solved in a direct manner.

The second inexpensive preconditioner is introduced for the Laplace operator ($\alpha_k = 1$, $\beta_k = 0$) in two dimensions : we choose R as the block diagonal matrix whose k^{th} block is

$$R^k \equiv \frac{h_k}{d_k}(I^k - P^k) + \Sigma^k, \tag{5.6}$$

where I^k and P^k have been introduced above and where Σ^k is the matrix corresponding to the Laplace-Beltrami operator $\Sigma_h^k \equiv -h_k \Delta_{\partial\Omega_k}$:

$$(\Sigma_h^k u_{kh}, v_{kh}) \equiv h_k \int_{\partial\Omega_k} \frac{d}{ds} u_{kh} \frac{d}{ds} v_{kh} ds, \qquad \forall u_{kh}, v_{kh} \in X_{kh}. \tag{5.7}$$

The choice of this preconditioner can be explained as follows : the term $\frac{h_k}{d_k}(I^k - P^k)$ approaches the Steklov-Poincaré operator for the lowest frequencies while the term Σ^k is used for the highest frequencies. The Steklov-Poincar'e operator is not so well approached in the intermediate frequencies. Quantitatively, we have the following result

Proposition 5.2 *The condition number $\kappa(\mathcal{R}^{-1}\mathcal{S})$ with respect to subspace V_B satisfies*

$$\kappa(\mathcal{R}^{-1}\mathcal{S}) \leq C \max_k \sqrt{\frac{d_k}{h_k}}. \tag{5.8}$$

Thus the condition number depends only on the maximal number of mesh points in one subdomain, and is much improved compared to the first preconditioner. Of course, this will be paid by more difficulty in solving the related linear systems.

The procedure for solving the preconditioning linear system is very close to the one discussed above for the first preconditioner: the matrix $\mathcal{R}$ is decomposed into $\mathcal{R} = \tilde{\mathcal{R}} + \mathcal{L}$, where $\mathcal{L}$ is a low rank matrix (rank K) and

$$\tilde{\mathcal{R}} \equiv \begin{pmatrix} \tilde{R} & B^T \\ B & 0 \end{pmatrix}, \tag{5.9}$$

and $\tilde{R}$ is the block diagonal matrix whose k^{th} block is

$$\tilde{R}^k \equiv \frac{h_k}{d_k} I^k + \Sigma^k. \tag{5.10}$$

Again the preconditioning system consists of solving twice a linear system with matrix $\tilde{\mathcal{R}}$ and once a coarse problem of dimension K. For the problems with $\tilde{\mathcal{R}}$, we eliminate first the unknowns non located at crosspoints and we are led to solving a small system whose dimension is proportional to the number of crosspoints. The main difficulty is to solve the systems with the blocks

$$
\begin{pmatrix}
\tilde{R}_e^{kl} & 0 & B_e^{kl\,T} \\
0 & \tilde{R}_e^{lk} & B_e^{lk\,T} \\
B_e^{kl\,T} & B_e^{lk\,T} & 0
\end{pmatrix},
\tag{5.11}
$$

with self explanory notations. Here, eliminating first U_e^{kl} and U_e^{lk} would lead to a linear system on Λ^{kl} with a dense matrix. The cost of solving this system would be proportional to the square of the number of unknowns. Therefore, we prefer instead solving directly the system, after having reordered carefully the unknowns. The reordering procedure, fully described in [2], permits to solve the system at an arithmetical cost proportional to the number of unknowns. Thus, here again, the preconditioning system can be solved with the optimal arithmetical complexity. However, in this case the programming effort is important, because reordering the unknowns is needed. Alternatively, a very close preconditioner to the latter can be designed (see [1]) when the lumped jump operator $\tilde{B}$ described in remark 1 is used, with a much easier practical implementation.

5.2   *Inner iterative procedure for better preconditioners*

The preconditioner will be introduced and analysed again for the Laplace operator in 2 dimensions. Let us choose

$$
R^k = (\Sigma^k)^{\frac{1}{2}},
\tag{5.12}
$$

where Σ^k is either the matrix

$$
\begin{pmatrix}
2 & -1 & & & & & -1 \\
-1 & 2 & -1 & & & & \\
& & \cdot & & \cdot & & \cdot \\
& & & \cdot & & \cdot & \cdot \\
& & & & -1 & 2 & -1 \\
-1 & & & & & -1 & 2
\end{pmatrix}
$$

for interior subdomains or a diagonal block of it otherwise. It is well known that the matrices R^k and S^k are spectrally equivalent. Thus the corresponding preconditioner $\mathcal{R}$ is also spectrally equivalent to $\mathcal{S}$ in the subspace of constraints. However solving the systems with the above mentioned preconditioner is not easier than solving the original system. Therefore, following [10], we are going to replace the above matrix $\mathcal{R}$ with another matrix $\hat{\mathcal{R}}$, spectrally equivalent to R, but leading to much cheaper implementation costs.

Let us introduce the matrix

$$
\mathcal{Q} = \begin{pmatrix} Q & B^T \\ B & 0 \end{pmatrix},
\tag{5.13}
$$

where Q is the block diagonal matrix whose k^{th} block is $Q^k \equiv h_k(I^k - P^k)$ if Ω_k is an internal subdomain and $h_k I^k$ otherwise. Note that $\mathcal{Q}$ is exactly the first preconditioner introduced in (5.1), for the special case of the Laplace operator.

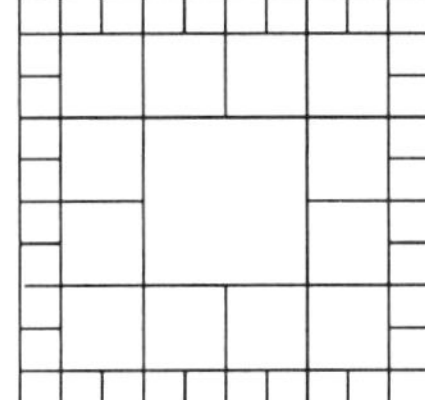

Figure 1 A smaller grid obtained by sparsening the initial one

Consider the eigenvalue problem : $\mu QW = \mathcal{R}W$. As in §5.1. it can easily be proved that the eigenvalues μ belong to segment $[c_1, c_2 \max_{1 \leq k \leq K} \frac{d_k}{h_k}]$ with positive constants c_1 and c_2 independant of d_k and h_k. Therefore, both matrices Q and $\mathcal{R}$ are symmetric and positive definite in subspace V_B and bounds on the spectrum of $Q^{-1}\mathcal{R}$ can be computed by a Lanczos type method. Then it follows immediately from [19],[14] that the Preconditioned Chebyshev iterative method in the subspace V_B can be applied for solving system $\mathcal{R}X = Y$. The Preconditioned Chebyshev iterative method can be represented in a compact form by $X^s = [I - P_s(Q^{-1}\mathcal{R})]\mathcal{R}^{-1}Y$ where $P_s(t)$ is the minimal polynomial of degree s related to a segment containing the eigenvalues of $Q^{-1}\mathcal{R}$ (see [19]). Finally a matrix $\hat{\mathcal{R}}$ can be defined by

$$\hat{\mathcal{R}} = \mathcal{R} \left[I - P_m(Q^{-1}\mathcal{R}) \right]^{-1} \tag{5.14}$$

with $m = O(\max_{1 \leq k \leq K} \sqrt{\frac{d_k}{h_k}})$. We have the following result :

Lemma 5.3 *Under the assumptions made, the matrix $\hat{\mathcal{R}}$ given by (5.17) is spectrally equivalent to matrix S, in the subspace of constraints V_B.*

The crucial point in solving the preconditioning problem is the product by matrix R, which can be achieved thanks to fast Fourier transforms.

An alternative choice of R^k is proposed in [13]: let $\hat{\mathcal{T}}_{kh}$ be a finite difference grid such that the trace of $\hat{\mathcal{T}}_{kh}$ on $\partial\Omega_k$ coincides with that of $\mathcal{T}_{kh}$. We assume that $\hat{\mathcal{T}}_{kh}$ has much less nodes that $\mathcal{T}_{kh}$, namely $O(\frac{d_k}{h_k})$ nodes. When $\frac{d_k}{h_k} \approx 2^p$, an example of such a grid for a square is given on the figure 1. Let $\hat{A}^k$ be the corresponding stiffness matrix. It can be factorized with $O(\frac{d_k}{h_k})$ arithmetical operations. Calling $\hat{S}^k$ the related Schur complement, we have that $\hat{S}^k$ is spectrally equivalent to S^k (see [18]) and that the product of a vector by $\hat{S}^k$ costs $O(\frac{d_k}{h_k})$ operations. It is thus possible to choose $R^k = \hat{S}^k$ instead of (5.12).

BIBLIOGRAPHY

[1]Y. Achdou, Yu.A. Kuznetsov, *Substructuring preconditioners for finite element methods on nonmatching grids* , East-West j. of Num. Math. 3, no 1, (1995) pp 1-28 .

[2]Y. Achdou, Yu.A. Kuznetsov, O.Pironneau, *Substructuring preconditioners for the* Q_1 *mortar element method.* , (1995), to appear in Numerische Mathematik.

[3] Y. Achdou, Y.Maday, O.Widlund, in preparation.

[4]Y. Achdou, O.Pironneau, *A fast solver for Navier-Stokes equations in the laminar regime using mortar finite element and boundary element method.*, to appear in SIAM journal of Numerical Analysis.

[5] G. Anagnostou, Y. Maday, C. Mavriplis, A. Patera, *On the mortar element method : generalization and implementation,* Proceedings of the third international conference on domain decomposition method for P.D.E. SIAM Philadelphia 1990.

[6] F.Ben Belgacem, *The mortar finite element method with Lagrange multipliers* , to appear.

[7]C. Bernardi, Y. Maday, A. Patera, *A new nonconforming approach to domain decomposition : the mortar element method* , Nonlinear partial differential equations and their applications, Pitman, H. Brezis, J.L. Lions eds. (1989).

[8] J.H. Bramble, J.E. Pasciak, *A preconditioning technique for indefinite systems resulting from mixed approximations of elliptic problems,* Mat. of Comp. 50, (1990) pp 1-18.

[9] J.H. Bramble, J.E. Pasciak, A.H. Schatz.*The construction of preconditioners for elliptic problems by substructuring, II,* Mat. of Comp. 49, (1987) pp 1-16.

[10] J.H. Bramble, J.E. Pasciak, A.H. Schatz.*The construction of preconditioners for elliptic problems by substructuring, III,* Mat. of Comp. 51, 181 (1988) pp 415-430.

[11] H. Elhman, G. Golub, *Inexact and preconditioned Uzawa algorithms for saddle point problems,* SIAM j. Numer. Anal. 31, (1991), pp 1645-1661.

[12] Yu.A. Kuznetsov. *Matrix iterative methods in subspaces,* in Proc. Int. Congress of Mathematicians, Warsaw, 1983, pp 1509-1521, North Holland, Amsterdam, (1984).

[13] Yu.A. Kuznetsov,*Efficient iterative solvers for elliptic finite element problems on nonmatching grids,* Rus. j. Num. Anal. Math. Modelling, vol 10, no 3, (1995), pp 187-211 .

[14]Yu.A. Kuznetsov, G.I.Marchuk. *Méthodes itératives et fonctionelles quadratiques,* Méthodes Mathématiques de l'Informatique-4 : sur les méthodes numériques en Sciences physiques et économiques. (J.L. Lions ,G.I.Marchuk eds) pp3-132, Dunod, Paris, 1974.

[15] P.Le Tallec. *Domain decomposition methods in computational mechanics.* Computational mechanics advances. 1,No 2 (1994) pp. 121-220.

[16] P.Le Tallec, T. Sassi. *Domain decomposition with non matching grids: Schur complement approach,* rapport CEREMADE 9323 Université de Paris Dauphine (1993).

[17] Y.Maday, O.Widlund, in preparation.

[18] S. Nepomniastchikh, *Mesh theorems of traces, normalizations of function traces and their inversion,* Sov. j. Num. Anal. Math. Modelling, vol 6, (1991), pp 223-242.

[19] R. Varga. *Matrix iterative analysis,* Prentice Hall series in Automatic computations, Englewood Cliffs, NJ (1962).

Solution of Singular Boundary Element Equations based on Domain Splitting

Ke CHEN

$\left(\textit{Department of Mathematics, University of Liverpool, England.}\right)$

1 Introduction

We consider the efficient solution of dense linear systems $A\underline{u} = \underline{f}$ by a preconditioned iterative method, where A is a $n \times n$ dense and unsymmetric matrix, and show that domain decomposition leads to a construction of practical preconditioners. Such linear systems arise from the solution of boundary element equations.

The boundary element methods (BEM), as a powerful alternative to finite element methods (FEM) and finite difference methods (FDM) for solving partial differential equations (PDE), are usually applied to PDE's with known free-space Green's functions; see [3]. However recent developments suggest that BEM can also be successfully applied to the solution of more general PDE's; see [11].

As far as the iterative solution is concerned, its success largely depends on the spectral properties of the integral operator or of the matrices of discrete linear systems. More precisely, if the underlying operator is smooth and compact, iterative methods can be very efficient without preconditioning; see [1], [4] and [5]. This paper addresses the case of a non-compact operator, where preconditioning is essential for iterative methods.

In the literature, work on preconditioning singular boundary element equations has mostly been based on 'algebraic considerations'. The underlying ideas fall into two main categories, a) design a preconditioner that can be inverted or solved easily, and contains or somehow represents the dominant part of contributions due to singular integrals; b) design a preconditioner that is sparse and somehow 'close' to the inverse of the coefficient matrix A. Essentially efficiency is the chief consideration and naturally the preconditioners suggested are often sparse; refer to [2], [6], [7], [9] and [10] among others.

In this paper, we propose to construct preconditioners for the singular integral

Domain Decomposition Methods in Sciences and Engineering, edited by R. Glowinski *et al.*
© 1997 John Wiley & Sons, Ltd.

operator equation first, based on boundary domain splitting, and then proceed with a construction of preconditioners for the dense linear systems. Using this new and general framework, we can show that several preconditioners in the literature are related. Our theory provides a justification for these preconditioners and points out a way toward new designs and further modifications. Some numerical results are reported.

2   Boundary elements and dense linear systems

Let $\Omega \in R^2$ denote a closed domain that may be interior and bounded, or exterior and unbounded, and $\Gamma = \partial\Omega$ be its (finite part) boundary that can be parameterised by $p = (x, y) = (x(s), y(s)), a \leq s \leq b$. Then a boundary integral equation that usually arises from reformulating a PDE in Ω can be written as

$$u(p) - \int_\Gamma \bar{k}(p, q)u(q)dS_q = f(p), \qquad p \in \Gamma, \tag{1}$$

or
$$u(s) - \int_a^b k(s, t)u(t)dt = f(s), \qquad s \in [a, b], \tag{2}$$

or simply
$$(I - \mathcal{K})u = f. \tag{3}$$

To solve the above equation numerically, we divide the boundary Γ (interval $[a, b]$) into m boundary elements (non-intersecting subintervals $E_i = [s_{i-1}, s_i]$). On each interval E_i, we may either approximate the unknown u by an interpolating polynomial of order r that leads to a collocation method, or apply a quadrature method of r nodes that gives rise to the Nyström method. Both discretization methods approximate equation (3) by

$$(I - \mathcal{K}_n)u_n = f, \tag{4}$$

where we can write

$$\mathcal{K}_n u = \mathcal{K}_n u_n = \sum_{j=1}^m \left[\sum_{i=1}^r w_i k(s, t_{ji})u_{ji} \right], \quad u_n(t_{ji}) = u(t_{ji}) = u_{ji}, \quad \text{and} \quad n = mr.$$

We use vector $\underline{u}$ to denote u_{ji}'s at all nodes. By a collocation step in equation (4), we obtain a linear system of equations

$$(I - K)\underline{u} = \underline{f}, \qquad \text{or} \qquad A\underline{u} = \underline{f}, \tag{5}$$

where matrices K and A are dense and unsymmetric (in general). The conditioning of A depends on the smoothness of kernel function $k(s, t)$. A strong singularity (as $t \to s$) leads to non-compactness of operator $\mathcal{K}$ and renders equation (5) difficult to solve by iterative methods without preconditioning.

To gain some insight into preconditioning, we may describe a BEM procedure as an interaction of three stages, each respectively characterized by

(1). Integral Operator $\mathcal{K}$

(2). Approximating Integrals $(\mathcal{K}_n u_n)(s)$

(3). Matrix Elements of K

Then stages $1 \rightarrow 2 \rightarrow 3$ represent the well-known (standard) BEM procedure. The simple fact is that any undesirable spectral properties of K (stage 3) must stem from the integral operator $\mathcal{K}$ (stage 1), although we need preconditioners for A (in stage 3) and we see that singularities tend to dominate integrals (in stage 2). Therefore we propose to exploit stages $3 \rightarrow 2 \rightarrow 1$ in search of a better preconditioned problem to work with or a general theoretical framework.

We remark that most preconditioners in the literature exploit stages $3 \rightarrow 2$ only.

3 Preconditioning techniques

We shall review three important preconditioners for solving $A\underline{u} = \underline{f}$ and illustrate each case by a 9×9 system.

3.1 Two grid based sparse column preconditioners

The success of this preconditioner, due to Yan [10], follows from the facts that a $n \times n$ sparse column matrix B with m nonzero long columns has its inverse of the identical sparsity and a system such as $B\underline{x} = \underline{y}$ for $x, y \in R^n$ can be solved in only $(n - m)m$ operations. For example, with $n = 9$,

$$
B = \begin{pmatrix}
\times & & \times & & & \times & & & \times \\
& \times & \times & & & \times & & & \times \\
& & \times & & & \times & & & \times \\
& & \times & \times & & \times & & & \times \\
& & \times & & \times & \times & & & \times \\
& & \times & & & \times & & & \times \\
& & \times & & & \times & \times & & \times \\
& & \times & & & \times & & \times & \times \\
& & \times & & & \times & & & \times
\end{pmatrix}.
$$

Formally, let $n = \eta m$ with η, m integers. Then from matrix $K = (k_1, k_2, \cdots, k_n)$, construct column vectors by

$$
k_i' = \begin{cases} k_i, & \text{if } i = \ell\eta, \quad 1 \leq \ell \leq m, \\ 0, & \text{otherwise,} \end{cases}
$$

and define a new matrix by $K' = (k_1', k_2', \cdots, k_n')$. Then we use $A' = (I + \eta K')$ as a preconditioner. It may be expected that A' is 'close' to $A = A_n$ because $A' \approx A_m$ which is the corresponding discrete matrix with m nodes.

3.2 Mesh neighbour based approximate inverses

The starting point in the mesh neighbour preconditioner of Vavasis [9] is that we hope to find a matrix such that $PA \approx I$ and this P should have its diagonal elements and immediate neighbours possess more importance as A usually comes from singular integral equations. In particular, assume that P is a quasi-tridiagonal matrix; for $n = 9$

we have

$$
P = \begin{pmatrix}
\times & \times & & & & & & & & \times \\
\times & \times & \times & & & & & & & \\
& \times & \times & \times & & & & & & \\
& & \times & \times & \times & & & & & \\
& & & \times & \times & \times & & & & \\
& & & & \times & \times & \times & & & \\
& & & & & \times & \times & \times & & \\
& & & & & & \times & \times & \times & \\
\times & & & & & & & \times & \times &
\end{pmatrix} . \tag{6}
$$

Then to find $P^T = (p_1, p_2, \cdots, p_n)$ in terms of $A = (a_1, a_2, \cdots, a_n)$, a direct approach is used. Let $PA = I \iff A^T P^T = I \iff A^T p_i = e_i$. Then, as p_i only has three nonzero positions $p_{i_1}, p_{i_2}, p_{i_3}$, an approximate solution is obtained by solving a 3×3 system

$$
\begin{pmatrix}
\times & \times & \times \\
\times & \times & \times \\
\times & \times & \times
\end{pmatrix}
\begin{pmatrix}
p_{i_1} \\
p_{i_2} \\
p_{i_3}
\end{pmatrix}
=
\begin{pmatrix}
0 \\
1 \\
0
\end{pmatrix} .
$$

3.3   Sparse LU decompositions

Similar to the idea of §3.2, the work of [2] assumes that preconditioner B should have its diagonal elements and immediate neighbours possess more importance for the same reason that A usually comes from singular integral equations. When $n = 9$, such a matrix B is of the same sparsity as P in equation (6).

As B^{-1} cannot maintain any sparsity property of B, it is proposed, in order to solve $B\underline{x} = \underline{y}$ for $x, y \in R^n$, to decompose B as $B = LU$, where L and U are sparse triangular matrices. To illustrate, for $n = 9$, we have

$$
B = LU =
\begin{pmatrix}
P_{1,1} & & & & & & & & \\
P_{2,1} & P_{2,2} & & & & & & & \\
& P_{3,2} & P_{3,3} & & & & & & \\
& & P_{4,3} & P_{4,4} & & & & & \\
& & & P_{5,4} & P_{5,5} & & & & \\
& & & & P_{6,5} & P_{6,6} & & & \\
& & & & & P_{7,6} & P_{7,7} & & \\
& & & & & & P_{8,7} & P_{8,8} & \\
P_{9,1} & P_{9,2} & P_{9,3} & P_{9,4} & P_{9,5} & P_{9,6} & P_{9,7} & P_{9,8} & P_{9,9}
\end{pmatrix}
\times
\begin{pmatrix}
1 & P_{1,2} & & & & & & & P_{1,9} \\
& 1 & P_{2,3} & & & & & & P_{2,9} \\
& & 1 & P_{3,4} & & & & & P_{3,9} \\
& & & 1 & P_{4,5} & & & & P_{4,9} \\
& & & & 1 & P_{5,6} & & & P_{5,9} \\
& & & & & 1 & P_{6,7} & & P_{6,9} \\
& & & & & & 1 & P_{7,8} & P_{7,9} \\
& & & & & & & 1 & P_{8,9} \\
& & & & & & & & 1
\end{pmatrix} .
$$

A general algorithm is given as follows $\qquad$ (**Calculation of LU**)

$$
\begin{array}{|l|}
\hline
P_{1,1} = B_{1,1}, \quad P_{2,1} = B_{2,1}, \quad P_{n,1} = B_{n,1}, \\
P_{1,2} = B_{1,2}/B_{1,1}, \quad P_{1,n} = B_{1,n}/B_{1,1}, \quad P_{2,2} = B_{2,2} - B_{2,1}P_{1,2}, \quad P_{n,2} = -B_{n,1}P_{1,2}, \\
s = B_{n,1}P_{1,n} \\
\hline
\text{FOR} \ \ i = 2 : n - 2 \\
\qquad \boxed{\begin{array}{ll}
P_{i,i+1} = B_{i,i+1}/P_{i,i}, & P_{i,n} = -B_{i,i-1}P_{i-1,n}/P_{i,i} \\
P_{i+1,i+1} = B_{i+1,i+1} - B_{i+1,i}P_{i,i+1}, & P_{n,i+1} = -P_{n,i}P_{i,i+1} \\
s = s + P_{n,i}P_{i,n} &
\end{array}} \\
\hline
\text{END} \\
P_{n,n-1} = P_{n,n-1} + B_{n,n-1}, \quad P_{n-1,n} = (B_{n-1,n} - B_{n-1,n-2}P_{n-2,n})/P_{n-1,n-1}, \\
P_{n,n} = B_{n,n} - P_{n,n-1}P_{n-1,n} - s. \\
\hline
\end{array}
$$

A related algorithm, needed at a preconditioning step, for solving a system $P\underline{x} = \underline{y}$ is

Forward substitution :

$$
\begin{array}{|l|}
\hline
\qquad\qquad z_1 = y_1/P_{1,1}, \qquad\qquad\qquad\qquad s = P_{n,1}, \\
\text{FOR } i = 2 : n - 1 \\
\qquad\qquad z_i = (y_i - P_{i,i-1}z_{i-1})/P_{i,i}, \qquad s = s + P_{n,i}z_i \\
\text{END} \\
\qquad\qquad z_n = (y_n - s)/P_{n,n}. \\
\hline
\end{array}
$$

Backward substitution :

$$
\begin{array}{|l|}
\hline
x_n = z_n, \qquad\qquad x_{n-1} = z_{n-1} - P_{n-1,n}x_n, \\
\text{FOR } i = n - 2 : -1 : 1 \\
\qquad\qquad x_i = z_i - P_{i,i+1}x_{i+1} - P_{i,n}x_n, \\
\text{END} \\
\hline
\end{array}
$$

In this paper, one of our aims has been to offer a new understanding of the above otherwise heuristic sparse preconditioners appeared in the literature. This is to be achieved by use of operator splittings or domain decomposition.

4 Operator and domain splitting

Here we introduce the idea of operator splitting that was originated in [7]. Use the partition of §2, $[a, b] = \bigcup_{i=1}^{m} E_i$. Accordingly we can partition variable u and vector $\underline{u}$ as follows $u = (u_1, u_2, \cdots, u_m)^T$ and $\underline{u} = (\underline{u}_1, \underline{u}_2, \cdots, \underline{u}_m)^T$.

Similarly operator $\mathcal{K}$ is partitioned into a matrix form and further we can observe that all singularities of $\mathcal{K}$ are contained in the following operator

$$
\bar{\mathcal{K}} = \begin{pmatrix}
\mathcal{K}_{1,1} & \mathcal{K}_{1,2} & & & & \mathcal{K}_{1,m} \\
\mathcal{K}_{2,1} & \mathcal{K}_{2,2} & \mathcal{K}_{2,3} & & & \\
& \mathcal{K}_{3,2} & \ddots & & \ddots & \\
& & & \ddots & \ddots & \mathcal{K}_{m-1,m} \\
\mathcal{K}_{m,1} & & & & \mathcal{K}_{m,m-1} & \mathcal{K}_{m,m}
\end{pmatrix}.
$$

The corresponding matrix out of K is

$$\bar{K} = \begin{pmatrix} K_{1,1} & K_{1,2} & & & & K_{1,m} \\ K_{2,1} & K_{2,2} & K_{2,3} & & & \\ & K_{3,2} & \ddots & & \ddots & \\ & & & \ddots & & K_{m-1,m} \\ K_{m,1} & & & K_{m,m-1} & K_{m,m} \end{pmatrix}.$$

Define operators $\mathcal{D} = I - \bar{\mathcal{K}}$ and $\mathcal{C} = \mathcal{K} - \bar{\mathcal{K}}$, and matrices $D = I - \bar{K}$ and $C = K - \bar{K}$. Then it can be shown that operator $\mathcal{D}$ is bounded and $\mathcal{C}$ is compact. So operator $\mathcal{D}^{-1}\mathcal{C}$ is also compact. Thus the solution of $A\underline{u} = \underline{f}$ is reduced to that of $[I - D^{-1}C]\underline{u} = D^{-1}\underline{f}$. Here D is in general a block quasi-tridiagonal matrix and the solution of $D\underline{x} = \underline{y}$ should similarly follow §3.3.

While it is now natural to use D as a preconditioner, we may conclude that an efficient preconditioner should contain D or its close approximation. Following [7], we may further show that $D_1 = diag(D)$ is also an efficient preconditioner.

5   Interpretation of preconditioners

We now use the theory of the above section to identify the operator splittings implied in the preconditioners of §3.

Firstly we see that the preconditioner of [10] uses the following splitting $\mathcal{K} = \mathcal{D} + \mathcal{C}$, where $\mathcal{D} = I - \tilde{\mathcal{K}}$ with $\tilde{\mathcal{K}} = (\tilde{\mathcal{K}}_{ij})$ and $\tilde{\mathcal{K}}_{ij} = \eta \int_{E_j/\eta} k(s,t)u_j(t)dt \approx \int_{E_j} k(s,t)u_j(t)dt = \mathcal{K}_{ij}$. Theoretically $\mathcal{C}$ is not compact but $\mathcal{C}$ should approach 0 if $m, n \to \infty$. Practically η should not be too large (then the method becomes more expensive if $m \approx n$).

Secondly, for the preconditioners of [9] and [2], the underlying splittings would be identical to that in the last section for piecewise constant approximations or the panel method (mid-point rule). The reason is that both were proposed based on assumptions on stage 2 (regarding singular integrals rather than operators).

6   Numerical results

The first problem to be tested has a weak singularity (see [10])

$$\text{Problem 1 :} \qquad u(s) + \gamma \int_{-\pi}^{\pi} u(s)b(s,t)dt = f(s), \qquad s \in [-\pi, \pi]$$

which arises from the solution of the exterior Neumann's problem of the Laplace equation (when $\gamma = 1$) over an elliptic boundary $p(s) = (\cos(s), \sin(s)/4)$. Here the kernel function is

$$b(s,t) = \frac{(p(t) - p(s)) \cdot n(p(t))|p'(t)|}{\pi|p(t) - p(s)|^2} = \frac{4}{\pi(17 - 15\cos(t+s))}.$$

The parameter γ is included to vary the difficulty of the problem. We specifically choose $f(s) = |\sin(s)| + 2\gamma\{4\cos(s)\log[(17 + 15\cos(s))/(17 - 15\cos(s))] + 17\sin(s)\tan^{-1}(15\sin(s)/8)\}/(15\pi)$ so that $u(s) = |\sin(s)|$.

Table 1   Convergence results of Problem 1 ($\gamma = 10$)

Method	Diagonal [7]	Inverse [9]	LU [2]	Mod LU [2]	Two grid [10]	Mod TG [10]
N = 16	6	6	23	26	7	13
N = 32	12	10	10	38	17	31
N = 64	17	15	10	45	34	65
N =128	18	19	13	45	41	80
N =256	19	20	13	41	50	87
N =512	18	20	14	40	53	89
N =1024	18	19	16	39	55	92

Table 2   Convergence results of Problem 2

Method	Diagonal [7]	Inverse [9]	LU [2]	Mod LU [2]	Two grid [10]	Mod TG [10]
N = 16	9	14	10	12	25	16
N = 32	10	18	19	16	50	22
N = 64	11	20	36	26	98	30
N =128	11	23	71	36	194	41
N =256	13	26	144	48	386	56
N =512	14	27	291	57	770	75
N =1024	14	30	599	62	1538	93

The second problem possesses a Cauchy singularity (see [7])

$$
\text{Problem 2}: \quad
\begin{cases}
\frac{1}{\pi}\int_{-1}^{1}\frac{w(t)\phi(t)}{t-x}dt + \int_{-1}^{1}\frac{(t^2-x^2)^2}{t^2+x^2}w(t)\phi(t)dt = f(x), \\
\frac{1}{\pi}\int_{-1}^{1}w(t)\phi(t)dt = 0, \qquad\qquad x \in (-1,1),
\end{cases}
$$

which has the exact solution $\phi(x) = x|x|$.

Both problems are discretized by the Nyström method using uniform nodes for Problem 1 and Chebyshev nodes for Problem 2. The conjugate gradient iterative method to the normal equation (CGN) is adopted; see [8]. The tolerance for residual errors is set to be TOL $= 10^{-J}$ where $J = 1 + \log(N)/\log(2)$. Results are shown in Tables 1–2 of comparisons of six preconditioners, where 'Mod' means a modified version and the modifications are based on discussions of last two sections.

Although further and more extensive tests are needed to compare these preconditioners, our preliminary conclusion is that for singular boundary element equations, the most robust preconditioner is that based on boundary domain decomposition [7].

Acknowledgement

The support for this research work from the Nuffield Foundation (UK) is gratefully acknowledged.

References

Amini S. and Chen K. (1989) Conjugate gradient method for second kind integral equations – applications to the exterior acoustic problem. *Engineering Analysis with Boundary Elements*, **6**(2), 72–77.

Amini S. and Maines N. (1994) Iterative solutions of boundary integral-equations. In : *Proceedings of the 14th conference on Boundary Element Methods*, ed. C. A. Brebbia, Ch.65, 193–200.

Brebbia C. A. et al (1984) *Boundary element techniques – theory and applications*, Springer-Verlag.

Hemker P. W. and Schippers H. (1981) Multigrid methods for the solution of Fredholm integral equations of the second kind. *Math. Comp.*, **36**, 215–232.

Hackbusch W. (1985) *Multigrid methods and applications*. Springer-Verlag.

Canning F. (1992) Sparse approximation for solving integral equations with oscillatory kernels. *SIAM J. Sci. Stat. Comp.*, **13**, 71–87.

Chen K. (1994) Efficient iterative solution of linear systems from discretizing singular integral equations. *Elec. Tran. Numer. Anal.*, **2**, 76–91.

Nachtigal N. M. et al (1992) How fast are nonsymmetric matrix iterations?. *SIAM J. Matr. Anal. Appl.*, **13**(3), 778–795.

Vavasis S. (1992) Preconditioning for boundary integral equations. *SIAM J. Matr. Anal. Appl.*, **13**(3), 905–925.

Yan Y. (1994) Sparse preconditioned iterative methods for dense linear systems. *SIAM J. Sci. Comp.*, **15**(5), 1190–1200.

Partridge P.W., Brebbia C. A. and Wrobel L. C. (1992) *The dual reciprocity boundary element method*. CMP Southampton, UK.

Some Recent Developments in Domain Decomposition Methods with Nonconforming Finite Elements

Jinsheng Gu [1] and Xiancheng Hu [2]

1 Introduction

Domain decomposition methods have recently become an important focus in the field of computational mathematics due to the development of parallel computers. The nonconforming finite element methods are effective for solving partial differential equations derived from mechanics and engineering [3, 4, 5, 14]. But, there has been no extensive study of domain decomposition methods with nonconforming finite elements which lack global continuity. Therefore, a rather systematic investigation on domain decomposition methods with nonconforming elements is presented in this paper.

It is well–known that extension theorems play key role in domain decomposition analysis, especially in the case of nonoverlapping subregions. We also know that extension theorems hold for conforming elements [1, 16, 17]. Hence the domain decomposition analysis can be performed for the conforming finite element discrete problems. When domain decomposition methods with nonconforming elements are studied, a core question in domain decomposition analysis is "Do the extension theorems hold for nonconforming finite elements?"

For this reason, we have established extension theorems for nonconforming elements based on conforming interpolation operators and further error estimates of

[1] Faculty 404, Department of Jet Propulsion, Beijing University of Aeronautics & Astronautics, Beijing 100083, P.R. China.
[2] Department of Applied Mathematics, Tsinghua University, Beijing 100084, P.R. China.

Domain Decomposition Methods in Sciences and Engineering, edited by R. Glowinski *et al.*
© 1997 John Wiley & Sons, Ltd.

nonconforming finite elements solution under weak conditions [6, 10, 12, 13]. The originality of the design of the nonoverlapping domain decomposition algorithms with nonconforming elements which are continuous at the midpoints of the element edges (such as the Crouzeix-Raviart elements), is that the internal cross points do not need to be handled. This leads to simplicity and high parallelism of our algorithms [6].

For the second order elliptic problems discretized by nonconforming element methods, all the domain decomposition algorithms, nonoverlapping or overlapping, are as efficient as their counterparts in the conforming cases, and even easier in implementation [6, 7]. For the Stokes problems discretized by the nonconforming mixed element methods [5, 14], several algorithms are presented and discussed in [11]. For fourth order elliptic problems, in the conforming case and in the Morley nonconforming discrete case, a series of algorithms have been developed. Many numerical results are consistent with the theoretical analysis; see [6].

The remainder of this paper consists of three sections. Extension theorems for nonconforming elements are presented in Sect. 2. Their applications and other results are indicated concisely in Sect. 3. Conclusions are given in Sect. 4.

2   Main results

We mainly consider the linear, selfadjoint Dirichlet elliptic problem of order $2m$ in variational form given by

$$u \in H_0^m(\Omega) : \quad a_\Omega(u,v) = (f,v), \quad \forall\, v \in H_0^m(\Omega). \tag{2.1}$$

Here, $m = 1, 2$, $\Omega \subset \Re^2$ is an open polygonal bounded domain and

$$(f,v) = \int_\Omega fv.$$

We assume that the bilinear form $a_\Omega(\cdot,\cdot)$, over Ω, satisfies the following standard conditions [17]:

$$\begin{cases} a_\Omega(w,v) = a_\Omega(v,w) \\ a_\Omega(v,\ v) \geq c\|v\|_{H^m(\Omega)}^2 \\ a_\Omega(w,v) \leq C\|w\|_{H^m(\Omega)}\|v\|_{H^m(\Omega)} \end{cases} \tag{2.2}$$

where c and C are positive constants.

Let $\Omega_h = \{e\}$ be a quasi–uniform mesh of Ω and let V_h be a finite element space associated with Ω_h. For (2.1), in the case $m = 1$, V_h is the Crouzeix–Raviart element space [5], or the piecewise quartic nonconforming rectangular element space [14], or the Wilson element space, or the Carey element space [3] or another nonconforming membrane element space. For (2.1), in the case $m = 2$, V_h is the Morley element space, or the Ziekenwicz element space, or the Adini element space, or another nonconforming plate element space[4]. Let

$$A(w,v) = \sum_{e \in \Omega_h} a_e(w,v)$$

and let

$$V_h^0 = \left\{ v \in V_h : \text{ the freedom of } v \text{ vanishes at each interpolation point } x \in \partial\Omega \right\}.$$

Then the nonconforming finite element discrete problem for (2.1) is

$$u_h \in V_h^0 : \quad A(u_h, v) = (f, v), \ \forall\, v \in V_h^0. \tag{2.3}$$

Let Γ be an open line segment in Ω such that $\Gamma \cap e = \emptyset$, $\forall\, e \in \Omega_h$ and let Ω be decomposed by Γ into two open subdomains, denoted by Ω_1 and Ω_2, which satisfies

$$\Omega_1 \bigcap \Omega_2 = \emptyset, \ \Omega_1 \bigcup \Omega_2 \bigcup \Gamma = \Omega, \ \overline{\Omega_1} \bigcup \overline{\Omega_2} = \overline{\Omega}.$$

For $k = 1, 2$, we make the following definitions:

$$A_k(w, v) = \sum_{e \in \Omega_k} a_e(w, v),$$

$$V_h^k = \left\{ v \in V_h^0 : \text{ the degrees of freedom of } v \text{ vanish at each nodal point } x \in \Omega \backslash \overline{\Omega_k} \right\},$$

$$V_h^{k,0} = \left\{ v \in V_h^0 : \text{ the degrees of freedom of } v \text{ vanish at each nodal point } x \in \Omega \backslash \Omega_k \right\},$$

$$\Phi_h = \Big\{ (v_1, v_2) : \ v_k \in V_h^k, \ A_k(v_k, w) = 0, \ \forall\, w \in V_h^{k,0}, \ k = 1, 2,$$

$$\text{the degrees of freedom of } \ v_1 \text{ and } v_2 \text{ are equal at each nodal point } x \in \Gamma \Big\}.$$

Theorem 2.1 ([6, 10, 12, 13]). *For the quasi–uniform mesh Ω_h, there exist two positive constants σ, τ, independent of the mesh parameter h, such that*

$$\tau A_2(v_2, v_2) \le A_1(v_1, v_1) \le \sigma A_2(v_2, v_2), \ \forall\, (v_1, v_2) \in \Phi_h$$

We can prove Theorem 2.1 by using a trace theorem, the regularity estimate of the elliptic problem, the additional finite element error estimate, the inverse inequality and using a conforming interpolation operator which forms a bridge between the nonconforming element space and the corresponding properly selected conforming element space. We omit its proof here. The interested reader are referred to [6, 10, 12, 13].

Theorem 2.1 plays a key role in the analysis of any two–subdomain nonoverlapping domain decomposition method for (2.3), see [7]. For example, the Dirichlet–Neumann alternating method, also known as the Marini–Quarteroni algorithm, can be applied to (2.3) with the same expression of the convergence factor, with σ, τ, in Theorem 2.1 as that in [15].

Theorem 2.1 is the so–called extension theorem in the two-subdomain case when (2.1) is a second order or fourth order problem. The extension theorems in the multi–subdomain case (with crosspoints), analogous to Lemma 3.5 [1] or Lemma 3.2 [17], have been established for second order problem, i.e. (2.1) with $m = 1$. It states that the strain energy of the discrete harmonic extension function of a subdomain does not exceed the sum of those of its adjacent (neighbouring) subdomains by at most a factor relevant to the maximum subdomain diameter (the coarse mesh parameter) H and the fine mesh parameter h. When V_h is a finite dimensional space whose elements (a function) are continuous at the mesh nodes, such as the Wilson element space or Carey element space (which are called *first kind nonconforming finite element space* for convenience), the factor is $\epsilon(1 + \ln \frac{H}{h})^2$. When V_h is a finite dimensional space whose element (a function) are continuous at each edge midpoint of $e \in \Omega_h$, such as the Crouzeix–Raviart space (which are called *second kind nonconforming finite element space*), the factor is $c(1+\ln \frac{H}{h}) \max(1+H^{-2}, 1+\ln \frac{H}{h})$. For the first kind nonconforming element space V_h, preconditioners can be constructed by substructuring which are as efficient as their counterparts in the conforming element case [6]. For the second kind nonconforming finite element space V_h, it is unnecessary and in fact impossible to calculate the values at the internal crosspoints. For this case, a number of simple and high parallel preconditioners and iterative domain decomposition algorithms have been developed [6].

3   Other results

1. In the two-subdomain nonoverlap case, the existing domain decomposition algorithms designed for the conforming finite element discrete problems (for (2.1) with $m = 1$) can be extended to the nonconforming case (for (2.1) with $m = 1, 2$) only after that the description has been changed [6,7,13]. In the multi–subdomain nonoverlapping case, we can construct preconditioners and iterative domain decomposition algorithms for the nonconforming finite element discrete problems (for (2.1) with $m = 1$ only) by revising properly those for the conforming element discrete problems [6]. Based on the extension theorems given above and other estimates, we can show that they are as efficient as their counterparts in the conforming element discrete case, and even easier in implementation.

2. The overlapping domain decomposition method (the parallel Schwarz alternating algorithm) for (2.3) has been studied when (2.1) is a second order or a fourth order problem. In each iteration, a coarse mesh problem is introduced and solved simultaneously with all the subproblems posed on the subdomains [8]. Its convergence is proved by using projection operator theory. When (2.1) is a second order problem, the convergence factor is independent of the fine mesh parameter h, and even of the coarse mesh parameter H, when the coarse mesh is properly employed.

3. The overlapping domain decomposition methods for second order and two dimensional nonselfadjoint elliptic problems discretized by the Crouzeix–Raviart elements have also been considered. We have shown that its convergence by the discrete maximum principle[9] under the condition that each internal angle of the triangular elements is no larger than $\frac{\pi}{2} - \alpha$ (α is a positive constant).

4. The nonconforming mixed finite element method has been applied to solving the following two dimensional stationary incompressible Stokes model problem with the kinetic viscosity coefficient 1

$$\begin{cases} -\Delta \mathbf{u} + \nabla p = f & \text{in } \Omega \\ \text{div } \mathbf{u} = 0 & \text{in } \Omega \\ \mathbf{u} = 0 & \text{on } \partial\Omega, \end{cases} \tag{3.1}$$

where Ω is a bounded polygonal domain, $\mathbf{u}$ is the velocity and p the pressure. The variational form of (3.1) is a saddle point problem. To discretize it, a piecewise constant element space and the second kind nonconforming finite element space are employed to approximate the pressure field and the velocity field, respectively;

see [5, 14]. An extension theorem for the discrete Stokes problem, analogous to Theorem 2.1, has been established based on the Brezzi–Babuska condition, the Stokes extension operator and Theorem 2.1. Then the two–subdomain nonoverlapping domain decomposition algorithms can be developed and analyzed correspondingly. As for the multi–subdomain cases, it can be handled using the techniques of Bramble, et al. [2] and the domain decomposition methods for the Laplace equation discretized by second kind nonconforming finite elements [11].

4   Conclusions

Through the systematic study of domain decomposition methods with nonconforming finite elements, we see that although the nonconforming finite element function lacks global continuity, a theoretical foundation can be established. The existing domain decomposition methods, developed in the conforming finite element discrete case, can be revised properly and extended to the nonconforming finite element discrete case. It is notable that some algorithms for the second kind nonconforming finite element discrete problems are simple and easy in implementation. But it remains an open problem how to design and analyze the nonoverlapping domain decomposition methods for fourth order elliptic problem in the multi–subdomain case.

REFERENCES

[1] Bramble J. H., Pasciak J. E. and Schatz A. H. (1986) The construction of preconditioners for elliptic problems by substructuring. I. *Math. Comp.* 47, 103–134.

[2] Bramble J. H., Pasciak J. E. and Schatz A. H. (1986) A preconditioning technique for indefinite systems resulting from mixed approximations of elliptic problems. *Math. Comp.* 50, 1–17.

[3] Carey G. F. (1976) An analysis of finite element equations and mesh subdivision. *Comput. Methods Appl. Mech. Engrg.* 9, 165–179.

[4] Ciarlet P. G. (1978) *The Finite Element Method for Elliptic Problems.* North–Holland, Amsterdam.

[5] Crouzeix M. and Raviart P. A. (1973) Conforming and nonconforming finite element methods for solving the stationary Stokes equations. *RAIRO Numer. Anal.* 7–R3, 33–76.

[6] Gu J. (1993) *Domain Decomposition Methods with Nonconforming Finite Elements.* Ph. D. thesis. (in Chinese). Dept. of Applied Math., Tsinghua University, China.

[7] Gu J. and Hu X. (1995) On domain decomposition methods in two–subdomain nonoverlap cases. *Chinese J. Num. Math. & Appl.* 17:1, 78–94.

[8] Gu J. and Hu X. (1994) On the overlap domain decomposition methods for elliptic problems in multi–subdomain case. *J. Tsinghua University.* 34:6, 50–57 (in Chinese).

[9] Gu J. and Hu X. (1995) Overlapping domain decomposition method for nonselfadjoint elliptic problems discretized by Crouzeix–Raviart elements. *Chinese J. Num. Math. & Appl.* 17:4.

[10] Gu J. and Hu X. (1995) Some estimates with nonconforming elements in the analysis of domain decomposition methods. UCD/CCM Report No. 39.

[11] Gu J. and Hu X. (1995) Domain decomposition methods for Stokes problem discretized by nonconforming mixed finite elements. *J. Beijing University of Aeronautics and Astronautics.* 21:4 (in Chinese).

[12] Gu J. and Hu X. (1994) On an essential estimate in the analysis of domain decomposition methods. *J. Comput. Math.* 12:2, 132–137.

[13] Gu J. and Hu X. (1994) Extension theorems for plate elements with applications. (Submitted to Math. Comp.). UCD/CCM Report No. 46.

[14] Han H. (1984) Nonconforming elements in the mixed finite element method. *J. Comput. Math.* 2:3, 223–233.

[15] Marini L. D. and Quarteroni A. (1989) A relaxation procedure for domain decomposition methods using finite elements. *Numer. Math.* 55, 575–598.

[16] Widlund O. B. (1987) An extension theorem for finite element spaces with three applications. In *Numerical Techniques in Continuum Mechanics*, Vol. 16. Braaunschweig Wiesbaden.

[17] Widlund O. B. (1988) Iterative substructuring methods: algorithms and theory for elliptic problems in the plane. in *Proceedings of 1st International Symposium on DDM.*

Schwarz Domain Decomposition Method for Multidimensional and Nonlinear Evolution Equations: Subdomains Have Overlaps

Qiming He and Lishan Kang

1 Introduction

Domain decomposition method is one of the most important approaches for solving partial differential equations numerically. This method has many merits: the size of a problem can be compressed by domain decomposition; various computing schemes can be used exploiting different geometric forms of the subdomains or the different features of the problems; parallel computing can be implemented on different subdomains, etc. Most present work focused on domain decomposition method for elliptic equations. Less effort has been devoloped to parabolic systems. In this article, we study the following initial-boundary value problem:

$$
\begin{cases}
\dfrac{\partial u}{\partial t} = Lu + f(t,x,u), & t \in (0,T),\ x \in \Omega, \\
u = 0, & t \in [0,T],\ x \in \partial\Omega, \\
u|_{t=0} = \varphi(x), & x \in \overline{\Omega},
\end{cases}
\tag{1}
$$

where $\Omega \subset R^m$ is a bounded convex region, with a boundary $\partial\Omega$ which is piecewise smooth, $u = (u_1,\cdots,u_J)^T$, $f = (f_1,\cdots,f_J)^T$, $\varphi = (\varphi_1,\cdots,\varphi_J)^T$, and the linear

[1] This work was supported in part by National Natural Science Foundation of China and National 863 High Technology Project of China.

[2] Institute of Software Engineering, Wuhan University, Wuhan 430072, People's Republic of China.

Domain Decomposition Methods in Sciences and Engineering, edited by R. Glowinski *et al.*

differential operator $L = (L_1, \cdots, L_J)^T$ is given by

$$L_j u = \sum_{k,l}^{m} D_k(a_{jkl}(x) D_l u_j) - c_j(x) u_j, \quad D_k = \frac{\partial}{\partial x_k}, \quad j = 1, 2, \cdots, J, \quad k = 1, 2, \cdots, m.$$

We suppose that the following conditions hold:

H_1) $a_{j,k,l}(x)$ is continuously differential on $\overline{\Omega}$, $c_j(x) \geq 0$ and is bounded on $\overline{\Omega}$, and

$$\sum_{k,l}^{m} a_{j,k,l}(x) \xi_k \xi_l \geq \gamma |\xi|^2, \quad \xi \in R^m, \quad \gamma > 0.$$

H_2) f is continuously differentiable for $(t, x, u) \in [0, T] \times \overline{\Omega} \times R^J$, and

$$\left| \frac{\partial f}{\partial t} \right| \leq b(1 + |u|), \quad \left| \frac{\partial f}{\partial u} \right| \leq b_1.$$

H_3) $\varphi(x) \in W_2^{(2)}(\Omega) \cap \dot{W}_2^{(1)}(\Omega)$.

Suppose that Ω is partitioned into $\Omega_1, \cdots, \Omega_I$, and that $\Omega_i \cap \Omega_{i_1} \neq \emptyset$ $(i \neq i_1)$, i.e. the subdomains overlap. $\partial\Omega_i$ stands for the boundary of Ω_i.

Let $t_n = n\Delta t (n = 0, 1, 2, \cdots, N)$, i.e. $N\Delta t = T$. Let $u(t_n, x) = u^n$, and discretize (1) with respect to t. We have

$$\begin{cases} \dfrac{u^{n+1} - u^n}{\Delta t} = L u^{n+1} + f^{n+1}, & x \in \Omega, \\ u^{n+1}|_{\partial\Omega} = 0, \\ u^0 = \varphi(x), \end{cases} \tag{2}$$

which can be written as

$$\begin{cases} (I + \Delta t L) u^{n+1} = u^n + \Delta t f^{n+1}, & x \in \Omega, \\ u^{n+1}|_{\partial\Omega} = 0, \end{cases} \tag{3}$$

where $f^{n+1} = f(t_{n+1}, x, u^{n+1})$.

The paper is organized as follows: Section 2 studies the existence for generalized solution of nonlinear elliptic equations (3); Section 3 discusses the Picard-Schwarz algorithm(named P-S algorithm in brief) for problem (3); the convergence of P-S algorithm is investigated in Section 4 and a priori estimates for semidiscrete solutions are presented in Section 5. The convergence of semidiscrete solutions is discussed in Section 6.

2   Existence of a generalized solution of the boundary value problem

We will now demonstrate that problem (3) has a generalized solution. The following lemma will be used in the proofs. Consider the transform:

$$z = T(y, \lambda),$$

where $y, z \in X$, X is a Banach space, $\lambda \in [a, b]$ is a parameter, and $[a, b]$ is a bounded interval. We assume that

(i) $T(y, \lambda)$ is defined for $y \in X, \lambda \in [a, b]$. For fixed $\lambda \in [a, b]$, $T(y, \lambda)$ is continuous in X. For y in a bounded set of X, $T(y, \lambda)$ is uniformly continuous with respect to λ.

(ii) For any fixed λ, $T(y, \lambda)$ is a compact transform.

(iii) There exists a constant $M > 0$ such that all possible solutions of $z - T(z, \lambda) = 0$, $\lambda \in [a, b]$, satisfy $\|z\| \leq M$.

(iv) Equation $z - T(z, a) = 0$ has a unique solution.

Then we have

Lemma 1(Leary-Schauder fixed point theorem). *Under the conditions* $(i) - (iv)$, *transform* $z = T(y, \lambda)$ *has a fixed point for* $\lambda \in [a, b]$, *i.e. equation* $z - T(z, \lambda) = 0$ *has solution for* $\lambda \in [a, b]$. *In particular, the solution of* $z - T(z, b) = 0$ *exists.*

Detailed proofs can be found in [2].

Now we start to prove that problem (3) has a generalized solution. For this purpose, we first consider the following boundary value problem containing the parameter $0 \leq \lambda \leq 1$:

$$\begin{cases} (I + \Delta t L)u^{n+1} = u^n + \lambda \Delta t f^{n+1}, & x \in \Omega, \\ u^{n+1}|_{\partial\Omega} = 0. \end{cases} \tag{4}$$

Taking $v \in \dot{W}_2^{(0)}(\Omega)$, and denoting $f(t_{n+1}, x, v) = h(x, v)$, we find using (Theorem 8. 3) that it is easy to see that the linear problem

$$\begin{cases} (I + \Delta t L)u^{n+1} = u^n + h(x, v), & x \in \Omega, \\ u^{n+1}|_{\partial\Omega} = 0 \end{cases} \tag{5}$$

has a unique generalized solution $u^{n+1} \in \dot{W}_2^{(1)}(\Omega)$. Thus we can define a transform from $v \in \dot{W}_2^{(0)}(\Omega)$ to $u^{n+1} \in \dot{W}_2^{(1)}(\Omega)$: $u^{n+1} = T(v, \lambda)$. What remains is to test this transform satisfies all conditions of Lemma 1; We these arguments. Hence a fixed point exists for $\lambda \in [0, 1]$, especially for $\lambda = 1$, problem (3) has a generalized solution $u^{n+1} \in \dot{W}_2^{(1)}(\Omega)$. Therefore we have

Theorem 1. *If conditions* $H_1) - H_3)$ *are satisfied, then the nonlinear boundary value problem (3) has a generalized solution* $u^{n+1} \in \dot{W}_2^{(1)}(\Omega)$.

3   P-S algorithm for the boundary value problem

The algorithm proposed in this section is a combination of a Picard iteration and Schwarz iteration, which is named the P-S algorithm. The procedure can be described as follows:

I. *Constructing the Picard Iteration.* Taking $u_0^{n+1} \in \dot{W}_2^{(1)}(\Omega)$, and making successive approximation for fixed n:

$$\begin{cases} (I + \Delta t L)u_{q+1}^{(n+1)} = u^n + \Delta t f(t_{n+1}, x, u_q^{n+1}), & x \in \Omega, \\ u_{q+1}^{(n+1)}|_{\partial\Omega} = 0, & q = 0, 1, 2, \cdots. \end{cases} \tag{6}$$

II. *Constructing Schwarz Iteration.* For every fixed q, (6) is a linear boundary value problem. Denoting by $u_{q+1}^{n+1} = v$, its Schwarz iteration can be specified as follows:

Step 1.   Taking an initial guess $v^{(0)} \in \dot{W}_2^{(1)}(\Omega)$, $p := 0$.

Step 2.   Solving the boundary value problems on subdomains Ω_i in parallel:

$$\begin{cases} (I + \Delta t L)v_i^{(p+1)} = u^n + \Delta t f(t_{n+1}, x, u_q^{n+1}), & x \in \Omega_i, \\ v_i^{(p+1)} = v^{(p)}, & x \in \partial\Omega_i, \ p = 0, 1, 2, \cdots, \ i = 1, 2, \cdots, I. \end{cases} \tag{7}$$

Step 3.   Continuing $v_i^{(p+1)}$ to Ω, i.e. setting

$$\hat{v}_i^{(p+1)} = \begin{cases} v_i^{(p+1)}, & x \in \Omega_i, \\ v_p, & x \in \Omega \setminus \Omega_i, \end{cases}$$

and then taking the mean value:

$$v^{(p+1)} = \frac{1}{I} \sum_{i=1}^{I} \hat{v}_i^{(p+1)},$$

and then setting $p := p + 1$ and going to Step 2. From I the Picard sequence (named the P-sequence) has been formed, and the Schwarz sequence (named the S-sequence) is given by II. In the next section we will prove that both sequences converge.

4   Convergence of the P-S iteration

The convergence of S-sequence can be obtained by using a method similar to that of [4]. Let $V_i = \dot{W}_2^{(1)}(\Omega_i)$, $V = \dot{W}_2^{(1)}(\Omega)$. Then, we have

Theorem 2.   *Suppose conditions $H_1) - H_3)$ are satisfied. If $V = \overline{\sum_{i=1}^{I} V_i}$, then the S -sequence $\{v^{(p)}\}$ converges to u_{q+1}^{n+1} in the $\dot{W}_2^{(1)}(\Omega)$ norm; if $V = \sum_{i=1}^{I} V_i$, then $\{v^{(p)}\}$ converges geometrically to u_{q+1}^{n+1} in the $\dot{W}_2^{(1)}(\Omega)$ norm.*

We now prove that P-sequence $\{u_{q+1}^{n+1}\}$ converges to u^{n+1}. As $q \to \infty$, we denote the error $e_q = u^{n+1} - u_q^{n+1}$. From (3) and (6), we have

$$\begin{cases} (I + \Delta t L)e_{q+1} = \Delta t(f(t_{n+1}, x, u^{n+1}) - f(t_{n+1}, x, u_q^{n+1})), \\ e_{q+1}|_{\partial\Omega} = 0. \end{cases} \tag{8}$$

It follows that

$$(e_{q+1}, v) + \Delta t a(e_{q+1}, v) = \Delta t(f(t_{n+1}, x, u^{n+1}) - f(t_{n+1}, x, u_q^{n+1}), v), \ \forall v \in \dot{W}_2^{(1)}(\Omega),$$

where

$$a(u, v) = \sum_{j=1}^{J} \int_{\Omega} \{\sum_{k,l} a_{jkl} D_k u_j D_l v_j + c_j u_j v_j\} dx.$$

Taking $v = e_{q+1}$ and making use of H_2), we can get

$$\|e_{q+1}\|_2^2 + \Delta t \|e_{q+1}\|_a^2 \le b_1 \Delta t \|e_q\|_2 \|e_{q+1}\|_2, \tag{9}$$

where $\| \cdot \|_a^2 = a(\cdot, \cdot)$. It follows that

$$\|e_{q+1}\|_2 \le b_1 \Delta t \|e_q\|_2.$$

By induction, we obtain

$$\|e_q\|_2 \le (b_1 \Delta t)^q \|e_0\|_2. \tag{10}$$

Then from (9) it follows that

$$\|e_{q+1}\|_2^2 + \Delta t \|e_{q+1}\|_a^2 \le \left(\frac{b_1 \Delta t}{2}\right)^2 \|e_q\|_2^2 + \|e_{q+1}\|_2^2).$$

Hence,

$$\|e_{q+1}\|_a^2 \le \frac{b_1^2}{4} \Delta t \|e_q\|_2^2$$
$$\le \frac{b_1}{2} \|e_q\|_2^2$$

for Δt small enough.

Notice that $\| \cdot \|_a^2 \ge \gamma \| \cdot \|_{\dot{W}_2^{(1)}(\Omega)}^2$. It follows that

$$\|e_{q+1}\|_{\dot{W}_2^{(1)}(\Omega)} \le \sqrt{\frac{b_1}{2\gamma}} \|e_q\|_2 \le \sqrt{\frac{b_1}{2\gamma}} (b_1 \Delta t)^q \|e_0\|_2. \tag{11}$$

(10) and (11) are the error estimates for the Picard iteration. It shows that the P-sequence is geometrically convergent in $\dot{W}_2^{(1)}(\Omega)$ norm.

Theorem 3. *If the conditions of Theorem 2 are satisfied, then the error estimates (10) and (11) hold, i.e. the P-sequence converges geometrically to the generalized solution of problem (3) in* $\dot{W}_2^{(1)}(\Omega)$ *norm.*

5   A priori estimates for the semidiscrete solution

In this section we will prove a priori estimates for the generalized solution of problem (3) in preparation for the convergence proofs in the next section.

First, from (3), we have

$$(u^{n+1}, v) + \Delta t a(u^{n+1}, v) = (u^n, v) + \Delta t (f^{n+1}, v), \ \forall v \in \dot{W}_2^{(1)}(\Omega). \tag{12}$$

Taking $v = u^{n+1}$ and applying conditions $H_1) - H_3$), and a computation, we can obtain

$$\|u^n\|_2 \le K_0. \tag{13}$$

Second, from (3), we have

$$(u^{n+1}, v) + \Delta t a(u^{n+1}, v) = (u^n, v) + \Delta t(f(t_{n+1}, x, u^{n+1}), v),$$
$$(u^n, v) + \Delta t a(u^n, v) = (u^{n-1}, v) + \Delta t(f(t_n, x, u^n), v).$$

Let $w^{n+1} = \frac{u^{n+1} - u^n}{\Delta t}$. It follows that

$$(w^{n+1} - w^n, v) + \Delta t a(w^{n+1}, v) = (f(t_{n+1}, x, u^{n+1}) - f(t_n, x, u^n), v), \qquad (14)$$

$n = 1, 2, \cdots, N - 1$. We define

$$w^0 = \frac{\varphi - u^{-1}}{\Delta t} = L\varphi + f(0, x, \varphi).$$

Then (14) is also definited for $n = 0$. Taking $v = w^{n+1}$ in (14), and by a computation, we can get

$$\|w^n\|_2 = \left\| \frac{u^n - u^{n-1}}{\Delta t} \right\|_2 \le K_1, \quad n = 1, 2, \cdots, N. \qquad (15)$$

Finally, we estimate $\|D_k u^n\|_2$. From (3), we have

$$(w^{n+1}, v) + a(u^{n+1}, v) = (f^{n+1}, v).$$

Taking $v = u^{n+1}$ and noting estimates (13) and (15), it can be established that

$$\|u^n\|_{\dot{W}_2^{(1)}(\Omega)} \le K_2, \qquad (16)$$

where the constants $K_0 - K_2$ are independent of Δt.

Theorem 4. *If the conditions of Theorem 3 are satisfied, then estimates (13), (15) and (16) hold.*

6   Convergence of semidiscrete solution

The following lemma will be used in our proofs of the rest of our theorems.

Lemma 2. *Under the conditions of Theorem 3, let*

$$W_{\Delta t}(t, x) = \frac{t - t_n}{\Delta t} u^{n+1} + \frac{t_{n+1} - t}{\Delta t} u^n, \ t_n \le t \le t_{n+1}, \ x \in \overline{\Omega}, \qquad (17)$$

and let $Q_T = \{0 \le t \le T, x \in \overline{\Omega}\}$. Then $\{W_{\Delta t}\}$ is strongly compact in $L_2(Q_T)$, $\{D_k W_{\Delta t}\}$ and and $\{D_t W_{\Delta t}\}$ are weakly compact in $L_2(Q_T)$.

The proof can be found in [5-6].

Now we prove that the semidiscrete solution $\{u^n\}$ is convergent (as $\Delta t \to 0$). Its limit is the generalized solution of problem (1). Let $Q_T^n = \{t_n < t \le t_{n+1}, x \in \overline{\Omega}\}$. Then, for $(t, x) \in Q_T^n$, we define

$$u_{\Delta t}(t, x) = u^{n+1}, \ \overline{u}_{k\Delta t}(t, x) = D_k u^{n+1} \ (k = 1, 2, \cdots, m), \ \tilde{u}_{\Delta t}(t, x) = \frac{u^{n+1} - u^n}{\Delta t}.$$

Thus these functions are all definited on Q_T. From Lemma 2, the following estimate holds

$$\sup_{0\leq t\leq T}\|u_{\Delta t}\|_2 + \sup_{0\leq t\leq T}\|\overline{u}_{k\Delta t}\|_2 + \sup_{0\leq t\leq T}\|\tilde{u}_{\Delta t}\|_2 \leq K_3, \quad k=1,2,\cdots,m, \tag{18}$$

where the constant K_3 is independent of Δt. Thus we can choose a subsequence $\{\Delta t_i\}$ such that $\{u_{\Delta t_i}\}$, $\{\overline{u}_{k\Delta t_i}\}$, $\{\tilde{u}_{\Delta t_i}\}$ converge weakly to $u(t,x)$, $\overline{D}_k(t,x)$, $\tilde{u}(t,x)$, respectively. Similar to [1], we can prove

$$\sup_{0\leq t\leq T}\|u\|_2 + \sup_{0\leq t\leq T}\|\overline{u}_k\|_2 + \sup_{0\leq t\leq T}\|\tilde{u}\|_2 \leq K_3, \quad k=1,2,\cdots,m,$$

and

$$\overline{u}_k(t,x) = D_k u, \quad \tilde{u}(t,x) = u_t.$$

Then using the definition of $u_{\Delta t}(t,x)$ and expression (17) of $W_{\Delta t}(t,x)$, it is not difficult to prove that $\{u_{\Delta t}\}$ is also strongly compact in $L_2(Q_T)$. Moreover, it and $\{u_{\Delta t_i}\}$ and $\{W_{\Delta t_i}\}$ converge strongly to the limit $u(t,x)$.

Now we prove that limit $u(t,x)$ is just the generalized solution of problem (1). Let $v(t,x)$ be a smooth function with compact support in Q_T. Then from (2) it is not difficult to show that

$$\int_0^T \{(\tilde{u}_{\Delta t}, v_{\Delta t}) + a(u_{\Delta t}, v_{\Delta t}) - (f_{\Delta t}, v_{\Delta t})\}dt = 0, \tag{19}$$

where $f_{\Delta t} = f(t_{n+1}, x, u_{\Delta t})$. We choose a subsequence $\{\Delta t_i\}$ $(i \to \infty$ as $\Delta t_i \to 0)$ and take the limit for

$$\int_0^T \{(\tilde{u}_{\Delta t_i}, v_{\Delta t_i}) + a(u_{\Delta t_i}, v_{\Delta t_i}) - (f_{\Delta t_i}, v_{\Delta t_i})\}dt = 0.$$

we obtain

$$\int_0^T \{(u_t, v) + a(u, v) - (f, v)\}dt = 0. \tag{20}$$

This shows that $u(t,x)$ is the generalized solution of problem (1) and it is not difficult to prove that it is unique. Hence $\{u^n\}$ converges strongly to the unique generalized solution $u(t,x) \in L_\infty((0,T), \dot{W}_2^{(1)}(\Omega)) \cap W_\infty^{(1)}((0,T), L_2(\Omega))$ of (1).

Theorem 5. *If the conditions of Theorem 4 are satisfied, then $\{u^n\}$ converges strongly to the generalized solution $u \in L_\infty((0,T), \dot{W}_2^{(1)}(\Omega)) \cap W_\infty^{(1)}((0,T), L_2(\Omega))$ of problem (1).*

We have carried our some numerical experiments using our proposed algorithm and our numerical results demonstrate that our method is effective for solving the nonlinear evolution equations. Because space limitations, we do not discuss these computations in detail in this paper.

REFERENCES

[1] Zhou Yulin(1991) *Applications of discrete function analysis to the finite deference method,* Inter. Acad. Publishers.

[2] Friedman A. (1964) *Partial differential equations of parabolic type,* Prentice-Hall.

[3] Gilbarg D. and Trudinger, N. (1977) *Elliptic partial differential equations of second order,* Springer-Verlag.

[4] Lu Tao, Shih T. M. and Liem C. B. (1992) *Domain decomposition method $-$new numerical techniques for solving PDE,* Science Press, Beijing (in Chinese).

[5] He Qiming and Kang Lishan(1994) The domain decomposition method for multidimensional and large scale nonlinear systems$-$case of subdomains overlapping, *Parallel Algorithms and Applications,* Vol. 4, No. 1-2, pp. 77-90.

[6] He Qiming and Evans D. J. (1995) Symmetric domain decomposition method for large scale and nonlinear evolution systems, *Parallel Algorithms and Applications,* Vol. 8, No. 1-2.

Chaotic Iterative Methods by Space Decomposition and Subspace Correction

Jianguo Huang

1 Preface

In this paper, we first construct an abstract chaotic algorithm based on the fundamental framework proposed in [2], [18]. We then prove the convergence property of the algorithm under realistic much receivable conditions and further present some convergence rate estimates with respect to the structure of space decomposition and the parameters of the inexact solvers. Finally, we apply the abstract theory to a concrete chaotic algorithm called S-CR method [12,13] and show the convergence rate can be estimated by $1 - (1 - \sqrt{1 - \frac{C\omega_1(2-\omega)}{(1+H^{-2})(1+log\frac{H}{h})^2}}\,)^{m-1}$, where m is the number of the subproblems, and ω_1 and ω, are the relaxation parameters, and H and h, the diameters of the coarse and finite element triangulation. These results guarantee the effectiveness of a number of chaotic algorithms when executed on message-passing distributed memory multiprocessor systems.

2 Description of the Algorithm

In the following, we shall discuss iterative algorithms to approximate the solution of a linear equation

$$Au = f \tag{2.1}$$

[1] Department of Applied Mathematics, Shanghai Jiaotong University, Shanghai 200240, P.R.China.

Domain Decomposition Methods in Sciences and Engineering, edited by R. Glowinski *et al.*
© 1997 John Wiley & Sons, Ltd.

where A is a symmetric positive definite (SPD) operator defined on a finite dimensional Hilbert space V with inner product $(\cdot, \cdot)$ and induced norm $\| \cdot \|$. (2.1) usually comes from the discretization of positive definite self-adjoint elliptic problem, with a proper boundary condition, by finite element method, finite difference method, etc.

Let V be decomposed into m subspaces $V_i \subset V$, $1 \leq i \leq m$, such that

$$V = \sum_{i=1}^{m} V_i, \qquad (2.2)$$

and, for simplicity, let $(\cdot, \cdot)_A = (A\cdot, \cdot)$ denote the new inner product on V with induced norm $\| \cdots \|_A$. For each k, we define $Q_k, P_k : V \to V_k$ as the orthogonal projection operators onto V_k associated with the inner product $(\cdot, \cdot)$ and $(\cdot, \cdot)_A$ respectively, and define $A_k : V_k \to V_k$ by

$$(A_k u_k, v_k) = (A u_k, v_k), \qquad u_k, v_k \in V_k. \qquad (2.3)$$

A_k can be regarded as a restriction of A on V_k and is SPD in the inner product $(\cdot, \cdot)$. It follows from the above definitions that

$$A_k P_k = Q_k A. \qquad (2.4)$$

Then, based on the framework in [2],[18], we construct the following chaotic algorithm:
Algorithm A. $u^0 \in V$ is an arbitrary initial value, assume $u^{k-1} \in V$ has been obtained. Then u^k is defined by

$$u^k = u^{k-1} + R_{\tau(k)} Q_{\tau(k)} (f - A u^{k-1}). \qquad (2.5)$$

Here we have enumerated the iterative solution in time order, R_i is the inexact solver of the subproblem defined on V_i, i.e. $R_i \simeq A_i^{-1}$, R_i is SPD in the inner product $(\cdot, \cdot)$, and $\tau(k)$ denotes the subscript of the subproblem used at the kth step. We assume that $\{\tau(k)\}_1^\infty$ include i, infinitely many times; here i is an arbitrary natural number, $1 \leq i \leq m$. It guarantees that iterated sequence $\{u^k\}$ is corrected by using the kth subproblem infinitely many times, which satisfies an intuitive convergence requirement. The above condition is named the admissible condition by L. Elsner et al [7],[8].

Let $E^k = u^k - u$ represent the error, then from (2.5) we have

$$E^k = (I - R_{\tau(k)} A_{\tau(k)} P_{\tau(k)}) E^{k-1}. \qquad (2.6)$$

3   Abstract Results

Let $\omega = \max_{1 \leq k \leq m} \rho(R_k A_k)$, $\omega_1 = \min_{1 \leq k \leq m} \rho(R_k A_k)$, where $\rho(A)$ denotes the spectral radius of A. In what follows, we assume that $0 < \omega < 2$. Then we have

Lemma 3.1 ([2],[18]) *Under the above condition,*

$$\|I - R_k A_k P_k\|_A \leq 1 \qquad (3.1)$$

where $1 \leq k \leq m$.

Lemma 3.2 *Let $M_1 = V_l$ be any element of $\{V_k\}_{k=1}^m$ (here $\{V_k\}_{k=1}^m$ is a finite set with sum V_k, $k = 1, 2, \cdots, m$ as its elements) and let M_2 be the sumspace of any subset of $\{V_k\}_{k=1}^m$ (i.e., $M_2 = \sum_{i=1}^l V_{t_i}$ for a subset $\{V_{t_i}\}_{i=1}^l \subset \{V_k\}_{k=1}^m$). Let $C_{M_1 M_2}$ denotes the positive constant, such that, for any $v \in M_1 + M_2$, there exist $v_i \in M_i$, $i = 1, 2$ satisfying that $v = v_1 + v_2$ and $[\|v_1\|_A^2 + \|v_2\|_A^2] \leq C_{M_1 M_2}^2 \|v\|_A^2$. Then we have*

$$\|v\|_A^2 \leq C_{M_1 M_2}^2 [\|P_1'v\|_A^2 + \|P_2'v\|_A^2], \qquad v \in M_1 + M_2,$$

where P_i' denote the orthogonal projection operators onto M_i with respect to inner product $(\cdot, \cdot)_A$, $i = 1, 2$, respectively.

This is just the variant of P.L. Lions' Lemma [15].

Lemma 3.3 *Let $M_1 = V_l$, M_2, P_2' and $C_{M_1 M_2}$ be defined as in Lemma 3.2. Then we have*

$$\|[I - R_l A_l P_l][I - P_2']v\|_A \leq \sqrt{1 - \frac{\omega_1(2 - \omega)}{C_{M_1 M_2}^2}} \|v\|_A, \quad v \in M_1 + M_2.$$

Lemma 3.4 *Let V_{k_1} and V_{k_2} be any two different elements of $\{V_k\}_{k=1}^m$. Then we have*

$$\|[I - R_{k_1} A_{k_1} P_{k_1}][I - R_{k_2} A_{k_2} P_{k_2}]v\|_A \leq \sqrt{1 - \frac{\omega_1(2 - \omega)}{9C_{k_1 k_2}^2}} \|v\|_A, \quad v \in V_{k_1} + V_{k_2},$$

where $C_{k_1 k_2}$ denotes the constant $C_{M_1 M_2}$ given in Lemma 3.2 for $M_i = V_{k_i}$, $i = 1, 2$.

Because of the finite number of choices of M_1, M_2, V_{k_1} and V_{k_2}, let $\sigma \in (0, 1)$ denote the maximum value of $\sqrt{1 - \frac{\omega_1(2-\omega)}{C_{M_1 M_2}^2}}$ and $\sqrt{1 - \frac{\omega_1(2-\omega)}{9C_{k_1 k_2}^2}}$ in all possible situations. Then from Lemma 3.3 and Lemma 3.4, we have

Lemma 3.5 *Let $M_1 = V_l$ be any element of $\{V_k\}_{k=1}^m$, M_2 be the sum of any subset of $\{V_k\}_{k=1}^m$, and let V_{k_1}, V_{k_2} be any two different elements of $\{V_k\}_{k=1}^m$. Then we have*

$$\begin{cases} \|[I - R_l A_l P_l][I - P_2']v\|_A \leq \sigma\|v\|_A, & v \in M_1 + M_2, \\ \|[I - R_{k_1} A_{k_1} P_{k_1}][I - R_{k_2} A_{k_2} P_{k_2}]v\|_A \leq \sigma\|v\|_A, & v \in V_{k_1} + V_{k_2}. \end{cases} \tag{3.2}$$

Lemma 3.6 *Let $\{t_1, t_2\}$ be an arbitrary subset of $\{1, 2, \cdots, m\}$, then for arbitrary K natural numbers $\alpha_k \in \{t_1, t_2\}, k = 1, 2, \cdots, K$, and $\{t_1, t_2\} \subset \{\alpha_1, \alpha_2, \cdots, \alpha_K\}$, we have*

$$\|\prod_{k=1}^K (I - R_{\alpha_k} A_{\alpha_k} P_{\alpha_k})v\|_A \leq \sigma\|v\|_A, \qquad v \in V_{t_1} + V_{t_2}, \tag{3.3}$$

where σ is defined as that in Lemma 3.5.

Lemma 3.6 follows from Lemma 3.1 and Lemma 3.5 easily.

Lemma 3.7 *For any integer l ($2 \leq l \leq m$), there exists a constant $\sigma^l \in (0, 1)$ such that, for arbitrary subset $\{t_1, t_2, \cdots, t_l\} \subset \{1, 2, \cdots, m\}$, arbitrary $\alpha_k \in \{t_1, t_2, \cdots, t_l\}$, $k = 1, 2, \cdots, K$ with $\{t_1, t_2, \cdots, t_l\} \subset \{\alpha_1, \alpha_2, \cdots, \alpha_K\}$, we have*

$$\|\prod_{k=1}^K (I - R_{\alpha_k} A_{\alpha_k} P_{\alpha_k})v\|_A \leq \sigma^l\|v\|_A, \quad v \in \sum_{k=1}^K V_{t_k}, \tag{3.4}$$

where $\sigma^2 = \sigma$ and $\sigma^{l+1} = \sigma + (1 - \sigma)\sigma^l$.

Proof. By induction. As $l = 2$, the result follows from Lemma 3.6. Assume the result is true for l $(2 \leq l < m)$, we want to prove the correctness for $l + 1$.

For arbitrary $\{t_1, t_2, \cdots, t_{l+1}\} \subset \{1, 2, \cdots, m\}$, arbitrary $\alpha_k \in \{t_1, t_2, \cdots, t_{l+1}\}$, $k = 1, 2, \cdots, K$ with $\{t_1, t_2, \cdots, t_{l+1}\} \subset \{\alpha_1, \alpha_2, \cdots, \alpha_K\}$, $v \in \sum_{k=1}^{l+1} V_{t_k}$, consider the estimate of $\prod_{k=1}^{K}(I - R_{\alpha_k} A_{\alpha_k} P_{\alpha_k})v$. Without loss of generality, we may assume that $\alpha_1 = t_1$, $t_1 \notin \{\alpha_2, \alpha_2, \cdots, \alpha_K\}$. Otherwise, through a search process, we can find some $i(1 \leq i \leq K)$, such that $\{t_1, t_2, \cdots, t_{l+1}\} \subset \{\alpha_i, \alpha_{i+1}, \cdots, \alpha_K\}$, and $\alpha_i \notin \{\alpha_{i+1}, \alpha_{i+2}, \cdots, \alpha_K\}$. We might as well suppose that $\alpha_i = t_1$. Then from Lemma 3.1

$$\|\prod_{k=1}^{K}(I - R_{\alpha_k} A_{\alpha_k} P_{\alpha_k})v\|_A \leq \|\prod_{k=i}^{K}[(I - R_{\alpha_k} A_{\alpha_k} P_{\alpha_k})v\|_A$$

which is converted to the estimate of the case for which the assumption holds.

Let $W = \sum_{k=2}^{l+1} V_{t_k}$, $e = \prod_{k=2}^{K}(I - R_{\alpha_k} A_{\alpha_k} P_{\alpha_k})v$. Then from the induction assumption, we have

$$\|e - P_{W^\perp}v\|_A = \|\prod_{k=2}^{K}(I - R_{\alpha_k} A_{\alpha_k} P_{\alpha_k})(v - P_{W^\perp}v)\|_A \leq \sigma^l \|v - P_{W^\perp}v\|_A, \qquad (3.5)$$

where $P_{W^\perp}$ denotes the orthogonal projection operator from V onto $W^\perp$ with respect to the inner product $(\cdot, \cdot)_A$.

Let $\eta = \frac{\|e - P_{W^\perp}v\|_A}{\|v - P_{W^\perp}v\|_A}$ ($\eta = 0$ as $v - P_{W^\perp}v = 0$), then $0 \leq \eta \leq \sigma^l$.

Introduce the auxiliary function

$$\begin{cases} v^* = P_{W^\perp}v + \dfrac{1}{\eta}(e - P_{W^\perp}v), & (\eta > 0), \\ v^* = v, & (\eta = 0). \end{cases} \qquad (3.6)$$

It is easy to see that

$$\begin{cases} \|v^*\|_A^2 = \dfrac{1}{\eta^2}\|e - P_{W^\perp}v\|_A^2 + \|P_{W^\perp}v\|_A^2 = \|v\|_A^2, \\ e = \eta v^* + (1 - \eta)P_{W^\perp}v^*. \end{cases} \qquad (3.7)$$

Then,

$$\|\prod_{k=1}^{K}(I - R_{\alpha_k} A_{\alpha_k} P_{\alpha_k})v\|_A = \|(I - R_{\alpha_1} A_{\alpha_1} P_{\alpha_1})e\|_A$$

$$= \|(I - R_{\alpha_1} A_{\alpha_1} P_{\alpha_1})[\eta v^* + (1 - \eta)P_{W^\perp}v*]\|_A$$

$$\leq \eta\|v\|_A + (1 - \eta)\sigma\|v\|_A \leq [\sigma + (1 - \sigma)\sigma^l]\|v\|_A.$$

The last inequalities follow from (3.7), Lemma 3.1 and Lemma 3.5. Here we also use the fact that $v^* \in V_{t_1} + W$. Lemma 3.7 is now proved.

Let $l = m$, we get the following result:

Lemma 3.8 *For arbitrary natural numbers $\alpha_i \in \{1, 2, \cdots, m\}$, $i = 1, 2, \cdots, K$, and $\{1, 2, \cdots, m\} \subset \{\alpha_1, \alpha_2, \cdots, \alpha_K\}$,*

$$\| \prod_{k=1}^{K} (I - R_{\alpha_k} A_{\alpha_k} P_{\alpha_k}) \|_A \leq \sigma^m < 1, \tag{3.8}$$

where $\sigma^m = 1 - (1 - \sigma)^{m-1}$.

Theorem 3.9 *Assume that V is split into m subspaces $\{V_k\}_{k=1}^{m}$ such that $V = \sum_{k=1}^{m} V_k$ and the inexact solvers R_k are chosen to be SPD on V_k with respect to the inner product $(\cdot, \cdot)$ for respectively, $k = 1, 2, \cdots, m$, which also satisfy*

$$0 < \max_{1 \leq k \leq m} \rho(R_k A_k) < 2.$$

Then Algorithm A is convergent. Furthermore, we have the following convergence rate estimates, i.e. if the iteration is corrected in each processor at least once in some iteration section, then the error will decrease according to the factor $1 - (1 - \sigma)^{m-1}$ after this iteration section, which is independent of the corrected times K and the concrete structure of this section.

Theorem 3.9 follows easily from Lemma 3.8.

4 An Application

Let Ω be a polygonal domain in R^2, for ease of exposition, and consider the problem

$$\begin{cases} -\displaystyle\sum_{i,j=1}^{2} \partial_i(a_{ij}\partial_j u) = f, & \text{in } \Omega, \\ u = 0, & \text{on } \partial\Omega. \end{cases} \tag{4.1}$$

Here $\partial_i = \frac{\partial}{\partial x_i}$, and we assume that the matrix of $(a_{ij})_{2\times 2}$ with continuously differential coefficients, is symmetric for each $x \in \Omega$ and there exist two constants C_0, C_1 such that

$$C_0|\xi|^2 \leq \sum_{i,j=1}^{2} a_{ij}(x)\xi_i\xi_j \leq C_1|\xi|^2, \quad x \in \bar{\Omega}, \quad \xi \in R^2. \tag{4.2}$$

The variational form of (4.1) is

$$\begin{cases} u \in H_0^1(\Omega), \\ a(u, v) = (f, v), & v \in H_0^1(\Omega). \end{cases} \tag{4.3}$$

Here $a(u, v) = \sum_{i,j=1}^{2} \int_\Omega a_{ij}\partial_i u \partial_j v\, dx$, and $(\cdot, \cdot)$ denotes the L^2-inner product on Ω with induced norm $\|.\|$.

Let $\{E_i\}_{i=1}^{m}$ be the coarse quasi-uniform triangulation of Ω with size H, i.e., there exist constants C_2, C_3 not depending on H such that each triangle E_i is contained in (respectively, contains) a disk of radius $C_3 H$ (respectively $C_2 H$). We further divide each E_i into smaller triangles to form a quasi-uniform finite element triangulation $\{K\}_{K \in T_h}$ of Ω with h as its diameter. Define

$$S^h(\Omega) = \{v \in C^0(\bar{\Omega}) : v|_K \text{ is linear}, K \in T_h\}, \qquad S_0^h(\Omega) = S^h(\Omega) \cap H_0^1(\Omega).$$

Then the finite element approximation of (4.3) reads as follows:

$$\begin{cases} u_h \in S_0^h(\Omega) \equiv V, \\ a(u_h, v) = (f, v), \qquad v \in V. \end{cases} \tag{4.4}$$

To give the chaotic algorithm in detail, we proceed to construct the following subdomains $\{\Omega_i\}_{i=1}^m$:

$$\Omega_i = \{\cup E_j : \bar{E}_j \cap \bar{E}_i \neq \emptyset, j = 1, 2, \cdots, m\}, \quad i = 1, 2, \cdots, m,$$

and define $V_i = H_0^1(\Omega_i) \cap V$, $i = 1, 2, \cdots, m$. Clearly, $V = \sum_{i=1}^m V_i$. Then we can apply Algorithm A of section 2 to solve (4.4), this is referred to as the S-CR method[12],[13] at this time. Here, the linear operator A in (2.1) is defined by $(Au, v) = a(u, v)$, $u, v \in V$.

Lemma 4.1 *For the S-CR method, the constant $C_{M_1 M_2}$ given in Lemma 3.2 can be estimated as follows:*

$$C_{M_1 M_2}^2 \leq C(1 + H^{-2})(1 + \log\frac{H}{h})^2, \tag{4.5}$$

where the constant C depends only on the constants C_0, C_1, C_2, C_3 given before and the domain Ω.

Proof. Without loss of generality, assume that $M_1 = V_1 = S_0^h(\Omega_1)$, $M_2 = \sum_{i=1}^l V_{t_i} = S_0^h(\tilde{\Omega}_1)$, where $\tilde{\Omega}_1 = \cup V_{t_i}$, for any open set F, $S^h(F) \equiv S^h(\Omega) \cap H_0^1(F)$, $S_0^h(F) \equiv S^h(F) \cap H_0^1(F)$. Then $M_1 + M_2 = S_0^h(\Omega_1 \cup \tilde{\Omega}_1)$. Let $\Omega' = \Omega_1 \cap \tilde{\Omega}_1$. Obviously Ω' is formed by some coarse triangles in $\{E_i\}_{i=1}^m$. For any $v \in M_1 + M_2$, let

$$v_1|_{\bar{\Omega}_1 \backslash \bar{\Omega}'} = v|_{\bar{\Omega}_1 \backslash \bar{\Omega}'}, \qquad v_2|_{\bar{\tilde{\Omega}}_1 \backslash \bar{\Omega}'} = v|_{\bar{\tilde{\Omega}}_1 \backslash \bar{\Omega}'}.$$

The construction of v_i on $\bar{\Omega}'$ is relatively complex. Define F^0 as the interior points set of F. For any coarse triangle E in Ω', if $\bar{E} \subset \bar{\Omega}' \backslash (\partial\Omega_1 \cap \partial\Omega')^0$, let $v_1 |_{\bar{E}} = v$; otherwise, E has at least one vertex on $\partial\Omega_1 \cap \partial\Omega'$, we then define E^i, $i = 1, 2, 3$ as its three vertices, E_{12}, E_{13}, E_{23}, the related edges; thus we have the following two cases:

(1). There exist two vertices, e.g. E^2, E^3 on $\partial\Omega_1 \cap \partial\Omega'$. We first construct two linear functions $g(x), g_1(x)$ on E, such that, $g(E^i) = v(E^i)$, $i = 1,2,3$, $g_1(E^1) = v(E^1)$, $g_1(E^2) = g_1(E^3) = 0$; and then construct the following an auxiliary function $\tilde{v}(x) \in S^h(E)$ such that

$$\begin{cases} a(\tilde{v}, w) = 0, \qquad w \in S_0^h(E), \\ \tilde{v}|_{E_{12}, E_{13}} = v - g, \\ \tilde{v}|_{E_{23}} = 0. \end{cases} \tag{4.6}$$

Finally, we define

$$v_1|_E = g_1 + \tilde{v}. \tag{4.7}$$

(2). There exists just one vertex e.g. E^3 on $\partial\Omega_1 \cap \partial\Omega'$. Let g_1 be the linear function on E such that $g_1(E^i) = v(E^i)$, $i = 1, 2$, $g_1(E^3) = 0$, and let g be defined as above. Let $\tilde{v}$ be the auxiliary function in $S^h(E)$ such that

$$\begin{cases} a(\tilde{v}, w) = 0, & w \in S_0^h(E), \\ \tilde{v}|_{\partial E} = v - g. \end{cases} \tag{4.8}$$

Then we define v_1 on E as (4.7) too. Thus, we get the construction of v_1 on $\bar{\Omega}'$. The function v_2 on $\bar{\Omega}'$ is defined as $v - v_1$. From the definitions above, we know that $v_i \in M_i$, $i = 1, 2$, and $v = v_1 + v_2$.

Now we proceed with the estimate of $C_{M_1 M_2}$. Clearly it suffices to consider the term

$$\int_E [\sum_{i,j=1}^2 a_{ij} \partial_i v_1 \partial_j v_1] dx$$

for E belonging to the above two cases, since $v_1|_E = v$ for other coarse triangle E. We only give the estimate for case (1). From (4.2), (4.6), (4.7), the extension Lemma and maximum norm estimates given in [1], [6], we know that

$$\int_E [\sum_{i,j=1}^2 a_{ij} \partial_i v_1 \partial_j v_1] dx$$
$$\leq C[\|\nabla g_1\|_{0,E}^2 + \|v - g\|_{H_{00}^{1/2}(E_{12})}^2 + \|v - g\|_{H_{00}^{1/2}(E_{13})}^2], \tag{4.9}$$

$$\|\nabla g_1\|_{0,E}^2 \leq C|v(E_1)|^2 \leq C[(1 + log\frac{H}{h})\|\nabla v\|_{0,E}^2 + H^{-2}\|v\|_{0,E}^2], \tag{4.10}$$

and

$$\|v - g\|_{H_{00}^{1/2}(E_{12})}^2 + \|v - g\|_{H_{00}^{1/2}(E_{13})}^2 \leq C(1 + log\frac{H}{h})^2\|\nabla v\|_{0,E}^2. \tag{4.11}$$

Here the definitions of the Sobolev's norms are the same as those in [1], [6]. From (4.8)-(4.10), we obtain

$$\int_E [\sum_{i,j=1}^2 a_{ij} \partial_i v_1 \partial_j v_1] dx \leq C[(1 + log\frac{H}{h})^2\|\nabla v\|_{0,E}^2 + H^{-2}\|v\|_{0,E}^2]. \tag{4.12}$$

For case (2), the application of the above technique will lead to the same estimate (4.12). Hence, from (4.2), (4.12) and Poincare's inequality [1],[6], we finally have

$$\|v_1\|_A^2 = a(v_1, v_1) \leq C(1 + log\frac{H}{h})^2(1 + H^{-2})\|v\|_A^2.$$

The Lemma then follows.

From Theorem 3.9 and Lemma 4.1, we easily have the following result:

Theorem 4.2 *Let the subspaces V_k be chosen as above, and the inexact solvers R_k be SPD with respect to the inner product $(\cdot, \cdot)(L^2$-inner product) which also satisfies*

$$0 < \omega_1 = \min_{1 \leq k \leq m} \rho(R_k A_k) \leq \omega = \max_{1 \leq k \leq m} \rho(A_k P_k) < 2.$$

Then the S-CR method is convergent. Furthermore, we have the following error estimate, i.e. if the iteration is corrected at each processor at least once in some iteration section, then the error will have decreased according to the factor $1 - (1 - \sqrt{1 - \frac{C\omega_1(2-\omega)}{(1+H^{-2})(1+log\frac{H}{h})^2}})^{m-1}$ *after this iteration section. Here, the constant* C *is independent of* H, h, ω_1, ω *and* m.

REFERENCES

[1] Bramble J. H. et al (1986) The construction of preconditioner for elliptic problems by substructuring. .I *Math.Comp.* 47:103-134.

[2] Bramble J. H. et al (1991) Convergence estimate for product iterative methods with application to domain decomposition *Math. Comp.* 57: 1-21.

[3] Chan T. F. (1988) Domain decomposition algorithms and computational fluid dynamics *Inter. Jour. of Supercomputer Appli.*, 2: 72-83.

[4] Ciarlet P. G. (1978) *The Finite Element Method for Elliptic Problems*, North-Holland, Amsterdam.

[5] Dryja M. and Widlund O. B. (1987) *An additive variant of the schwarz alternating method for the case of many subregions*, Courant institute of mathematical science, Technical Report 339.

[6] Dryja M. and Widlund O. B. (1989) Towards a unified theory of domain decomposition algorithms for elliptic problems. *Domain decomposition methods for partial differential equations*, Philadelphia.

[7] Elsner L. et al (1988) Convergence and property of ART and SSOR algorithms. *Numer. Math.* 59: 91-106.

[8] Elsner L. et al (1990) On the convergence of asynchronous paracotractions with applications to tomographic reconstruction from incomplete data. *LAA* 130: 65-82.

[9] Golub G. H. and VanLoan P. H. (1989) *Matrix Computation*, John Hopkins Univ. Press, USA.

[10] Hageman L. and Young D. (1981) *Applied Iterative Methods*, Academic Press, USA.

[11] Hachbusch W. (1985) *Multi-grid Methods and Applications*, Springer-Verlag, Berlin, Heidelberg.

[12] Kanq L. S. (1984) *Asynchronous Parallel Algorithms for Mathematics-Physics Problems*(In Chinese), Science Press, Beijing.

[13] Kang L. S. (1987) *Parallel Algorithms and Domain Decomposition*, Wuhan University Press, Wuhan.

[14] Kang L. S. (1990) *Split Methods for Solving Multi-dimentional PDE*(In Chinese), Shanghai Science and Technology Press, Shanghai.

[15] Lions P. L. (1988,1989) On the schwarz alternating method (1),(2). In: *R.Glowinski et al ed., International Symposium on Domain Decomposition Methods for Partial Differential equations*, SIAM, Philadelphia.

[16] Varga R. S. (1962) *Matrix Iterative Analysis, Prentice-hall*,U.S.A.

[17] Xu J. (1989) *Theory of multilevel methods*, Ph.D. Thesis, Cornell University.

[18] Xu J. (1992) Iterative methods by space decomposition and subspace correction. *SIAM Review*, 34: 581-613.

Domain Decomposition Methods to Penalty Combinations for Singularity Problem

ZI-CAI LI

Abstract

An embedding technique is presented in this paper to implement penalty combinations into parallel by the iterative substructuring method. Penalty combinations using a penalty integral are much simpler than the direct constraints to match different admissible functions. In combinations, the Ritz-Galerkin method is used in the singular subdomains Ω^+ where exist solution singularities, and the singular particular solutions are chosen to be admissible functions so that only a few of them are needed to cope well with the difference grids in the finite difference methods used in the rest of the solution domain. Consequently, while applying the iterative substructuring methods, we may attain the singular domain Ω^+ to the interface of two subregions in the domain decomposition methods, and regard Ω^+ as a "fat" interface or called an interface "zone". The matrix contribution resulting from Ω^+ can be included into the preconditioner matrix due to a few of unknown coefficients to be sought. The new embedding technique may reform penalty combination easily into parallel computing by the existing, iterative substructuring methods. Such an technique has been proven to be effective, by a brief analysis and numerical experiments of Motz's problem given in this paper.

1.1 Penalty Combinations of the Ritz-Galerkin and Finite Difference Methods

Parallel penalty combinations are presented in this paper for solving singularity problems, to join the combined methods with the domain decomposition methods (simply written DDMs). In this paper the penalty combinations to combine the finite difference method with the Ritz-Galerkin method, to regain the superconvergence rates $O(h^{2-\delta})$, where $\delta(> 0)$ is an arbitrarily small number. Usually, the singular domain should be chosen as one subregion, where the mixed Neumann-Dirichlet problem in

---

^{0} Department of Applied Mathematics, National Sun-Yat Sen University, Kaohsiung, Taiwan 80424, R.O.C.

Domain Decomposition Methods in Sciences and Engineering, edited by R. Glowinski *et al.*

DDMs is solved. However, since only a few unknown coefficients are used with the total number L, having an asymptotic relation:

$$L = O(|lnh|), \tag{1.1}$$

where h is the maximal mesh spacing of difference grids, the singular subdomain may be regarded as a "fat" interface or called an interface "zone" in the renovated domain decomposition methods. Indeed, a few more unknown coefficients in the preconditioner matrices will not cause much effort in computation in the inverse preconditioner matrices.

Consider the Poisson equation with the Dirichlet boundary condition

$$-\Delta u = -\left(\frac{\partial^2 u}{\partial x^2} + \frac{\partial^2 u}{\partial y^2}\right) = f(x, y), \quad (x, y) \in \Omega, \tag{1.2}$$

$$u = 0, \quad (x, y) \in \partial\Omega, \tag{1.3}$$

where Ω is a polygon domain, and the function f is smooth enough. For simplicity, we assume that only one singular point of the solution $u(x, y)$ exists in $\bar{\Omega} = \Omega \cup \partial\Omega$. Let Ω be divided by a piecewise straight line Γ_0 into two subdomains Ω^+ and Ω^-. The Ritz-Galerkin method is used in $\Omega^+ \cup \partial\Omega$ including the singular point, and the finite difference method is used in Ω^-. The subdomain Ω^- is again split by quasiuniform difference grids into small rectangles $\square_{ij}$ and triangles $\triangle_{ij}$. Denote $u_{i,j} = u(x_i, y_i)$, as the solution on the difference nodes (i, j). The finite difference method can be regarded as a special kind of finite element method if using piecewise bilinear and linear interpolatory functions and if using special integration rules to approximate the integrals involved (also see [Li(86,95)]).

Assume that the solution u in Ω can be spanned by $u = \Psi_0 + \sum_{i=1}^{\infty} D_i \Psi_i$, where D_i are the expansion coefficients, and $\Psi_i (i = 1, 2, \cdots \infty)$ are complete and linearly independent basis functions that are known. Then the admissible functions in combinations of the RG-FDMs are written as:

$$v = \begin{cases} v^- &= v_1, \text{ in } \Omega^-, \\ v^+ &= \Psi_0 + \sum_{i=1}^{L} \tilde{D}_i \Psi_i . \text{ in } \Omega^+, \end{cases} \tag{1.4}$$

where $\tilde{D}_i$ are unknown coefficients to be sought. If the particular solutions of (1.2) and (1.3) are chosen as Ψ_i, their total number will greatly decrease for a given accuracy of solutions.

Since there occurs discontinuity of solutions on Γ_0, i.e., $v^+ \neq v^-$ on Γ_0, we define another space

$$H = \{v \in L^2(\Omega), v \in H^1(\Omega^-), \text{ and } v \in H^1(\Omega^+)\}, \tag{1.5}$$

where $H^1(\Omega)$ is the Sobolev space. Let $V_h (\subseteq H)$ denote a finite dimensional collection of the function v in (1.4) satisfying (1.3). We will couple v^+ and v^- by the following integrals

$$\hat{D}(u, v) = \frac{P_c}{h^\sigma} \int_{\Gamma_0} (u^+ - u^-)(v^+ - v^-) d\ell, \tag{1.6}$$

where $P_c(> 0)$ the penalty constant, and $\sigma(\geq 0)$ the penalty power. Hence, we lead to the penalty combination of RG-FDMs, to seek a solution $u_h \in V_h$ such that

$$a_h(u_h, v) = f_h(u, v), \quad \forall v \in V_h \quad where \tag{1.7}$$

$$a_h(u, v) = \widehat{\iint_{\Omega^-}} \nabla u \nabla v \, ds + \iint_{\Omega^+} \nabla u \nabla v \, ds + \frac{P_c}{h^\sigma} \hat{\int_{\Gamma_0}} (u^+ - u^-)(v^+ - v^-) d\ell \tag{1.8}$$

In virtual computation, the integrals $D(u, v)$ on Γ_0 are evaluated approximately by integration rules. When $\sigma \geq 4$, the superconvergence rates of error norms $\overline{\|\epsilon\|}_h = O(h^{2-\delta})$, and $\overline{\|\epsilon\|}_h = O(h^{3/2})$ in the solution derivatives can be achieved by penalty combinations for the quasiuniform difference grids: $S_1 = \cup_{ij} \square_{ij}$, and $S_1 = (\cup_{ij} \square_{ij}) \cup (\cup_{ij} \triangle_{ij})$ respectively.

1.2   Domain Decomposition Methods to Penalty Combination

We will still follow the typical approaches of Bjorstad and Widlund (86), Bramble, Pasciak and Schatz (86), Manteuffel and Parter (90) and Morz (89). Let Ω be also split into Ω_I and Ω_{II} by Γ in DDMs:

$$\bar{\Omega} = \partial\Omega \cup \Omega_I \cup \Omega_{II} \cup \Gamma, \tag{1.9}$$

where Ω_I, Ω_{II} and Γ are two subregions and their interface. To distinguish partitions between DDMs and combinations, we call *subregions* Ω_I, Ω_{II} and *interface* Γ in DDMs, but *subdomains* Ω^+, Ω^- and *common boundary* Γ_0 in combinations respectively, where

$$\bar{\Omega} = \partial\Omega \cup \Omega^+ \cup \Omega^- \cup \Gamma_0. \tag{1.10}$$

The key treatments of the DDMs applied to the combinations are how to embed Ω^+, Ω^- and Γ_0 into Ω_I, Ω_{II} and Γ.

It should be noted from Eq. (1.1) that the number of the coefficients D_l is much less than that of the variables $v_{ij} \in \{\Gamma \cap \partial\Omega^+\}$. Hence we may simply attain the singular domain Ω^+ to interface Γ, and call the interface "zone" Γ^* instead,

$$\Gamma^* = \Gamma \cup \Gamma_0 \cup \Omega^+. \tag{1.11}$$

So we have

$$\Omega^+ \subset \Gamma^* \quad and \quad \Gamma \subset \Gamma^*. \tag{1.12}$$

The subdomain Ω^- is also divided into Ω_I and Ω_{II}.

Below let us describe more in details the renovated DDMs of combinations. Denote the variables by x_1, x_2 and x_3 in Ω_I, Ω_{II} and their interface Γ^*, respectively. Then the unknown expansion coefficients $\{D_i\}$ are included in x_3. Denote $\tilde{x} = (x_1, x_2, x_3)^T$. The equations (1.7) are then reduced to a linear algebraic equation system

$$A\tilde{x} = \tilde{b}, \tag{1.13}$$

where $\tilde{b} = (b_1, b_2, b_3)^T$, and

$$A = \begin{bmatrix} A_{11} & 0 & A_{13} \\ 0 & A_{22} & A_{23} \\ A_{13}^T & A_{23}^T & A_{33} \end{bmatrix}. \tag{1.14}$$

In (1.14), A_{11}, A_{22} and A_{33} are the coefficient matrices from Ω_I, Ω_{II} and Γ^*. Also A_{13} denotes the relation matrices between x_1 and x_3, and A_{23} denotes those between x_2 and x_3.

By using the block Gaussian elimination, we can again reduce (1.13) to a system of algebraic equations of x_3 only:

$$S\, x_3 = \hat{b}_3 , \tag{1.15}$$

where the Schur matrix

$$S = A_{33} - A_{13}^T A_{11}^{-1} A_{13} - A_{23}^T A_{22}^{-1} A_{23}, \tag{1.16}$$

and the vector $\hat{b}_3 = b_3 - A_{13}^T A_{11}^{-1} b_1 - A_{23}^T A_{22}^{-1} b_2$. We may solve x_3 from (1.15) by the preconditioner conjugate gradient method (PCGM) (see [Golub and Loan (89)]). Then x_1 in Ω_I and x_2 in Ω_{II} can be solved in parallel. For the penalty combinations, we choose the preconditioning matrix

$$S_1 = A_{33}^{(1)} - A_{13}^T A_{11}^{-1} A_{13}, \tag{1.17}$$

where $A_{33}^{(1)}$ excludes the matrix components of the contribution of Ω_{II} to A_{33}, but includes that of Ω^+ when using (1.12) in the embedding techniques.

The penalty integral

$$\hat{D}(v, v) = \frac{P_c}{h^\sigma} \int_{\Gamma_0} (v^+ - v^-)^2 \, d\ell , \tag{1.18}$$

plays a role to match the admissible functions v^+ and v^-. The variables v_{ij} on Γ_0 and the coefficients $\{D_\ell\}$ are also included in x_3. We then have [Bjorstad and Widlund (86)]

$$\begin{bmatrix} A_{11} & 0 & A_{13} \\ 0 & A_{22} & A_{23} \\ A_{13}^T & A_{23}^T & A_{33} \end{bmatrix} \begin{bmatrix} A_{11} & 0 & A_{13} \\ 0 & A_{22} & A_{23} \\ A_{13}^T & 0 & A_{33}^{(1)} \end{bmatrix}^{-1} \begin{bmatrix} 0 \\ 0 \\ y \end{bmatrix} = \begin{bmatrix} 0 \\ 0 \\ SS_1^{-1}y \end{bmatrix} , \tag{1.19}$$

where y are the sequential solutions of x_3 on the interface boundary (or "zone"). The operation involving the inverse matrix can be carried out by two steps [Bramble, Pasciak and Schatz (86)].

Step I. Solve the Dirichlet problem on $\Omega_{II} \cup \Gamma$:

$$A_{22}x_2 + A_{23}y = 0. \tag{1.20}$$

Step II. Solve the Neumann-Dirichlet problem on $\Omega_I \cup \Gamma$

$$\begin{bmatrix} A_{11} & A_{13} \\ A_{13}^T & A_{33}^{(1)} \end{bmatrix} \begin{bmatrix} x_1 \\ x_3 \end{bmatrix} = \begin{bmatrix} 0 \\ y \end{bmatrix} . \tag{1.21}$$

In fact, the preconditioner matrix S_1 in (1.17) is the Schur matrix of the above equation.

1.3 Analysis of Preconditioning Condition Number

Let the matrices A and B be positive definitive and symmetric, and B be the preconditioner matrix. Then the preconditioning number is defined by the following ratios of the maximal and minimal eigenvalues of $B^{-1}A$.

$$Con.\,(B^{-1}A) = Con.\,(A\,B^{-1}) = \lambda_{\max}(B^{-1}A)/\lambda_{\min}(B^{-1}A)\,, \qquad (1.22)$$

$$\lambda_{\max}(B^{-1}A) = \max_{\|x\|\neq 0}\frac{A(x,x)}{B(x,x)}\,, \quad \lambda_{\min}(B^{-1}A) = \min_{\|x\|\neq 0}\frac{A(x,x)}{B(x,x)}\,, \qquad (1.23)$$

where $A(x,x) = x^T A x$, and the Euclidean norm $\|x\| = (\sum_{i=1}^{n} x_i^2)^{1/2}$.

Consider the Poisson equation on Ω^- in Fig. 1.1b with the mixed Neumann-Dirichlet condition:

$$-\Delta u = f \text{ in } \Omega^-\,, u = 0 \text{ on } \partial\Omega \setminus \Gamma_0, \qquad \frac{\partial u}{\partial n} = 0 \text{ on } \Gamma_0, \qquad (1.24)$$

where Γ_0 is the common boundary of Ω^+ and Ω^-. Suppose that the finite difference method (or the finite element method) is used in Ω^-. Let A_a and B_a be the associated matrix and the preconditioner matrix in the traditional DDMs. Many reports on DDMs such as [Bjorstad and Widlund (86), Bramble et al. (86), Manteuffel and Parter (90) and Mroz (89)] have obtained the $B - D$ bounds

$$\alpha_0 B_a(x_a, x_a) \le A_a(x_a, x_a) \le \alpha_1 B_a(x_a, x_a), \qquad (1.25)$$

where $\alpha_0 = 1$, $\alpha_1 = C(1 + \ell n^2 h)$, and C is a bounded constant independent of h. We can prove the following theorem.

Theorem 1. *Let the $B - D$ condition hold. Then for the DDMs of penalty combination (1.7) by the embedding technique (1.12), there exist the bounds*

$$Con.(S_1^{-1}S) \le \alpha_1/\alpha_0\,. \qquad (1.26)$$

This theorem displays that the renovated DDMs in this paper will not cause deterioration of preconditioning condition numbers of the combinations, compared with the existing DDMs.

1.4 Numerical Experiments for Motz's Problem

Consider Motz's problem which solves the Laplace equation on a rectangle Ω $(-1 < x < 1, 0 < y < 1)$

$$\Delta u = \frac{\partial^2 u}{\partial x^2} + \frac{\partial^2 u}{\partial y^2} = 0 \text{ in } \Omega, \qquad (1.27)$$

with the mixed Neumann-Dirichlet boundary conditions

$$u|_{x<0 \wedge y=0} = 0, \qquad u|_{x=1} = 500, \qquad \frac{\partial u}{\partial y}\Big|_{y=0} = \frac{\partial u}{\partial y}\Big|_{y=0 \wedge x>0} = \frac{\partial u}{\partial x}\Big|_{x=-1} = 0. \qquad (1.28)$$

Note that there exists an angular singularity at the origin (0,0) due to the intersection point of the Neumann-Dirichlet boundary conditions.

We split Ω by Γ_0 into two subdomains Ω^+ and Ω^-, where the singular subdomain Ω^+ is a small rectangle $(-\frac{1}{2} < x < \frac{1}{2}, \; 0 < y < \frac{1}{2})$, and Ω^- is the rest of Ω (see Fig. 1.2a). The Ritz-Galerkin method and the finite difference method are used in Ω^+ and Ω^- respectively, with the admissible functions [Li(89,95)]

$$v = \left\{ \begin{array}{ll} v^- & \text{in } \Omega^- \\ v^+ = \sum_{\ell=0}^{L} D_\ell r^{\ell+\frac{1}{2}} \cos(\ell + \frac{1}{2})\theta, \end{array} \right. \tag{1.29}$$

where (r, θ) are the polar coordinates, and D_ℓ are the coefficients to be sought. The subdomain Ω^- is again divided into small squares as shown in Fig. 1.2b.

For DDMs of penalty combinations, we choose Fig. 1.3 as the computational model, where Ω^- is divided into three subregions

$$\Omega^- = \Omega_I^- \cup \Omega_{II,1}^- \cup \Omega_{II,2}^-, \quad \Omega_I = \Omega_I^-, \quad \Omega_{II} = \Omega_{II,1}^- \cup \Omega_{II,2}^- \tag{1.30}$$

and the interface "zone" Γ^* in (1.12), where Γ is the common boundary Ω_I and Ω_{II}. Choose the zero initial values: $D_\ell^{(0)} = 0$ and $v_{ij}^{(0)} = 0$ on $\Gamma \cup \Gamma_0$. Then the sequences $D_\ell^{(k)}$ and $v_{ij}^{(k)}$ on $\Gamma \cup \Gamma_0$ $D_\ell^{(k)}$ can be obtained from the DDMs. To measure the iterative errors we may compute the following sequential errors,

$$\Delta_{\max}^{(k)} = \max \left\{ \max_{(i,j) \in \Gamma \cup \Gamma_0} |u_{ij}^{(k)} - u_{i,j}^{(k-1)}|, \max_\ell |D_\ell^{(k)} - D_\ell^{(k-1)}| \right\} \tag{1.31}$$

Table 1 presents the sequential errors (1.31) until the reduced ratios $\Delta_{maix}^{(k)}/\Delta_{max}^{(1)} \leq 10^{-6}/2$. It can be seen that only 7–8 iterations are needed in the renovated DDMs. Table 2 provides the approximate coefficients in the iterations. The above computational results show an effectiveness of parallel penalty combinations for singularity problems.

Acknowledgements

This work was supported in part by the research grants from the National Science Council of Republic of China under Grant No.NSC83-0208-M110-001 and No.NSC84-2121-M110-003-MS.

Table 1.1 Error norms in iterations the by Penalty Combination using preconditioner S_1 with $P_c = 1$ and $\sigma = 4$ by $\Delta_{max}^{(k)}/\Delta_{max}^{(k)} \leq 10^{-6}/2$

Iteration number	Δ_{Max}				
	$MS = 2$	4	6	8	10
1	401.1	403.3	403.8	403.9	403.9
2	80.17	78.29	77.88	79.73	79.65
3	5.802	7.444	7.300	7.467	7.436
4	0.3281	0.2061	0.1640	0.1760	0.1917
5	$0.5317 * 10^{-1}$	0.1460	0.1282	$0.2953 * 10^{-1}$	$0.2902 * 10^{-1}$
6	$0.1186 * 10^{-1}$	$0.1309 * 10^{-2}$	$0.3653 * 10^{-2}$	$0.1517 * 10^{-2}$	$0.1433 * 10^{-2}$
7	$0.4412 * 10^{-4}$	$0.1914 * 10^{-3}$	$0.2300 * 10^{-3}$	$0.6330 * 10^{-3}$	$0.6161 * 10^{-3}$
8	/	$0.1162 * 10^{-4}$	$0.3964 * 10^{-4}$	$0.3516 * 10^{-5}$	$0.3794 * 10^{-5}$

Table 1.2 Leading coefficients by the DDMs of Penalty Combinations as $Ms = 8$,$L + 1 = 6$,$Pc = 1$ and $\sigma = 4$.

Iteration number	$\tilde{D}_0$	$\tilde{D}_1$	$\tilde{D}_2$	$\tilde{D}_3$	$\tilde{D}_4$	$\tilde{D}_5$
1	403.8726	9.8300	34.6974	-45.6345	0.5574	-16.3848
2	404.9640	89.5558	24.3091	-3.1285	5.9365	3.7177
3	401.0731	87.7007	16.8425	-8.6168	1.6305	0.7451
4	401.0927	87.6446	16.9661	-8.7928	1.7377	0.6990
5	401.0925	87.6467	16.9644	-8.7924	1.7295	0.7285
6	401.0926	87.6469	16.9636	-8.7939	1.7306	0.7295
7	401.0926	87.6470	16.9636	-8.7940	1.7312	0.7299
8	401.0926	87.6470	16.9636	-8.7940	1.7312	0.7299
Com. Coeffs.	401.0926	87.6470	16.9636	-8.7939	1.7312	0.7299
True Coeffs.	401.1625	87.6559	17.2379	-8.0712	1.4403	0.3311

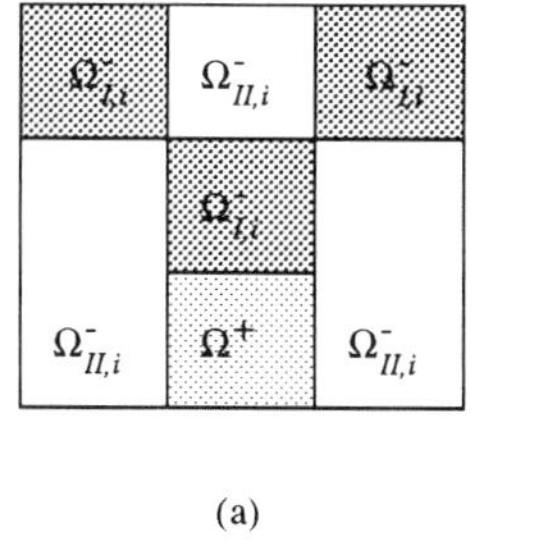

(a)

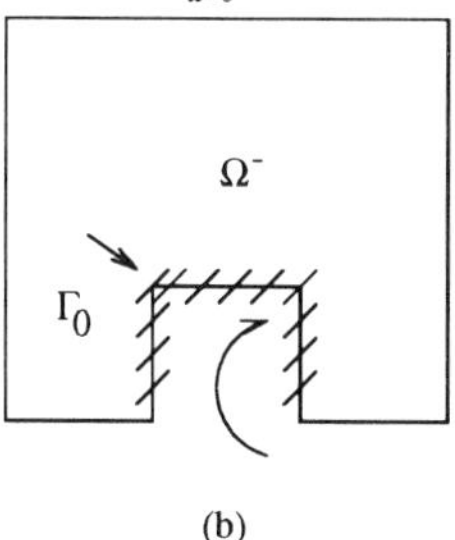

(b)

Figure 1.1 The partition of the DDMs when $\Omega^+ \subset \Gamma^*$

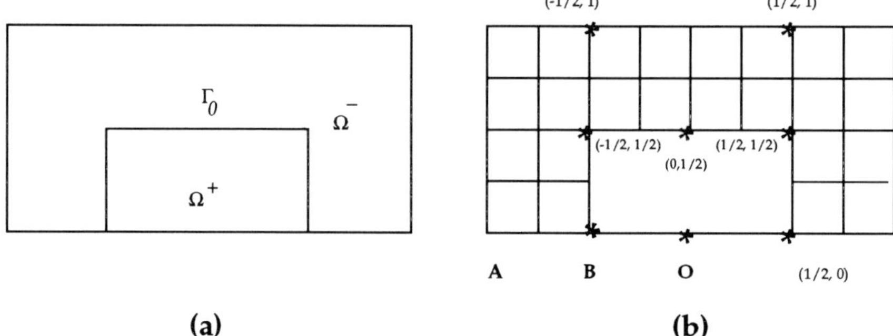

Figure 1.2   Partitions of Motz's problem in the combinations and partition in DDMs

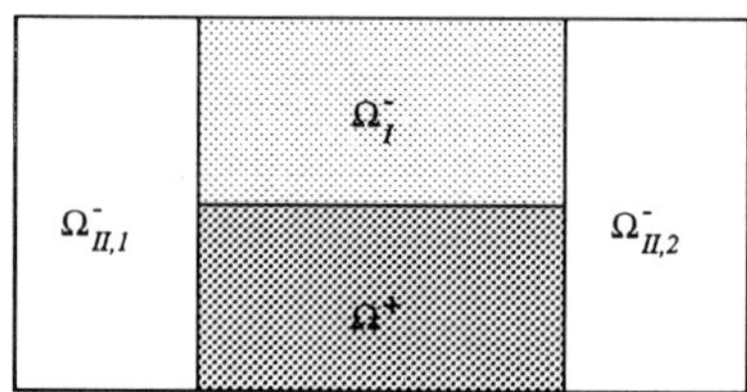

Figure 1.3   Partitions of Motz's problem in the DDMs while $\Omega^+ \subset \Omega_I$,

References

P.E. Bjorstad and O.B. Widlund, Iterative methods for the solution of elliptic equations on the regions partitioned into substructure, SIAM J. Numer. Anal., Vol. 23, pp. 1097–1120, 1986.

J.H. Bramble, J.W. Pasciak and A.H. Schatz, An iterative method for elliptic problems on regions partitioned into substructures, Math. Comp., Vol. 46, pp. 361–369, 1986.

G.H. Golub and C.F. van Loan, Matrix Computations (2nd Edition), The Johns Hopkins University Press, 1989.

T.A. Manteuffel and S.C. Parter, Preconditioning and boundary conditions, SIAM J. Numer. Anal., Vol. 27, pp. 656–694, 1990.

M. Mroz, Domain decomposition method for elliptic mixed boundary methods, Computing, Vol. 42, pp. 45–59, 1989.

Z.C. Li, An approach combining the Ritz-Galerkin and finite difference methods, Numer. Methods for PDE, Vol. 5, pp. 279–295, 1989.

Z.C. Li, Superconvergence of penalty combinations of the Ritz-Galerkin and finite difference method, Technical Report, Department of Applied Mathematics, National Sun Yat-sen University, Kaohsiung, Taiwan, 1995.

Domain Decomposition Methods with Strip Substructures

Monika Mróz [1]

Abstract: Domain Decomposition Method for finite element discretization of planar elliptic variational problems with regular and discontinuous coefficients is analyzed. The domain is divided into strip shaped subdomains. We construct Neumann – Dirichlet substructuring algorithms. The approximate solution is obtained iteratively by solving local problems associated with each strip and the global problem associated with the coarse triangulation. Convergence of algorithms is almost optimal with respect to the parameters of triangulations and independent of the jumps of coefficients.

1 Introduction

Domain Decomposition Method for finite element discretization of elliptic problems with discontinuous coefficients is analyzed. We introduce two level nested triangulation. The domain is divided into strip shaped subdomains. We construct Neumann – Dirichlet substructuring algorithm with coarse space.

The partition of the domain into such subdomains has several advantages. The bandwidth of local matrices is narrow, which minimizes computations and memory requirements. Also the structure of local problems is useful for vectorization of an algorithm.

The substructuring preconditioners are usually constructed on subdomains (boxes) defined by the coarse triangulation, cf. Bramble, Pasciak, Schatz (1986). The convergence rate of iterative method which is depends on the condition number of preconditioned system, is bounded polylogarithimcally in H/h. Here H and h denote parameters of coarse and fine triangulations. Our algorithms have similar convergence properties. The condition number of preconditioned system for strips is proportional to $(1 + \ln(H/h))$. The related numerical experiments are included in Mróz (1995).

[1] Institute of Applied Mathematics and Mechanics, Warsaw University,Banacha 2, 02-097 Warsaw, Poland, Email `monika@appli.mimuw.edu.pl`. This work has been supported by Polish Scientific Grant 211669101

Domain Decomposition Methods in Sciences and Engineering, edited by R. Glowinski *et al.*

The algorithms presented in this paper can be extended to 3–D case. Also they are applicable to parabolic and nonlinear problems. The Neumann–Dirichlet preconditioner can be used as inexact solver in Additive Schwarz Method. In such situation we obtain optimal convergence and much simpler implementation. We can construct substructuring preconditioner for boxes. Each strip is divided into boxes and for each strip we apply our Neumann–Dirichlet preconditioner. The convergence of such method is the same as for strips. All these extensions are analyzed in Mróz (1995).

2 Model problem

We consider the problem of finding an approximate solution of the following elliptic, boundary value problem.

For given a bilinear form $a(\cdot,\cdot)$ and linear functional $l(\cdot)$ on $H_0^1(\Omega)$ we want to find $u \in H_0^1(\Omega)$ such that

$$a(u,v) = l(v) \quad \forall v \in H_0^1(\Omega) \,, \tag{1}$$

where Ω is a Lipschitz bounded domain in R^2. For simplicity of presentation we assume that Ω is a polygon.

We will distinguish two cases for $a(\cdot,\cdot)$.

bilinear form with regular coefficients

The bilinear form in this case is as follows:

$$a(u,v) = \sum_{i=1}^{2} \int_{\Omega} \frac{\partial u}{\partial x_i} \frac{\partial v}{\partial x_i}) dx \,. \tag{2}$$

bilinear form with discontinuous coefficients

We consider the variational problem of the form (1) up to replacing $a(\cdot,\cdot)$ by the form

$$a^{\rho}(u,v) = \int_{\Omega} \rho(x) \sum_{i=1}^{2} \frac{\partial u}{\partial x_i} \frac{\partial v}{\partial x_i} dx \,. \tag{3}$$

The function $\rho(x)$ is piecewise constant i.e

$$\rho(x) = \rho_i > 0 \,, \quad x \in \Omega_i \,, \tag{4}$$

where Ω_i denote the strip shape subdomains consisting of elements Ω^j, defined in Section 2. The jumps of coefficients between subdomains may be large. This model problem can be applied to the case when the function $\rho(x)$ varies moderately on each subdomain and is discontinuous between subdomains. In this case the coefficient is merely equal to the mean value of $\rho(x)$ on the subdomain.

Let $l(v)$ denote the linear form defined by

$$l(v) = (f,v)_{L^2(\Omega)} = \int_{\Omega} fv dx \,. \tag{5}$$

3   Finite Element approximation

A two level triangulation is defined on the domain Ω. First, we construct a coarse triangulation Ω_H that consists of shape regular (cf. Ciarlet (1978)), nonoverlaping triangles Ω^j of diameter of order H. In second step, we further divide each element of the triangulation Ω_H into smaller, shape regular triangles of diameter $O(h)$. They form the fine triangulation Ω_h.

Spaces of piecewise linear, continuous functions on Ω_H and Ω_h are denoted by $V^H(\Omega)$ and $V^h(\Omega)$. The restriction to subspaces of functions vanishing on $\partial\Omega$ is denoted by $V_0^h(\Omega)$ and $V_0^H(\Omega)$ respectively. The corresponding approximate problem for (1) is then:

Find $u^* \in V_0^h(\Omega)$ such that

$$a(u^*, v) = l(v) \quad \forall v \in V_0^h(\Omega) . \tag{6}$$

Let $\{\phi_j^h\}$ be the set of standard, piecewise linear, nodal basis functions, thus $V_0^h(\Omega) = span\{\phi_j^h\}$. In this basis, the discrete variational problem (6) can be rewritten as a system of linear equations

$$A\underline{u} = \underline{f} , \tag{7}$$

where coefficients $A_{jl} = a(\phi_j^h, \phi_l^h)$ and $f_j = l(\phi_j^h)$ The matrix A is positive definite and symmetric. The condition number of A is proportional to h^{-2}.

In the same manner, we can formulate the discrete variational problem for the bilinear form $a^\rho(\cdot, \cdot)$ from (3). The condition number of A^ρ is proportional to $\frac{\max_j \rho^j}{\min_j \rho^j} h^{-2}$.

4   Neumann–Dirichlet preconditioner – regular coefficients

In this section we construct a Neumann–Dirichlet preconditioner for the problem (2) with regular coefficients.

The domain Ω is divided into N strips Ω_i, $i = 1, \ldots, N$ called strips. We assume that the boundary of each strip consists only of boundaries of elements from the coarse triangulation Ω_H and there are no nodes of Ω_H inside the strip. The strip Ω_i has common boundary only with at most two neighboring strips. This common interface between two strips is called Γ_i,

$$\Gamma_i = \partial\Omega_i \cap \partial\Omega_{i+1} .$$

Every point of Γ_i belongs to exactly two strips. The boundary of each strip consists of the two interface lines Γ_{i-1} and Γ_i and parts of $\partial\Omega$.

Let us denote the odd strips by Dirichlet superscript Ω_i^D and even strips by Neumann superscript Ω_i^N. The bilinear forms $a_i(\cdot, \cdot)$ represents restrictions of $a(\cdot, \cdot)$ to Ω_i. In order to distinguish the form $a_i(\cdot, \cdot)$ defined on Dirichlet or Neumann type strips we will add suitable superscript to this notation. Thus bilinear forms $a_i^D(\cdot, \cdot)$ and $a_i^N(\cdot, \cdot)$ are defined on Dirichlet or Neumann type strips respectively.

We define the local orthogonal projection P_i of the space $V^h(\Omega_i) \cap V_0^h(\Omega)$ onto $V_0^h(\Omega_i)$ by

$$a_i(P_i u, v) = a_i(u, v) \quad \forall v \in V_0^h(\Omega_i) , \tag{8}$$

and the local discrete harmonic function $H_i u$ with respect to bilinear form $a_i(\cdot, \cdot)$ by

$$a_i(H_i u, v) = 0 \qquad \forall v \in V_0^h(\Omega_i) , \tag{9}$$
$$H_i u = u \qquad \text{on } \partial\Omega_i .$$

For any function $V^h(\Omega_i) \cap V_0^h(\Omega)$ we have

$$u = P_i u + H_i u .$$

Thus

$$a(u, u) = \sum_{even\ i} a_i^N(u, u) + \sum_{odd\ i} \left(a_i^D(P_i u, P_i u) + a_i^D(H_i u, H_i u) \right) . \tag{10}$$

Dryja and Proskurowski (1985) constructed the Neumann–Dirichlet preconditioner by omitting the last term of representation (10). Then the upper estimate of $a(u, u)$ by this preconditioner depends on H^{-2} since the Trace and Extension Lemmas were used for strip shape subdomains. In order to avoid such dependence, the mechanism of global transportation of information should be introduced into the definition of the preconditioner (cf. Widlund (1988). To meet this requirement we include the term $a(I_H u, I_H u)$ to the definition of the preconditioner, where I_H denotes the nodal value interpolation operator from $V_0^h(\Omega)$ onto $V_0^H(\Omega)$, defined by

$$(I_H u)(x) = u(x) , \tag{11}$$

if x is a node of the triangulation Ω_H.

We are ready to define a bilinear form $b(\cdot, \cdot)$ that corresponds to the Neumann–Dirichlet preconditioner,

$$b(u, v) = \sum_{odd\ i} a_i^D(\mathcal{P}_i(u - I_H u), \mathcal{P}_i(v - I_H v)) +$$
$$+ \sum_{even\ i} \tilde{a}_i^N(u - I_H u, v - I_H v) + a(I_H u, I_H v) . \tag{12}$$

where

$$\tilde{a}_i^N(u, v) = a_i^N(u, v) + H^{-2}(u, v)_{L^2(\Omega_i^N)} . \tag{13}$$

Theorem 1 *For any function $u \in V_0^h(\Omega)$, the following inequalities holds*

$$m \left(1 + \ln \frac{H}{h}\right)^{-1} b(u, u) \leq a(u, u) \leq M\, b(u, u) , \tag{14}$$

where the positive m and M are independent of H and h

The proof of this theorem is given in Mróz (1995).

To solve the linear system (7), we use a preconditioned gradient method (PCG) (cf. Concus, Golub, O'Leary (1976)). In each step of the PCG method the system corresponding to the preconditioner is to be solved:

Find $u \in V_0^h(\Omega)$ such that

$$b(u, v) = g(v) \quad \forall v \in V_0^h(\Omega) . \tag{15}$$

The finite element space $V_0^h(\Omega)$ is decomposed,

$$V_0^h(\Omega) = \tilde{V}_0^h(\Omega) \oplus V_0^H(\Omega) \,,$$

where $\tilde{V}_0^h(\Omega)$ is a subspace of $V_0^h(\Omega)$ consisting of functions vanishing at the nodes of the coarse triangulation Ω_H.

The algorithm of solving the problem (15) is as follows.

Algorithm 2

1. *Construct the system for the nodal basis functions ϕ_j^h connected with the nodes of the interiors of Ω_i^D. For such functions $I_H\phi_j^h = 0$, hence (15) is reduced to separate subproblems of finding local projections $\mathcal{P}_i w$, $i = 1, 3, \ldots$ by solving*

$$a_i^D(\mathcal{P}_i w, \phi_j^h) = g(\phi_j^h) \quad \forall \phi_j^h \in V_0^h(\Omega_i^D)\,.$$

   These systems correspond to solving the subproblems individually for each strip Ω_i with homogeneous Dirichlet boundary conditions on interface lines Γ_i.

2. *We build the system associated with $\overline{\Omega}_i^N$ excluding nodes of the coarse triangulation Ω_H. For such basis functions $I_H\phi_j^h = 0$, thus we search for w on Ω_i^N $i = 2, 4, \ldots$ by solving*

$$\tilde{a}_i^N(w, \phi_j^h) = g(\phi_j^h) - \sum_{odd\ i} a_i^D(\mathcal{P}_i w, \phi_j^h) \quad \forall \phi_j^h \in \tilde{V}^h(\Omega_i^N)$$

   These systems correspond to solving the subproblems individually for each strip Ω_i^N with Neumann boundary conditions on interface lines Γ_i, excluding the nodes of the triangulation Ω_H, where we impose homogeneous Dirichlet boundary conditions.

3. *Construct the system for basis function ϕ_j^H from $V_0^H(\Omega)$. Note that $\phi_j^H - I_H\phi_j^H = 0$, thus in order to find the interpolation $I_H u$ we solve the global system*

$$a(I_H u, \phi_j^H) = g(\phi_j^H) \quad \forall \phi_j^H \in V_0^H(\Omega)$$

4. *In Step 1 the projections $\mathcal{P}_i w$ have been computed, so now we find the solution u on Ω_i^D. The discrete harmonic function $\mathcal{H}_i w$ is obtained from the system.*

$$a_i^D(\mathcal{H}_i w, \phi_j^h) = 0 \quad \forall \phi_j^h \in V_0^h(\Omega_i^D)\,,$$

   and then $w = \mathcal{H}_i w + \mathcal{P}_i w$ or we can solve the system

$$a_i^D(w, \phi_j^h) = g(\phi_j^h) \quad \forall \phi_j^h \in V_0^h(\Omega_i^D)\,.$$

   The boundary conditions are to be imposed so as to fix the values of w on the interface lines Γ_i as calculated in the previous Step. In the second case we do not need to keep the values of $\mathcal{P}_i w$ in all nodes of Ω_i after Step 1 is completed.

   The solution u on Ω is obtained from the formula

$$u = w + I_H u \,.$$

This algorithm admit quite high level of coarse-grained parallelism. Steps 1, 2 and 4 consist of solving a number of small separate systems. Furthermore Steps 2 and 3 may be performed concurrently.

5   Neumann–Dirichlet preconditioner – discontinuous coefficients

In this section we consider the differential problem with coefficients constant on each strip Ω_i

We will use the notation and definitions introduced in previous Section. The restriction of the bilinear form $a^\rho(\cdot,\cdot)$ to Ω_i is denoted by $a_i^\rho(\cdot,\cdot)$. The definition of projection $P_i^\rho u$ and discrete harmonic function $H_i^\rho u$ is the same as in (8) and (9) with respect to the bilinear form $a_i^\rho(\cdot,\cdot)$. The idea of construction of the Neumann–Dirichlet preconditioner is similar to that for regular coefficients. In order to avoid the dependence on jumps of coefficients, the weights are introduced into Neumann problems on even (Neumann type) strips. We first define the function u^ρ on $\partial\Omega$ and on Γ_i, $i = 1, \ldots, n-1$

$$\begin{aligned}
u^\rho(x) &= (\rho_i + \rho_{i+1})^{1/2}(u(x) - I_H u(x)) && x \in \Gamma_i\,, \\
u^\rho(x) &= 0 && x \in \partial\Omega\,.
\end{aligned} \tag{16}$$

The function $\tilde{\mathcal{H}}_i u^\rho \in V^h(\Omega_i^N)$ is a local discrete harmonic function defined on Ω_i^N with respect to the bilinear form

$$\tilde{a}_i^N(u,v) = (\nabla u, \nabla v)_{L^2(\Omega_i^N)} + H^{-2}(u,v)_{L^2(\Omega_i^N)}\,, \tag{17}$$

i.e.

$$\begin{aligned}
\tilde{a}_i^N(\tilde{\mathcal{H}}_i u^\rho, v) &= 0 && \forall v \in V_0^h(\Omega_i^N)\,, \\
\tilde{\mathcal{H}}_i u^\rho(x) &= u^\rho(x)\,, && x \in \partial\Omega_i^N\,.
\end{aligned}$$

The bilinear form $b^\rho(\cdot,\cdot)$ that corresponds to the Neumann–Dirichlet preconditioner for the case with piecewise constant coefficients is defined by

$$\begin{aligned}
b^\rho(u,v) &= \sum_i \rho_i(\nabla\mathcal{P}_i(u - I_H u), \nabla(v - I_H v))_{L^2(\Omega_i)} + \\
&+ \sum_{even\ i} \tilde{a}_i^N(\tilde{\mathcal{H}}_i u^\rho, \tilde{\mathcal{H}}_i v^\rho) + a^\rho(I_H u, I_H v)\,.
\end{aligned} \tag{18}$$

Theorem 2 *For any function $u \in V_0^h(\Omega)$, the following inequalities hold*

$$m\left(1 + \ln\frac{H}{h}\right)^{-1} b^\rho(u,u) \le a^\rho(u,u) \le M\, b^\rho(u,u), \tag{19}$$

provided coefficients of the bilinear form $a^\rho(\cdot,\cdot)$ are constant on strip Ω_i, (see (4)), the positive constants m and M are independent of H, h and the jumps of ρ_i.

This theorem is proved in Mróz (1995).

6 Neumann-Dirichlet preconditioner as inexact solver in ASM

In this section we apply the Neumann–Dirichlet preconditioner to Additive Schwarz Method as inexact solver for local problems. Such approach allow us to construct a structural algorithm with optimal estimates on convergence. We use the framework of ASM developed by Dryja and Widlund (1990). The algorithm is presented for the case of regular coefficients.

The space $V_0^h(\Omega)$ is represented as a sum of two spaces

$$V_0^h(\Omega) = V_0 + V_1 = V_0^H(\Omega) + V_0^h(\Omega). \tag{20}$$

Let us define inner local products $b_i(\cdot,\cdot)$, $i = 0,1$

$$\begin{aligned}
b_0(u,v) &= a(u,v), & u,v &\in V_0^H(\Omega), \\
b_1(u,v) &= \sum_{odd\ i} a_i^D(\mathcal{P}_i u, \mathcal{P}_i v) + \sum_{even\ i} \tilde{a}_i^N(u,v) & u,v &\in V_0^h(\Omega),
\end{aligned} \tag{21}$$

where $\tilde{a}_i^N(u,v)$ was introduced in (13).

Let $\mathcal{T}_i$ denote the approximate projections from $V_0^h(\Omega)$ to V_i with respect to bilinear form $b_i(\cdot,\cdot)$

$$b_i(\mathcal{T}_i u, v) = a(u,v) \quad \forall v \in V_i \tag{22}$$

If the operator $\mathcal{T} = \mathcal{T}_0 + \mathcal{T}_1$ is invertible then (6) is equivalent the following auxiliary problem:

Find $u \in V_0^h(\Omega)$ which satisfies

$$\mathcal{T} u = g \tag{23}$$

where the right hand side g has to be chosen so that the auxiliary equation (23) has the same solution as (6).

Theorem 3 *The operator* $\mathcal{T} : V_0^h(\Omega) \to V_0^h(\Omega)$ *is symmetric and the following estimates hold*

$$m\,a(u,u) \le a(\mathcal{T} u, u) \le M\,a(u,u) \quad \forall u \in V_0^h(\Omega), \tag{24}$$

where constants m and M are independent of H, h.

The proof of this theorem can be found in Mróz (1995).

In each step of PCG method (cf. Concus, Golub O'Leary (1976)) we have to calculate the function $w \in V_0^h(\Omega)$

$$w = \mathcal{T} u = w_0 + w_1, \quad w_i = \mathcal{T}_i u,$$

where $u \in V_0^h(\Omega)$ is a given function. The algorithm of finding $w_0 = \mathcal{T}_0 u$ involves solving the global problem defined on the coarse space $V_0^H(\Omega)$, see Step 3 of Algorithm 1. Let us now outline the algorithm of finding $w_1 = \mathcal{T}_1 u$.

Algorithm 2

1. *Construct the system for the nodal basis functions ϕ_j^h associated with the nodes of interiors of Ω_i^D*

$$a_i^D(\mathcal{P}_i w_1, \phi_j^h) = a(u, \phi_j^h) \quad \forall \phi_j^h \in V_0^h(\Omega_i^D).$$

2. *Build the system for the nodal basis functions ϕ_j^h associated with the nodes of $\overline{\Omega}_i^N$. Thus we compute w_1 on $\overline{\Omega}_i^N$ $i = 2, 4, \ldots$ by solving*

$$\tilde{a}_i^N(w_1, \phi_j^h) = a(u, \phi_j^h) - \sum_{odd\ i} a_i^D(\mathcal{P}_i w_1, \phi_j^h) \quad \forall \phi_j^h \in V^h(\Omega_i^N) \cap V_0^h(\Omega)$$

It reduces to solving the subproblems in each subdomain Ω_i^N with Neumann boundary conditions on interface lines Γ_i, and homogeneous Dirichlet boundary conditions on $\partial\Omega$.

3. *In Step 1 the projections $\mathcal{P}_i w_1$ have been computed, so now we find the function w_1 on Ω_i^D. This is done in the same way as in Step 4 of Algorithm 1 .*

References

Bramble J. H., Pasciak J. E., Schatz A. H. (1986) An iterative method for elliptic problems on regions partitioned into substructures. *Math. Comp.* 46, 361–369.

Ciarlet P. G. (1978) *The Finite Element Method for Elliptic Problems.* North–Holland.

Concus P., Golub G. H., O'Leary D. P.(1976) A generalized conjugate gradient method for the numerical solution of elliptic PDE. In *Sparse Matrix Computations* (Ed. J. R. Bunch and D. J. Rose). Academic Press, N.J., pp. 309–332.

Dryja M., Proskurowski W. (1985) Capacitance matrix method using strips with alternating Neumann and Dirichlet boundary conditions. *Applied Numer. Math.* 1.

Dryja, M., Widlund, O. B. (1990) Towards a Unified Theory of Domain Decomposition Algorithms for Elliptic Problems, in *Third International Symposium on Domain Decomposition Methods for Partial Differential Equations, held in Houston, Texas, March 20-22, 1989* (ed. T. Chan and others) SIAM, Philadelphia, PA.

Mróz M. (1995) Domain Decomposition Methods with Strip Substructures for Finite Element Problems, PHD Thesis, Institute of Applied Mathematics and Mechanics, Warsaw University.

Widlund O. B. (1988) Iterative substructuring methods: algorithms and theory for problems in the plane. *Domain Decomposition Methods for PDEs* (Ed. R. Glowinski and others) SIAM, Philadelphia.

PROXIMAL DOMAIN DECOMPOSITION ALGORITHMS AND APPLICATION TO ELLIPTIC PROBLEMS

by

Said Oualibouch and Noureddine El Mansouri

Institut d'Informatique et I.A., Université de Neuchâtel,
Rue Emile-Argand 11, CH-2007 Neuchâtel.
(Switzerland)
e-mail: said.oualibouch@info.unine.ch

Abstract: Many problems in mathematical economics, engineering and mechanics reduce to large, sparse, unstructured, and poorly conditioned optimization problems. Because of their size and lack of structure, these problems are hard to solve by classical and global methods which are not adapted to modern parallel computers. In this setting, decomposition methods are very attractive, because the solution of the global problem can be reduced to the iterative solution of many subproblems of smaller size. These techniques are designed for exploiting the full power of modern parallel computers, because of the built-in parallelism of the algorithms and the character of the associated data. In this paper, we are interested in linear and nonlinear elliptic problems, and we present two additive algorithms based on the proximal techniques. They yield two overlapping additive domain decomposition methods: *Proximal-Jacobi method* and *Proximal Schwarz methods*. These methods can also be viewed as regularized versions of Additive Schwarz methods. In order to validate these algorithms, some numerical results are given for the homogeneous Dirichlet problem.

Keywords: Domain Decomposition Methods, Additive Schwarz Methods, Proximal Techniques.

0.1   INTRODUCTION

The main characteristic of decomposition algorithms in convex programming is the splitting of a large-scale problem into a set of reduced size subproblems which may be solved either in parallel or in sequence. Very often, the structure of system induces a splitting of the variable set in disjoint subsystems yielding some coupling constraints. In our approach, the subsystems may overlap, but the advantage is that there are no artificial coupling constraints. The efficiency of the decomposition methods depends not only on the specific properties of the objective functions as convexity, smoothness or separability but mostly on the degree of the coupling between the subsystems.

Besides the motivation of reducing the size of the problems appears the possibility of decentralizing the optimal decision among the local subproblems as in an ideal hierarchical organization. It is well-known, see [1], that most classical approaches perform only a partial decentralization and need a heavy coordination upper level to build a solution from the local proposals. This is due to the lack of uniqueness of subproblems solutions which in turn is a

Domain Decomposition Methods in Sciences and Engineering, edited by R. Glowinski *et al.*

direct consequence of the non-smoothness of the coordination function. It is then natural to introduce regularizing terms in the decentralized process to cope with both non-uniqueness and non-smoothness as in the proximal point algorithm [11].

In this paper we are interested in solving the following convex problem:

$$\min_{v \in K} f(v), \quad K \subset H \tag{0.1}$$

where the function f is proper convex lower-semi-continuous, H a finite dimensional Hilbert space H and K a closed convex subset of H such that $\operatorname{ri} \operatorname{dom} f \cap \operatorname{ri} K \neq \emptyset$. We suppose that $H = \sum_{i=0}^{m} H_i$ and $K = \sum_{i=0}^{m} K_i$, $K_i \subset H_i$. The problem (0.1) is equivalent to the following one:

$$\min_{v_i \in H_i, \ i=0,...,m} F(\sum_{i=0}^{m} v_i) \tag{0.2}$$

where $F = f + \chi_K$ with χ_K is the $\{0, \infty\}$-indicator function of the subset K. In this case, we can use parallel methods (at least m processors) to solve the problem (0.2).

To solve a partial differential equation or a variational inequality often is equivalent to minimizing an energy function. In practice, there are different ways to decompose this energy function and to decompose the space of minimization, see [5], [13] and the references therein. In order to illustrate some of the possible ways, let us consider the following homogeneous Dirichlet problem:

$$\begin{cases} -\Delta u &= f \quad \text{in } \Omega \\ u &= 0 \quad \text{on } \partial\Omega \end{cases} \tag{0.3}$$

Where Ω is a bounded domain in $\mathbb{R}^n$ and $\partial\Omega$ its boundary. The problem (0.3) can be reduced to the following one

$$\min_{v \in H} F(v) \tag{0.4}$$

where the function F is defined by: $F(v) = \int_\Omega (\frac{1}{2}|\nabla v|^2 - fv)dx$ and $H = H_0^1(\Omega) \cap V_h$, V_h being the finite element space. Space decomposition (SD) can be done in different ways. For example, the finite element space itself is the linear span of the finite element basis, therefore it can be easily regarded as the sum of subspaces. The multilevel method is another way to decompose a finite element space. In the overlapping DDM, we decompose a domain Ω into overlapping subdomains Ω_i, $i = 1, ..., m$, this means that $\Omega = \cup_{i=1}^{m} \Omega_i$, and for each Ω_i, there exists a subdomain Ω_j such that $\Omega_i \cap \Omega_j \neq \emptyset$. If the subdomains overlap uniformly, it is known that, see [6],

$$H_0^1(\Omega) = H_0^1(\Omega_1) + H_0^1(\Omega_2) + ... + H_0^1(\Omega_m) \tag{0.5}$$

The central idea to construct parallel methods by SD derives from the observation that the space H can be decomposed into the sum of smaller and simpler subspaces as in (0.5), then the minimization problem (0.4) can be replaced by a problem similar to (0.2) where $H_i = H_0^1(\Omega_i) \cap V_h$. For a general SD, the minimizer of (0.2) may not be unique, but it has been proved that several methods converge for (0.2), for example the sequential Gauss-Seidel method and the parallel Jacobi method, see [12]. In the overlapping DDM case, the Schwarz alternating method is nothing else but the Gauss-Seidel method. Applying the Gauss-Seidel and Jacobi methods to some linear and nonlinear problems, several sequential and parallel overlapping DDM's can be got, see [12], [13].

The rest of this paper is organized as follow: the section 0.2 will give an overview of proximal techniques. In section 0.3, we will present two versions of additive proximal domain decomposition methods. Finally, we will conclude with some numerical results and remarks.

0.2 PROXIMAL TECHNIQUES

Let denote the inner product on H by $\langle .,. \rangle$ and the induced norm by $\|.\|$. Let $T : H \rightrightarrows H$ be a set-valued map (operator) on H, we define its graph by: $\mathrm{Gr}(T) = \{(x,y) \in H \times H | w \in Tx\}$, the inverse T^{-1} of T is the operator defined by: $\mathrm{Gr}(T^{-1}) = \{(x,y) \in H \times H | (y,x) \in \mathrm{Gr}(T)\}$. The operator T is said to be monotone if $\langle x - x', y - y' \rangle \geq 0$ for all (x,y) and (x',y') in $\mathrm{Gr}(T)$. T is said to be maximal monotone if it is monotone and its graph is not properly contained in the graph of any other monotone operator. Several authors have extensively studied the theory of maximal monotone operators in Hilbert spaces and applications, among others Brezis [3], Dolezal [4] and Aubin and Ekeland [2].

Many problems from mathematical programming, complementarity, mathematical economics and other fields can be formulated in the way of the fundamental problem of finding an element $z \in H$ such that

$$0 \in Tz \tag{0.6}$$

For example, if T is the subdifferential operator ∂F of F ($F \not\equiv +\infty$), then Minty [9] have shown that T is maximal monotone, and the problem of minimizing the function F is equivalent to that of finding a zero of T, i.e. $F(z) = \min F(x)$ means that $0 \in Tz$. The problem (0.6) is equivalent to that of finding a fixed point of the resolvent operator $J_\lambda^T = (I + \lambda T)^{-1}$ (to simplify, we denote this operator by J_λ), $\lambda > 0$, i.e. find $z \in H$ such that: $z = J_\lambda z$.

As early as 1962, Minty [10] pointed out that, when the operator T is maximal monotone, its resolvent (the Moreau-Yosida resolvent) J_λ is single-valued on $\mathbb{R}^n$ and non-expansive. This result suggests that a solution to the inclusion $0 \in T(z)$ may be iteratively approximated using the classical iteration $z^{(k+1)} = J_\lambda z^{(k)}$. One could modify this scheme by iteratively varying the scalar λ and by choosing the iterator $z^{(k+1)}$ to be an approximate solution to the equation $(I + \lambda T)z = z^{(k)}$, i.e., $z^{(k+1)} \approx J_\lambda z^{(k)}$. The proximal point algorithm precisely applies these ideas. The algorithm, starting from any point $z^{(0)}$, generates a sequence $\{z^{(k)}\}$ in H as follows:

$$x^{(k+1)} = J_{\lambda_k} x^{(k)} \qquad (0.7)$$

where $\{\lambda_k\}$ is some sequence of positive real numbers. A wide variety of global convergence results for the proximal point algorithm can be found in literature. As early as 1970 and 1972, Martinet [7] [8] proved the convergence of the exact proximal point algorithm for certain special cases of the operator T with fixed $\lambda_k \equiv \lambda$. The first theorem on the convergence of the general proximal point algorithm was proved by Rockafellar in 1976 [11]. His theorem not only insures the global convergence under a mild approximating rule, but also describes the global behavior if the inclusion $0 \in T(z)$ has no solution. The convergence rate of the proximal point algorithm depends on the properties of the operator T, the choice of the sequence $\{\lambda_k\}$, and the accuracy of the approximation $z^{(k+1)} \approx J_{\lambda_k} z^{(k)}$.

Consider the general form of the convex optimization problem (0.4). One method to solve (0.4) is to regularize the objective function by using the proximal regularization, already introduced in this document. Given a real positive parameter λ, we recall that a proximal approximation (regularization) of F is defined by:

$$F_\lambda(x) = \inf_u \{F(u) + \frac{1}{2\lambda}\|u - x\|^2\} \qquad (0.8)$$

This function is convex and differentiable and when minimized possesses the same set of minimizers and the same optimal value as problem (0.4). When we consider the right part of (0.8) and seek its optimality condition, we have

$$\begin{aligned}
0 \in \partial F(u^*) + (1/\lambda)(u^* - x) \quad &\Leftrightarrow \quad 0 \in \lambda\partial F(u^*) + (u^* - x) \\
&\Leftrightarrow \quad x \in (I + \lambda\partial F)(u^*) \\
&\Leftrightarrow \quad u^* = (I + \lambda\partial F)^{-1}(x)
\end{aligned}$$

This is the motivation of the terminology "proximal regularization". In such case $T = \partial F$ and the proximal point iteration (0.7) is equivalent to:

$$x^{(k+1)} = \text{Argmin}\{F(x) + (1/2\lambda_k)\|x - x^{(k)}\|^2\} \qquad (0.9)$$

If the function F is separable and consequently the variables $(x_i)_{i=0}^m$ are independent, the space H can be expressed as a product space, i.e. $H = \prod_{i=0}^m H_i$ and $T = \prod_{i=0}^m T_i$, where T_i is a maximal monotone operator on H_i, then we have $(I + \lambda T)^{-1} = \prod_{i=0}^m (I + \lambda T_i)^{-1}$. This property is fundamental when we are interested in a parallel implementation of a proximal decomposition algorithm. In this case, observe that we can substitute the proximal step (0.9) by $m + 1$ elementary proximal steps that can be executed simultaneously each on its own processor.

0.3 ADDITIVE PROXIMAL ALGORITHMS

In the case where the function F is not supposed to be separable but
the variables are independent, Martinet [7] proposed the following sequential
algorithm: at iteration k, we solve sequentially for $j = 0, ..., m$

$$x_j^{k+1} = \arg\min_{x_j \in H_j} \{F(\sum_{i=0}^{j-1} x_i^{k+1} + x_j + \sum_{i=j+1}^{m} x_i^k) + \frac{1}{2\lambda}\|x_j - x_j^k\|^2\} \qquad (0.10)$$

Martinet [7] has established the convergence of this regularized relaxation
method when F has continuous partial derivatives. But in our application
we don't assume the separability of the function nor the independence of the
variables, i.e, the subspaces H_i, $i = 0, ..., m$ are not orthogonal to each other.
In such case, Tai [12] proposed a parallel Jacobi-like iteration and established
its convergence when F has Lipschitz continuous and coercive derivative. In
this paper, we refer to this algorithm by Jacobi-Schwarz. This Jacobi-Schwarz
is given by:
Do in parallel for $j = 0, ..., m$:

> *Step 0:*   Choose $u_j^0 \in H_j$ and $\alpha_j > 0$, such that $\sum_j \alpha_j \leq 1$.
>
> *Step 1:* $u_j^{k+1/2} = \arg\min_{u_j \in H_j}\{F(\sum_{i=0}^{j-1} u_i^k + u_j + \sum_{i=j+1}^{m} u_i^k)\}$ (0.11)
>
> *Step 2:* $u_j^{k+1} = u_j^k + \alpha_j(u_j^{k+1/2} - u_j^k)$ (0.12)

In our approach, we have no assumptions about the smoothness of the objective,
we overcome this thanks to the proximal regularization of the functions to be
minimized in the right side of (0.11).

0.3.1 Parallel Proximal-Jacobi Algorithm (PJ)

Do in parallel for $j = 0, ..., m$:

> *Step 0:*   Choose $u_j^0 \in H_j$ and α_j, such that $\sum_j \alpha_j \leq 1$.
>
> *Step 1:* $u_j^{k+1/2} = \arg\min_{u_j \in H_j}\{F(\sum_{i=0}^{j-1} u_i^k + u_j + \sum_{i=j+1}^{m} u_i^k)$
> $$+ \frac{1}{2\lambda}\|u_j - u_j^k\|^2\} \qquad (0.13)$$
>
> *Step 2:* $u_j^{k+1} = u_j^k + \alpha_j(u_j^{k+1/2} - u_j^k)$ (0.14)

The minimizer of (0.2) may not be unique, therefore in the convergence analysis
of our algorithm, we will only prove that $u^{n+1} = \sum_{i=0}^{m} u_i^{n+1}$ converges to the
minimizer of (0.2)

Before giving some convergence results, we recall some definitions in convex
analysis.

Theorem 1 *Assume that F is coercive, convex lower semi-continuous on H
and $\sum_j \alpha_j \leq 1$, $0 < \alpha_i < 1$. Then the sequence $\{u^n = \sum_{i=0}^{m} u_i^n\}_n$ has the
convergence property.*

Proof: Let us begin by proving the convergence of the sequence $\{F(u^n)\}_n$,

$$
\begin{aligned}
F(u^{n+1}) &= F(\textstyle\sum_{i=0}^m u_i^{n+1}) \\
&= F(u^n + \textstyle\sum_{i=0}^m \alpha_i(u_i^{n+1/2} - u_i^n)) \\
&= F(\textstyle\sum_{i=0}^m \alpha_i(u^n + u_i^{n+1/2} - u_i^n) + (1 - \sum_{i=0}^m \alpha_i)u^n) \\
&\leq \textstyle\sum_{i=0}^m \alpha_i F(u^n + u_i^{n+1/2} - u_i^n) + (1 - \sum_{i=0}^m \alpha_i)F(u^n) \\
&\leq \textstyle\sum_{i=0}^m \alpha_i(F(u^n) - \frac{1}{2\lambda}\|u_i^n - u_i^{n+1/2}\|) + (1 - \sum_{i=0}^m \alpha_i)F(u^n)
\end{aligned}
$$

$$
F(u^{n+1}) \leq F(u^n) - \frac{1}{2\lambda}\sum_{i=0}^m \alpha_i\|u_i^n - u_i^{n+1/2}\|^2 \tag{0.15}
$$

Therefore, we proved that the sequence $\{F(u^n)\}_n$ is a decreasing sequence bounded below by $F(u)$, u is the minimizer of (0.2). So $\{F(u^n)\}_n$ is a convergent sequence. From the last inequality (0.15), we can easily deduce that the sequence $\{u^{n+1} - u^n\}$ vanishes as $n \to \infty$. F is coercive, so the sequences $\{u^n\}$ and $\{u^{n+1/2}\}$ have a limit point u^∞. If we consider the optimality condition of the minimization problem in the right side of (0.13), it can be seen that $y_i^{n+1/2} = \partial F_i^n(u_i^{n+1/2}) = (u_i^n - u_i^{n+1/2})/\lambda$, where $F_i^n(v_i) = F(u^n - u_i^n + v_i)$. And From (0.15), we can also deduce that for each $i = 0, ..., m$, $y_i^{n+1/2}$ vanishes as $n \to \infty$. Finally, the limit point of u^n is a minimizer of the problem (0.2).

Remark 1 *When we assume that the function F is differentiable and its gradient is uniformly continuous, the proof of convergence of the PJ algorithm is similar to that of Jacobi-Schwarz algorithm, proposed by X.-C. Tai in [12].*

0.3.2   Proximal Schwarz Algorithm (PS)

The basic iteration of the algorithm considered in this section is similar to that multiplicative one proposed by Martinet [7], as described in (0.10). The difference is that our algorithm is additive and the separability of the problem is not necessary for us, in other words, the subdomains may overlap. This parallel Proximal-Schwarz Algorithm is described bellow:

Do in parallel for $j = 0, ..., m$:

Step 0:   Choose $u_j^0 \in H_j$.

Step 1: $u_j^{k+1/2} = \arg\min_{u_j \in H_j}\{F(\sum_{i=0}^{j-1} u_i^k + u_j + \sum_{i=j+1}^m u_i^k)\}$ (0.16)

The convergence of this algorithm is proven without using any assumption about the smoothness of the objective function.

0.4   NUMERICAL RESULTS AND CONCLUSIONS

We have implemented these algorithms for a homogeneous Dirichlet problem (0.3) on a 2D domain decomposed into two overlapped subdomains, i.e., two subspaces. We take $\alpha_1 = \alpha_2 = 0.45$ and for each value of λ, we count the number of iterations until $\|u^n - \overline{u}\| < 10^{-12}$ ($\overline{u}$ is the exact solution). The obtained numerical results show that the parameter λ influences the rate of convergence of the Proximal-Jacobi (PJ) and Proximal-Schwarz (PS) algorithms. Figure 0.1 shows the behavior of these algorithms as well as our reference algorithm Jacobi-Schwarz (JS). We notice that, for our example, the algorithm PS is more efficient for a good choice of λ.

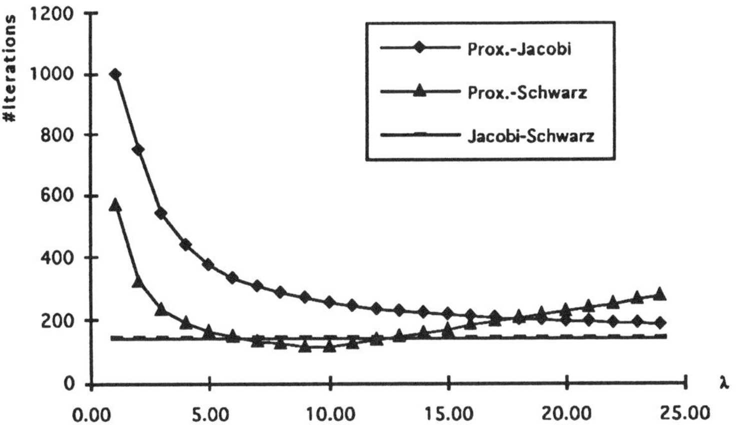

Figure 0.1   Behavior of the proposed algorithms in function of λ

To sum up, in this paper we have proposed some new methods that are useful for both linear and nonlinear elliptic problems. The numerical results show that for certain values of the parameter λ, the proximal regularization is a good way to accelerate the Jacobi-like domain decomposition methods. The corresponding theoretical results and a more detailed proof of theorem 1 and some applications to nonlinear elliptic problems will be submitted elsewhere.

REFERENCES

[1] Arrow K. and Hurwics L. (1960) Decentralization and computation in resource allocation. *Essays in Economics and Econometrics*, R. Phouts ed., Chapell Hill.

[2] Aubin J.P. and Ekeland I. (1984) Applied Nonlinear Analysis, *Wiley Interscience*, New York.

[3] Brezis H. (1973) Opérateurs Maximaux Monotones. *North Holland*, Amsterdam.

[4] Dolezal V. (1979) Monotone Operators and Applications in Control and Network Theory, *Elsevier*, Amsterdam.

[5] Le Tallec P. (1994) Domain decomposition methods in computational mechanics. *Computational Mechanics Advances, North-Holland*, J.T. ODEN ed., 1: 121–220.

[6] Lions P.L. (1988) On the schwarz alternating method. 1. *First Intenational Symposium on Domain Decomposition Methods for Partial Differential Equations*, R. Glowinski and G. H. Golub and G. A. Meurant and J. Periaux ed., SIAM.

[7] Martinet B. (1971) Minimisation d'une fonctionnelle dans un espace produit par une méthode de relaxation. *Revue francaise d'informatique et de recherche operationnelle*, R-3: 121–126.

[8] Martinet B. (1972) Determination approchée d'un point fixe d'une application pseudo-contractante. *Compte rendu de l'accademie des sciences, Paris*, Serie A 274: 163–165.

[9] Minty G.J. (1964) On the monotonicity of the gradient of a convex function. *Pacific J. Math.*, 14: 243–247.

[10] Minty G.J. (1962) Monotone nonlinear operators in Hilbert space. *Duke Mathematics Journal*, 29: 341–346.

[11] Rockafellar R.T. (1976) Monotone operators and proximal point algorithm. *SIAM J. Control*, 14:5: 877–898.

[12] Tai X.-C. (1994) Parallel function decomposition and space decomposition methods, Part 1: Theory. *Report 94, Bergen University*.

[13] Xu J.C. (1992) Iterative methods by space decomposition and subspace correction. SIAM Review, 34: 4: 581–613.

Stability of implicit extrapolation methods

Ulrich Rüde[1]

1 Introduction

Multilevel methods decompose the solution space in a nested sequence of subspaces. These subspaces are then used to construct a multigrid or multilevel preconditioned conjugate gradient method. Thus the multilevel structure is the basis for the efficient solution of a partial differential equation (PDE). In this setting, the solution is defined in the topmost space and the multilevel structure is just used to accelerate some iterative solution method.

Besides this *algebraic* perspective, the multilevel structure may be also used to improve the accuracy of the discretization itself. Under certain conditions, the nested mesh and space structure can be exploited by various *extrapolation* schemes.

Extrapolation results for finite difference methods can be found in [MS83] and for finite elements in [BLR86]. All *explicit* extrapolation techniques rely on the existence of *global error expansions* for the approximate solutions, typically of the form

$$u_h - u = \sum_{i=1}^{k} h^{\alpha_i} e_i + R_{k+1}, \tag{1}$$

where u_h and u are the numerical and the true solution of the differential equation, e_i are functions independent of h, and R_{k+1} is a remainder term. The parameter h denotes the mesh width and can be interpreted as identifying a single space V_h in the nested multilevel structure. Once an expansion of the form (1) has been proven to

[1] Institut für Mathematik, Universität Augsburg, D-86135 Augsburg, Germany, email: `ruede@math.uni-augsburg.de`

Domain Decomposition Methods in Sciences and Engineering, edited by R. Glowinski *et al.*

exist, and when the exponent α_1 of the leading term has been identified, then a linear combination of u_h and u_{2h} of the form

$$\bar{u}_h = \frac{2^{\alpha_1}}{2^{\alpha_1} - 1} u_h - \frac{1}{2^{\alpha_1} - 1} u_{2h}$$

will lead to an approximation where the dominating error term is eliminated. Using this scheme recursively, leads to the well-known *Richardson extrapolation table.*

Extrapolation is only computationally feasible, when the coefficients α_i are known, and when they form a quickly growing sequence. If the α_1 do not grow quickly, so that the expansion has many terms of almost equal order, extrapolation will be less attractive, since many linear combinations are needed before the approximation order is significantly increased.

Unfortunately, these situations are typical in many practical PDE problems when reentrant corners, interfaces with jumps in the coefficients or rough data may lead to *singularities*, each of which may create its own error contributions to the error expansion with a new set of exponents α_i.

In the case of reentrant corner singularities, the α_i are still known and a local extrapolation technique has been proposed in [BR88], based on meshes refined locally according to the interior angles.

In this paper we will discuss another variant of extrapolation, where the Richardson principle is applied indirectly. Such *implicit extrapolation* methods have been introduced in [JR94, Rüd91a]. They are closely related to the so-called τ-extrapolation in multigrid methods, see e.g. [Hac85]Chapter 14.1.3.

Implicit extrapolation is based on a local element-by-element analysis of numerical quadrature and differentiation rules. Thus they primarily depend on the smoothness of the shape functions rather than the global regularity of the solution. The regularity of the solution is of course required to justify the use of high order (polynomial) approximations in general. Implicit extrapolation has the advantage that it is formally not applied to the solution itself, but only to the finite element functions approximating the solution. Thus the specific analysis is independent of the solution properties. Whether high order finite elements are suitable to approximate a given solution can be decided independently. Furthermore, implicit extrapolation is suitable for nonuniform grids as they may occur in a local refinement context.

Using extrapolation in a multilevel context is especially attractive, when the extrapolation to obtain higher order discretizations is efficiently integrated with a fast multilevel solver. For integrating Richardson extrapolation with a full multigrid method and a comparison with multigrid τ-extrapolation, see e.g. Lin and Schüller [SL85]. The implicit extrapolation which is the main topic of this paper is most efficiently implemented by the multigrid τ-extrapolation algorithm. Due to the space limitations in this paper we refer for all algorithmic details to [JR94, Rüd91a]. We remark only that implicit extrapolation can be implemented by a trivial change of a conventional (FAS) multigrid method which consists of only a multiplication by an additional extrapolation factor in the fine-to-coarse transfer. Thus integrating implicit extrapolation into an existing multilevel algorithm is easy, and the basic solver remains untouched, since the higher order accuracy is obtained by a defect correction-like iteration.

2   The implicit extrapolation method

For exposition, we consider the simplest case of an elliptic PDE in one dimension

$$u'' = f, \quad u(0) = u(1) = 0. \tag{2}$$

We will use the equivalent formulation as a minimization problem

$$\min_{u \in H_0^1(0,1)} E(u), \quad \text{where } E(u) = \int_0^1 (u'(x))^2 - 2u(x)f(x) \, dx, \tag{3}$$

where $H_0^1(0,1)$ denotes the the usual Sobolev space of order 1 enforcing homogeneous Dirichlet boundary conditions. Next, we introduce the mesh $0 = x_0 < x_1 < \ldots < x_n = 1$, and discretize (3) directly, by representing the continuous function u by the vector $u_h = (u_0^h, \ldots, u_n^h)^T$. With the mesh widths $h_i = x_{i+1} - x_i$ and the midpoints $x_{i+1/2} = 1/2(x_{i+1} + x_i)$, we may replace

$$u'(x_{i+1/2}) \approx \frac{u(x_{i+1}) - u(x_i)}{h_i}$$

and the integration by the *midpoint sum*

$$\int_0^1 f(x) \, dx \approx \sum_{i=0}^{n-1} h_i f(x_{i+1/2})$$

for the numerical approximation to $(u'(x))^2$ and the *trapezoidal sum*

$$\int_0^1 f(x) \, dx \approx \sum_{i=0}^{n-1} \frac{h_i}{2} (f(x_{i+1}) + f(x_i))$$

for approximating the integral over $u(x)f(x)$. These approximations combined, lead to

$$E(u) \approx E_h(u_h) = \sum_{i=0}^{n-1} h_i \left[\frac{1}{2} \left(\frac{u_{i+1} - u_i}{h_i} \right)^2 + \frac{f_{i+1} u_{i+1} + f_i u_i}{2} \right].$$

The corresponding normal equations are

$$\begin{aligned} u_0 &= 0 \\ \frac{-u_{i+1} + u_i}{h_{i+1}} + \frac{u_i - u_{i-1}}{h_i} + \frac{h_{i+1} + h_i}{2} f_i &= 0 \quad \text{for } i = 1, \ldots, N-1 \\ u_N &= 0 \end{aligned}$$

Thus we have recovered the discretization by second order finite differences, or, equivalently, the discretization by piecewise linear finite elements with lumped mass matrix.

The basic reason for deriving the discrete system in this form is the existence of asymptotic error expansions for

$$E(u) - E_h(u) = e_2 h^2 + \ldots + e_{2k} h^{2k} + R_{2k+1} \tag{4}$$

when u is sufficiently smooth. This result holds even for a nonuniform basic discretization $x_0, \ldots, x_n$, if only the refined grids are constructed by recursively inserting the interval midpoints. This result is proved in [Lyn68]. Based on the expansion (4) we may now consider extrapolated functionals of the form

$$\bar{E}_h(u_h) = \frac{4}{3} E_h(u_h) - \frac{1}{3} E_{2h}(u_h), \quad \bar{\bar{E}}_h(u_h) = \frac{16}{15} \bar{E}_h(u_h) - \frac{1}{15} \bar{E}_{2h}(u_h), \quad \text{etc.} \quad (5)$$

Here we interpret u_h simply as a vector of values, and note that $E_{2h}(u_h)$ only uses every second value in u_h.

Obviously, each of the extrapolated functionals defines a new system of normal equations for u_h. In section 3 we will give conditions, when the higher order representation of the functional results in an improved accuracy for u_h. Before we continue this argument, we mention that results analogous to (4) hold in two space dimensions, as proved in [Rüd93] for the case of triangular meshes. Of course this is a crucial result, since the main interest here is in methods which generalize to higher space dimensions. In [JR94] it is furthermore shown that the implicit extrapolation method is equivalent to using higher order finite elements in a special case.

3   Stability of implicit extrapolation

In the above section we have introduced the *implicit extrapolation principle* for deriving higher order difference discretizations for elliptic PDE. In this section, we will interpret u_h as a finite element approximation with lowest oder, that is piecewise linear finite elements. By the extrapolation, as in (5), we construct higher order representations of the integrals. These extrapolated quadrature rules are defined for the continued refinement of a basic mesh $x_0, x_1, \ldots, x_n$ by recursively inserting the interval midpoints. If the basic mesh is associated with the space of piecewise linear finite element functions V_h, it would seem natural to identify each of the refined meshes with the corresponding (hierarchical) finite element spaces $V_h \subseteq V_{h/2} \subseteq V_{h/4} \subseteq \ldots$ of piecewise linear functions.

However, since the basic integration/differentiation rules are already correct in all these spaces, their approximation properties cannot be improved by extrapolation. If there is a positive effect of the implicit extrapolation, it cannot be directly understood within the h-refinement space structure.

Therefore, we consider a p-refinement where the basic space of piecewise linear approximations V_h is enlarged by piecewise polynomials of increasing order. Thus the first refinement from space V_h to $\bar{V}_{h/2}$ with a first step of extrapolation corresponds to adding quadratic functions in each interval (x_i, x_{i+1}). The next step to $\bar{V}_{h/4}$ introduces two additional degrees of freedom in each element and this corresponds to adding cubic and 4th order basis functions. In the next level $\bar{V}_{h/8}$ another four degrees of freedom are added and thus all polynomials up to degree 8.

At the same time as we add p-refinement, the extrapolation according to 5 increases the accuracy of the integration. Unfortunately, this increase of accuracy does not keep up with the rapid growth of the spaces. Only in the first step, when quadratic shape functions are added, the extrapolation provides sufficiently accurate integration rules. In the second step of extrapolation, only the functions up to degree 3 are integrated exactly However in this step cubic *and* 4th order functions are added to the solution space. Thus there is a mismatch between the functions in the finite element space and the accuracy of the extrapolated quadrature rules.

In conventional finite element analysis, the need for numerical quadrature to compute the stiffness matrix (and right hand side) is rarely made explicit. The finite element space is basic, and the use of another space (say of nodal values) for numerical quadrature is usually not explicit in the basic analysis. Here, in the context of implicit extrapolation, the converse is true. We directly work in the space used for quadrature and our interpretation of this as a p-version finite element space is just an artifact.

Generally speaking, we are faced with the following situation.

- The continued refinement has produced a large space V which we identify with a p-refinement of our basic discretization. The associated finite element problem is denoted by

$$\min_{u \in V} E(u), \quad \text{where } E(u) = a(u, u) - 2(f, u). \tag{6}$$

- In the space V, however, we cannot assume that we represent the functionals correctly. Our extrapolation procedure falls behind in accuracy, so what we really solve is not (6) but

$$\min_{\tilde{u} \in V} \tilde{E}(\tilde{u}), \quad \text{where } \tilde{E}(\tilde{u}) = \tilde{a}(\tilde{u}, \tilde{u}) - 2(f, \tilde{u}) \tag{7}$$

- Though we cannot say, how accurate $\tilde{E}$ is with respect to E in the full space V, we have constructed the method such, that V has a subspace W, where both functionals agree well with each other, say

$$|E(w) - \tilde{E}(w)| \leq \epsilon ||w||^2 \text{ for all } w \in W. \tag{8}$$

In W we consider the equation defined by the original functional

$$\min_{w \in W} E(w) \tag{9}$$

The best we can hope for in this situation is that our computed solution $\tilde{u}$ of problem (7) is close to the solution w of problem (9). This is shown in the following theorem.

Theorem 3.1 (Stability of implicit extrapolation) *Let V be a Hilbert space with norm $|| \cdot ||$ and bilinear forms $a(\cdot, \cdot)$ and $\tilde{a}(\cdot, \cdot)$, $f \in V^*$ with positive constants c_1, c_2 such that*

$$c_1 ||v||^2 \leq a(u, u) \leq c_2 ||v||^2.$$

Let u, $\tilde{u}$, and w be defined by (6), (7), and (9), respectively. Assume that (8) holds and that

$$\tilde{a}(v, v) \geq a(v, v) \text{ for all } v \in V. \tag{10}$$

Then there exists $c > 0$ such that

$$||w - \tilde{u}|| \leq c(\sqrt{\epsilon}||w|| + ||w - u||).$$

Proof.

$$
\begin{aligned}
\tilde{a}(w - \tilde{u}, w - \tilde{u}) &= \tilde{a}(w, w - \tilde{u}) - \tilde{a}(\tilde{u}, w - \tilde{u}) \\
&= \tilde{a}(w, w - \tilde{u}) - a(u, w - \tilde{u}) \\
&= \tilde{a}(w, w - \tilde{u}) - a(w, w - \tilde{u}) + a(w - u, w - \tilde{u}) \\
&= (\tilde{a} - a)(w, w - \tilde{u}) + a(w - u, w - \tilde{u}) \\
&\leq \sqrt{(\tilde{a} - a)(w, w)\,(\tilde{a} - a)(w - \tilde{u}, w - \tilde{u})} + \sqrt{a(w - u, w - u)\,a(w - \tilde{u}, w - \tilde{u})} \\
&\leq \sqrt{(\tilde{a} - a)(w, w)\,\tilde{a}(w - \tilde{u}, w - \tilde{u})} + \sqrt{a(w - u, w - u)\,\tilde{a}(w - \tilde{u}, w - \tilde{u})}.
\end{aligned}
$$

Therefore

$$
\sqrt{\tilde{a}(w - \tilde{u}, w - \tilde{u})} \leq \sqrt{(\tilde{a} - a)(w, w)} + \sqrt{a(w - u, w - u)}.
$$

and thus there exists $c > 0$ with

$$
||w - \tilde{u}|| \leq c(\sqrt{\epsilon}||w|| + ||w - u||).
$$

$\square$

This stability theorem requires condition (10), which states that we must not underestimate the true energy $E(u)$ in our numerical analogue $\tilde{E}(u)$. In many practical situations, this condition is violated as soon as we attempt a second step of extrapolation Thus the potential increase in accuracy is not reflected in the solution.

A computational remedy has been suggested in [Rüd91b]. Obviously it is sufficient so solve (7) subject to the constraint $\tilde{u} \in W$. If such a constraint is used to make V coincide with W, the instability cannot occur.

However, condition (10) indicates that we need not strictly impose $\tilde{u} \in W$, but it suffices to correct a possible underestimate of the energy of solution components which do not lie in W. Since the orthogonal complement of W in V consists of all polynomials of degree larger than some k_0 and smaller than k_1, it is simple to construct such a correction. Thus (7) becomes modified to

$$
\min_{\tilde{u} \in V} \tilde{E}(\tilde{u}) + \rho D(\tilde{u}), \tag{11}
$$

where ρ is a sufficiently large real parameter and $D(\tilde{u})$ is a quadratic form which vanishes on all (piecewise) polynomials of degree $\leq k_0$ and is positive for all (piecewise) polynomials of degree $> k_0$ and $\leq k_1$. Thus the *accuracy condition* (8) is maintained, and clearly, for some ρ large enough, the *stability condition* (10) will be satisfied, too. Consequently, an ansatz of the form (11) where $\tilde{E}$ is constructed by implicit extrapolation may obtain any approximation order, provided that the *penalty term D* is chosen appropriately.

$D(\tilde{u})$ can e.g. be constructed by computing the finite differences estimating the kth derivatives for $k_0 < k \leq k_1$ in each element. Thus $D(\tilde{u})$ is a local operator, effecting only nodal values within a single element. Computational experiments showing the effectiveness of this technique are given in [Rüd91b].

4   A numerical experiment for a singular interface problem

The implicit extrapolation technique can be generalized to two and three space dimensions. We will now present a two-dimensional numerical experiment in a nontrivial situation. We consider the diffusion equation

$$\nabla \cdot \alpha(x,y)\nabla u(x,y) \;=\; 0 \text{ in } (0,1)^2 \tag{12}$$

$$u(x,y) \;=\; g(x,y) \text{ on } \partial(0,1)^2, \tag{13}$$

where the coefficient has jumps across the lines $x = 1/2$ and $y = 1/2$. In particular, we assume

$$\alpha(x,y) = \left\{ \begin{array}{ll} 1 & \text{in } (0,1/2) \times (1/2,1) \cup (1/2,1) \times (0,1/2) \\ 3 & \text{in } (0,1/2)^2 \cup (1/2,1)^2 \end{array} \right. .$$

We choose the boundary data such that the solution has the form

$$u(x,y) = u(r,\phi) = \Phi(\phi)r^{(2/3)},$$

where (r,ϕ) are polar coordinates with respect to the point $(x,y) = (1/2,1/2)$, and where $\Phi(\phi)$ is of the form

$$\Phi(\phi) = \sin(2/3\phi + \beta)$$

in each of the four quadrants of the domain, with β chosen such that the interface conditions for the normal derivatives are satisfied. The solution is depicted in Figure 1. Since the singularity is of the same type as in the case of an L-shaped domain,

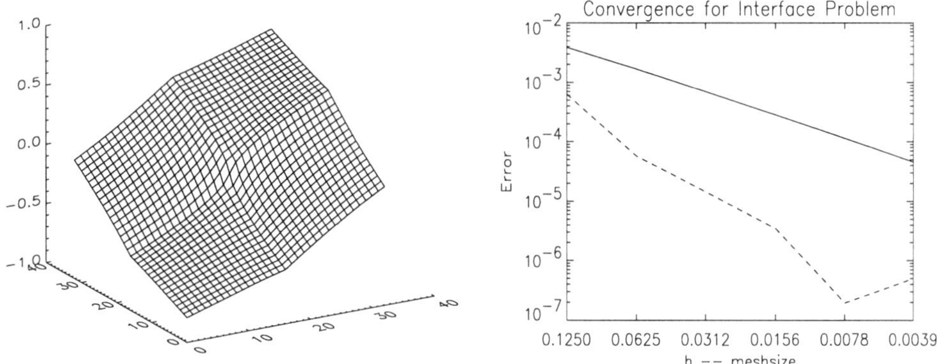

Figure 1   Solution of problem (12,13) and convergence rates.

we expect poor convergence. It has been shown, that in a point-wise sense and for a uniform discretization with piecewise linear triangular elements, we can only expect a convergence rate of $O(h^{4/3})$ This is reflected in the upper (unbroken) line of Figure 1.

A conventional technique to improve on this is to use local refinement to better resolve the singularity. An alternative is based on extrapolation, as in [BR88] or [Rüd88]. This is the approach we will take, focussing on implicit extrapolation.

Thus we solve a system of the form (4), however, we must adapt the extrapolation parameter to the nonstandard leading term in the expansion. We thus minimize $(2^{4/3} - 1)^{-1}(2^{(4/3)}E_h(u_h) - E_{2h}(u_h))$. The convergence rate for the corresponding solution is shown in the lower (broken) line of Figure 1.

The error is visualized in Figure 2. The effect of implicit extrapolation is clearly

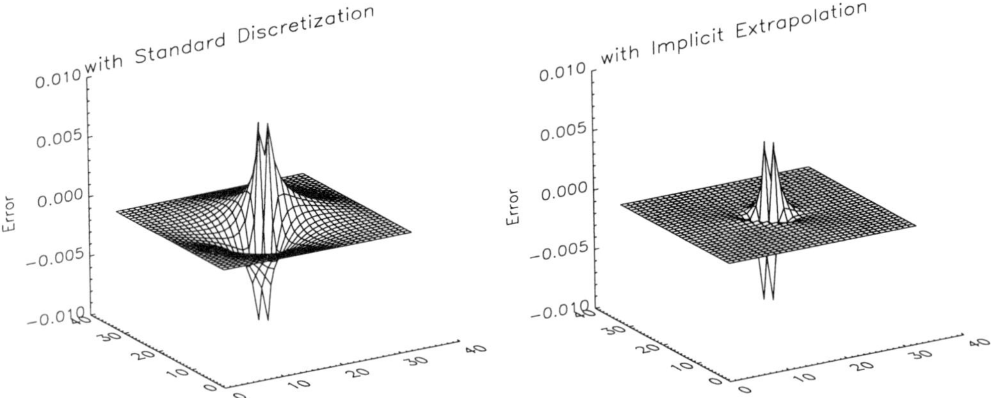

Figure 2   Error for standard discretization and implicit extrapolation.

visible. Since we have not refined locally, the poor resolution of the singularity is not significantly improved. However, with implicit extrapolation, the global spreading of the error, the so-called *pollution effect* is fully suppressed. This is also clear form the convergence graph in Figure 1, where it is shown that implicit extrapolation has recovered $O(h^2)$ convergence away from the singularity.

5   Conclusions

In this paper we have briefly presented the implicit extrapolation method. We have proved a stability condition and have suggested computational techniques to satisfy the stability condition in a general setting. A numerical example has shown, how the method can be used even in the case of singular interface problems. Many more aspects of the method, in particular its combination with splitting extrapolation (see [TmSbL90]) and the so-called sparse grid techniques remain to be analyzed in the future. Some preliminary results are given in [Rüd91b].

References

[BLR86] Blum H., Lin Q., and Rannacher R. (1986) Asymptotic error expansions and Richardson extrapolation for linear finite elements. *Numer. Math.* 49: 11–37.

[BR88] Blum H. and Rannacher R. (1988) Extrapolation techniques for reducing the pollution effect of reentrant corners in the finite element method. *Numer. Math.* 52: 539–564.

[Hac85] Hackbusch W. (1985) *Multigrid Methods and Applications.* Springer Verlag, Berlin.

[JR94] Jung M. and Rüde U. (1994) Implicit extrapolation methods for multilevel finite element computations. In Manteuffel T. (ed) *Preliminary Proceedings of the Colorado Conference on Iterative Methods, Breckenridge, Colorado, April 4-10, 1994.* To appear in S.J.Sci.Comput., 17(1), 1996.

[Lyn68] Lyness J. N. (1968) The calculation of the Stieltjes' integral. *Num. Math.* 12: 252–265.

[MS83] Marchuk G. and Shaidurov V. (1983) *Difference Methods and their Extrapolations.* Springer, New York.

[Rüd88] Rüde U. (1988) On the accurate computation of singular solutions of Laplace's and Poisson's equation. In McCormick S. F. (ed) *Multigrid Methods: Theory, Applications, Supercomputing: Proceedings of the Third Copper Mountain Conference on Multigrid Methods, April 5-10, 1987.* Marcel Dekker, New York, NY, USA.

[Rüd91a] Rüde U. (1991) Adaptive higher order multigrid methods. In Hackbusch W. and Trottenberg U. (eds) *Proceedings of the Third European Conference on Multigrid Methods, October 1-4, 1990,* pages 339–351. Birkhäuser, Boston, MA, USA. International Series of Numerical Mathematics, Vol. 98.

[Rüd91b] Rüde U. (September 1991) Extrapolation and related techniques for solving elliptic equations. Bericht I-9135, Institut für Informatik, TU München.

[Rüd93] Rüde U. (1993) Extrapolation techniques for constructing higher order finite element methods. Bericht I-9304, Institut für Informatik, TU München.

[SL85] Schüller A. and Lin Q. (December 1985) Efficient high order algorithms for elliptic boundary value problems combining full multigrid techniques and extrapolation methods. Arbeitspapiere der GMD 192, Gesellschaft für Mathematik und Datenverarbeitung.

[TmSbL90] Tao L., min Shih T., and bo Liem C. (August 1990) An analysis of splitting extrapolation for multidimensional problems. *Systems Science and Mathematical Sciences* 3(3): 261–272.

Substructure Preconditioners for Nonconforming Plate Elements

Zhongci Shi and Zhenghui Xie

1 Introduction

In this paper, we generalize the BPS algorithm [1] to nonconforming element approximations of the biharmonic equation. We construct a preconditioner for the Morley element by substructuring on the basis of a space decomposition. The space decomposition is introduced by partitioning discrete biharmonic functions into low and high frequency components through intergrid transfer operators between coarse and fine meshes and a conforming interpolation operator. The method leads to a preconditioned system with the condition number bounded by $C(1 + \log^2 H/h)$ in the case with interior cross points, and by C in the case without interior cross points, where H is the subdomain size and h is the mesh size. These techniques are applicable to other nonconforming plate elements and are well suited to parallel computation.

For conforming element discrete problems of a second order elliptic equation, Bramble et al [1] and Widlund [7] have obtained certain preconditioners which are easily inverted in parallel and can reduce the condition number of a discrete system from $O(h^{-2})$ to $O(1 + \log^2 H/h)$. The main idea is a decomposition given by $v = \Pi_H v + (v - \Pi_H v)$, where Π_H is the interpolation operator on coarse meshes and the nodal parameters of $v - \Pi_H v$ vanishes on the coarse mesh nodes, and an extension theorem. Gu and Hu [3] have obtained a similar result for Wilson nonconforming element. Zhang [9] has constructed preconditioners for certain conforming plate elements on the basis of a space decomposition by adding certain vertex spaces. However, for Morley element, since the finite element spaces are not nested, and the functions have bad discontinuities, a space decomposition similar to those mentioned above does not hold.

We introduce a conforming interpolation operator for the Morley element and related intergrid transfer operators, and then construct a space decomposition to overcome these difficulties. Brenner [2] has introduced a conforming interpolation operator E_h by taking averages of the nodal parameters associated with the function and its first derivatives among the relevant elements, and taking zero as the nodal

[1] National Laboratory of Scientific & Engineering Computing
Chinese Academy of Sciences, P.O.Box 2719, Beijing 100080, China.
e-mail: shi@lsec.cc.ac.cn   szc@lsec.cc.ac.cn

Domain Decomposition Methods in Sciences and Engineering, edited by R. Glowinski *et al.*

parameters associated with its second-order derivatives, in order to deal with an overlapping domain decomposition method. To be suited to a parallel computation in the substructure preconditioning, we modify Brenner's approach so that the nodal parameters of $E_h v_h$ depend only on those of v_h on the boundaries of substructures. Zhang [9] on the other hand, has defined an interpolation operator for certain conforming plate elements by setting the nodal parameters for second-order derivatives to zero. We use it to define the intergrid transfer operators I_h from coarse to fine meshes and I_H from fine to coarse meshes. Then we generalize the BPS algorithms and Widlund theory of substructure preconditioning to nonconforming plate elements.

2   A Preconditioning Algorithm

Let Ω be a bounded polygonal domain in R^2. Let J_h and J_H be quasi-uniform triangulations of Ω with h and H as mesh parameters respectively. Assume that J_h can be obtained by refining J_H, so that J_H and J_h form a two-level triangulations on Ω and the nodes of J_H are those of J_h. Let $S^h(\Omega)$ be the Morley element space [5] and let $S_0^h(\Omega)$ be a subspace of $S^h(\Omega)$ with nodal parameters vanishing at the boundary nodes. The Morley element discrete problem is : Find $u_h \in S_0^h(\Omega)$ such that

$$a_h(u_h, v_h) = (f, v_h), \quad \forall v \in S_0^h(\Omega),$$

where

$$a_h(u, v) = \sum_{T \in J_h} \sum_{|\alpha|=2} \int_T D^\alpha u D^\alpha v \, dx, \quad (f, v) = \int_\Omega f v \, dx.$$

Let $J_H = \{\Omega_k\}_{k=1}^N$. The vertices of J_H will be labeled by v_j (ordered in some way) and Γ_{ij} will denote the edge with endpoints v_i and v_j. $S_0^h(\Omega_j)$ will denote the subspace of $S_0^h(\Omega)$ consisting of functions with nodal parameters vanishing on $\bar{\Omega} \backslash \Omega_j$. In addition, $S^h(\Omega_j)$ will be the set of functions which are restrictions, of those in $S_0^h(\Omega)$, to $\bar{\Omega}_j$. In what follows, c and C (with or without subscript) will denote generic positive constants which are independent of H, h and Ω_k.

We construct our preconditioner B through its corresponding bilinear form $B(\cdot, \cdot)$ defined on $S_0^h(\Omega) \times S_0^h(\Omega)$.

We decompose functions in $S_0^h(\Omega)$ as follows:

Write $w = w_P + w_H$, where $w_P \in S_0^h(\Omega_1) \oplus \cdots \oplus S_0^h(\Omega_N)$ satisfies

$$a_h^k(w_P, \phi) = a_h^k(w, \phi), \quad \forall \phi \in S_0^h(\Omega_k), \text{ for each } k,$$

where

$$a_h^k(u, v) = \sum_{T \in J_h, T \subset \Omega_k} \sum_{|\alpha|=2} \int_T D^\alpha u D^\alpha v \, dx.$$

Notice that w_P is determined on Ω_k by the nodal parameters of w on Ω_k and that

$$a_h^k(w_H, \phi) = 0 \text{ for all } \phi \in S_0^h(\Omega_k).$$

Thus on each Ω_k, w is decomposed into a function w_P whose nodal parameters vanish on $\partial \Omega_k$ and a function $w_H \in S^h(\Omega_k)$ which satisfies the above homogeneous equations and has the same nodal parameters as w at $\bigcup_k \partial\Omega_k$. We shall refer to such a function w_H as "discrete a_h^k–biharmonic".

We note that the above decomposition is orthogonal with respect to the inner-product $a_h(\cdot, \cdot)$ and hence, $a_h(w, w) = a_h(w_P, w_P) + a_h(w_H, w_H)$.

To define the bilinear form $B(\cdot, \cdot)$, we introduce a linear interpolation operator E_h, and intergrid transfer operators I_h and I_H. The conforming relative of Morley element is the Argyris quintic element. Let $AR^h(\Omega)$ and $AR^H(\Omega)$ be the Argyris quintic element space associated with J_h and J_H, respectively; see [2].

For an arbitrary vertex p of J_h, we assign to it one of its adjacent edge midpoints e_p. If $p \in \bigcup \Gamma_{ij}$, we assign to it e_p which belongs to $\bigcup \Gamma_{ij}$. If $p \in \partial\Omega$, we assign to it e_p which belongs to $\partial\Omega$. For $v \in S_0^h(\Omega)$, we define $E_h v \in AR^h(\Omega)$ such that

$$
\begin{aligned}
E_h v(p) &= v(p), \quad \forall \text{ verties } p \\
\partial_n E_h v(m) &= \partial_n v(m), \quad \forall \text{ midpoint } m \\
D^\alpha E_h v(p) &= 0, \quad |\alpha| = 2;
\end{aligned}
\tag{2.1}
$$

and

$$
\begin{aligned}
\partial_x E_h v(p) &= \partial_n v(e_p) \cos\beta + \frac{v(p) - v(a)}{l_{ap}} \sin\beta, \\
\partial_y E_h v(p) &= \partial_n v(e_p) \sin\beta + \frac{v(a) - v(p)}{l_{ap}} \cos\beta,
\end{aligned}
\tag{2.2}
$$

where $n = (\cos\beta, \sin\beta), s = (-\sin\beta, \cos\beta)$ are the unit normal and tangential vector respectively, and l_{ap} is the length of the segment ap (cf. Figure 2.1). We note that (2.1) is defined as in Brenner [2] but that (2.2) is different.

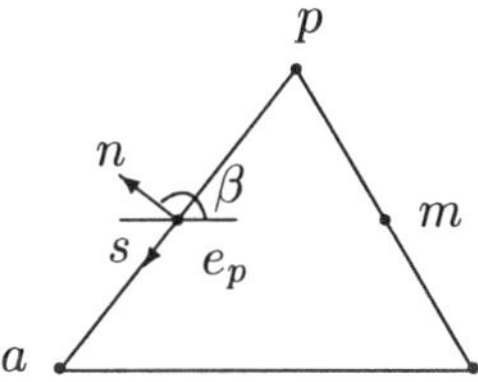

Figure 2.1

From the definition of the conforming interpolation operator E_h, we can see that nodal parameters of $E_h v_h$ on $\cup\Gamma_{ij}$ depend only on those of v_h on $\cup\Gamma_{ij}$. This property is important in our discussion.

The intergrid transfer operator $I_H : AR^h(\Omega) \longrightarrow AR^H(\Omega)$ is defined by (cf.[9])

$$
\begin{cases}
D^\alpha I_H v(p) = D^\alpha v(p), \text{ for } |\alpha| \leq 1 \\
D^\alpha I_H v(p) = 0, \text{ for } |\alpha| = 2 \\
\partial_n I_H v(m) = \partial_n v(m), \text{ for all internal midpoints } m \in J_H,
\end{cases}
$$

for $v \in AR^h(\Omega)$. The intergrid transfer operator $I_h : AR^H(\Omega) \longrightarrow AR^h(\Omega)$ is defined similarly.

Now we construct a preconditioner. From the decomposition of Argyris quintic element space

$$
AR^h(\Omega) = I_h I_H AR^h(\Omega) \oplus AR^h(\Omega)',
$$

we decompose $w_H \in S^h(\Omega_k)$ into

$$w_H = w_E + w_V,$$

where $w_V \in S^h(\Omega_k)$ is a discrete a_h^k–biharmonic function such that the nodal parameters of $E_h w_V$ on $\partial \Omega_k$ are those of $I_h I_H E_h w_H$ along each Γ_{ij}. Thus w_E is a discrete a_h^k–biharmonic function in Ω_k for each k such that the nodal parameters of $E_h w_E$ vanish at all nodes of coarse meshes. Let

$$\bar{S}_0^h(\Omega) = \{v_h \in S_0^h(\Omega); \text{ nodal parameters of } E_h v_h|_{\Gamma_{ij}} = \text{ those of } I_h I_H E_h v_h|_{\Gamma_{ij}}\}$$

for all Γ_{ij}. Then, we have a space decomposition

$$S_0^h(\Omega) = \bar{S}_0^h(\Omega) \oplus (S_0^h(\Omega))'.$$

Using this decomposition, we now define the bilinear form $B(\cdot, \cdot)$ as follows

$$
\begin{aligned}
B(w, \phi) = {} & a_h(w_P, \phi_P) \\
& + \sum_{\Gamma_{ij}} \left\{ \langle \partial_s \bar{w}_E, \partial_s \bar{\phi}_E \rangle_{H_{00}^{1/2}(\Gamma_{ij})} + \langle \partial_n \bar{w}_E, \partial_n \bar{\phi}_E \rangle_{H_{00}^{1/2}(\Gamma_{ij})} \right\} \\
& + \sum_{\Gamma_{ij}} \{ (w_V(v_i) - w_V(v_j) - D\bar{w}_V(v_i)(v_i - v_j)) \\
& \quad \cdot (\phi_V(v_i) - \phi_V(v_j) - D\bar{\phi}_V(v_i)(v_i - v_j))H^{-2} \\
& \quad + (D\bar{w}_V(v_i) - D\bar{w}_V(v_j))(D\bar{\phi}_V(v_i) - D\bar{\phi}_V(v_j)) \} \\
& + \sum_{T \in J_H} \sum_m (\partial_n(w_V - (w_V)_I)(m))(\partial_n(\phi_V - (\phi_V)_I)(m)),
\end{aligned}
$$

(2.3)

where and from now on $\bar{v} = E_h v$, and $\langle \cdot, \cdot \rangle_{H_{00}^{1/2}(\Gamma_{ij})}$ means $H_{00}^{1/2}(\Gamma_{ij})$-inner product which is defined by

$$
\begin{aligned}
\langle v, w \rangle_{H_{00}^{1/2}(\Gamma_{ij})} = {} & \int_{\Gamma_{ij}} \int_{\Gamma_{ij}} \frac{(v(x) - v(y))(w(x) - w(y))}{|x - y|^2} ds(x) ds(y) \\
& + \int_{\Gamma_{ij}} v(x) w(x) \left(\frac{1}{|x - v_i|} + \frac{1}{|x - v_j|} \right) ds(x), \quad v, w \in H_{00}^{1/2}(\Gamma_{ij}).
\end{aligned}
$$

We shall demonstrate how the linear system $Bw = g$ can be solved efficiently.

Given g, the problem of solving $Bw = g$ reduces to finding the functions w_P and w_H. The function w_P restricted to Ω_k satisfies

$$a_h^k(w_P, \phi) = (g, \phi) \text{ for all } \phi \in S_0^h(\Omega_k). \tag{2.4}$$

Thus it can be obtained by solving in parallel the corresponding biharmonic Dirichlet problem (2.4) on each subdomain. With w_P known, we are left with the equation

$$
\sum_{\Gamma_{ij}} \left\{ \langle \partial_s \bar{w}_E, \partial_s \bar{\phi}_E \rangle_{H_{00}^{1/2}(\Gamma_{ij})} + \langle \partial_n \bar{w}_E, \partial_n \bar{\phi}_E \rangle_{H_{00}^{1/2}(\Gamma_{ij})} \right\}
$$

$$
+ \sum_{\Gamma_{ij}} \{ (w_V(v_i) - w_V(v_j) - D\bar{w}_V(v_i)(v_i - v_j))
$$

$$
\cdot (\phi_V(v_i) - \phi_V(v_j) - D\bar{\phi}_V(v_i)(v_i - v_j)) H^{-2} \tag{2.5}
$$

$$
+ (D\bar{w}_V(v_i) - D\bar{w}_V(v_j))(D\bar{\phi}_V(v_i) - D\bar{\phi}_V(v_j)) \}
$$

$$
+ \sum_{T \in J_H} \sum_m (\partial_n(w_V - (w_V)_I)(m))(\partial_n(\phi_V - (\phi_V)_I)(m))
$$

$$
= (g, \phi) - a_h(w_P, \phi).
$$

(The last equality holds since $a_h(w_P, \phi_H) = 0$). Notice that the value of $(g, \phi) - a_h(w_P, \phi)$ for each ϕ depends only on the nodal parameters of $\bar{\phi}$ on all Γ_{ij}. From the definition of the interpolation operator E_h, we see that the value of $(g, \phi) - a_h(w_P, \phi)$ for each ϕ depends only on the nodal parameters of ϕ on all Γ_{ij}. Thus (2.5) gives rise to a set of equations which can be treated as follows: for each Γ_{ij}, choose ϕ in a subspace of $S_0^h(\Omega)$ such that the nodal parameters of $\bar{\phi}$ vanish in the all interior mesh points of every Ω_k and on all other Γ_{ij}. Thus, on this subspace, (2.5) decouples into independent problems of finding $\bar{w}_E \in AR_0^h(\Gamma_{ij})$, $I_H \bar{w}_E = 0$ given by

$$
\langle \partial_s \bar{w}_E, \partial_s \bar{\phi} \rangle_{H_{00}^{1/2}(\Gamma_{ij})} + \langle \partial_n \bar{w}_E, \partial_n \bar{\phi} \rangle_{H_{00}^{1/2}(\Gamma_{ij})}
$$

$$
= (g, \phi) - a_h(w_P, \phi), \ \forall \phi \in S_0^h(\Omega), \ I_H \bar{\phi} = 0, \ \bar{\phi} \in AR_0^h(\Gamma_{ij}) \tag{2.6}
$$

for each Γ_{ij}. Note that these are local problems with unknowns corresponding to the nodes on Γ_{ij} and may be solved in parallel.

Next we solve for $\bar{w}_V$ on the edges. We consider the subspace $\{\phi$; nodal parameters of $\bar{\phi}|_{\Gamma_{ij}} = $ those of $I_h I_H \bar{g}|_{\Gamma_{ij}}, g \in S_0^h(\Omega)\}$. Then, (2.5) reduces to

$$
\sum_{\Gamma_{ij}} \{ (w_V(v_i) - w_V(v_j) - D\bar{w}_V(v_i)(v_i - v_j))
$$

$$
\cdot (\phi_V(v_i) - \phi_V(v_j) - D\bar{\phi}_V(v_i)(v_i - v_j)) H^{-2}
$$

$$
+ (D\bar{w}_V(v_i) - D\bar{w}_V(v_j))(D\bar{\phi}_V(v_i) - D\bar{\phi}_V(v_j)) \} \tag{2.7}
$$

$$
+ \sum_{T \in J_H} \sum_m (\partial_n(w_V - (w_V)_I)(m))(\partial_n(\phi_V - (\phi_V)_I)(m))
$$

$$
= (g, \phi) - a_h(w_P, \phi).
$$

The nodal parameters of $\bar{w}_V$ at nodes of $T \in J_H$ determine those of w_V on all edges Γ_{ij}, and hence $w_H = w_E + w_V$ is known on all edges Γ_{ij}.

The last step consists of determining w_H in each Ω_k so that

$$
a_h^k(w_H, \phi) = 0 \text{ for } \phi \in S_0^h(\Omega_k). \tag{2.8}
$$

This problem is similar to (2.4), which can also be solved in parallel on each subdomain. Hence the solution of $Bw = g$ is determined by $w = w_P + w_H$.

We summarize the process by outlining the steps for obtaining the solution of

$$B(w, \phi) = (g, \phi) \text{ for all } \phi \in S_0^h(\Omega),$$

and hence for computing the action of B^{-1}.

Algorithm.

1. Find w_P by solving biharmonic Dirichlet problems on the subdomains. The solution of each individual Dirichlet problem on subdomains may be done in parallel.

2. Find $\bar{w}_E$ on Γ_{ij} by solving a one-dimensional equation on each Γ_{ij}; this may be done in parallel.

3. Find $\bar{w}_V$ on $\bigcup \Gamma_{ij}$ by solving a coarse mesh equation and then extending it to all edges Γ_{ij} by operator I_h.

4. Find w_H by extending the nodal values of $w_E + w_V$ on $\cup \Gamma_{ij}$ to all subdomains. As in step 1, the solution may be done in parallel.

3   Estimates of the Condition Number

We have the following theorem.

Theorem. *There are positive constants λ_0, λ_1 and C such that*

$$\lambda_0 B(w, w) \leq a_h(w, w) \leq \lambda_1 B(w, w), \quad \forall w \in S_0^h(\Omega),$$

where $\lambda_1/\lambda_0 \leq C(1 + \log^2 H/h)$. If all of the nodes of Ω_k lie on $\partial\Omega$, then $\lambda_1/\lambda_0 \leq C$.

The proof can be found in [6,8]. It means that the condition number grows at most like $(1 + \log^2 H/h)$ as h tends to zero. Therefore the preconditioned iteration converges rapidly.

Remark. We can easily get similar results for many other nonconforming plate elements [4].

REFERENCES

[1] Bramble J. H., Pasciak J. E. and Scharz A. H. (1986) The construction of preconditioners for elliptic problems by substructuring I. *Math.Comp.*, 47(175): 103–134.

[2] Brenner S. C. (1994) A two-level additive Schwarz preconditioner for nonconforming plate elements. *Domain Decomposition Methods in Scientific and Engineering Computing, Proceeding of DDM7*, (David E. Keyes, Jinchao Xu ed), 9–14.

[3] Gu J. and Hu X. (1994) On an essential estimate in the analysis of domain decomposition methods. *Journal of Comput. Math.*, 12(2): 132–137.

[4] Lascaux P. and Lesaint P. (1975) Some nonconforming finite elements for the plate bending problem. *RAIRO Anal. Numer.*, 9: 9–53.

[5] Shi Z. C. (1990) On the error estimate for Morley element. *Mathematica Numerica Sinica*, 12: 113–118.

[6] Shi Z. C. and Xie Z. H. (1995) Substructure preconditioners for nonconforming plate elements, preprint.

[7] Widlund O. B. (1988) Iterative substructuring methods: Algorithms and theory for elliptic problem in the plane. In *the First International Šymposium on Domain Decomposition Methods for Partial Differential Equations,* G. R. Glowinski et al. (eds.), SIAM Philadelphia, 113–128.

[8] Xie Z. H.(1996) *Domain Decomposition and Multigrid Methods for Nonconforming Plate Elements,* Ph.D Dissertation of Institute of Computational Mathematics & Scientific/Enginering Computing, Academia Sinica, April.

[9] Zhang X. (1991) Studies in Domain Decomposition : Multilevel methods and the biharmonic Dirichlet problem. Technical Report 584, Courant Institute, Now York University, September.

Domain Decomposition with Patched Subgrids

K.H. TAN[1] and M.J.A. BORSBOOM [2]

1 ABSTRACT

This paper describes a domain decomposition method for advection-diffusion problems on a two-dimensional domain that has been subdivided into a number of non-overlapping sub-domains, each covered by its own structured grid. The discretization of the problem's PDE over the entire domain is based on local finite difference discretizations on each subgrid, together with several so-called coupling equations that are needed to couple the discrete solution at subdomain interfaces. The choice of coupling equations depends on the discrete problem at hand and should result in a fast converging overall iterative procedure, at the same time maintaining sufficient accuracy of the solution across the interfaces. We show how both requirements can be fulfilled within the approach of optimized parameterized interface conditions when the grid at the interfaces is non-smooth, thereby extending our previous work on regularly patched Cartesian grids.

2 INTRODUCTION

One of the main issues in domain decomposition is the specification of suitable interface conditions at the artificially introduced internal boundaries of subdomains.

[1] Delft Hydraulics, Dept S&O, P.O. Box 152, 8300 AD Emmeloord, the Netherlands,
`Kian.Tan@wldelft.nl`
[2] Delft Hydraulics, Dept S&O, P.O. Box 152, 8300 AD Emmeloord, the Netherlands,
`Mart.Borsboom@wldelft.nl`

Domain Decomposition Methods in Sciences and Engineering, edited by R. Glowinski *et al.*

Obviously, the order of accuracy of the discretization inside the subdomains should be retained over the interfaces. Also physical requirements like mass conservation should be fulfilled. At the same time however, these internal boundary conditions should be designed for fast convergence of the overall iterative solution procedure that pieces together the global solution from the solutions of the different subproblems.

In this paper we propose a method for constructing such proper interface conditions in grid patching for the coupling of the solution of non-overlapping subdomains, each having its own grid. They can be regarded as discretizations of so-called coupling equations, involving suitably chosen combinations of several derivatives which are expressed in either a global coordinate system or a local computational coordinate system. The performance of the resulting domain decomposition algorithm is illustrated for the two-dimensional model problem

$$\frac{\partial c}{\partial t} + v_1 \frac{\partial c}{\partial x} + v_2 \frac{\partial c}{\partial y} = D(\frac{\partial^2 c}{\partial x^2} + \frac{\partial^2 c}{\partial y^2}) \text{ on } \Omega \ , D > 0 \ , \quad c = g \text{ on } \Gamma_i \ . \tag{1}$$

Here, Γ_i denotes those parts of the boundary where Dirichlet inflow conditions are imposed. On the remaining part of the boundary $\delta\Omega \backslash \Gamma_i$ we specify outflow conditions, based on one-sided discretizations of the advection part of (1).

3   THE MODEL PROBLEM ON NON-OVERLAPPING SUBDOMAINS

Consider problem (1) on a number of naturally ordered parallelograms Ω_k,

$$\Omega = \bigcup_{k=1}^{K} \Omega_k \qquad \Omega_k \cap \Omega_{k+1} = \Gamma_{kk+1} = \Gamma_{k+1k} \qquad k = 1, \ldots, K - 1 \ .$$

An example of a strip decomposition which we consider is given in Figure 1.

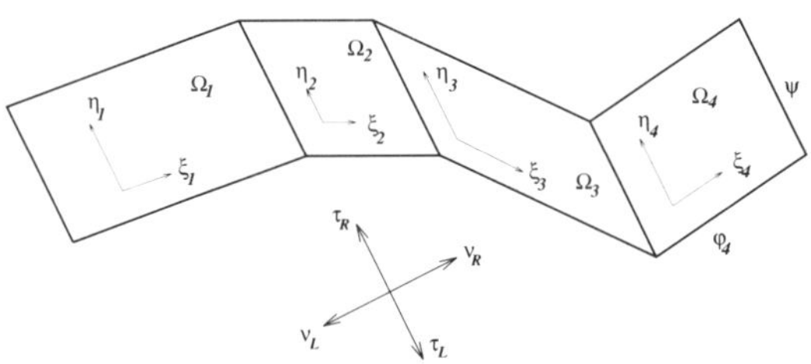

Figure 1   Example of a strip decomposition.

Prior to the discretization, (1) is transformed on each subdomain to a local coordinate system. We consider on each subdomain k the linear transformation $\xi_k = \xi_k(x, y)$, $\eta_k = \eta_k(x, y)$, that transforms Ω_k to a rectangle in computational space:

$$J_k = \left(\begin{array}{cc} \frac{\partial x}{\partial \xi_k} & \frac{\partial y}{\partial \xi_k} \\ \frac{\partial x}{\partial \eta_k} & \frac{\partial y}{\partial \eta_k} \end{array} \right) = \left(\begin{array}{cc} h_{\xi_k} & 0 \\ 0 & h_{\eta_k} \end{array} \right) \left(\begin{array}{cc} \cos \phi_k & \sin \phi_k \\ -\sin \psi & \cos \psi \end{array} \right) \ , \tag{2}$$

with $-\frac{\pi}{2} < \phi_k - \psi < \frac{\pi}{2}$. The resulting transformed equations are then discretized in the local, i.e. per subdomain, computational coordinate system, for which uniform Cartesian grids are employed. Without loss of generality we may assume that each local Cartesian grid has mesh size 1, since the actual mesh size in physical space can be included in the transformation via scalars h_{ξ_k} and h_{η_k}.

To keep our discussion of coupling clear, we consider only grid lines in ξ_k-direction that connect at the interfaces, i.e. $h_{\eta_k} = h_\eta$, $\forall k$. To facilitate the specification of the discrete equations holding at the subdomain boundaries, the local grids extend beyond both the physical and internal boundaries, these boundaries lying *in between* the outer two layers of grid points. See Figure 2, where the bold lines indicate the boundaries of subdomains. The □- and •-points indicate the locations where respectively boundary conditions and coupling conditions are discretized (cf. Section 4).

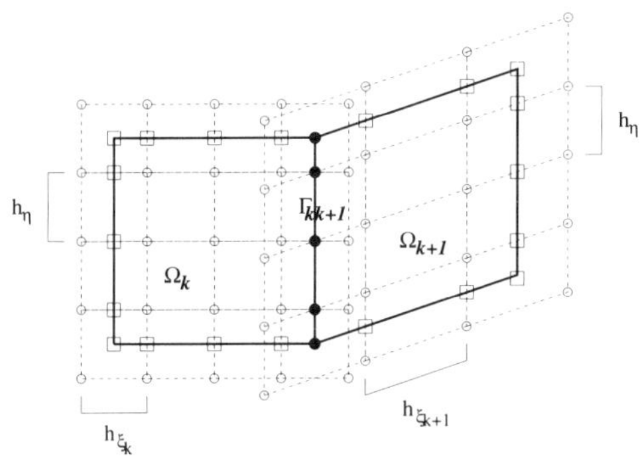

Figure 2   Two subdomains and two local grids.

For later purposes it is convenient to define transformations with respect to the normal and tangential direction at interfaces as well. Let ν_{kk+1} and ν_{k+1k} denote the unit outward normal (at the interface) to Ω_k and Ω_{k+1}. Tangential vectors τ_{kk+1} and τ_{k+1k} are defined as the vectors that are obtained by rotating ν_{kk+1} and ν_{k+1k} counterclockwise. For our decomposition, we have that $\forall k$, $\nu_{kk+1} = \nu_R$, $\tau_{kk+1} = \tau_R$, $\nu_{k+1k} = \nu_L$, and $\tau_{k+1k} = \tau_L$. The local coordinate transformations $\xi_k(\nu_R, \tau_R), \eta_k(\nu_R, \tau_R)$ and $\xi_k(\nu_L, \tau_L), \eta_k(\nu_L, \tau_L)$ are still defined by transformation matrices of the form (2), but with the rotation angles ϕ_k and ψ replaced by $\hat{\phi}_k = \phi_k - \psi$, $\hat{\psi} = 0$ and $\bar{\phi}_k = -\pi + \phi_k - \psi$, $\bar{\psi} = \pi$, respectively.

The transformed equations that are discretized per subdomain k are:

$$\frac{\partial c_k}{\partial t} + v_1^k \frac{\partial c_k}{\partial \xi_k} + v_2^k \frac{\partial c_k}{\partial \eta_k} = D_{\xi\xi}^k \frac{\partial^2 c_k}{\partial \xi_k{}^2} + D_{\eta\eta}^k \frac{\partial^2 c_k}{\partial \eta_k{}^2} + D_{\xi\eta}^k \frac{\partial^2 c_k}{\partial \xi_k \eta_k} \quad \text{on } \Omega_k \, , \qquad (3)$$

with transformed velocities and diffusion coefficients v_1^k, v_2^k, $D_{\xi\xi}^k$, $D_{\eta\eta}^k$, and $D_{\xi\eta}^k$. Its discretization is given in Section 4.

3.1   *Coupling equations*

The key point of this paper is the construction of equations that provide an accurate and fast coupling of the individual solutions c_k at the interfaces Γ_{kk+1}. From a continuous point of view, it is sufficient to impose (see e.g. [QL88])

$$\frac{\partial c_k}{\partial \nu_{kk+1}} = \frac{\partial c_{k+1}}{\partial \nu_{kk+1}} \quad \text{on } \Gamma_{kk+1} \qquad k = 1,\ldots,K-1 \,, \qquad [\frown] \qquad (4a)$$

$$c_{k+1} = c_k \quad \text{on } \Gamma_{k+1k} \qquad k = 1,\ldots,K-1 \,. \qquad [\frown] \qquad (4b)$$

The multi-domain formulation (3)–(4b) is then equivalent with the original problem
(1), since $c|_{\Omega_k} = c_k$. Note that for each interface $\Gamma_{kk+1} = \Gamma_{k+1k}$ two conditions are
formulated. For comparison with the discrete conditions that follow, we have listed
them as conditions for either subdomain k ($[\frown]$) or subdomain $k+1$ ($[\frown]$). For obvious
reasons these conditions are called coupling conditions, or interface conditions. As
indicated in e.g. [Tan92] it is allowed to use more general conditions

$$\Phi_{kk+1}(c_k) = \Phi_{kk+1}(c_{k+1}) \quad \text{on } \Gamma_{kk+1} \qquad k = 1,\ldots,K-1 \,, \qquad [\frown] \qquad (5a)$$

$$\Phi_{k+1k}(c_{k+1}) = \Phi_{k+1k}(c_k) \quad \text{on } \Gamma_{k+1k} \qquad k = 1,\ldots,K-1 \,, \qquad [\frown] \qquad (5b)$$

as long as they imply conditions (4a) and (4b). This seems to be an academic matter
only, would it not be that these conditions have a significant influence on the rate
of convergence of the solution procedure for the global discrete problem. This has
motivated the authors to develop coupling equations involving more general operators:

$$\begin{aligned}
\Phi_{ij} \equiv \quad &\zeta_{ij}\left(I + h_\eta \beta_{ij}\frac{\partial}{\partial \tau_{ij}} + (h_\eta)^2 \delta_{ij}\frac{\partial^2}{\partial \tau_{ij}{}^2} \right) + \\
&h_{\xi_j}\alpha_{ij}\frac{\partial}{\partial \nu_{ij}}\left(I + h_\eta \gamma_{ij}\frac{\partial}{\partial \tau_{ij}} + (h_\eta)^2 \epsilon_{ij}\frac{\partial^2}{\partial \tau_{ij}{}^2} \right) \,,
\end{aligned} \qquad (6)$$

where i, j are either $k, k+1$ or $k+1, k$, depending on the direction in which the coupling
operator is used. Scalars h_{ξ_j} and h_η are the mesh size coefficients in transformations
(2).

In equation (6), ζ_{ij}, α_{ij}, β_{ij}, γ_{ij}, δ_{ij}, and ϵ_{ij} are coupling *functions* from $\Gamma_{ij} \to \mathbb{R}$,
not coefficients. It is clear that these functions have to satisfy certain conditions to
guarantee equivalence of the multi-domain problem with the single-domain problem.
The imposed conditions should still imply (4a)–(4b), which is already obtained as soon
as Φ_{kk+1} and Φ_{k+1k} are distinctive enough.

In the remaining of this paper we assume the coupling functions to be optimized
for convergence speed following the strategy of [Tan95]. This optimization of coupling
functions turns out to provide equivalence automatically.

4   The model problem on virtually overlapping local grids

Having specified the entire continuous multi-domain problem we now consider its
discretization on local grids of size $(n_{x_k} + 2) \times (n_y + 2)$. We refer to Figure 2 for
a schematic diagram of the location of grid points in the two-subdomain case. The
discrete multi-domain problem consists of:

- a standard 9-point discretization based on central differences of the
  transformed equations (3) at the $n_{\xi_k} \times n_\eta$ 'inner' grid points of each
  subdomain;

- a discretization of the boundary conditions at the □-points with 3×2- or 2×3-point (at the corners 2×2-point) computational molecules;
- discrete equivalents of coupling equations, applied at the •-points;
- backward Euler time discretization.

4.1   discrete coupling equations

Obviously, the spatial discretization of (1) in the interior grid points is second-order accurate. Our aim is to formulate discrete analogues of coupling equations that do not affect this accuracy. Hereto we employ a similar two-step discretization as used for (1). We first transform the continuous coupling equations (5a)–(5b) to the local coordinate systems and then discretize the result with 2×3 stencils.

Note that because of the choice of the stencil, we are unable to discretize all terms of the transformed coupling equations. So we (have to) neglect all terms containing second derivatives in ξ_k and ξ_{k+1}, which includes certain mixed derivatives.

As a consequence, discretized coupling equations contain both transformation and discretization errors. In order to quantify them, let $\bar{c}_k$ and $\bar{c}_{k+1}$ be the exact solutions of the discrete coupling equations, satisfying:

$$(\Phi_{kk+1} + \tilde{T}_{\Phi_{kk+1}} + \tilde{D}_{\Phi_{kk+1}})(\bar{c}_k) = (\Phi_{kk+1} + \check{T}_{\Phi_{kk+1}} + \check{D}_{\Phi_{kk+1}})(\bar{c}_{k+1}) , \quad [\curvearrowleft] \qquad (7a)$$

$$(\Phi_{k+1k} + \check{T}_{\Phi_{k+1k}} + \check{D}_{\Phi_{kk+1}})(\bar{c}_{k+1}) = (\Phi_{k+1k} + \tilde{T}_{\Phi_{k+1k}} + \tilde{D}_{\Phi_{k+1k}})(\bar{c}_k) . \quad [\curvearrowright] \qquad (7b)$$

The T-terms and D-terms represent transformation and discretization errors.

We will first address the size of the transformation errors.

Proposition 4.1 *Let* Φ_{kk+1} *be of the form (6)* ($\nu_{kk+1} = \nu_R$, $\tau_{kk+1} = \tau_R$)*. Consider the coordinate systems* $(\xi_k(\nu_R, \tau_R), \eta_k(\nu_R, \tau_R))$ *and* $(\xi_{k+1}(\nu_R, \tau_R), \eta_{k+1}(\nu_R, \tau_R))$ *which specify the angles* $\hat{\phi}_k$ *and* $\hat{\phi}_{k+1}$ *between de interface and the* ξ_k *and* ξ_{k+1} *axis. Then*

$$\tilde{T}_{\Phi_{kk+1}}(c_k) = -h_{\xi_{k+1}}(h_\eta)^2 \alpha_{kk+1}\epsilon_{kk+1} \tan \hat{\phi}_k \cdot \frac{\partial^3 c_k}{\partial \tau_R^3} , \qquad (8)$$

$$\check{T}_{\Phi_{kk+1}}(c_{k+1}) = -h_{\xi_{k+1}}(h_\eta)^2 \alpha_{kk+1}\epsilon_{kk+1} \tan \hat{\phi}_{k+1} \cdot \frac{\partial^3 c_{k+1}}{\partial \tau_R^3} . \qquad (9)$$

A similar proposition holds for Φ_{k+1k}.

This proposition shows that transformation errors increase with increasing skewness of the ξ-coordinate direction with respect to ν_R, as expected. However, for reasonable grid connections, i.e. when $\hat{\phi}_k - \hat{\phi}_{k+1} \approx 0$, transformation errors almost entirely cancel out, since they are made in *both* left- and right-hand side of the coupling equations.

Even when the transformation errors do not cancel, they can still be acceptable. From (8) and (9) it follows that these errors are of third or second order in the grid spacing, depending on whether the leading term of the coupling operators is zeroth or first order. The former applies when the coupling condition has an important Dirichlet component, while the latter applies when the coupling consists of Neumann plus higher derivatives. Of course, this second or third order transformation error is only acceptable when coefficients like $\epsilon_{kk+1}(\tan \hat{\phi}_{k+1} - \tan \hat{\phi}_k) < \mathcal{O}(1)$ (cf. (8)–(9)).

As for the discretization errors we remark that the D-terms in (7a)–(7b) are $\mathcal{O}((h_{\xi_k})^2) + \mathcal{O}((h_\eta)^2)$. The same important remark as for the transformation errors can be made here. Discretization errors in coupling equations tend to cancel out when grid connections are not too irregular, i.e. when $h_{\xi_k} - h_{\xi_{k+1}} \approx 0$. Note that it is the difference between the angle of subgrids at both sides of the interface that is important in the transformation errors, while it is change in the mesh size in ξ-direction that is important in the discretization errors.

Summarizing, the discretized interface conditions provide a second-order accurate coupling between the solutions per subdomain, which error tends to cancel out for small differences in subgrid angles and mesh sizes. This brings us to the conclusion that for the type of decomposition and discretization considered in this paper, the coupling at interfaces is an entirely transparent process. The discrete solution behaves as if no interfaces would have been present. This is also illustrated by taking $\phi_k = \phi$, $h_{\xi_k} = h_\xi$, $\forall k$, in which case overlapping grid points actually coincide. Then our approach automatically reduces to our coupling technique for regularly patched grids ([TB93]), which can be regarded as nothing else but a preconditioning technique for the discrete system of equations, without altering its solution.

We remark that the cancellation of discretization and transformation errors in interface conditions is entirely due to the special extension of the subgrids over the interface with the interface itself in between. Note also that our discrete coupling, by the simple fact of discretizing (5a)–(5b) automatically includes some form of interpolation.

5   Numerical experiments

The entire set of equations to be solved each time step can be represented by a block tridiagonal system $Bc = f$. Its principal submatrices B_{kk} represent the discretization of (3) in the interior of each of the subdomains, together with the discretized boundary conditions and the left-hand sides of the subsequent discrete coupling equations. Off-diagonal blocks only contain the right-hand sides of the discrete coupling equations. This system $Bc = f$ is solved by GCR [EES83], applied to the right-preconditioned system $BM^{-1}Mc = f$. The preconditioner M is taken as $M = \text{blockdiag}(B_{11}, \ldots, B_{KK})$. ILU(2)-preconditioned BiCGSTAB(4) ([SF93]) is used for solving up to sufficient precision the subsystems $B_{kk}u_k = r_k$ that occur during the GCR process.

The first example deals with a simple advection problem on $\Omega = [-1,0] \times [-1,1] + [0, \cos\phi_2] \times [x\sin\phi_2 - 1, x\sin\phi_2 + 1]$ with velocity field $(v_1, v_2) = (10, 10)$ and negligible diffusion $D = 10^{-7}$ on two grids, one of which is rotated slightly with respect to the other, $\phi_1 = \psi = 0$, $\phi_2 = \arctan(0.5)$. To study the effect of interface coupling on the spatial discretization error, we take a small time step $\Delta t = 10^{-4}$. A cone $c(x, y) = 10^{-8\left((x+\frac{1}{2})^2 + (y+\frac{1}{2})^2\right)}$ is taken as initial condition.

The error in the numerical solution consists of both space and time discretization errors made in the interior of the subdomains, and coupling errors made at the interface. The error introduced at the interface can be visualized by following the peak of the initial condition while it propagates from the first to the second subdomain and plotting the error in the numerical solution at the exact spot of the peak. Bilinear interpolation is used when this spot does not coincide with a grid point, resulting in an interpolated value $c^*(t)$ (which explains the 'hops' in the figures). Although this adds interpolation errors to these figures, tendencies for the development of the discretization error can still clearly be noticed in the figures. Figure 3 shows the error

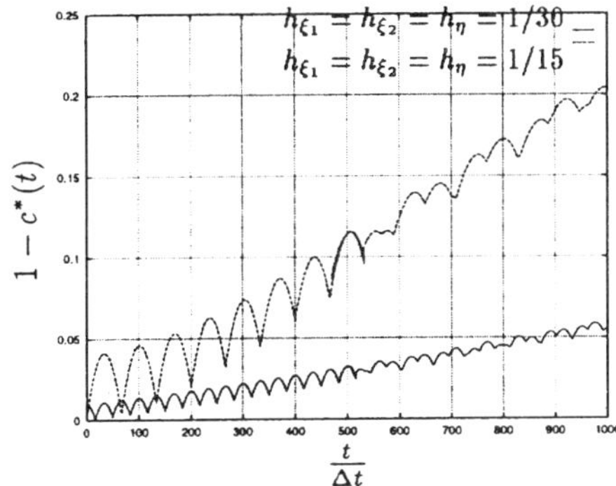

Figure 3   Error development at peak spot, $\Delta t = 10^{-4}$.

development at the peak for this Example 1, for two different mesh sizes. We clearly observe that at the interface virtually no error is introduced, since no jump takes place at $\frac{t}{\Delta t} = 500$. We also recognize the second-order accuracy of the scheme in space. Note that the error in both subdomains grow equally fast, although the space discretization error is expected to be smaller in subdomain 2 where the grid has been rotated in the direction of the flow.

In addition we verify the transparency of the coupling by comparing the accuracy of the discrete solution when increasing the number of interfaces for a problem of fixed size. The advection problem of Example 1 is solved on a decomposition of 20 thin strips, each strip covered by a grid of 5×62 points, which includes a row of virtual points at all sides. We set $\phi_{2k+1} = 0$ for the odd-numbered subdomains and $\phi_{2k} = \arctan(0.5)$ for the even-numbered subdomains. This gives rise to a 'stairway' pattern of subdomains, the solution on which (Figure 4) should be comparable to the previously computed numerical solution for $h = \frac{1}{30}$ since the number of grid points at which the PDE is discretized in the interior of subdomains is the same. The accuracy of the solution on 2 and 20 strips turned out to be indeed comparable; the difference in the error development at the peak spot was less than 5 %.

6   Concluding remarks.

We have presented a patched-grid domain decomposition method in which subgrids may vary per subdomain. The patching of the solution at subdomain interfaces is provided for by optimized interface conditions. Such optimized interface conditions have been designed to meet accuracy requirements as well as convergence rate requirements. The method is discussed here for non-overlapping strip decompositions

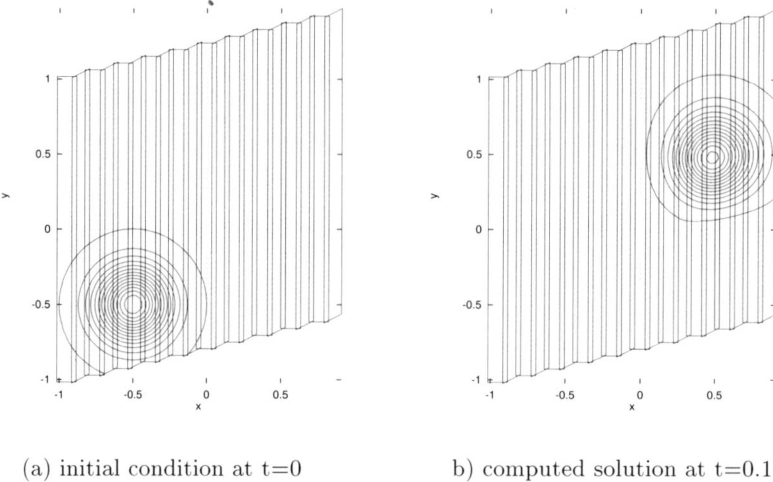

(a) initial condition at t=0   b) computed solution at t=0.1

Figure 4   Multiple interfaces.

and simple structured grids only, but can be extended to any decomposition into
non-overlapping subdomains and any grid transformation, due to the formulation of
the coupling conditions in terms of local normal and tangential derivatives at the
interfaces, which are then transformed to the local computational coordinate system.

REFERENCES

[EES83] Eisenstat S. C., Elman H. C., and Schultz M. H. (April 1983) Variational
iterative methods for nonsymmetric systems of linear equations. *SIAM J. Numer.
Anal.* 20(No. 2): 345–357.

[QL88] Quarteroni A. and Landriani G. S. (June 1988) Iteration by subdomains
in numerical fluid dynamics. 3rd German-Italian Symposium on Applications of
Mathematics in Technology.

[SF93] Sleijpen G. L. G. and Fokkema D. R. (1993) BiCGstab(ℓ) for linear equations
involving unsymmetric matrices with complex spectrum. *ETNA* 1: 11–32.

[Tan92] Tang W. P. (March 1992) Generalized Schwarz splittings. *SIAM J. Sci.
Stat. Comput.* 13(2): 573–595.

[Tan95] Tan K. H. (May 1995) *Local Coupling in Domain Decomposition.* PhD
thesis, University Utrecht, faculty Mathematics and Informatics.

[TB93] Tan K. H. and Borsboom M. J. A. (1993) On generalized Schwarz coupling
applied to advection-dominated problems. In Keyes D. E. and Xu J. C. (eds)
*Domain Decompositions Methods for Partial Differential Equations, Proc. 7th.
Int. Symp. on*, pages 125–130. AMS.

Domain Decomposition Methods for Unbounded Domains

Dehao Yu

1 Introduction

In many fields of scientific and engineering computing it is necessary to solve boundary value problems of partial differential equations over unbounded domains. There the standard techniques such as the finite element method will meet some difficulties, even if they are very effective for bounded domains. The standard results for convergence and error estimates of finite element methods are not valid on unbounded domains. No doubt it would be feasible to restrict the domain within some bounded subdomain simply, but the limit of finite element solutions is not necessarily the solution of original problem. In order to get enough accuracy a very large bounded subdomain should be taken, and a very high cost of computation must be paid. Then in the recent twenty years many new methods for solving problems over unbounded domains, such as the infinite element method [1, 11, 13], the boundary element method [2, 5, 7, 17], the coupling method of finite elements and boundary elements [8, 12, 14], the finite element method with approximate conditions on an artificial boundary [3, 15], etc., have been developed. However, each of them has its own advantages and disadvantages. We believe that the coupling of the finite element method and the natural boundary element method has some advantages over other coupling methods [14, 16]. All these methods lead to complicated linear algebraic equations.

The domain decomposition methods are important computational techniques developed rapidly in recent years [6, 10]. A domain is divided into subdomains, a large, difficult and complicated problem is reduced to some small, easy and simple subproblems. But up to now most of the theoretical analyses and algorithms are only valid for bounded domains. In fact, it is more important to develop domain decomposition method for unbounded domains. But when an unbounded domain is divided into subdomains, there is still at least one unbounded subdomain. In this case, the finite element method by itself is not enough; the boundary reduction method,

[1] The project supported by the National Natural Science Fundation of China and partially by the Research Grant Council of Hong Kong.
[2] State Key Laboratory of Scientific and Engineering Computing, Institute of Computational Mathematics and Scientific/Engineering Computing, Chinese Academy of Science, Beijing 100080, P. R. China

Domain Decomposition Methods in Sciences and Engineering, edited by R. Glowinski *et al.*
© 1997 John Wiley & Sons, Ltd.

which is an effective measure for handling some problems on unbounded domains, will be needed [9, 20].

There are many different ways to do the boundary reduction. The natural boundary reduction suggested and developed by K. Feng and D. Yu is one of them, and it has many distinctive advantages [4, 17]. Based on the natural boundary reduction, an overlapping and a non-overlapping domain decomposition methods for unbounded domains are discussed in this paper. Using one or two circles as artificial boundaries, we can divided an unbounded domain with a closed boundary into non-overlapping or overlapping subdomains. For the interior, small, bounded subdomain, the standard finite element method can be used without any difficulty. In the exterior circular subdomain, the results on the natural boundary reduction [17] can be applied directly. Then a Schwarz alternating method and a Dirichlet-Neumann method for unbounded domains are developed. Especially, for Poisson equation, some careful estimates of contraction factors and convergence rates are given. The theoretical and numerical results show that their convergences are very fast [19, 20].

In this paper, a harmonic boundary value problem is discussed in detail. Since many results on the natural boundary reduction of the biharmonic, plane elasticity and Stokes equations are also given in [17], developing domain decomposition methods for those problems over unbounded domains is also possible.

Steklov-Poincare operators play an important role in domain decomposition methods. The relationship between Steklov-Poincare operators and natural integral operators, and the inverse formulas of Steklov-Poincare operators are discussed in [21].

2 Schwarz Alternating Method Based on Natural Boundary Reduction

Let Ω be an unbounded domain with a closed curve Γ_0 as its boundary. The circles Γ_1 and Γ_2 are in Ω and their radii $R_1 > R_2 > 0$. Let Ω_1 be the bounded domain between Γ_0 and Γ_1, and Ω_2 the unbounded domain outside Γ_2. Then the original problem is decomposed into two subproblems over Ω_1 and Ω_2 with $\Omega_1 \cap \Omega_2 \neq \emptyset$. Taking some initial boundary value on Γ_1, e.g. zero, combining it with the given boundary condition on Γ_0, we solve the problem over Ω_1, get the value of solution on Γ_2, and then solve the problem over Ω_2, get the value of solution on Γ_1, and then solve the problem over Ω_1 again, and so on. This is a Schwarz alternating algorithm: the standard finite element method is applied to Ω_1, and the Poisson integral formula, which is obtained by the natural boundary reduction, is used for Ω_2.

Consider an exterior boundary value problem of the harmonic equation

$$-\Delta w = 0, \text{ in } \Omega, \quad w = g, \text{ on } \Gamma_0. \tag{1}$$

where Ω is the domain exterior to the closed curve Γ_0. The problem is equivalent to

$$\begin{cases} -\Delta u &= f, & \text{in} & \Omega, \\ u &= 0, & \text{on} & \Gamma_0 \end{cases} \tag{2}$$

for some proper f. With the condition that u is bounded in infinity, (2) has unique solution. Define the following Schwarz alternating algorithm:

$$\begin{cases} -\Delta u_1^{(k)} &= f, \quad &\text{in} \quad &\Omega_1, \\ u_1^{(k)} &= 0, \quad &\text{on} \quad &\Gamma_0, \quad k = 1, 2, \cdots \\ u_1^{(k)} &= u_2^{(k-1)}, \quad &\text{on} \quad &\Gamma_1, \end{cases} \tag{3}$$

and

$$\begin{cases} -\Delta u_2^{(k)} &= f, \quad &\text{in} \quad \Omega_2, \\ u_2^{(k)} &= u_1^{(k)}, \quad &\text{on} \quad \Gamma_2, \end{cases} \quad k = 1, 2, \cdots, \tag{4}$$

where $u_2^{(0)} \in H^{\frac{1}{2}}(\Gamma_1)$ is arbitrarily given, e. g. $u_2^{(0)} = 0$. The solution of problem (2) is in the space

$$V = \{v \in W_0^1(\Omega) | v = 0 \quad \text{on} \quad \Gamma_0\}.$$

Let

$$V_1 = \{v \in H^1(\Omega_1) | v = 0 \quad \text{on} \quad \Gamma_0 \cup \Gamma_1\}, \quad V_2 = \{v \in W_0^1(\Omega_2) | v = 0 \quad \text{on} \quad \Gamma_2\},$$

then

$$u_1^{(k)} - u_2^{(k-1)} \in V_1, \qquad u_2^{(k)} - u_1^{(k)} \in V_2.$$

From the bilinear form

$$D(u, v) = \int_\Omega \nabla u \cdot \nabla v dx, \tag{5}$$

the inner product $(u, v)_1$ and the norm $\| \cdot \|_1$ in V can be defined. Then (3) and (4) are equivalent to the variational problems

$$\begin{cases} \text{Find } u_1^{(k)} \in V_1 + u_2^{(k-1)} \quad \text{such that} \\ D(u_1^{(k)} - u, v_1) = 0, \qquad \forall v_1 \in V_1 \end{cases} \tag{6}$$

and

$$\begin{cases} \text{Find } u_2^{(k)} \in V_2 + u_1^{(k)} \quad \text{such that} \\ D(u_2^{(k)} - u, v_2) = 0, \qquad \forall v_2 \in V_2, \end{cases} \tag{7}$$

respectively. Let $P_{V_i^\perp} : V \to V_i^\perp$, $i = 1, 2$, denote the projectors in $(\cdot, \cdot)_1$,

$$e_i^{(k)} = u - u_i^{(k)}, \qquad i = 1, 2$$

be the errors, we have

$$\begin{cases} e_1^{(k)} &= P_{V_1^\perp} e_2^{(k-1)}, \\ e_2^{(k)} &= P_{V_2^\perp} e_1^{(k)}, \end{cases} \quad k = 1, 2, \cdots. \tag{8}$$

Then,

$$\begin{cases} e_1^{(k+1)} &= P_{V_1^\perp} P_{V_2^\perp} e_1^{(k)}, \quad &k = 1, 2, \cdots, \\ e_2^{(k+1)} &= P_{V_2^\perp} P_{V_1^\perp} e_2^{(k)}, \quad &k = 0, 1, \cdots. \end{cases} \tag{9}$$

Theorem 1

$$\lim_{k \to \infty} \|e_i^{(k)}\|_1 = 0, \qquad i = 1, 2, \tag{10}$$

and there is a constant $\alpha \in [0,1)$ such that

$$\|P_{V_1^\perp} P_{V_2^\perp}\| \le \alpha, \qquad \|P_{V_2^\perp} P_{V_1^\perp}\| \le \alpha, \tag{11}$$

Furthermore,

$$\|e_1^{(k)}\|_1 \le \alpha^{k-1}\|e_1^{(1)}\|_1, \qquad \|e_2^{(k)}\|_1 \le \alpha^k \|e_2^{(0)}\|_1. \tag{12}$$

Theorem 1 shows that the above Schwarz alternating method converges geometrically.

It is difficult to estimate the contraction factor for a general unbounded domain Ω. Here let Ω be an exterior domain of a circle Γ_0 with radius R_0, and $R_0 < R_2 < R_1$.

Let $\gamma' : W_0^1(\Omega_2) \to H^{\frac{1}{2}}(\Gamma_1)$ and $\gamma'' : H^1(\Omega_1) \to H^{\frac{1}{2}}(\Gamma_2)$ be the Dirichlet trace operators, $P_1 : H^{\frac{1}{2}}(\Gamma_1) \to H_0^1(\Omega_1) = \{v \in H^1(\Omega_1) | v = 0 \text{ on } \Gamma_0\}$ and $P_2 : H^{\frac{1}{2}}(\Gamma_2) \to W_0^1(\Omega_2)$ be the Poisson integral operators. Then,

$$\gamma'' P_1 \gamma' P_2 : H^{\frac{1}{2}}(\Gamma_2) \to H^{\frac{1}{2}}(\Gamma_2),$$

$$\gamma' P_2 \gamma'' P_1 : H^{\frac{1}{2}}(\Gamma_1) \to H^{\frac{1}{2}}(\Gamma_1).$$

Theorem 2 *The operators $\gamma'' P_1 \gamma' P_2$ and $\gamma' P_2 \gamma'' P_1$ are contraction mappings:*

$$\|\gamma'' P_1 \gamma' P_2 f\|_{\frac{1}{2},\Gamma_2} \le \delta \|f\|_{\frac{1}{2},\Gamma_2}, \qquad \forall f \in H^{\frac{1}{2}}(\Gamma_2), \tag{13}$$

$$\|\gamma' P_2 \gamma'' P_1 g\|_{\frac{1}{2},\Gamma_1} \le \delta \|g\|_{\frac{1}{2},\Gamma_1}, \qquad \forall g \in H^{\frac{1}{2}}(\Gamma_1), \tag{14}$$

where $0 < \delta < 1$, and we can take

$$\delta = \max\left(\frac{\ln(R_2/R_0)}{\ln(R_1/R_0)}, \left(\frac{R_2}{R_1}\right)^2\right). \tag{15}$$

From theorem 2, we can see that, the contraction factor δ only depends on R_0, R_1, and R_2, and the larger the ratio R_1/R_2, the smaller the factor δ. From the proof of the theorem, we can also see that $\delta \to 0$ when $R_2/R_0 \to 1$, and that

$$\frac{|\alpha_n^*|}{|\alpha_n|} = \frac{R_2^{2|n|} - R_0^{2|n|}}{R_1^{2|n|} - R_0^{2|n|}} \le \left(\frac{R_2}{R_1}\right)^{2|n|}, \qquad n = \pm 1, \pm 2, \cdots, \tag{16}$$

where α_n and α_n^* are coefficients of Fourier expansion of f and $\gamma'' P_1 \gamma' P_2 f$ (or g and $\gamma' P_2 \gamma'' P_1 g$), respectively. This means that the contraction factor is exponentially attenuate with the frequency.

Applying theorem 2 to the above alternating algorithm, we have

$$\begin{cases} \|u_2^{(k)} - u\|_{1,\Omega_2} \le C\delta^k \to 0, \\ \|u_1^{(k)} - u\|_{1,\Omega_1} \le C\delta^k \to 0 \end{cases} \tag{17}$$

when $k \to \infty$, where C depends on u, Ω_1, Ω_2 and the initial value on the artificial boundary.

For proofs of these theorems are given in [20].

3   A Dirichlet-Neumann Method Based On Natural Boundary Reduction

Consider the Dirichlet problem for the Poisson's equation

$$\begin{cases} -\Delta u & = & f, & \text{in} & \Omega, \\ u & = & g, & \text{on} & \Gamma_0, \end{cases} \tag{18}$$

where Ω is an unbounded domain with a closed curve Γ_0 as its boundary. Adding the proper boundary condition at infinity, (18) has unique solution.

Let circle Γ_1, with radius R_1, encircle Γ_0 such that dist $(\Gamma_1, \Gamma_0) > 0$. Then Ω is divided into an interior subdomain Ω_1 and an exterior subdomain Ω_2. Γ_1 is the artificial boundary. We define the following Dirichlet-Neumann alternating domain decomposition method.

Step 1. Choose original $\lambda^0 \in H^{\frac{1}{2}}(\Gamma_1)$, $n = 0$.

Step 2. Solve a Dirichlet problem in Ω_2:

$$\begin{cases} -\Delta u_2^n & = & f, & \text{in} & \Omega_2, \\ u_2^n & = & \lambda^n, & \text{on} & \Gamma_1. \end{cases} \tag{19}$$

Step 3. Solve a mixed boundary value problem in Ω_1:

$$\begin{cases} -\Delta u_1^n & = & f, & \text{in} & \Omega_1, \\ \frac{\partial u_1^n}{\partial n_1} & = & -\frac{\partial u_2^n}{\partial n_2}, & \text{on} & \Gamma_1, \\ u_1^n & = & g, & \text{on} & \Gamma_0. \end{cases} \tag{20}$$

Step 4. Set $\lambda^{n+1} = \theta_n u_1^n + (1 - \theta_n)\lambda^n$, on Γ_1.

Step 5. Set $n = n + 1$, goto step 2.

By the theory of the natural reduction [17], the solution of (19) is given by following the Poisson integral formula:

$$u_2^n(r, \varphi) = \frac{r^2 - R_1^2}{2\pi} \int_0^{2\pi} \frac{\lambda^n(\varphi')}{R_1^2 + r^2 - 2R_1 r \cos(\varphi - \varphi')} d\varphi' + \iint_{\Omega_2} G(p, p') f(p') dp', \tag{21}$$

and there is a natural integral equation on Γ_1:

$$\frac{\partial u_2^n}{\partial n_2}(\varphi) = -\frac{1}{4\pi R_1} \int_0^{2\pi} \frac{\lambda^n(\varphi')}{\sin^2 \frac{\varphi - \varphi'}{2}} d\varphi' + \iint_{\Omega_2} [\frac{\partial}{\partial n} G(p, p')] f(p') dp', \tag{22}$$

where $G(p, p')$ is the Green function for Ω_2:

$$G(p, p') = \frac{1}{4\pi} \ln \frac{R_1^4 + r^2 r'^2 - 2rr' R_1^2 \cos(\varphi - \varphi')}{R_1^2[r^2 + r'^2 - 2rr' \cos(\varphi - \varphi')]}.$$

Then in fact we need not solve (19) by the finite or boundary element method in step 2. Applying (22), $\frac{\partial u_2^n}{\partial n_2}$ can be found directly from λ^n, and obviously, the computation is fully parallel. As for the numerical computation of the hypersingular integral given by (22), see [17, 18]. Since Ω_1 is a small bounded subdomain, it is no difficulty to solve (20) by the standard finite element method in step 3.

Let $e_i^n = u - u_i^n$, $i = 1, 2$, $\mu = u|_{\Gamma_1} - \lambda^n$, then the errors e_1^n and e_2^n satisfy

$$\begin{cases} -\Delta e_2^n &= 0, \quad \text{in} \quad \Omega_2, \\ e_2^n &= \mu^n, \quad \text{on} \quad \Gamma_1 \end{cases} \tag{23}$$

and

$$\begin{cases} -\Delta e_1^n &= 0, \quad \text{in} \quad \Omega_1, \\ \frac{\partial e_1^n}{\partial n_1} &= -\frac{\partial e_2^n}{\partial n_2}, \quad \text{on} \quad \Gamma_1, \\ e_1^n &= 0, \quad \text{on} \quad \Gamma_0. \end{cases} \tag{24}$$

Moreover, it is important to choose θ_n properly. If θ_n is badly chosen, the algorithm will diverge.

By the standard domain decomposition theorey [10] this Dirichlet-Neumann method is equivalent to a precondition Richardson iteration method. Then the convergence analysis can be reduced to the estimation of the characteristic values of $S_1^{-1}S$, i.e., estimating upper and lower bounds for

$$\frac{(S\mu, \mu)}{(S_1\mu, \mu)} = 1 + \frac{(S_2\mu, \mu)}{(S_1\mu, \mu)} = 1 + \frac{a_2(H_2\mu, H_2\mu)}{a_1(H_1\mu, H_1\mu)}. \tag{25}$$

Here $S = S_1 + S_2$ is the Steklov-Poincare operator on Γ_1, H_1 and H_2 are harmonic extension operators of functions on Γ_1 to Ω_1 and Ω_2, respectively,

$$a_i(u, v) = \iint_{\Omega_i} \nabla u \nabla v \, dx dy, \qquad i = 1, 2. \tag{26}$$

Theorem 3 *If circle Γ_0' with radius $R_0' < R_1$ is the smallest circle which encloses surrounds Γ_0 and has the same center as Γ_1, then*

$$a_2(H_2\lambda, H_2\lambda) \le a_1(H_1\lambda, H_1\lambda) \le \frac{R_1^2 + (R_0')^2}{R_1^2 - (R_0')^2} a_2(H_2\lambda, H_2\lambda), \tag{27}$$

$$\frac{1}{\sqrt{2}} \|\lambda\|_{\frac{1}{2}, \Gamma_1}^2 \le a_1(H_1\lambda, H_1\lambda) \le \frac{R_1^2 + (R_0')^2}{R_1^2 - (R_0')^2} \|\lambda\|_{\frac{1}{2}, \Gamma_1}^2, \tag{28}$$

$$\frac{1}{\sqrt{2}} \|\lambda\|_{\frac{1}{2}, \Gamma_1}^2 \le a_2(H_2\lambda, H_2\lambda) \le \|\lambda\|_{\frac{1}{2}, \Gamma_1}^2, \tag{29}$$

where

$$\lambda \in H_0^{\frac{1}{2}}(\Gamma_1) = \{\mu \in H^{\frac{1}{2}}(\Gamma_1), \int_{\Gamma_1} \mu ds = 0\}.$$

Theorem 4 *Under the assumption of theorem 3 our Dirichlet-Neumann method and the corresponding preconditioned Richardson iteration method converge when $0 < \theta < 1$. In particular, when $\theta = \frac{R_1^2 + (R_0')^2}{2R_1^2 + (R_0')^2}$, the contraction factor satisfes $\delta \le \frac{(R_0')^2}{2R_1^2 + (R_0')^2}$, the convergence rate $-\ln \delta \ge \ln[2(\frac{R_1}{R_0'})^2 + 1]$, the condition number*

$$cond \ (S_1^{-1}S) \le 1 + (\frac{R_0'}{R_1})^2 < 2.$$

4   Conclusion

1. The boundary reduction is a forceful means for handling problems over unbounded domains. Based on the boundary reduction and the boundary element methods, some domain decomposition methods, where subdomains are overlapping or non-overlapping, are developed. These methods are applicable to unbounded domains and have the advantages of both the finite and the boundary element methods, while standard domain decomposition method can only be used for bounded domain.

2. Comparing with other kinds of boundary reduction, the natural boundary reduction has many advantages. Since a circle can be taken as an artificial boundary, the Poisson integral formulas and the natural boundary integral equations can be applied directly in the exterior subdomain Ω_2, where we do not need solve any equations, and a full parallel computation can be implemented. It is totally different from the finite element computation.

3. Only the subproblem over interior subdomain Ω_1 should be solved by the finite element method. Many available standard programs can be used. Since Ω_1 can be taken quite as small as possible, only few elements are necessary for solving the subproblem over Ω_1. It is much simpler than the coupling method of FEM and BEM.

4. Both the theoretical analysis and the numerical experiment show that these methods are feasible and converges very quickly, provided that the ratio R_1/R_2 (for overlapping subdomains) or R_1/R_0' (for non-overlapping subdomains) is not too close to 1. The larger the ratio, the faster the convergence.

5. These methods are suitable not only for solving Poisson equation, but also for solving biharmonic, plane elasticity and Stokes equations. Since many results of the natural boundary reduction for those equations are already given in [17], the extension of these methods is not difficult. Moreover, they can also be extended to more general problems when some more general boundary reduction is used.

REFERENCES

[1]   Feng Kang (1980) Differential vs. integral equations and finite vs. infinite elements. *Math. Numer. Sinica*, 2(1):100-105.

[2]   Feng Kang and Yu De-hao (1983) Canonical integral equations of elliptic boundary value problems and their numerical solution. *Proc. of China-France Symp. on FEM (Beijing, 1982)*, Science Press, Beijing, 211-252.

[3]   Feng Kang (1984) Asymptotic radiation conditions for reduced wave equation. *J. Comp. Math.*, 2(2):130-138.

[4]   Feng Kang and Yu De-hao (1994) A theorem for the natural integral operator of harmonic equation. *Math. Numer. Sinica*, 16(2):221-226.

[5]   Giroire J. and Nedelec J. C. (1978) Numerical solution of an exterior Neumann problem using a double layer potential. *Math. of Comput.*, 32(144): 973-990.

[6]   Glowinski R., Golub G. H., Meurant G. A. and Periaux J. eds. (1988) *Proc. of 1st International Symposium on Domain Decomposition Methods for Partial Differential Equations*, SIAM, Philadelphia, PA.

[7] Hsiao G. C., Wendland W. L. (1985) On a boundary integral method for some exterior problems in elasticity. Proc. Tbilisi University UDK 539. 3, *Math. Mech. Astron.*, 257: 31-60.

[8] Hsiao G. C. (1988) *The coupling of BEM and FEM-a brief review, Boundary Elements X,* Vol. 1:431-446.

[9] Hsiao G. C., Khoromskij B. N., Wendland W. L. (1994) Boundary integral operators and domain decomposition, Preprint 94-11, Math. Institut A, Univ. Stuttgart.

[10] Lu Tao, Shih T. M. and Liem C. B. (1992) *Domain Decomposition Methods -New Numerical Techniques for solving PDE,* Science Press, Beijing.

[11] Thatcher R. W. (1978) On the finite element for unbounded regions, *SIAM J. Numer. Anal.*, 15(3).

[12] Wendland W. L. (1986) On asymptotic error estimates for the combined BEM and FEM, *Innovative Numerical Methods in Engrg.*, 88:55-70.

[13] Ying Long-an (1978) The infinite similar element method for calculating stress intensity factors. *Scientia Sinica*, 21(1): 19-43.

[14] Yu De-hao (1983) Coupling canonical boundary element method with FEM to solve harmonic problem over cracked domain. *J. Comp. Math.*, 1(3): 195-202.

[15] Yu De-hao (1985) Approximation of boundary conditions at infinity for harmonic equation. *J. Comp. Math.*, 3(3): 219-227.

[16] Yu De-hao (1991) A direct and natural coupling of BEM and FEM, Boundary Elements XIII. *Computational Mechanics Publications, Southampton,* 995-1004.

[17] Yu De-hao (1993)*Mathematical Theory of Natural Boundary Element Method,* Science Press, Beijing.

[18] Yu De-hao (1993) The numerical computation of hypersingular integrals and its application in BEM. *Advances in Engineering Software*, 18:103-109.

[19] Yu De-hao (1994) The domain decomposition method of alternative FEM and natural BEM over unbounded domain. *Proc. of the 6th China-Japan Symposium on Boundary Element Methods,* International Academic Publishers, Beijing, 3-8.

[20] Yu De-hao (1994) A domain decomposition method based on natural boundary reduction over unbounded domain. *Math. Numer. Sinica,* 16(4): 448-459.

[21] Yu De-hao (1995) On relationship between Steklov-Poincare operators and natural integral operators and Green functions. *Math. Numer. Sinica,* 17(3):331-341.

An Additive Schwarz Algorithm for a Variational Inequality

Shuzi Zhou

1 Introduction

Let Ω be a bounded polygonal domain, V be a subspace of the Sobolev space $H^k(\Omega), a(\cdot, \cdot)$ be a continuous, coercive and symmetric bilinear form on $V \times V, f \in V^*$. For simplicity, we assume that the elements of V satisfy homogeneous boundary condition on $\partial\Omega$. Consider the variational inequality: find $u \in K$ such that

$$a(u, v - u) \geq f(v - u), \quad \forall v \in K \tag{1}$$

where

$$K = \{v \in V : v \geq \phi \text{ in } \Omega, \phi \in H^1(\Omega), \phi \leq 0 \text{ on } \partial\Omega\} \tag{2}$$

or

$$K = \{v \in V : \phi \leq v \leq \psi \text{ in } \Omega, \phi, \psi \in H^1(\Omega), \phi \leq 0 \leq \psi \text{ on } \partial\Omega\}. \tag{3}$$

Assume that $V^h \subset H_0^1(\Omega)$ is the finite element approximation of V and that the set of nodal parameters includes the value of the function in V_h at the nodes. Suppose $\phi, \psi \in C^0(\bar{\Omega})$. The finite element approximation of problem (1), (2) or problem (1), (3) is: find $u_h \in K^h$ such that

$$a_h(u_h, v - u_h) \geq f_h(v - u_h), \quad \forall v \in K^h, \tag{4}$$

where

$$K^h = \{v \in V^h : v \geq \phi \text{ at all the nodes }\}, \tag{5}$$

[0] Research supported by NNSF of China
[1] Department of Applied Mathematics, Hunan University, Changsha, Hunan, 410082, P. R. China

or

$$K^h = \{v \in V^h : \phi \leq v \leq \psi \text{ at all the nodes }\}, \tag{6}$$

$a_h(.,.)$ is a continuous, coercive and symmetric form on $V^h \times V^h$, $f_h \in (V^h)^*$. In [2] a multiplicative Schwarz algorithm for solving problem (1), (2) and problem (1), (3) with $k = 1$ was studied. In [5] a multiplicative Schwarz algorithm for solving problem (1), (3) with $k = 2$ was proposed. Combining the methods of [2, 5] and the so-called average method of [3, 4], mainly for solving problem (1), (3) with $k = 1$, an additive Schwarz algorithm was given in [5]. A differential additive Schwarz algorithm has been discussed for solving (4) in [3, 4] provided that K^h is a cone. Based on the work just mentioned, we propose an additive Schwarz algorithm for solving problem (4), (5) and problem (4), (6) in more general cases. We give the algorithm and the convergence theorem and get so-called finite step convergence for coincident components.

A similar algorithm has been studied in [8] and [9]. But they require a special choice of the initial value. In [8], monotone convergence was proven and in [9] the h independent convergence rate has been obtained.

2   An Additive Schwarz Algorithm

We use the two-level triangulation of Ω, given by Dryja and Widlund (ref. [1]). In this way, we get overlapping open subregions $\Omega_i, i = 1, \cdots, m$. Let $V_i = V^h \cap H_0^1(\Omega_i)$. The algorithm is defined as follows.

  <u>Algorithm I</u>
  <u>Step 1</u>. Given $\omega_i > 0, i = 1, \cdots, m$ with $\Sigma_{i=1}^m \omega_i = 1$. Take $u^0 \in K^h$ and $n := 0$;
  <u>Step 2</u>. For $i = 1, \cdots, m$ solve the following subproblems: find $u^{n_i} \in K_i^n$ such that

$$a_h(u^{n,i}, v - u^{n,i}) \geq f_h(v - u^{n,i}), \quad \forall v \in K_i^n,$$

where $K_i^n = (u^n + v_i) \cap K^h$,
  <u>Step 3</u> $u^{n+1} = \sum_{i=1}^m \omega_i u^{n,i}$;
  <u>Step 4</u> $n := n + 1$, go to step 2.

In order to prove the convergence of Algorithm I, we consider the following variational inequality: find $u^* \in K^h$, such that

$$a_h(u^*, v - mu^*) \geq f_h(v - mu^*), \quad \forall v \in K_1^* + \cdots + K_m^*, \tag{7}$$

where $K_i^* = (u* + V_i) \cap K^h, i = 1, \cdots, m$.

Lemma 1. *Problem (7) is equivalent to Problem (4), (5) or Problem (4), (6).*

  Proof. It is easy to prove that the solution of Problem (4), (5) (or (4), (6)) satisfies (7). Now we prove that any solution of (7) solves (4), (5) (or (4), (6)). Assume that u^* solves (7). It is sufficient to prove, for any $v \in K^h$, that

$$a_h(u^*, v - u^*) \geq f_h(v - u^*). \tag{8}$$

Let Ω_ε be the open subset of Ω with $\bar{\Omega}_\varepsilon \subset \Omega$, containing all the interior nodes. Then, there exists $\{\theta_i\}$ such that $0 \leq \theta_i \leq 1$, $\theta_i \in C_0^\infty(\Omega_i)$ and $\sum_{i=1}^m \theta_i = 1$ on $\bar{\Omega}_\varepsilon$. Therefore, we have

$$v = \sum_{i=1}^m I^h(\theta_i v) \text{ on } \bar{\Omega}$$

where I^h is the V^h–interpolation operator. Obviously $v_i = I^h(\theta_i v) \in V_i$ and

$$v - u^* = \sum_{i=1}^{m} v_i - u* = \sum_{i=1}^{m} w_i - mu^*, \tag{9}$$

where $w_i = u^* + I^h(\theta_i(v - u^*))$. It is easy to see that $w_i \in K_i^*$. Hence we have, by (7), that

$$a_h(u^*, \sum_{i=1}^{m} w_i - mu^*) \geq f_h(\sum_{i=1}^{m} w_i - mu^*),$$

which combined with (9) yields (8). The lemma is proved.

Let $J_h(v) = a_h(v, v)/2 - f_h(v)$. Then $u_h, u^{n,i}$ are, respectively, the solutions of the following problems:

$$u^h \in K^h, \quad J^h(u_h) = \min_{v \in K^h} J_h(v),$$

$$u^{n,i} \in K_i^n, \quad J_h(u^{n,i}) = \min_{v \in K_i^n} J_h(v).$$

By using the lemma and the strict convexity of $J(v)$, we obtain the following convergence theorem.

Theorem 1. *The sequence $\{u^n\}$ produced by Algorithm I converges to u_h in V^h. Moreover we have $u^{n,i} \to u_h, (n \to \infty), i = 1, \cdots, m$.*

Assume that $k = 1$ and that V^h is a Lagrange finite element space. Then Problem (4), (5) is equivalent to a linear complementary problem which has the following matrix form:

$$U \geq \Phi, \ AU - F \geq 0, \ (U - \Phi)^T(AU - F) = 0, \tag{10}$$

where $U, \Phi \in R^N$, their components $\{U_j\}, \{\Phi_j\}$ are respectively the values of u_h, φ at the interior nodes, $A = (a_{ij})_{i,j=1}^N, a_{ij} = a_h(\varphi_i, \varphi_j), F = (F_j)_{j=1}^N, F_j = f(\varphi_j)$, and $\{\varphi_j\}$ is the basis of V^h. Then K^h corresponds to

$$C = \{V \in R^N : V_j \geq \Phi_j; j = 1 \cdots, N\}.$$

Let $I = \{1, \cdots, N\}$. Denote the index set of the nodes in Ω_i by I_i. The matrix form of Algorithm I is as follow.

Algorithm I*

<u>Step 1</u> Given $\omega_i > 0, 1, \cdots, m$ with $\sum_{i=1}^{m} \omega_i = 1$. Take $U^0 \in C. n := 0$;

<u>Step 2</u> For $i = 1, \cdots, m$, solve the subproblems : find $U^{n,i} \in R^N$ such that $U^{n,i} \geq \Phi$ and

$$U_j^{n,i} = U_j^n \text{ for } j \in I/I_i,$$

$$(AU^{n,i} - F)_j \geq 0, (U_j^{n,i} - \Phi_j)(AU^{n,i} - F)_j = 0 \text{ for } j \in I_i;$$

<u>Step 3</u> $U^{n+1} = \sum_{i=1}^{m} \omega_i U^{n,i}$;

<u>Step 4</u> $n := n + 1$, go to Step 2.

Assume that U is the solution of (10) and let $J = \{j \in I : U_j = \Phi_j\}$. We call J the coincident set and U_j the coincident component if $j \in J$. We say that problem

(10) is nondegenerate if $j \in J$ implies $(AU - F)_j > 0$. By Theorem 1 we know that $U^n \to U(n \to \infty)$ and $U^{n,i} \to U(n \to \infty), i = 1, \cdots, m$. It is not difficult to show that the coincident components are reached by the corresponding components of $U^{n,i}$ within a finite number of iteration steps, just as the following theorem states.

Theorem 2. *If (10) is nondegenerate, then there exists a positive integer n_0 such that for $n \geq n_0$*

$$U_j^{n,i} = \Phi_j, \forall j \in J \cap I_i,$$
$$U_j^{n,i} > \Phi_j, \forall j \in (I \backslash J) \cap I_i.$$

A counterexample shows that U^n does not have this property of $U^{n,i}$. Now we consider the so-called average with variable weights (ref. [4]). Assume $Q \in \Omega$. We say that Q is a k-point if Q belongs to at most k subregions.

Let

$$\Omega(i_1, \cdots, i_k) = \Omega_{i_1} \bigcap \cdots \bigcap \Omega_{i_k}.$$

We construct a function $\bar{u}^n$ as follows: for a k-point $Q \in \Omega(i_1, \cdots, i_k)$ define

$$\bar{u}^{n+1}(Q) = \frac{1}{k} \sum_{j=1}^{k} u^{n,i_j}(Q),$$

where $u^{n,i}$ is defined by Algorithm I with $\omega_i = \frac{1}{m}, i = 1, \cdots, m$. Then it is easy to see that the following relationship between u^n and $\bar{u}^n$ holds:

$$u^{n+1}(Q) = \frac{k}{m}\bar{u}^{n+1}(Q) + \frac{m-k}{m}u^n(Q) \text{ for } Q \in \Omega(i_1, \cdots, i_k).$$

Therefore we know that $\bar{u}^n \to u(n \to \infty)$ by Theorem 1. Moreover we can show

Theorem 3. *If (10) is nondegenerate and $\bar{U}^n$ is the vector corresponding to $\bar{u}^n$ then there exists a positive integer n_0 such that for $n \geq n_0$*

$$\bar{U}_j^n = \Phi_j, \forall j \in J,$$
$$\bar{U}_j^n > \Phi_j, \forall j \in I \backslash J.$$

For Problem (4), (6) we can establish similar definitions and conclusions.

REFERENCES

[1] Dryja M. (1989) in *Proceedings of DDM2* (eds T. F. Chan, R. Glowinski, J. Periaux& O. B. Widlund), SIAM, 168-172.

[2] Lions P. L. (1988) in *Proceedings of DDM1* (eds. R. Glowinski, G. H. Golub, G. A. Murant &J. Periaux), SIAM, 1-40.

[3] Lu T. (1991) *Sys. Sci. &Math. Sci,* 4: 340-349.

[4] Lu T., Shih T. M. and Lilm C. B. (1992) Domain Decomposition Method (in Chinese), Academic Press, Beijing.

[5] Scarpini F. (1990) *Calcolo,* 27:57-72.

[6] Xu X. J. and Shen S. M. (1994) *J. Numer. Math. of Chinese Universities,* 16: 186-194.

[7] Bedea L. (1995) *SIAM J. Numer. Anal.* 28: 179-204.

[8] Kuznetsov Y., Neittaanmäki P. and Tarvainen P. (1995) in *Proceedings of DDM7* (eds. D. E. Keyes & J. C. Xu), AMS.

[9] Zen J. P. (1995), Monotone Convergence and Convergenve rate Analysis of DDM for the Solution of LCP, preprint.

Part II Domain Decomposition and Multi-Level Methods

Additive Schwarz Methods without Subdomain Overlap and with New Coarse Spaces

Petter E. Bjørstad [1] Maksymilian Dryja [2] Eero Vainikko [3]

1   Introduction

This chapter develops two additive Schwarz methods with new coarse spaces. The methods are designed for elliptic problems in 2 and 3 dimensions with discontinuous coefficients. The methods use no explicit overlap of the subdomains, subdomain interaction is via the coarse space. The first method has a rate of convergence proportional to $(H/h)^{1/2}$ when combined with a suitable Krylov space iteration. This rate is independent of discontinuities in the coefficients of the equation. The method has good parallelization properties and does not require a coarse grid triangulation, that is, one is free to use arbitrary, irregular subdomains. The second method uses a diagonal scaling in addition to the standard coarse space. This method is not as robust and flexible as the first method. A suitable model problem that we will consider has the variational form:

[1] Para//ab, Institutt for Informatikk, University of Bergen, N-5020 Bergen, Norway. http://www.ii.uib.no/~petter, Email: petter@ii.uib.no. This research was partially supported by NFR grant 27625.

[2] Department of Mathematics, Warsaw University, Banacha 2, 02-097 Warsaw, Poland. Email: dryja@mimuw.edu.pl. This research was partially supported by the NSF under grant NSF-CCR-9204255, and in part by the Colorado School of Mines.

[3] Para//ab, Institutt for Informatikk, University of Bergen, N-5020 Bergen, Norway.

Domain Decomposition Methods in Sciences and Engineering, edited by R. Glowinski *et al.*
© 1997 John Wiley & Sons, Ltd.

Find $u^* \in V(\Omega)$ such that

$$\sum_{i=1}^{N} \int_{\Omega_i} \rho_i \nabla u^* \cdot \nabla v \, dx = \int_{\Omega} f v \, dx \quad \forall v \in V(\Omega), \tag{1}$$

in an appropriate Sobolev space $V(\Omega)$. Here ρ_i are positive constants and $\Omega = \cup_{i=1}^{N} \Omega_i$. We are interested in the discontinuous coefficient case, i.e., we have (possibly large) jumps in the values of ρ_i across subdomain boundaries.

This problem is quite common in practice where, for example, the coefficients ρ_i represent physical (e.g., material) properties that change across the domain of interest. An example of this is petroleum reservoir simulation [BK95], where the reservoir consists of inhomogeneous rock where the permeability can change by orders of magnitude across relatively small sections. The freedom to make the subdomain boundaries follow the material discontinuities may be quite important in this and similar applications.

The problem has attracted much attention, [Sko92] and [BS92] developed parallel algorithms showing experimentally that the Additive Schwarz algorithm converged very satisfactorily when using small or minimal overlap between the subdomains. The paper [BX91] contains estimates for the weighted L^2-projection indicating a lack of stability and in [Xu91] there are counter examples showing that these estimates are sharp. The report [DSW94] discusses many domain decomposition algorithms for this problem in 3 dimensions and characterizes a class of distributions of the coefficients ρ_i for which the L^2-projection is stable. This class, called a quasi-monotone distribution, requires a monotone path from all subdomains to the subdomain having the largest coefficient, traversing (through the faces) the subdomains that have a common node [4], see Figure 1. For more references the reader should consult the bibliographies given in the above mentioned papers.

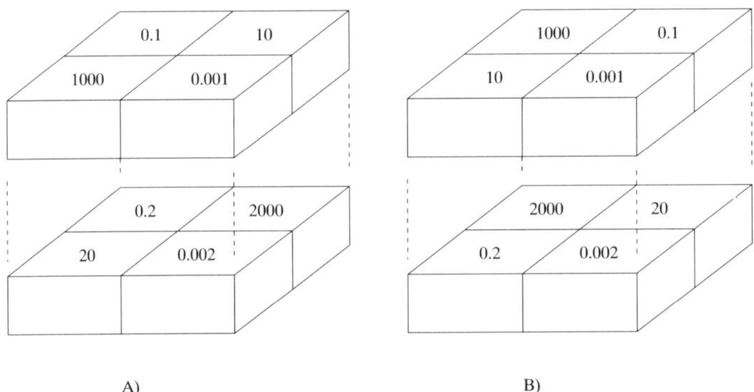

Figure 1   In A) the distribution is not quasi-monotone as there is no monotone path from 1000 to 2000. In B) we show a quasi-monotone distribution.

---

[4] If there is a path through edges only, we have a weak quasi-monotone distribution. This situation improves our estimates, but is not discussed further here.

2 Definition of a Discrete Model Problem

Let Ω be a polyhedral region in R^d, $d = 2, 3$. We consider two triangulations of Ω, a coarse and a fine, with elements Ω_i and e_i each having diameters H_i and h_i respectively. We assume that the fine triangulation is a refinement of the coarse subdivision. We further require a constant α_i, independent of the triangulation, such that

$$\hat{h}_i = \inf_{j: e_j \cap \partial\Omega_i} h_j \geq \alpha_i \sup_{j: e_j \cap \partial\Omega_i} h_j, \tag{2}$$

for each substructure Ω_i. This defines $\hat{h}_i$, the diameter of the smallest element touching the boundary in substructure Ω_i and requires the substructure boundary elements to be quasi-uniform. Note that elements in the interior of a substructure are not subject to this restriction, but we require all elements of the fine triangulation to be shape regular. We note that our assumptions can be relaxed to only require shape regularity near subdomain boundaries, at the expense of a more technical exposition. In order to keep the notation simple the reader should note that we often use the quantity h_i under the (limited) quasi-uniform assumption stated in (2).

We have no such assumption on the coarse triangulation, and note that our results are valid for an irregular coarse partitioning into subdomains as long as each element Ω_i with diameter H_i is a union of the fine triangles e_i. In order to keep the presentation simple we will only discuss the case where V^h is a finite element space of piecewise linear, continuous functions defined on the fine triangulation and vanishing on $\partial\Omega$, the boundary of Ω. Our discrete problem is of the form:

Find $u \in V^h(\Omega)$ such that

$$a(u, v) = f(v) \quad \forall v \in V^h(\Omega), \tag{3}$$

where

$$a(u, v) = \sum_{i=1}^{N} \int_{\Omega_i} \rho_i \nabla u \cdot \nabla v \, dx, \quad f(v) = \int_{\Omega} f v \, dx \tag{4}$$

and ρ_i are positive constants.

3 The Additive Average Method

We decompose the space V^h into

$$V^h = V_0 + V_1 + \cdots + V_N,$$

where, for $i = 1, \cdots, N$, $V_i = H_0^1(\Omega_i) \cap V^h$ and zero outside of Ω_i. The space V_0, which we call the coarse space is defined as the range of an interpolation-like operator I_A, $V_0 = \text{Range}(I_A)$. For $u \in V^h$ on $\bar{\Omega}_i$ we define $I_A u$ as follows:

$$I_A u = \begin{cases} u(x), & x \in \partial\Omega_{ih} \\ \bar{u}_i, & x \in \Omega_{ih}, \end{cases} \tag{5}$$

where $\partial\Omega_{ih}$ and Ω_{ih} denote the nodal points of $\partial\Omega_i$ and Ω_i respectively. Let n_i denote the number of nodes on $\partial\Omega_i$. For any function $u \in V^h$ we define the quantity $\bar{u}_i$ for each $\partial\Omega_i$ by

$$\bar{u}_i = \frac{1}{n_i} \sum_{x \in \partial\Omega_{ih}} u(x), \tag{6}$$

the average of the nodal values of u that are on the boundary of substructure Ω_i. Due to this construction and in order to distinguish this method from other additive Schwarz methods, we will denote it the Additive Average method.

We now state a lemma about the properties of our interpolation-like operator I_A which will be useful for the analysis of the proposed algorithm.

Lemma 1. For any $u \in V^h$:

$$a(I_A u, I_A u) = \sum_i \rho_i |I_A u|^2_{H^1(\Omega_i)} \le C \sum_i \rho_i \frac{H_i}{\hat{h}_i} |u|^2_{H^1(\Omega_i)} \tag{7}$$

and

$$\sum_i \rho_i \|u - I_A u\|^2_{L^2(\Omega_i)} \le C \sum_i \rho_i H_i^2 |u|^2_{H^1(\Omega_i)}, \tag{8}$$

where C is a constant independent of H_i and h_i and of the jumps in ρ_i. Here $\hat{h}_i$ is defined in (2).

Proof: First, consider the interpolation error $\|u - I_A u\|^2_{L^2(\Omega_i)}$. We have in d dimensions ($d = 2, 3$),

$$\begin{aligned} \|u - I_A u\|^2_{L^2(\Omega_i)} &\le 2 \left\{ \|u\|^2_{L^2(\Omega_i)} + \int_{\Omega_i} (I_A u)^2 dx \right\} \\ &\le C \left\{ \|u\|^2_{L^2(\Omega_i)} + H_i^d \, \bar{u}_i^2 \right\}. \end{aligned} \tag{9}$$

We can estimate $\bar{u}_i^2$ by

$$\bar{u}_i^2 \le C H_i^{1-d} \|u\|^2_{L^2(\partial\Omega_i)},$$

since

$$\bar{u}_i^2 = \frac{1}{n_i^2} \Big(\sum_{x \in \partial\Omega_{ih}} u(x) \Big)^2 \le \frac{1}{n_i} \sum_{x \in \partial\Omega_{ih}} u(x)^2 \tag{10}$$

and $1/n_i \le C(h_i/H_i)^{d-1}$ because of assumption (2).

Using this, we obtain an intermediate inequality which is needed in the proof of both (7) and (8),

$$\|u - I_A u\|^2_{L^2(\Omega_i)} \le C \left\{ \|u\|^2_{L^2(\Omega_i)} + H_i \|u\|^2_{L^2(\partial\Omega_i)} \right\}. \tag{11}$$

We now proceed to prove (7).

$$\begin{aligned} |I_A u|^2_{H^1(\Omega_i)} &= |I_A u - \bar{u}_i|^2_{H^1(\Omega_i)} \\ &\le C \sum_{e_j \subset \bar{\Omega}_i} h_j^{-2} \|I_A u - \bar{u}_i\|^2_{L^2(e_j)} \\ &\le C h_i^{d-2} \sum_{x \in \partial\Omega_{ih}} (u(x) - \bar{u}_i)^2 \\ &\le C h_i^{d-2} \sum_{x \in \partial\Omega_{ih}} u^2(x), \end{aligned}$$

where we have used the inverse inequality and the properties of I_A. The last inequality follows from the observation that,

$$\sum_{x\in\partial\Omega_{ih}} (u(x) - \bar{u}_i)^2 = \sum_{x\in\partial\Omega_{ih}} (u(x) - \bar{u}_i)u(x) = \sum_{x\in\partial\Omega_{ih}} (u^2(x) - n_i\bar{u}_i^2) \leq \sum_{x\in\partial\Omega_{ih}} u^2(x).$$

From this we conclude that

$$|I_A u|^2_{H^1(\Omega_i)} \leq \frac{C}{h_i}\|u\|^2_{L^2(\partial\Omega_i)}. \tag{12}$$

Using the trace inequality

$$\|u\|^2_{L^2(\partial\hat{\Omega})} \leq C\left\{|u|^2_{H^1(\hat{\Omega})} + \|u\|^2_{L^2(\hat{\Omega})}\right\}, \tag{13}$$

for a reference element $\hat{\Omega}$, of unit diameter, see for example Theorem 1.5.1.10 in [Gri85], we can write:

$$\begin{aligned}
|I_A u|^2_{H^1(\Omega_i)} &\leq Ch_i^{-1}\|u\|^2_{L^2(\partial\Omega_i)} \\
&\leq Ch_i^{-1}H_i^{d-1}\|u\|^2_{L^2(\partial\hat{\Omega})} \\
&\leq Ch_i^{-1}H_i^{d-1}\left\{|u|^2_{H^1(\hat{\Omega})} + \|u\|^2_{L^2(\hat{\Omega})}\right\} \\
&\leq Ch_i^{-1}H_i^{d-1}\left\{\frac{1}{H_i^{d-2}}|u|^2_{H^1(\Omega_i)} + \frac{1}{H_i^d}\|u\|^2_{L^2(\Omega_i)}\right\} \\
&\leq C\frac{H_i}{\hat{h}_i}|u|^2_{H^1(\Omega_i)},
\end{aligned} \tag{14}$$

where we have used Poincare's inequality on the second term in the last step. Summation over all elements Ω_i proves (7), the first part of the lemma.

In order to prove the second part of Lemma 1, we apply the result from (14) to the second term of (11) and Poincare's inequality on the first term concluding that

$$\|u - I_A u\|^2_{L^2(\Omega_i)} \leq CH_i^2|u|^2_{H^1(\Omega_i)}.$$

Summation over all subdomains Ω_i completes the proof of Lemma 1.

Introduce bilinear forms $b_i(u, v)$ on $V_i \times V_i$ of the form

$$b_i(u, v) = a(u, v) \quad i = 1, \cdots, N$$

and the bilinear form on the coarse space

$$b_0(u, v) = \sum_i \rho_i \hat{h}_i^{d-2} \sum_{x\in\partial\Omega_{ih}} (u(x) - \bar{u}_i)(v(x) - \bar{v}_i). \tag{15}$$

Note that for $v = \phi$, a basis function associated with a nodal point $x \in \partial\Omega_i$, we have

$$\sum_{x\in\partial\Omega_{ih}} (u(x) - \bar{u}_i)(\phi(x) - \bar{\phi}_i) = \sum_{x\in\partial\Omega_{ih}} (u(x) - \bar{u}_i)\phi(x),$$

since

$$\sum_{x \in \partial \Omega_{ih}} (u(x) - \bar{u}_i)\bar{\phi}_i = 0.$$

Let

$$T = T_0 + T_1 + \cdots + T_N, \tag{16}$$

where

$$b_i(T_i u, v) = a(u, v) \quad \forall\, v \in V_i, \quad i = 0, \cdots, N.$$

The discrete problem (3) can be replaced by the equation

$$Tu = g,$$

where $g = \sum_{i=0}^{N} g_i$, and $g_i = T_i u$ is the solution of

$$b_i(g_i, v) = f(v) \quad \forall\, v \in V_i.$$

We refer the reader to [SBG95] for a detailed explanation of this reformulation.
We now state the main result of this paper:

Theorem 1 *The operator T is self-adjoint positive definite in V^h with the scalar product $a(u, v)$ and for any $u \in V^h$*

$$\gamma_0 \delta^{-1} a(u, u) \leq a(Tu, u) \leq \gamma_1 a(u, u),$$

where γ_0 and γ_1 are positive constants independent of h_i and H_i and of the jumps of ρ_i. Here $\delta = \max_i H_i/\hat{h}_i$ and $\hat{h}_i$ is defined in (2).

Proof: Using the general framework of Additive Schwarz methods (Chapter 5 in the book [SBG95]), we need to check three key assumptions.

Assumption 1 *Let C_0 be the minimum constant such that for all $u \in V^h$ there exists a representation $u = \sum_i u_i$, $u_i \in V_i$, with*

$$\sum_i b_i(u_i, u_i) \leq C_0^2 a(u, u).$$

Let $u_0 = I_A u$ and $w = u - u_0$. Note that w vanishes on each $\partial \Omega_i$. In light of this let us define $u_i \in V_i$ by the nodal values of w at all nodal points $x \in \Omega_{ih}$ and zero otherwise. Note that

$$u = \sum_{i=0}^{N} u_i.$$

We then have

$$\begin{aligned}
\sum_{i=1}^{N} b_i(u_i, u_i) &= \sum_{i=1}^{N} \rho_i |u - u_0|_{H^1(\Omega_i)}^2 \\
&= a(u - u_0, u - u_0) \\
&\leq 2a(u, u) + 2a(u_0, u_0).
\end{aligned}$$

We now apply Lemma 1 to the second term above and obtain

$$\sum_{i=1}^{N} b_i(u_i, u_i) \leq C\delta a(u, u). \tag{17}$$

It remains to handle the case $i = 0$, that is, u_0. We use the same inequalities as in the proof of the first part of Lemma 1 to estimate $b_0(u_0, u_0)$,

$$b_0(u_0, u_0) = \sum_i \rho_i \hat{h}_i^{d-2} \sum_{x \in \partial\Omega_{ih}} (u(x) - \bar{u}_i)^2 \leq C \sum_i \frac{\rho_i}{\hat{h}_i} \|u\|_{L^2(\partial\Omega_i)}^2. \tag{18}$$

The inequality in (14) provides a bound on $\|u\|_{L^2(\partial\Omega_i)}^2$,

$$b_0(u_0, u_0) \leq C \sum_i \rho_i \frac{H_i}{\hat{h}_i} |u|_{H^1(\Omega_i)}^2 \leq C\delta a(u, u).$$

Combining (17) and the previous inequality gives a final bound on the energy of our decomposition of u

$$\sum_{i=0}^{N} b_i(u_i, u_i) \leq C\delta a(u, u),$$

which verifies Assumption 1.

Assumption 2 *Define $0 \leq \mathcal{E}_{ij} \leq 1$ to be the minimal values that satisfy*

$$|a(u_i, u_j)| \leq \mathcal{E}_{ij}[a(u_i, u_i)]^{1/2}[a(u_j, u_j)]^{1/2} \quad \forall u_i \in V_i, \ u_j \in V_j, \ i, j = 1, \ldots, N.$$

Define $\rho(\mathcal{E})$ to be the spectral radius of $\mathcal{E} = \{\mathcal{E}_{ij}\}$. Note that we do not include the subspace V_0.

In our case $\rho(\mathcal{E}) = 1$ since $\mathcal{E}_{ii} = 1$ and $\mathcal{E}_{ij} = 0$ when $i \neq j$.

Assumption 3 *Let $\omega > 0$ be the minimum constant such that*

$$a(u, u) \leq \omega b_i(u, u) \qquad \forall u \in V_i, \ i = 0, \cdots, N.$$

First, for $i = 1, 2, \ldots, N$, we can take $\omega = 1$ since $b_i(u, v) = a(u, v)$. To handle the case $i = 0$ we use the inverse inequality,

$$\begin{aligned}
a(u_0, u_0) &= \sum_{i=1}^{N} \rho_i |u_0|_{H^1(\Omega_i)}^2 \\
&= \sum_{i=1}^{N} \rho_i |u_0 - \bar{u}_i|_{H^1(\Omega_i)}^2 \\
&\leq C \sum_{i=1}^{N} \rho_i \hat{h}_i^{d-2} \sum_{x \in \partial\Omega_{ih}} (u_0(x) - \bar{u}_i)^2 \\
&= C b_0(u_0, u_0),
\end{aligned}$$

which shows that ω can be taken equal to C.

Note that in the case when $b_0(u, v)$ is the exact bilinear form, that is, if $b_0(u, v) = a(u, v)$ then $\omega = 1$ and we have $\gamma_1 = 2$ in the upper bound of Theorem 1.

4   An Additive Diagonal Scaling Method

This method is a modification of the method in the previous section. We decompose our domain into non-overlapping subdomains as before. We use the classical coarse space, i.e., piecewise linear elements defined on the substructures, but complement this by yet another decomposition derived from solving a diagonal problem along all interior interfaces. That is, we define a local problem for each node on the interior interfaces (corresponding to a local subdomain equal to the support of the nodal basis function). This is just a special diagonal scaling, hence the name of the method. For a quasi-monotone distribution of the ρ_i the estimate of the resulting condition number for the preconditioned system is of the same order as for the method in Section 3 while in the general (non quasi-monotone) case it is worse by a factor H/h in 3 dimensions. Our theory for this method requires the triangulation of each substructure Ω_i to be quasi-uniform with parameter h_i. We also note that the method requires the substructures to form a regular decomposition, hence it is potentially less versatile.

We proceed to describe and analyze this method using the same notation and framework as before. The decomposition of the space V^h is of the form

$$V^h = V_{-1} + V_0 + V_1 + \ldots + V_N.$$

Here V_i, for $i = 1, \ldots, N$, are the same as in Section 3. The coarse space $V_0 = V^H$, is a space of piecewise linear functions on the coarse triangulation which vanish on $\partial\Omega$ (we assume that the coarse triangulation with triangular elements is shape regular). The space V_{-1} is called a diagonal space and is the restriction of V^h to $\Gamma = \cup_i \partial\Omega_i$. Functions in this space vanish at all nodal points outside Γ.

The bilinear form $b_i(u, v) : V_i \times V_i \to R$, for $i = 0, 1, \ldots, N$, is given by

$$b_i(u, v) = a(u, v) \quad u, v \in V_i,$$

while for $i = -1$

$$b_{-1}(u, v) = \sum_{i=1}^{N} \rho_i h_i^{d-2} \sum_{x \in \partial\Omega_{ih}} u(x)v(x). \tag{19}$$

The operators $T_i : V^h \to V_i$, $i = -1, 0, \ldots, N$, are defined by the bilinear forms above and

$$T = T_{-1} + T_0 + \cdots + T_N.$$

Theorem 2 *The operator T is self-adjoint positive definite in V^h with the scalar product $a(u, v)$ and for any $u \in V^h$*

$$\gamma_2 \delta^{-1} a(u, u) \leq a(Tu, u) \leq \gamma_3 a(u, u),$$

where

$$\delta = \max_i \begin{cases} \frac{H_i}{h_i}, & \Omega \subset R^d, d = 2, 3 \text{ when } \rho_i \text{ is quasi} - \text{monotone} \\ \frac{H_i}{h_i} \log \frac{H_i}{h_i}, & \Omega \subset R^2 \text{ when } \rho_i \text{ is not quasi} - \text{monotone} \\ \left(\frac{H_i}{h_i}\right)^2, & \Omega \subset R^3 \text{ when } \rho_i \text{ is not quasi} - \text{monotone}, \end{cases}$$

and γ_2 and γ_3 are positive constants independent of h_i and H_i and of the jumps of ρ_i.

Proof:

We need to check the same three assumptions as in Section 3.

Assumption 1:

Let $u_0 = Q_H^\rho u$ be the L_2- projection with weights ρ_i from V^h on $V_0 = V^H$, that is,

$$\sum_{i=1}^{N} \rho_i (u_0, v)_{L^2(\Omega_i)} = \sum_{i=1}^{N} \rho_i (u, v)_{L^2(\Omega_i)} \quad \forall v \in V^H.$$

It is known that

$$\sum_{i=1}^{N} \rho_i \|u - u_0\|_{L^2(\Omega_i)}^2 \leq C \sum_{i=1}^{N} \rho_i H_i^2 |u|_{H^1(\Omega_i)}^2, \tag{20}$$

in the case when ρ_i has a quasi-monotone distribution, see [DSW94], while

$$\sum_{i=1}^{N} \rho_i \|u - u_0\|_{L^2(\Omega_i)}^2 \leq \begin{cases} C \sum_{i=1}^{N} \rho_i H_i^2 \log \frac{H_i}{h_i} |u|_{H^1(\Omega_i)}^2, & \Omega \subset R^2 \\ C \sum_{i=1}^{N} \rho_i H_i^2 \frac{H_i}{h_i} |u|_{H^1(\Omega_i)}^2, & \Omega \subset R^3, \end{cases} \tag{21}$$

in the general case, see [BX91] and [Xu91].

Let $w = u - u_0$ and on each Ω_i we define

$$w = w_I + w_B,$$

where $w_I(x) = w$ for all $x \in \Omega_{ih}$ and zero on $\partial\Omega_i$. Note that $w_B \in V_{-1}$. Let $u_i = w_I$ on Ω_i and zero outside of Ω_i, for $i = 1, 2, \ldots, N$, and $u_{-1} = w_B$. We see immediately that $u = \sum_{i=-1}^{N} u_i$.

Again using (14) we obtain,

$$h_i^{d-2} \sum_{x \in \partial\Omega_{ih}} u_{-1}^2(x) \leq C h_i^{-1} \|u - u_0\|_{L^2(\partial\Omega_i)}^2 \tag{22}$$

$$\leq C \frac{H_i}{h_i} \left\{ |u - u_0|_{H^1(\Omega_i)}^2 + H_i^{-2} \|u - u_0\|_{L^2(\Omega_i)}^2 \right\}.$$

We now sum over substructures applying (20) or (21) (depending on the distribution of the ρ_i) to the last term. The projection $u_0 = Q_H^\rho$ is L_2 and H^1 stable in the case of quasi-monotone coefficients ρ_i, otherwise the extra factor $\log(H_i/h_i)$ or H_i/h_i from (21) appears. Hence, with δ defined in Theorem 2,

$$b_{-1}(u_{-1}, u_{-1}) \leq C \delta a(u, u). \tag{23}$$

Using the same arguments we prove that

$$b_0(u_0, u_0) \leq C \delta^{1/2} a(u, u). \tag{24}$$

We now consider the estimates for $i = 1, \ldots, N$. We have

$$b_i(u_i, u_i) = b_i(w_I, w_I) \leq 2 b_i(w, w) + 2 b_i(w_B, w_B).$$

By using the inverse inequality and (22) we estimate the last term as follows:

$$b_i(w_B, w_B) \leq C h_i^{d-2} \sum_{x \in \partial \Omega_{ih}} \rho_i (u - u_0)^2(x) \leq C \rho_i \frac{H_i}{h_i} \|u - u_0\|_{H^1(\Omega_i)}^2.$$

Hence,

$$\sum_{i=1}^{N} b_i(u_i, u_i) \leq C(\delta^{1/2} a(u, u) + a(w_B, w_B)) \leq C \delta a(u, u). \tag{25}$$

Adding (23), (24), and (25) we obtain

$$\sum_{i=-1}^{N} b_i(u_i, u_i) \leq C \delta a(u, u),$$

which verifies Assumption 1.

Assumption 2:

Note that we can take V_{-1} and V_0 as the coarse spaces. In this case $\rho(\mathcal{E}) = 1$ as in Section 3. This slightly modifies the standard theory regarding Assumption 3, as we get 2ω instead of ω, see [SBG95] for details.

Assumption 3:

This assumption is verified by noting that we have $\omega = 1$ for $i = 0, 1, \ldots N$, while $\omega \leq C$ for the case $i = -1$. In this case we use the same arguments as in the proof of Theorem 1.

5   Implementation

In this section we briefly discuss implementation of the Additive Average method, in particular the issues relevant for a parallel implementation.

The subdomain solution can proceed completely in parallel, consider therefore the coarse problem,

$$\begin{pmatrix} A_a & A_{ab} \\ A_{ab}^T & A_b \end{pmatrix} \begin{pmatrix} u_a \\ u_b \end{pmatrix} = \begin{pmatrix} 0 \\ v_b \end{pmatrix}. \tag{26}$$

We note that one extra unknown per subdomain, corresponding to the average values $\bar{u}_i$ from (6), has been introduced in the vector u_a. The vector u_b has one component for each node on all subdomain boundaries. The subscripts a and b denote average and boundary respectively. The linear system has a very special structure that can be exploited in the solution algorithm. First, from (15) we note that each block in the coarse matrix will contain a diagonal scaling of the form $\hat{h}_i^{d-2} \rho_i$. The block A_a is diagonal with entries n_i (ignoring the scaling), the number of nodes belonging to each substructure boundary. Similarly, the block A_{ab} contains the nonzero entry, -1, in such a way that the first block equation just expresses the relation (6) between average and boundary values. The block A_b is also diagonal and its values correspond to the number of subdomains that share a given node on the substructure boundaries.

Additive algorithms normally proceed by restricting the current vector to the coarse space, perform a coarse space solution and interpolate this solution back to the fine grid. In our case the restriction operator is implemented as follows. First, the values of all interior nodes in a substructure are added and the result is stored in the vector v_a. Next, we compute the right hand side vector v_b according to the formula

$$v_b := v_b - A_{ab}^T A_a^{-1} v_a. \tag{27}$$

This computation is the transpose of the interpolation-like operator defined in (5).

We solve the coarse matrix problem (26) by first forming the Schur complement

$$S = A_a - A_{ab} A_b^{-1} A_{ab}^T.$$

Due to the special structure of the blocks we can explicitly compute S quite easily or multiply S with a vector in an inner iterative procedure. We next solve the Schur complement system

$$S u_a = -A_{ab} A_b^{-1} v_b$$

and then compute the values of the internal boundary nodes u_b by

$$u_b = A_b^{-1}(v_b - A_{ab}^T u_a).$$

The resulting values u_a and u_b are then added to the solutions from the subdomains in the standard additive fashion. This outline shows how the special structure of our coarse problem results in an efficient implementation and an overall algorithm very suitable for parallel computation.

Table 1   Computational complexity of the preconditioning step in the Additive
   Average method compared with the Additive Schwarz method with minimal
   overlap

Task	2-dimensions		3-dimensions	
	Average	Additive	Average	Additive
Subdomain (size)	$(n-1)^2$	$(n+1)^2$	$(n-1)^3$	$(n+1)^3$
Coarse problem (size)	m^2	$(m-1)^2$	m^3	$(m-1)^3$
Transfers (flops)	$2(n-1)^2 + 20n$	$34(n+1)^2$	$2(n-1)^3 + 30n^2$	$126(n+1)^3$
Communication (words)	2	$8(n+1)$	2	$12(n+1)^2$
Communication (startups)	2	10	2	14

6   Complexity

We compare the complexity of the preconditioning step in one iteration of a Krylov subspace method when using the classical Additive Schwarz method and our new Additive Average method. We consider only Additive Schwarz with minimal overlap, that is, each subdomain extends a distance h into its neighbor making the total overlap area have width $2h$. We give the required number of operations per subdomain in a way which can be implemented on a parallel machine where each subdomain is mapped to a (possibly virtual) processor. Additionally, a full Krylov step requires some vector operations and a residual calculation (a distributed matrix vector product), but this has the same cost in the two methods.

For simplicity, we consider the case where we have m^d subdomains each having n^d rectangular grid blocks in d dimensions ($d = 2, 3$). This means that we have $(n - 1)^d$ interior nodal points in each subdomain. We ignore the (setup) cost of forming the Schur complement described in the previous section. Table 1 shows that each subdomain problem is slightly smaller since we have no overlap in the new method, while the coarse problem has dimension m^d, an increase from $(m - 1)^d$ for the Additive Schwarz method. One should also note that each subdomain problem in the new methods have a constant coefficient ρ_i, while in the Additive Schwarz method there are jumps in the coefficients near the subdomain boundaries because of the overlap. This may make the subdomain problems easier to solve with our new approach. The two coarse problems have similar nonzero structure. A potentially more significant difference can be seen in the reduced cost of performing the interpolation and restriction (called 'transfer' in Table 1) as well as the reduced communication requirement. The term $2(n - 1)^2$ reflects the need to add all interior unknown values and later add the computed average value to all interior nodes, altogether two additions per nodal point. The term $20n$ estimates the work along the boundary of each subdomain according to the equations in the previous section and similarly $30n^2$ in 3 dimensions. The reduced communication reflects the fact that only one average value must be communicated (both ways) between the coarse solver and each subdomain. The Additive Schwarz algorithm has two phases of communication of similar size; after the subdomain solution and in the residual calculation. In the Additive Average method the first phase is very small and a full iteration of a parallel code will therefore have both the number of messages and the size of the messages reduced by almost 50 percent. We will describe actual parallel performance in a forthcoming report.

7 Numerical Results

We have carried out numerical experiments comparing the new Additive Average method and the Additive Diagonal Scaling method with the standard Additive Schwarz method. In all tests we consider both a quasi-monotone distribution and a distribution which is not quasi-monotone. The domain and the distributions are derived from Figure 1 by translation of the upper half in two dimensions, while the full 3-dimensional piece is translated in order to build the 3-dimensional examples. That is, the value of ρ is constant in each subdomain. The algorithms have been implemented as described, for convenience we use an (accurate) iterative solver for subdomain and coarse grid problems. We further used the constant 1.7 instead of h^{d-2} in (15) and (19) when $d = 2$ since this value produced slightly better results.

First consider Table 2. We compare the two distributions and a reference case where $\rho_i = 1$ in two dimensions. Note the distinct jump in the condition number for the Additive Diagonal Scaling method in the non quasi-monotone case, while the Additive Average method shows no such dependence. We note that the Additive Schwarz method works well for all cases, with a very low condition number for the quasi-monotone case considered in this example. The factor H/h changes by a factor 2 in the two parts of the table. We observe that both the Additive Average method and the Additive Diagonal Scaling method reflect this while the Additive Schwarz method only displays a clear H/h dependence in the non quasi-monotone case.

Table 2 Comparison of the methods in 2 dimensions. The number of iterations required to reduce the residual by 10^{-6} and a condition number estimate is listed.

	4×4 Subdomains, 4×4 Blocks			8×8 Subdomains, 8×8 Blocks		
Method	Average	Diagonal	Additive	Average	Diagonal	Additive
Overlap	0	0	2	0	0	2
$\rho = 1$	22 (11.2)	17 (6.41)	15 (5.34)	34 (25.4)	27 (13.4)	18 (7.64)
q-mon.	25 (11.6)	19 (6.71)	16 (4.68)	37 (25.8)	29 (13.7)	15 (4.64)
Not q-mon.	24 (11.7)	28 (18.0)	16 (5.11)	42 (25.7)	58 (50.8)	21 (9.88)

Table 3 Comparison of the methods in 3 dimensions. The number of iterations required to reduce the residual by 10^{-6} and a condition number estimate is listed.

	4^3 Subdomains, 4^3 Blocks			8^3 Subdomains, 8^3 Blocks		
Method	Average	Diagonal	Additive	Average	Diagonal	Additive
Overlap	0	0	2	0	0	2
$\rho = 1$	19 (7.95)	16 (5.47)	19 (10.4)	32 (21.3)	27 (13.5)	26 (18.7)
q-mon.	23 (8.46)	20 (6.29)	21 (9.13)	37 (21.6)	33 (15.1)	24 (11.8)
Not q-mon.	23 (8.47)	34 (53.2)	23 (12.8)	39 (21.9)	134 (351)	44 (51.9)

Table 3 contains a similar comparison in the 3 dimensional case. We observe qualitatively the same effects. The condition number estimates grow somewhat faster than the factor of two change in H/h, but the reader should note that the problem changes in both Tables 2 and 3, that is, the number of coefficient jumps corresponds to the number of subdomains.

The substantial increase in the condition number from a quasi-monotone case to a non quasi-monotone distribution is not always followed by a similar increase in computational work. As an example, using the Additive Diagonal Scaling method with 4^3 subdomains and 8^3 blocks per subdomain, the condition number changes from 52.9 to 237, while the required number of iterations only increases from 53 to 60. If we exclude the two smallest eigenvalues and compute the resulting 'effective condition numbers', we get 32.3 and 38.0 respectively, thus explaining the small change in iteration count.

In the non quasi-monotone case we note that the Additive Schwarz method has a factor 4 increase in its condition number (when H/h changes by a factor 2), while it performs very well in the other cases. A standard estimate for the required number of iterations thus predicts that the computational effort may increase by as much as a factor of two which is indeed the case here. The Tables show that the particular distribution of coefficients may have a significant influence on the actual rate of convergence of two of the algorithms and this effect is in agreement with the theoretical predictions.

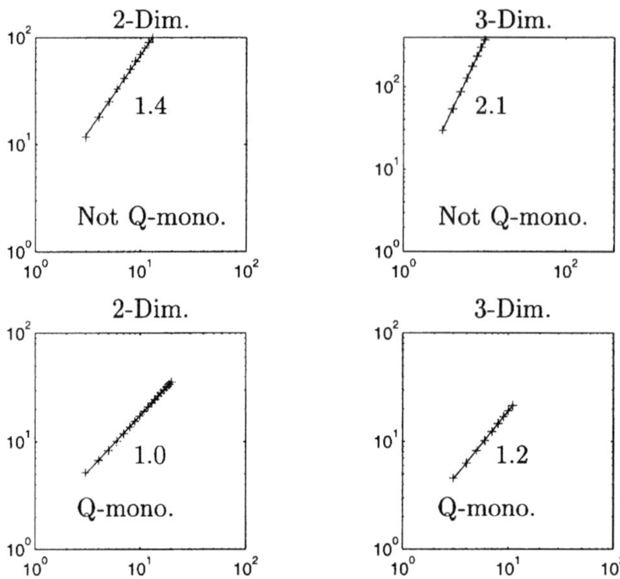

Figure 2   The Additive Diagonal Scaling method in 2 and 3 dimensions. The condition number of the iteration operator is plotted as a function of the ratio H/h. The least square fitted lines have slopes as indicated on the plots.

Figures 2 and 3 study the H/h behavior of the methods in more detail and compare this with the Additive Schwarz method in Figure 4. All computations are based on a domain subdivided into 4×4 subdomains in 2 dimensions and similarly into $4 \times 4 \times 4$ subdomains in 3 dimensions. We then change the ratio H/h by decreasing h. The log-log plots enable us to get an estimate for the slope β assuming that the condition number behaves like $(H/h)^\beta$. In Figure 2 we consider the Additive Diagonal Scaling method. The bottom part of the figure displays the quasi-monotone case and we observe a perfect slope of 1.0 in 2 dimensions, while the slope is near 1.2 in 3 dimensions. Both the actual condition numbers (crosses) and a line corresponding to a best least squares approximation are shown. In the non quasi-monotone case Theorem 2 predicts a slope near 2 in 3 dimensions and considerably better behavior in 2 dimensions, and we compute the values 2.1 and 1.4 respectively.

Figure 3 shows the same information for the Additive Average method. In this case there should be no sensitivity to the distributions and this is reflected fully in our computations. The slope is near 1.1 in 2 dimensions and near 1.2 in 3 dimensions. According to the theory presented the slopes should not exceed 1. The reason for this difference will be further investigated.

Finally, in Figure 4 we plot the behavior of the standard Additive Schwarz method with minimal overlap on the same test problems. In 2 dimensions our quasi-monotone problem converges with virtually no dependence on H/h. We therefore also include the case $\rho_i = 1$ (Poisson's equation) where the slope is about 0.7. The non quasi-monotone

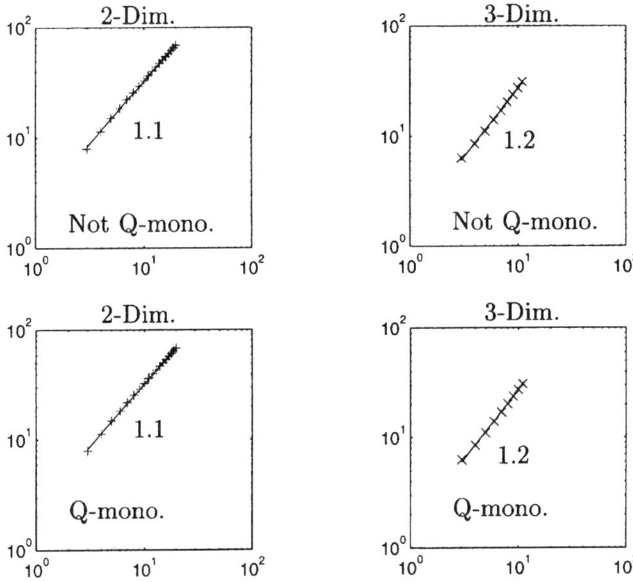

Figure 3   The Additive Average method in 2 and 3 dimensions. The condition number of the iteration operator is plotted as a function of the ratio H/h. The least square fitted lines have slopes as indicated on the plots.

case is also better than this having a slope of about 0.3. In 3 dimensions the overall picture looks similar, the non quasi-monotone case has a slope of 0.9, while the quasi-monotone distribution shows a slope of 0.3 (0.8 for a Poisson problem). This behavior is better than expected and different from the growth observed in Table 3. In that table the number of subdomains was allowed to grow (and therefore the number of jumps in the coefficients). The two cases are therefore not directly comparable, but a more detailed understanding of the Additive Schwarz method in this context is needed.

8   Conclusion

We have presented and analyzed two additive Schwarz methods, the Additive Average method and the Additive Diagonal Scaling method and compared them with the standard Additive Schwarz method with minimum overlap. Both methods behave well and are easy to implement. The Additive Average method has more flexibility since no regularity of the division into subdomains is required. We can also prove a satisfactory bound on the condition number for this method. The bound does not depend on assumptions of a quasi-monotone distribution of the coefficients. An increase in the condition number like H/h is predicted and this is often acceptable since a scaled up problem tend to be divided into more subdomains (keeping the ratio H/h bounded) in order to run efficiently on a scalable parallel computer.

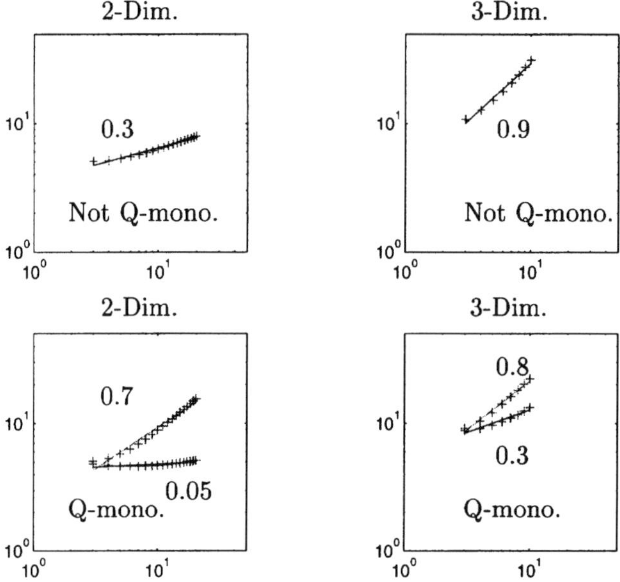

Figure 4   The Additive Schwarz method in 2 and 3 dimensions. The condition number of the iteration operator is plotted as a function of the ratio H/h. The least square fitted lines have slopes as indicated on the plots.

Our numerical experiments confirm the important difference that many methods exhibit when changing from a quasi-monotone to a non quasi-monotone distribution. Thus, this characterization of distributions where the weighted L^2 projection is stable is directly reflected in the behavior of some algorithms. However, it should be mentioned that a non quasi-monotone distribution is quite special. Both theory and computational experiments show that many methods are quite good even in the case where we only have a weak quasi-monotone distribution, that is, one can traverse through edges as well as faces in Figure 1. Additionally, there are cases where the growth in the condition number not necessarily implies a corresponding increase in the number of iterations.

The experiments further confirm that the Additive Average method is not affected by such a change in the distribution of the coefficients, but our estimates of the relevant condition number show a slightly faster growth with H/h than predicted by theory. The cause of this difference is at the present not fully determined and will be subject to further research.

Finally, it should be noted that the Additive Schwarz method shows better than predicted performance in both 2 and 3 dimensions on our test problems with discontinous coefficients. This observation has been reported earlier (see, for example [BS92]), but it is still not fully explained by theory.

References

[BK95] Bjørstad P. E. and Kårstad T. (1995) Domain decomposition, parallel computing and petroleum engineering. In Keyes D. E., Saad Y., and Truhlar D. G. (eds) *Domain-Based Parallelism and Problem Decomposition Methods in Computational Science and Engineering*, chapter 3, pages 39–56. SIAM.

[BS92] Bjørstad P. E. and Skogen M. (1992) Domain decomposition algorithms of Schwarz type, designed for massively parallel computers. In Keyes D. E., Chan T. F., Meurant G. A., Scroggs J. S., and Voigt R. G. (eds) *Fifth International Symposium on Domain Decomposition Methods for Partial Differential Equations*, pages 362–375. SIAM, Philadelphia, PA.

[BX91] Bramble J. H. and Xu J. (1991) Some estimates for a weighted L^2 projection. *Math. Comp.* 56: 463–476.

[DSW94] Dryja M., Sarkis M., and Widlund O. B. (March 1994) Multilevel Schwarz methods for elliptic problems with discontinuous coefficients in three dimensions. Technical Report 662, Department of Computer Science, Courant Institute. To appear in Numer. Math., 1995.

[Gri85] Grisvard P. (1985) *Elliptic Problems in Nonsmooth Domains*. Pitman, Boston.

[SBG95] Smith B. F., Bjørstad P., and Gropp W. (1995) *Domain Decomposition: Parallel Multilevel Methods for Elliptic Partial Differential Equations*. Cambridge University Press.

[Sko92] Skogen M. D. (February 1992) *Schwarz Methods and Parallelism*. PhD thesis, Department of Informatics, University of Bergen, Norway.

[Xu91] Xu J. (1991) Counter examples concerning a weighted L^2 projection. *Math. Comp.* 57: 563–568.

Multilevel domain decomposition and multigrid methods for unstructured meshes: algorithms and theory

Tony F. Chan, Susie Go [1] and Jun Zou [2]

We will summarize some of our recent theoretical and numerical results on domain decomposition and multigrid methods for second order elliptic problems on unstructured meshes. We first present a general framework for convergence analyses applicable to unstructured meshes, which can be viewed as a natural extension of the one formulated by Xu [Xu92] for structured meshes. Then this framework is applied to two level and multilevel Schwarz methods for elliptic problems on unstructured meshes. As we allow general coarse grids whose boundaries may be non-matching to the boundary of the fine grid, special treatments are needed to implement different types of boundary conditions. We will propose a couple of such treatments. Finally, numerical results for domain decomposition and multigrid methods on unstructured meshes are presented to show similar convergence properties as we expect for standard structured meshes.

---

[1] Dept. of Math., Univ. of Calif. at Los Angeles, Los Angeles, CA 90024-1555, USA. e-mail: chan@math.ucla.edu, sgo@math.ucla.edu.

[2] Dept. of Math., Univ. of Calif. at Los Angeles, Los Angeles, CA 90024-1555, USA, and Computing Center, the Chinese Academy of Sciences, Beijing 100080, P. R. China. e-mail: zou@math.ucla.edu, zou@math.cuhk.hk.

Domain Decomposition Methods in Sciences and Engineering, edited by R. Glowinski *et al.*

1 Introduction

Recently, unstructured finite element meshes have become very popular in scientific computing, cf. Barth [Bar92] and Mavriplis [Mav92], primarily because of their flexibility in adapting to complicated geometries and the resolution of fine scale structures in the solution. Since no natural coarser grids exist as in structured meshes, practical multilevel domain decomposition and multigrid algorithms must allow coarser grids which are non-quasi-uniform and with boundaries and interior elements which are not necessarily matching to that of the fine mesh. Therefore, the traditional solvers have to be modified so that their efficiency will not be adversely affected by the lack of structure.

In this paper, we first propose a general framework for convergence analyses applicable to unstructured meshes, which can be viewed as a natural extension of the one formulated by Xu [Xu92] for structured meshes. Then this framework is applied to two level and multilevel Schwarz methods for elliptic problems on unstructured meshes. Very general meshes and subdomains are allowed: neither the fine mesh nor the coarse mesh need to be quasi-uniform, the subdomains can be of arbitrary shapes and sizes, and the coarse mesh need not be nested to, or cover the same physical domain as the fine mesh. Some existing related works on unstructured meshes can be found in Chan, Smith and Zou[CZ94a, CSZ94, CZ95, CZ94b], Cai [Cai95], Bramble-Pasciak-Xu [BPX91], Douglas-Douglas [DD93], Bank-Xu [BX, BX94].

We subsequently describe how to create a coarse grid hierarchy by successive coarsening of a fine grid using a maximum independent set approach. As we allow completely non-matching coarse grids to fine grids, special treatments are needed for different types of boundary conditions. We will propose a couple of such treatments to ensure that a proper sequence of coarse subspaces exists for the domain decomposition or multigrid methods. Then we discuss how to implement interpolation operators from coarse grids to finest grids.

Finally, numerical experiments on domain decomposition and multigrid methods on unstructured meshes will be presented to demonstrate similar convergence properties as we expect for standard structured meshes.

The paper is arranged as follows: In Section 2, we formulate the general framework for the convergence analysis for the additive type domain decomposition methods, and a perturbation extension of the result will be presented afterwards for the application to non self-adjoint parabolic problems. In Section 3 we introduce the fine and coarse finite element spaces, domain decompositions, the L^2-optimal approximation and H^1-stability of the standard finite element interpolant and the Clément interpolant. Section 4 will be devoted to the application of the abstract convergence theory developed in Section 2 to second order elliptic problems. Section 5 discusses how to generate a coarse grid sequence and to implement boundary conditions for Neumann boundary part, and Section 6 shows some numerical experiments using domain decomposition and multigrid methods. The results in Sections 2-4 are a summary of some of the main results in Chan-Zou [CZ95].

2   Convergence theory for additive preconditioners

Let V, and V^k, $0 \le k \le p$ be finite dimensional vector spaces with inner products $(\cdot, \cdot)$ and $(\cdot, \cdot)_k$, resp. The spaces V^k are not necessarily subspaces of V. The space V^0 is special, usually referring to the coarse grid space.

Given a symmetric positive definite (SPD) operator A on V and $f \in V$, we are interested in solving the equation $Au = f$ on V, which arises from the discretization of elliptic or parabolic problems by using finite element methods. As A is ill-conditioned, our goal is to find a good preconditioner M for A such that MA is better conditioned than A, and the action of M is inexpensive to calculate. Then one can use iterative methods, like *Conjugate Gradient* method, for $MAu = Mf$ instead of $Au = f$.

We will study preconditioners of the following additive type:

$$M = \sum_{k=0}^{p} I_k R_k Q_k, \tag{2.1}$$

where the "interpolation" operators $I_k : V^k \to V$ are linear, and the "projection" operators Q_k are the adjoints of I_k defined by

$$(Q_k u, v_k)_k = (u, I_k v_k), \ \forall\, u \in V, \ \ v_k \in V^k, \tag{2.2}$$

and $R_k : V^k \to V^k$ are given SPD operators, approximating the inverses of the restrictions of A on V^k in some sense. It is easy to verify that M is an SPD operator on V.

We remark that the preconditioner form (2.1) is a natural extension of the one introduced by Xu [Xu92] with nested subspaces. The awareness of this general form (2.1) was due to [GO93], see also [CSZ94], [SBG95] and [Hol95].

Following the framework of Xu [Xu92] for structured meshes in which all V^k are subspaces of V, one can bound the condition number of MA for the present unstructured cases in terms of three parameters K_0, ω_0 and α_0 defined as follows:

(P1) For any $u \in V$, there exist $u_k \in V^k$ ($0 \le k \le p$) such that $u = \sum_{k=0}^{p} I_k u_k$ and

$$\sum_{k=0}^{p} (R_k^{-1} u_k, u_k)_k \le K_0 (Au, u).$$

(P2) For any $u_k \in V^k$, $k = 0, 1, \cdots, p$,

$$(AI_k u_k, I_k u_k) \le \omega_0 (R_k^{-1} u_k, u_k)_k.$$

(P3) For any $u \in V$ and $u_k \in V^k$ ($1 \le k \le p$),

$$\sum_{k=1}^{p} (Au, I_k u_k) \le \alpha_0^{\frac{1}{2}} (Au, u)^{\frac{1}{2}} \left(\sum_{k=1}^{p} (AI_k u_k, I_k u_k) \right)^{\frac{1}{2}}.$$

We have the following bound for the condition number of $\kappa(MA)$ (see [CZ95] for the proof):

Theorem 1 *Under the assumptions* **(P1)** *-* **(P3)**, $\kappa(MA) \leq \omega_0(\alpha_0 + 1)K_0$.

Remark 2.1 **(P1)** *and* **(P2)** *are natural extensions of assumptions from the theory for structured meshes by Xu [Xu92], where all the spaces V^k, $0 \leq k \leq p$, are assumed to be subspaces of V.* **(P1)** *means that any function in V can be decomposed into a sum of functions in spaces V^k and this partition is stable with the perturbed "energy" norm in some sense.* **(P2)** *is equivalent to $\lambda_{\max}(R_k A_k) \leq \omega_0$ where $A_k = Q_k A I_k$ is the "restriction" of A on V^k, it means that the approximation of R_k to the inverse of A_k cannot be "too bad".* **(P3)** *is a condition on the "local" properties of V^k $(1 \leq k \leq p)$ in some sense, i.e. the image spaces $I_k V_k$ of V^k under the mapping I_k cannot overlap one another too much in the fine space V.*

Note that our **(P3)** *is not the extension of the corresponding assumption used in [Xu92]. It might be replaced by the extension of the so-called strengthened Cauchy-Schwarz inequality in [Xu92] for nested subspaces with identity operators I_i $(1 \leq i \leq p)$:*

(P3*) *Let $\varepsilon_{ij} \in (0, 1]$ be the smallest constants that satisfy*

$$(AI_i u_i, I_j u_j) \leq \varepsilon_{ij}(AI_i u_i, I_i u_i)^{\frac{1}{2}}(AI_j u_j, I_j u_j)^{\frac{1}{2}}, \quad \forall u_i \in V^i, u_j \in V^j, i, j = 1, \cdots, p.$$

It is easy to verify that **(P1)**, **(P2)** *and* **(P3*)** *imply* **(P3)**. *Thus* **(P3)** *is a weaker assumption than* **(P3*)**. *We prefer* **(P3)** *to* **(P3*)** *as* **(P3)** *is more convenient to check than* **(P3*)** *for the non-nested subspaces.*

2.1 Multilevel additive preconditioners for SPD operators

Let V and V^l $(0 \leq l \leq L)$ be defined as in Section 2, and furthermore, we assume that for each $l : 1 \leq l \leq L$, the space V^l can be decomposed into a sum of subspaces V_k^l $(1 \leq k \leq N_l)$. Then the multilevel additive preconditioners for the given SPD operator A is defined as follows

$$M = \sum_{l=0}^{L} \sum_{k=1}^{N_l} I_k^l R_k^l Q_k^l \tag{2.3}$$

where the "interpolation" operators $I_k^l : V_k^l \to V$ are linear, and the "projection" operators Q_k^l are the adjoints of I_k^l defined by

$$(Q_k^l u, v_k^l)_l = (u, I_k^l v_k^l), \ \forall u \in V, \ v_k^l \in V_k^l, \tag{2.4}$$

and $R_k^l : V_k^l \to V_k^l$ are given SPD operators, approximating the inverses of the restrictions of A on V_k^l in some sense. It is easy to verify that M is an SPD operator on V. Note that for $l = 0$, we adopt the notation $N_0 = 1, I_k^0 = I^0, Q_k^0 = Q^0, R_k^0 = R^0$.

As in the last section, the condition number of MA can be bounded in terms of three parameters K_0, ω_0 and α_0 defined as follows:

(P1$'$) For any $u \in V$, there exist $u_k^l \in V_k^l$ ($0 \leq l \leq L$, $1 \leq k \leq N_l$) such that $u = \sum_{l=0}^{L} \sum_{k=1}^{N_l} I_k^l u_k^l$ and

$$\sum_{l=0}^{L} \sum_{k=1}^{N_l} ((R_k^l)^{-1} u_k^l, u_k^l)_k \leq K_0(Au, u).$$

(P2$'$) For any $u_k^l \in V_k^l$, $0 \leq l \leq L$, $1 \leq k \leq N_l$,

$$(A I_k^l u_k^l, I_k^l u_k^l) \leq \omega_0((R_k^l)^{-1} u_k^l, u_k^l)_l.$$

(P3$'$) For any $u \in V$ and $u_k^l \in V_k^l$, $0 \leq l \leq L$, $1 \leq k \leq N_l$,

$$\sum_{l=0}^{L} \sum_{k=1}^{N_l} (Au, I_k^l u_k^l) \leq \alpha_0^{\frac{1}{2}} (Au, u)^{\frac{1}{2}} \left(\sum_{l=0}^{L} \sum_{k=1}^{N_l} (A I_k^l u_k^l, I_k^l u_k^l) \right)^{\frac{1}{2}}.$$

Similar to Theorem 1, we have

Theorem 2 *Under the assumptions* **(P1$'$)** *-* **(P3$'$)**, $\kappa(MA) \leq \omega_0(\alpha_0 + 1)K_0$.

2.2 *Additive preconditioners for small perturbations of SPD operators*

The results of this section are applicable to general non-symmetric parabolic problems, cf. [CZ94b, CZ95]. Let V be a finite dimensional space with the scalar product $(\cdot, \cdot)$, and E a non-symmetric operator on V which is a small perturbation of the SPD operator A, that is, $E = A + B$, and we solve the equation $Eu \equiv (A + B)u = f$ on V. Our goal is to find a good preconditioner M for the non-symmetric operator E. Then we can use iterative methods, like *GMRES* or *BiCGSTAB*, to solve $MEu = Mf$ instead of $Eu = f$. Let us consider the *GMRES* method. It is known (cf. [EES83]) that the convergence rate of *GMRES* depends on the following two parameters:

$$\beta_1 = \min_{u \neq 0} \frac{(u, MEu)_A}{(u, u)_A}, \quad \beta_2 = \max_{u \neq 0} \frac{\|MEu\|_A}{\|u\|_A}. \tag{2.5}$$

If $\beta_1 > 0$, *GMRES* converges, and at the mth iteration the residual is bounded as (cf. [EES83])

$$\|Mf - MEu^m\|_A \leq \left(1 - \frac{\beta_1^2}{\beta_2^2} \right)^{m/2} \|Mf - MEu^0\|_A.$$

Let the spaces V^k ($1 \leq k \leq p$), the scalar products $(\cdot, \cdot)_k$, linear operators $I_k : V^k \rightarrow V$, the adjoints Q_k of I_k and the SPD operators $R_k : V^k \rightarrow V^k$ be defined as in Section 2. Then we define the preconditioner M as in (2.1) by $M = \sum_{k=0}^{p} I_k R_k Q_k$ for operator E. Note that we still use an SPD preconditioner M even though E is non-symmetric.

We introduce two assumptions for the perturbation operator B:

(P4) For any $u \in V$ and $u_k \in V^k$, $1 \leq k \leq p$,

$$\sum_{k=1}^{p}(Bu, I_k u_k) \leq \alpha_1^{\frac{1}{2}}(Au, u)^{\frac{1}{2}}\left(\sum_{k=1}^{p}(AI_k u_k, I_k u_k)\right)^{\frac{1}{2}}.$$

(P5) There exists a constant $\mu_1 \in (0, 1)$ such that for any $u, v \in V$,

$$|(Bu, v)| \leq \mu_1 \|u\|_A \|v\|_A.$$

We have the following estimates about two parameters β_1 and β_2 which determine the convergence rate of *GMRES* iteration for solving $MEu = Mf$ (see [CZ95] for the proof):

Theorem 3 *If in addition to* **(P1)** *-* **(P5)***, we assume further that*

$$\mu_1^2 + \alpha_1 \leq \frac{(1 - \mu_1)^2}{2\omega_0 K_0}, \tag{2.6}$$

then we have

$$\beta_1 = \min_{u \neq 0}\frac{(u, ME)_A}{(u, u)_A} \geq \frac{(1 - \mu_1)^2}{4K_0}, \quad \beta_2 = \max_{u \neq 0}\frac{\|MEu\|_A}{\|u\|_A} \leq 2\omega_0 \alpha_0^{\frac{1}{2}}(1 + \alpha_0 + \alpha_1 + \mu_1^2)^{\frac{1}{2}}.$$

3　Finite elements and domain decompositions

For an open bounded domain Ω in $\subset R^d$ ($d = 2, 3$), suppose we are given a family of shape regular (not necessarily quasi-uniform) triangulations $\{\mathcal{T}^h\}$ on Ω, consisting of simplices. We will not discuss the effects of approximating Ω but always assume that the triangulations $\{\mathcal{T}^h\}$ of Ω are exact, i.e., Ω is either a polygon or a polyhedron, and $\Omega = \Omega^h \equiv \cup_{\mathcal{T}^h \in \mathcal{T}^h}\mathcal{T}^h$. Let V^h be a piecewise linear finite element subspace of $H_0^1(\Omega)$ defined on $\mathcal{T}^h$.

Decompose the domain Ω into p non-overlapping subdomains $\tilde{\Omega}^k$ ($1 \leq k \leq p$), then extend each $\tilde{\Omega}^k$ to a larger one Ω^k such that the distance between $\partial\Omega^k$ and $\partial\tilde{\Omega}^k$ is bounded from below by $\delta_k > 0$. We allow each Ω^k to be of quite different size and of quite different shape from other subdomains. Then we define the subspaces $\{V^k\}_{k=1}^{p}$ of V^h corresponding to the subdomains $\{\Omega^k\}_{k=1}^{p}$ by $V^k = \{v \in V^h; \ v = 0 \ \ \text{on} \ \ \Omega\backslash\Omega^k\}$.

We introduce also a coarse grid $\mathcal{T}^H$ which forms a σ_0-shape regular triangulation of Ω, but has nothing to do with $\mathcal{T}^h$, i.e., none of the nodes of $\mathcal{T}^H$ need to be nodes of $\mathcal{T}^h$. Let Ω^0 be the coarse grid domain, i.e. $\Omega^0 = \cup_{\mathcal{T}^H \in \mathcal{T}^H}\mathcal{T}^H$.

Denote by V^0 (resp. $\hat{V}^0$) the subspace of $H_0^1(\Omega^0)$ (resp. $H^1(\Omega^0)$) consisting of piecewise linear functions defined on $\mathcal{T}^H$. Note that Ω^0 usually does not match with Ω, and $V^0 \not\subset V^h$.

In addition, we need to impose a few reasonable assumptions on the coarse grid Ω^0. Roughly speaking, we assume that for each coarse element τ^H, all its neighboring fine elements having non-empty intersections with τ^H form a subregion whose measure can be bounded by a constant times the one of τ^H; the coarse grid part outside the fine grid is of the fine element sizes while the fine grid part outside the coarse grid is of the coarse element sizes. See detailed assumptions in [CZ95].

3.1 H^1-stability and L^2-optimal approximation of linear interpolants

As the coarse space V^0 is non-nested to the fine space V^h for our interest, the convergence proof for the domain decomposition methods requires the existence of an operator I_0 mapping the non-nested coarse space V^0 to a nested subspace of V^h satisfying the following H^1-stability and L^2-optimal approximation properties: for any coarse grid function $u \in V^0$,

$$|I_0 u|_{1,\Omega} \leq C|u|_{1,\Omega^0}, \quad \|u - I_0 u\|_{0,\Omega} \leq Ch|u|_{1,\Omega^0}$$

where Ω^0 is the coarse grid domain. More specifically, we require these two properties to hold locally in order to deal with general unstructured meshes.

There exist many options for such grid-transfer operators. But as this grid-transfer operator I_0 enters the algorithm, we want it to be as simple as possible.

Standard and modified finite element interpolants The simplest one is obviously the standard finite element interpolant Π_h corresponding to the fine space V^h. To make $I_0 V^0$ a subspace of the fine space V^h, we need some special treatment on the part of the coarse grid boundary close to the fine grid boundary part which assumes Neumann boundary conditions. If one has pure Dirichlet boundary conditions, then the zero extension operator outside of the coarse grid domain should be the most natural and also effective option, cf. [CS94, CSZ94]. If one has other types of boundary conditions, one choice is to require that the coarse grid covers the fine grid Neumann boundary part, then Π_h is well-defined everywhere in the fine grid domain, cf. [CSZ94, CZ95, CZ94b]. It is shown that this interpolant has the required two properties.

Another choice is to define the operator I_0 in the interior part of the coarse grid domain by the standard interpolant Π_h but outside of the coarse grid domain by other simple linear interpolation, see Section 5 for more details.

Clément's interpolant For the proof of the convergence of domain decomposition methods, or more specifically for the verification of the partition assumption **(P1)** in Theorem 1, we need another grid-transfer operator R_H mapping the fine space V^h to the coarse space V^0 and R_H must also possess the two properties of the H^1-stability and L^2-optimal approximation: for any fine grid function $u \in V^h$,

$$|R_H u|_{1,\Omega^0} \leq C|u|_{1,\Omega}, \quad \|u - R_H u\|_{0,\Omega^0} \leq C H |u|_{1,\Omega}.$$

However, as we are dealing with unstructured meshes which can be non-quasi-uniform, R_H should be defined completely locally. We know the standard finite element interpolant corresponding to the coarse space V^0 is defined locally, but it does not possess the required two properties. Two satisfactory operators are Clément interpolant [Cle75] and Scott-Zhang interpolant [SZ90], cf. [CSZ94, CZ95, CZ94b].

4   Two level and multilevel additive Schwarz method for elliptic problems

In this section, we apply the general theory of Section 2 to the following second order elliptic problems:

$$-\sum_{i,j=1}^{d} \frac{\partial}{\partial x_i}\left(a_{ij}\frac{\partial u}{\partial x_j}\right) + b\,u = f, \quad \text{in } \Omega \tag{4.1}$$

with some mixed boundary conditions on $\partial\Omega$. Here $\Omega \subset R^d$ ($d = 2, 3$) as described in Section 3, $(a_{ij}(x))$ is symmetric, uniformly positive definite, and $b(x) \geq 0$ in Ω.

The weak formulation of the above problem is: Find $u \in H_0^1(\Omega)$ such that

$$A_\Omega(u, v) = (f, v), \ \forall\, v \in H_0^1(\Omega)$$

with

$$A_\Omega(u, v) = \int_\Omega \left(\sum_{i,j=1}^{d} a_{ij} \frac{\partial u}{\partial x_j} \frac{\partial v}{\partial x_i} + b\,uv \right) dx.$$

The finite element problem is: Find $u \in V^h$ such that

$$A_\Omega(u, v) = (f, v), \ \forall\, v \in V^h. \tag{4.2}$$

4.1   Two level additive Schwarz method

Based on the local finite element spaces V^k ($1 \leq k \leq p$) and the coarse space $V^0 = V^H$ defined in Section 3, Schwarz methods are preconditioning for the linear system (4.2) that are built using local and coarse grid solves.

We define scalar products $(\cdot, \cdot)_k = (\cdot, \cdot)_{0,\Omega^k}$ on V^k for $1 \leq k \leq p$ and $(\cdot, \cdot) = (\cdot, \cdot)_{0,\Omega}$ on V^h, and then define an SPD operator A on V^h and a coarse operator A_0 by

$$(Au, v) = A_\Omega(u, v), \ \forall\, u, v \in V^h; \quad (A_0 u, v)_0 = A_{\Omega^0}(u, v), \ \forall\, u, v \in V^0$$

and local operators A_k, $1 \leq k \leq p$ by

$$(A_k u, v)_k = A_{\Omega^k}(u, v), \ \forall\, u, v \in V^k.$$

Since $V^k \subset V^h$, $1 \leq k \leq p$, we define $I_k : V^k \to V^h$ to be the natural injection operator. Note that $V^0 \not\subset V^h$. Define $I_0 : V^0 \to V^h$ to be the standard interpolant Π_h discussed in Section 3.1. One may also use other choices of I_0, e.g., the Clément interpolant R_h^0. Choose the local solvers R_k, $0 \leq k \leq p$ to be exact solvers, i.e. $R_k^{-1} = A_k$. Then the preconditioner M in (2.1) for A is: $M = \sum_{k=0}^{p} I_k A_k^{-1} Q_k$.

The following theorem gives the bound of the condition number $\kappa(MA)$. These results were proved in our previous work [CZ94a, CSZ94], but only global bounds were obtained there.

Theorem 4 *Under the assumptions* **(A1)** *-* **(A6)**, *we have*

$$\kappa(MA) \leq C \max_{1 \leq k \leq p} \frac{H_k^2}{\delta_k^2}.$$

Outline of the proof. By Theorem 1, it suffices to verify the three assumptions **(P1)**, **(P2)** and **(P3)**. **(P2)** is a direct consequence of the Cauchy-Schwarz inequality and the two properties of H^1-stability and the L^2-optimal approximation for the interpolants I_k discussed in Section 3.1, which gives the bound $\omega_0 = O(1)$. **(P3)** follows easily from the Cauchy-Schwarz inequality and the assumption that any point in Ω belongs to only a fixed number of subdomains, which gives us a bound $\alpha_0 = O(1)$. For **(P1)**, we use the Clément interpolant R_H and the partition of unity associated with the subdomains to define the required partition. Then **(P1)** can be proved using the two properties of H^1-stability and the L^2-optimal approximation for the interpolants I_k and the Clément interpolant R_H, giving a bound $K_0 = \max_k H_k^2/\delta_k^2$.

4.2 Multilevel additive Schwarz method for elliptic problems

Let $\Omega \subset R^d$ $(d = 2, 3)$ be a convex polygonal or polyhedral domain. Consider the same elliptic problem as defined in (4.1) and its finite element discretization (4.2). We will construct multilevel additive type preconditioners for the finite element system.

Let $\{\mathcal{T}^l\}_{l=0}^L$ be a not necessarily nested sequence of shape regular triangulations on Ω with h_l the maximum diameter of all elements in $\mathcal{T}^l$. $\mathcal{T}^L = \mathcal{T}^h$ is the finest triangulation on which the finite element space V^h and in return the finite element problem are defined. Denote the coarser domains corresponding to the coarser triangulations $\mathcal{T}^l$, $0 \leq l \leq L - 1$ by Ω^l.

For $0 \leq l < L$, let $V^l \subset H^1(\Omega^l)$ be the piecewise linear finite element space defined on $\mathcal{T}^l$ with proper boundary conditions imposed.

Assume that for each level $l = 0, 1, \cdots, L$, $\{\Omega_k^l\}_{k=1}^{N_l}$ is an overlapping domain decomposition of Ω^l, obtained by extending a given non-overlapping subdomain covering $\{\tilde{\Omega}_k^l\}_{k=1}^{N_l}$ of Ω^l such that $\text{dist}(\partial\tilde{\Omega}_k^l, \partial\Omega_k^l \cap \Omega^l) \geq \delta_l > 0$, $1 \leq k \leq N_l$. δ_l is called the l-th level overlapping ratio. Here the boundaries of the subdomains Ω_k^l are required to align with the boundaries of the l-th level elements in $\mathcal{T}^l$. We also impose some assumptions on each coarse grid similar to the ones discussed in Section 3, see [CZ95] for details.

For each coarser space V^l $(1 \leq l \leq p)$, we define a mapping $I_l : V^l \to V^h$ to be the standard finite element interpolant Π_h with proper modifications on boundary nodes, and for each subdomain Ω_k^l on l-th level, define a local subspace by

$$V_k^l = \{v \in V^l; \quad v = 0 \quad \text{on} \quad \partial\Omega_k^l \cap \Omega^l\} \subset V^l$$

and a prolongation operator $I_k^l : V_k^l \to V^h$ to be I_l, but I_k^l's adjoint $Q_k^l : V^h \to V_k^l$ by

$$(Q_k^l u, v_k^l)_l = (u, I_k^l v_k^l), \ \forall\, u \in V^h, \ v_k^h \in V_k^l$$

where $(\cdot, \cdot)_l = (\cdot, \cdot)_{0,\Omega^l}$ is the scalar product in $L^2(\Omega^l)$.

Furthermore, we define local operators $A_k^l : V_k^l \to V_k^l$ by

$$(A_k^l u, v)_l = A_{\Omega_k^l}(u, v), \ \forall\, u, v \in V_k^l,$$

and let $R_k^l = (A_k^l)^{-1}$ for simplicity of exposition, then we may construct the additive Schwarz preconditioner as in (2.3) by

$$M = \sum_{l=0}^L \sum_{k=1}^{N_l} I_k^l R_k^l Q_k^l \equiv \sum_{l=0}^L \sum_{k=1}^{N_l} I_k^l (A_k^l)^{-1} Q_k^l.$$

For the condition number $\kappa(MA)$, we have

Theorem 5 *Under the assumptions* **(H1)** *-* **(H4)**, $\kappa(MA) \lesssim \rho^2 L^2$, *where* $\rho = \max_{1 \leq l \leq L}(h_l + h_{l-1})/\delta_l$.

Outline of the proof. By Theorem 2, it suffices to verify the three assumptions **(P1$'$)**, **(P2$'$)** and **(P3$'$)**. To this aim, we need the two properties of the H^1-stability and L^2-optimal approximation for the interpolants I_k and the Clément interpolant discussed in Section 3.1; and moreover, we need also the H^1-stability and L^2-stability of the interpolants I_k.

To define the proper partition in **(P1$'$)**, we use the orthogonal projections P^l : $H^1(\Omega) \to I_l V^l$ defined by

$$A_\Omega(P^l u, v) = A_\Omega(u, v), \ \forall\, u \in H^1(\Omega), v \in I_l V^l,$$

and the following property of projections P^l which can be proved by using the Aubin-Nitsche trick, Sobolev extension theorem and Clément's interpolant: for $0 \leq l < L$,

$$\|v - P^l v\|_{0,\Omega} \lesssim h_l \|v\|_{1,\Omega}, \ \forall\, v \in V^h.$$

Using these results, we can show that $K_0 = O(\rho^2 L)$, $\omega_0 = O(1)$ and $\alpha_0 = O(L)$.

Remark 4.1 *The condition number bound given in Theorem 5 grows like* L^2. *It is known that in the structured case, one can remove this dependence on* L *and obtain optimal condition number (cf. Zhang [Zha92] and Oswald [Osw92]). At this point, we do not know how to obtain a similar optimal bound for our unstructured case.*

4.3   *Additive Schwarz method for non-symmetric parabolic problems*

The abstract framework of Section 2.2 can be applied to general non-symmetric second order parabolic problems to obtain similar optimal convergence results as in structured meshes. We remark that symmetric positive definite solvers can be used both for local subproblems and for the global coarse problem. We refer to [CZ94b, CZ95] for more details.

5   Treatment of Neumann boundary conditions

Practical multilevel algorithms using unstructured grids require some method to produce the coarse grid hierarchy (since it not naturally obtained from the fine grid problem), along with the associated interpolation and restriction operators. In [CS94], we generate a sequence of coarse grids by recursively coarsening a fine, unstructured grid using a maximal independent set approach [Gui93]. We observed that the performance of multilevel methods using grids generated by this method performed as well as standard multilevel methods on a structured mesh, but the performance of the methods deteriorated considerably when a mixed boundary condition was used instead of a purely Dirichlet boundary condition.

The convergence rate proofs in [CSZ94] for domain decomposition methods with non-matching grids using interpolations with zero extension required the assumptions that the coarse grid covers all of the Neumann boundary and that no coarse grid element lies completely outside the fine grid. The motivation for this was that with zero extension outside the coarse grid, corrections were being improperly made for the parts of the fine grid which lie outside the coarse domains. While this would not have a serious effect for problems with Dirichlet boundary conditions, it would slow down the method for problems with Neumann boundary conditions. Since the maximal independent set approach for grid coarsening generally creates coarse grid domains which are subsets of the fine grid, this suggests that the observed deterioration in the rate of convergence for mixed boundary condition problems may be remedied by either creating coarse grid domains which completely cover the fine grid domain or improving the transfer matrices used in the multilevel methods to get from one level to the next. We will discuss these two approaches next.

5.1 Modifying coarse grid boundaries

A sequence of coarse grids is generated by finding the maximal independent set of boundary nodes by eliminating every other boundary node and then finding the maximal independent set of the interior nodes. The resulting vertex set is then triangulated using a triangulation algorithm. Since this approach for grid coarsening generally creates coarse grid domains which do not cover the fine grid domain, we modify the coarse grids so that they do not violate this condition. In our implementation, we physically move the boundary nodes of the coarse grids in a systematic way so that if a fine grid node to be eliminated is exterior to the coarse grid boundary, the positions of one or more nearby coarse boundary nodes are adjusted so that the fine grid node will be interior to the new coarse grid boundary (see Figure 1). For our purposes, the boundary adjustment algorithm need only be applied to the edges where a Neumann boundary condition occurs and is not necessary for edges with a Dirichlet boundary condition. For practical purposes however, we applied the boundary adjustment algorithm to all edges regardless of the boundary condition.

5.2 Approximate interpolation/restriction operators

An alternative to modifying the coarse grid domains is to instead improve the interpolation and restriction operators used to transfer information between different levels. The interpolation matrices used in [CS94] were formed by taking each fine node and searching for the coarse grid element in which it lies, then interpolating with the coarse nodes which make up that element. If no such coarse grid element can be found, then zero weights were set for all coarse nodes. This results in a zero extension for all fine nodes exterior to the coarse domain. The restriction matrices were taken to be the transposes of these interpolation matrices.

Instead of zero extension for fine nodes exterior to the coarse grid domain, we have modified the current interpolation matrices for these points by selecting the nearest coarse boundary edge and interpolating with the two coarse nodes which make up

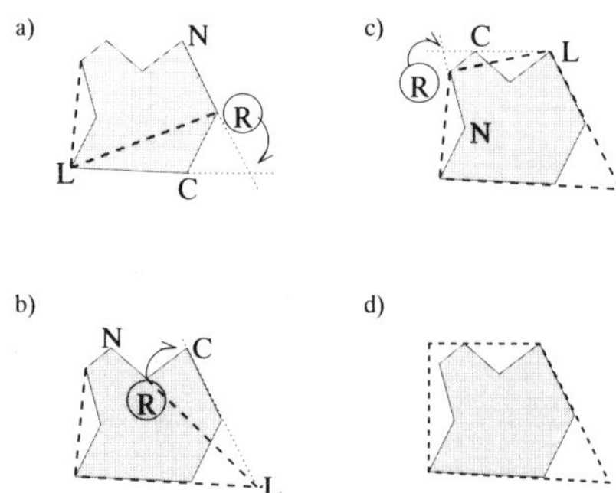

Figure 1   Modifying the coarse grid boundaries. Shaded region is the fine grid
domain, dashed line is the coarse grid boundary, dotted lines show coarse grid
boundary adjustment. L and R are coarse boundary nodes, while C is the fine grid
boundary node to be eliminated. Modify coarse edges by moving R.

that edge. Values at these exterior fine nodes were approximated with the value at the
nearest point on this coarse boundary edge (see Figure 2). This idea was motivated
by Bank and Xu's [BX94] coarsening algorithm for hierarchical basis methods.

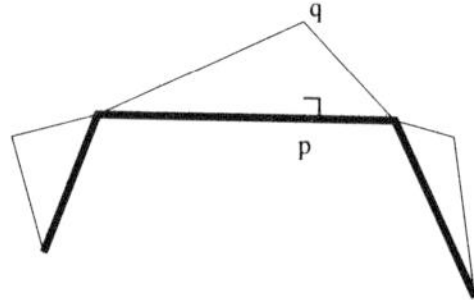

Figure 2   Values on a fine grid point outside the coarse domain, q, are
approximated with the nearest point, p, on the nearest coarse boundary edge
(thick line).

6   Numerical results

In this section, we provide some numerical results of domain decomposition and
multigrid experiments on unstructured grids for the Poisson equation with the airfoil
mesh (from T. Barth and D. Jesperson of NASA Ames) shown in Figure 3 as our fine
grid domain. All numerical experiments were performed using the Portable, Extensible
Toolkit for Scientific Computation (PETSc) of Gropp and Smith [GS], running on a
Sun SPARC 20. Piecewise linear finite elements were used for the discretizations and
the resulting linear system was solved using either two-level overlapping Schwarz or
multigrid preconditioning with full $GMRES$ as an outer accelerator. We compared two

different triangulation algorithms: Cavendish [Cav74] and Baker [Bak94] algorithms. The coarse grid hierarchy of the airfoil mesh triangulated with Baker's algorithm is shown in Figure 4 where G^2 refers to the first coarsening of the fine grid, G^1 is the coarsening of G^2, and G^0 is the coarsening of the G^1.

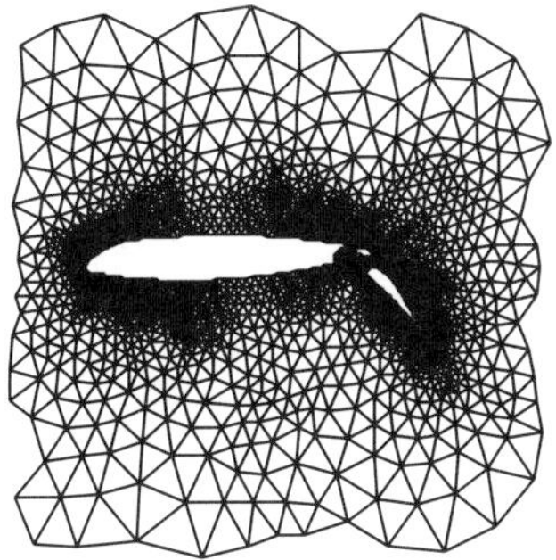

Figure 3 The *airfoil* grid with 4253 unknowns.

In our first experiment, we solve a mildly varying coefficient problem:

$$\frac{\partial}{\partial x}((1+xy)\frac{\partial u}{\partial x}) + \frac{\partial}{\partial y}((1+\sin(4x+4y)\frac{\partial u}{\partial y}) = x^2\sin(3y)$$

with either a purely Dirichlet boundary condition or a mixed boundary condition: Dirichlet for $x \leq 0.2$ and homogeneous Neumann for $x > 0.2$. For this problem, the Dirichlet condition was $u = x^2\sin(4y)$.

We solved this problem using two-level additive and multiplicative Schwarz preconditioning with the fine grid domain partitioned into 32 subdomains using the Recursive Spectral Bisection method as in [CS94]. The initial iterate is set to be zero and the iteration is stopped when the discrete norm of the residual is reduced by a factor of 10^{-5}.

The results are summarized in Table 1. For these results, the coarse grids used were triangulated using Cavendish's algorithm. The column labeled "unmodified boundaries" shows the number of iterations until convergence for coarse grid domains which do not cover the fine domain with both non-zero extension of fine nodes in $G^{k+1}\backslash G^k$, and zero extension (in parentheses). The column labeled "modified boundaries" shows the results for coarsening with coarse grid domains covering fine grid domains.

The second experiments show the results for the Poisson equation using multigrid preconditioning. The same two kinds of boundary conditions were used, but with a homogeneous Dirichlet condition instead. A V-cycle multigrid method with pointwise Gauss-Seidel smoothing and 2 pre and 2 post smoothings per level was used with the same initial iterate and the stopping criterion was reduced to 10^{-6}. The results of the multigrid experiments on the airfoil grid are summarized in Table 2, with both Cavendish triangulated coarse grids and Baker triangulated grids. The method of triangulation seemed to have little effect on the convergence rates.

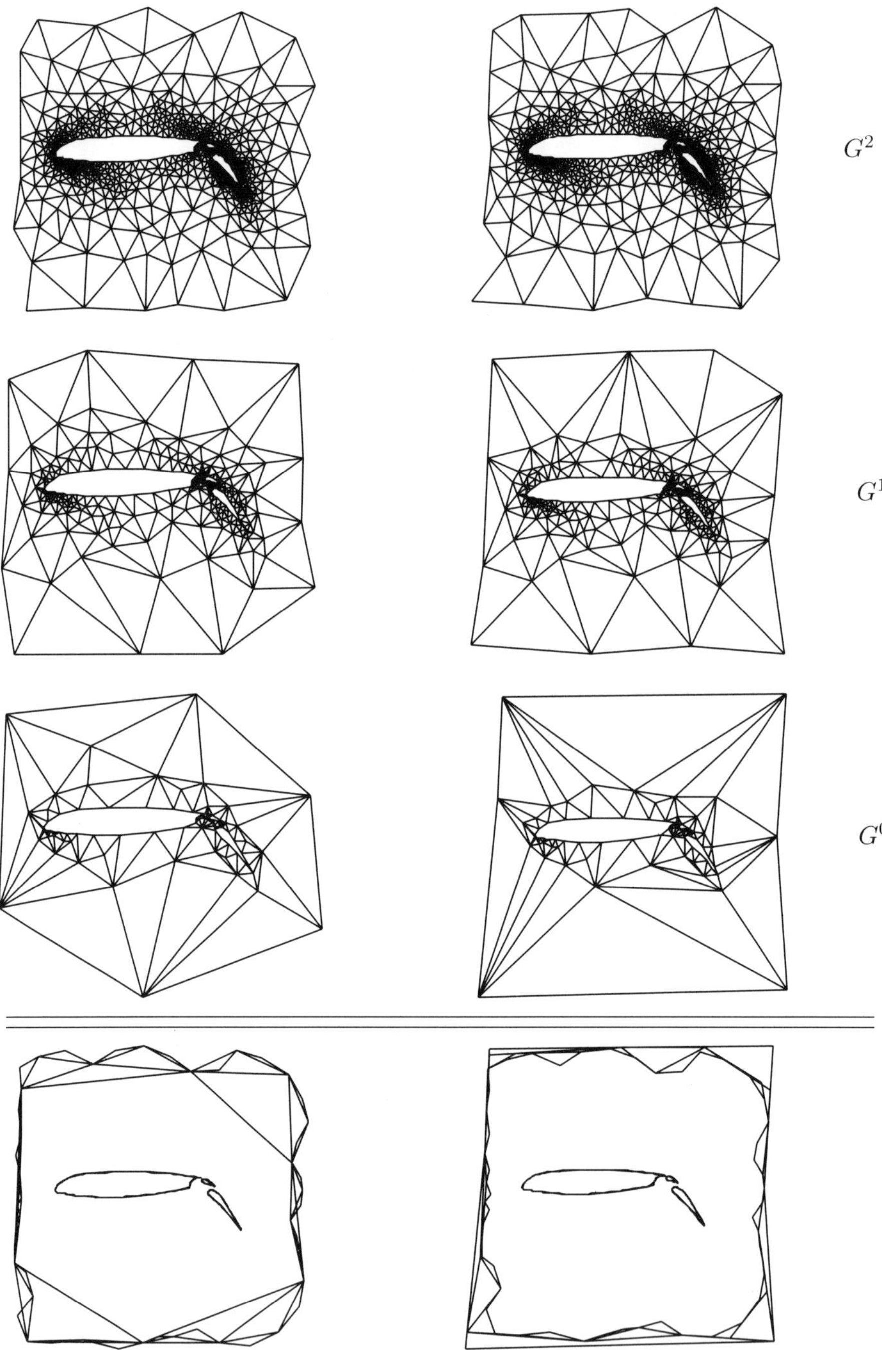

G^2

G^1

G^0

Figure 4 *Airfoil* grid hierarchy with unmodified boundaries (left) and modified boundaries (right).

Table 1 Multiplicative DD iterations for the *airfoil* grid with 4253 unknowns. *
indicates identical results since no coarse grid was used.

Overlap (# elements)	Coarse grid	Unmodified Boundaries		Modified Boundaries	
		Dir BC	Mixed BC	Dir BC	Mixed BC
0	None	23 (*)	73 (*)	*	*
	G^2	9 (9)	9 (17)	9	10
	G^1	15 (12)	17 (24)	15	17
	G^0	20 (20)	21 (33)	19	24
1	None	13 (*)	39 (*)	*	*
	G^2	6 (6)	7 (9)	7	8
	G^1	9 (9)	10 (16)	10	10
	G^0	12 (12)	13 (19)	12	15
2	None	10 (*)	30 (*)	*	*
	G^2	5 (5)	6 (9)	6	7
	G^1	8 (7)	8 (12)	8	8
	G^0	9 (9)	11 (15)	9	13

In addition to the airfoil domain, we ran experiments on an annulus domain [CS94] for comparison. Multigrid results on the annulus grid are summarized in Tables 3–5.

The numerical results show the significance of the assumption that when interpolations with zero extension are used, the coarse grid must cover the Neumann boundary of the fine grid problem; when the coarse grid domains do not cover the Neumann boundary, the convergence rates deteriorate noticeably.

The multigrid experiments on the annulus grid show that in both approaches used to treat Neumann boundary conditions, we obtained methods which were nearly mesh-size independent.

The zero extension transfer operator is the simplest to implement and effective for Dirichlet boundary conditions. If there are Neumann boundaries, then the improved interpolation approach seems to work well and is not too difficult to implement. It can also be used with independently generated coarse grids. The modified boundary approach is equally effective but can be less robust.

Acknowledgments: We would like to thank Barry Smith for his suggestions and patient guidance in using the PETSc software.

This work was partially supported by the NSF under contract ASC 92-01266, ONR under contract ONR-N00014-92-J-1890, and the Army Research Office under contract DAAL-03-91-C-0047 (Univ. Tenn. subcontract ORA4466.04 Amendment 1 and 2).

Table 2   MG iterations for the *airfoil* grid with 4253 unknowns.

Cavendish Triangulation Results

MG levels	Unmodified Boundaries			Modified Boundaries		
	nodes	Dir BC	Mixed BC	nodes	Dir BC	Mixed BC
2	1170	5 (5)	6 (8)	1170	5	6
3	338	5 (5)	6 (9)	342	5	7
4	95	5 (5)	7 (9)	98	5	8

Baker Triangulation Results

MG levels	Unmodified Boundaries			Modified Boundaries		
	nodes	Dir BC	Mixed BC	nodes	Dir BC	Mixed BC
2	1170	4 (4)	5 (8)	1170	4	6
3	336	4 (4)	6 (9)	333	5	7
4	98	5 (5)	7 (9)	98	5	7

Table 3   MG iterations for the *annulus* grid with 8448 unknowns.

MG levels	Unmodified Boundaries			Modified Boundaries		
	nodes	Dir BC	Mixed BC	nodes	Dir BC	Mixed BC
2	2175	5 (5)	8 (48)	2175	5	8
3	574	5 (5)	8 (58)	572	6	8
4	158	5 (5)	8 (50)	156	6	8

Table 4   MG iterations for the *annulus* grid with 2176 unknowns.

MG levels	Unmodified Boundaries			Modified Boundaries		
	nodes	Dir BC	Mixed BC	nodes	Dir BC	Mixed BC
2	575	5 (5)	7 (20)	575	5	7
3	159	5 (5)	7 (20)	159	5	7
4	47	5 (5)	7 (20)	47	5	7

Table 5   MG iterations for the *annulus* grid with 576 unknowns.

MG levels	Unmodified Boundaries			Modified Boundaries		
	nodes	Dir BC	Mixed BC	nodes	Dir BC	Mixed BC
2	159	4 (4)	6 (11)	159	4	7
3	47	4 (4)	6 (12)	47	4	7
4	15	4 (4)	6 (12)	15	4	7

References

[Bak94] Baker T. (November 1994) Triangulations, mesh generation and point placement strategies. In D.Caughey (ed) *Computing the Future*, pages 101–115. Ithaca, NY.

[Bar92] Barth T. (March 1992) Aspects of unstructured grids and finite-volume solvers for the Euler and Navier-Stokes equations. In *Special course on unstructured grid methods for advection dominated flows*. Special course at the VKI, Belgium.

[BPX91] Bramble J. H., Pasciak J. E., and Xu J. (1991) The analysis of multigrid algorithms with nonnested spaces or noninherited quadratic forms. *Math. Comp.* 56(193): 1 – 34.

[BX] Bank R. and Xu J.A hierarchical basis multigrid method for unstructured meshes. In Hackbusch W. and Wittum G. (eds) *Tenth GAMM-Seminar Kiel on Fast Solvers for Flow Problems*. Vieweg-Verlag, Braunschweig. To appear.

[BX94] Bank R. and Xu J. (1994) An algorithm for coarsening unstructured meshes. Technical report, Dept. of Math., Univ. of Calif. at San Diego.

[Cai95] Cai X.-C. (1995) The use of pointwise interpolation in domain decomposition methods with nonnested meshes. *SIAM J. Sci. Comp.* 16(1).

[Cav74] Cavendish J. (1974) Automatic triangulation of arbitrary planar domains for the finite element method. *Int'l J. Numer. Meth. Engineering* 8: 679–696.

[Cle75] Clément P. (1975) Approximation by finite element functions using local regularization. *R.A.I.R.O. Numer. Anal.* R-2: 77–84.

[CS94] Chan T. F. and Smith B. (1994) Domain decomposition and multigrid methods for elliptic problems on unstructured meshes. In Keyes D. and Xu J. (eds) *Domain Decomposition Methods in Science and Engineering, Proceedings of the Seventh International Conference on Domain Decomposition, October 27-30, 1993, The Pennsylvania State University*. American Mathematical Society, Providence.

[CSZ94] Chan T. F., Smith B., and Zou J. (February 1994) Overlapping Schwarz methods on unstructured meshes using non-matching coarse grids. Technical Report 94-8, Department of Mathematics, University of California at Los Angeles.

[CZ94a] Chan T. F. and Zou J. (1994) Additive Schwarz domain decomposition methods for elliptic problems on unstructured meshes. *Numerical Algorithms* 8: 329–346.

[CZ94b] Chan T. F. and Zou J. (April 1994) Domain decomposition methods for non-symmetric parabolic problems on unstructured meshes. Technical Report 94-22, Department of Mathematics, University of California at Los Angeles.

[CZ95] Chan T. F. and Zou J. (1995) A convergence theory of multilevel additive Schwarz methods on unstructured meshes. Technical report, Department of Mathematics, University of California at Los Angeles.

[DD93] Douglas C. and Douglas J. (1993) A unified convergence theory for abstract multigrid or multilevel algorithms, serial and parallel. *SIAM J. Numer. Anal.* 30: 136–158.

[EES83] Eisenstat S. C., Elman H. C., and Schultz M. H. (1983) Variational iterative methods for nonsymmetric systems of linear equations. *SIAM J. Numer. Anal.* 20 (2): 345–357.

[GO93] Griebel M. and Oswald P. (1993) Remarks on the abstract theory of additive and multiplicative Schwarz algorithms. Technical Report TUM-19314, SFB-Bericht Nr. 342/6/93 A, Technical University of Munich.

[GS] Gropp W. and Smith B. F.Portable, Extensible Toolkit for Scientific Computation (PETSc). Available via anonymous ftp at `info.mcs.anl.gov` in the directory `pub/pdetools`.

[Gui93] Guillard H. (1993) Node-nested multi-grid method with Delaunay coarsening. Technical report, INRIA, Sophia Antipolis, France.

[Hol95] Holst M. (1995) An algebraic Schwarz theory. Technical report, Applied Math, California Institute of Technology.

[Mav92] Mavriplis D. (July 1992) Unstructured mesh algorithms for aerodynamic calculations. Technical Report 92-35, ICASE, NASA Langley, Virginia.

[Osw92] Oswald P. (1992) On discrete norm estimates related to multilevel preconditioners in the finite element method. In Ivanov K. and Sendov B. (eds) *Proc. Int. Conf. Theory of Functions, Varna 91*, pages 203–241.

[SBG95] Smith B., Bjørstad P., and Gropp W. (1995) *Domain decomposition: parallel multilevel methods for elliptic partial differential equations.* To appear.

[SZ90] Scott L. R. and Zhang S. (1990) Finite element interpolation of nonsmooth functions satisfying boundary conditions. *Math. Comp.* 54: 483–493.

[Xu92] Xu J. (December 1992) Iterative methods by space decomposition and subspace correction. *SIAM Review* 34: 581–613.

[Zha92] Zhang X. (1992) Multilevel Schwarz methods. *Numer. Math.* 63(4): 521–539.

Adaptive Monotone Multigrid Methods for some Non–Smooth Optimization Problems

RALF KORNHUBER[1]

1   INTRODUCTION

The weak solution of an elliptic selfadjoint boundary value problem is obtained by minimizing the corresponding quadratic energy functional $\mathcal{J}$. We will consider the more general problem

$$\mathcal{J}(u) + \phi(u) = \ \min$$

with ϕ denoting a convex functional which is piecewise quadratic but not differentiable. Such non–smooth optimization problems are modeling physical phenomena involving a change of phase. Obstacle problems or time–discretized two–phase Stefan problems are typical examples (see e.g. [Cra88, DL72, HS90] for further applications). The continuous problem is discretized by piecewise linear finite elements with respect to a sequence of triangulations resulting from the successive adaptive refinement of a given initial mesh. A corresponding adaptive algorithm has been described in [Koron]. In this paper we will concentrate on the efficient solution of the nonlinear discrete problems arising on each refinement level.

The most delicate question in constructing a multigrid method for a nonlinear problem is how to represent the nonlinearity on the coarse grids. This process usually involves some kind of linearization. Unfortunately, the computed corrections

---

[1] Weierstraß–Institut Berlin, Mohrenstraße 39, D–10117 Berlin, Germany
email: kornhuber@wias-berlin.de

Domain Decomposition Methods in Sciences and Engineering, edited by R. Glowinski *et al.*

may exceed the region in which the actual linearization is valid. This problem is often remedied by *a posteriori* damping of the coarse grid correction [Hac85]. The appropriate selection of damping parameters is a non–trivial task [HR89, Hop93]. The basic idea of *monotone multigrid methods* to be presented here is first to find out a neighborhood of the actual iterate in which the actual linearization is valid and then to constrain the coarse grid correction to this neighborhood. In this way, we ensure monotonically decreasing energy in course of the iteration. It turns out that such kind of *local linearization* is equivalent to the damping of the inaccessible nonlinear coarse grid correction. Suitable damping parameters are implicitly incorporated in the constraints. This approach provides globally convergent methods and we can prove asymptotic multigrid convergence rates. In comparison with previous multigrid algorithms, monotone multigrid methods turned out to be superior both from a theoretical and from a numerical point of view [Kor94, Korara].

As proofs of the basic convergence results have been already presented elsewhere [Kor94, Korara], we will try to give an algorithmically oriented presentation here. In this way, we hope to simplify further generalizations of the underlying ideas and the implementation in existing multigrid codes. A more detailed description will be contained in a forthcoming work [Korarb].

2   DISCRETIZATION OF THE CONTINUOUS PROBLEM

Let Ω be a bounded, polygonal domain in the Euclidean space $I\!\!R^2$. We consider the optimization problem

$$u \in H : \quad \mathcal{J}(u) + \phi(u) \leq \mathcal{J}(v) + \phi(v), \quad v \in H, \tag{1}$$

on a closed subspace $H \subset H^1(\Omega)$. For simplicity, we select $H = H_0^1(\Omega)$ corresponding to homogeneous Dirichlet boundary conditions. Other boundary conditions of Neumann or mixed type and the case of three space dimensions can be treated in a similar way [BEK93]. The quadratic functional $\mathcal{J}$,

$$\mathcal{J}(v) = \tfrac{1}{2}a(v,v) - \ell(v), \tag{2}$$

is induced by a continuous, symmetric, and H–elliptic bilinear form $a(\cdot, \cdot)$ and a bounded, linear functional ℓ. H is equipped with the energy norm $\| \cdot \| = a(\cdot, \cdot)^{1/2}$. The convex functional ϕ of the form

$$\phi(v) = \int_\Omega \Phi(v(x)) \, dx, \tag{3}$$

is generated by a scalar function $\Phi : I\!\!R \to I\!\!R \cup \{+\infty\}$. We assume that Φ is convex and piecewise quadratic,

$$\Phi(z) = \tfrac{1}{2}b_i z^2 - f_i z + c_i, \quad \theta_i \leq z \leq \theta_{i+1}, \tag{4}$$

on a partition $-\infty \leq \theta_0 < \theta_1 < \ldots < \theta_N < \theta_{N+1} \leq +\infty$ of the closed interval $K \subset I\!\!R$ bounded by θ_0, θ_{N+1} and that $\Phi(z) = \infty$ holds, if $z \notin K$. To make sure that $0 \in K$, we assume $\theta_0 \leq 0 \leq \theta_{N+1}$. The convexity implies that Φ is continuous on K but the derivative Φ' may be discontinuous at the transition points θ_i, $i = 1, \ldots, N$. From the assumptions on Φ, the functional ϕ is convex, lower semi–continuous, and proper (i.e. $\phi(v) > -\infty$ and $\phi \not\equiv +\infty$). Hence, it follows from well–known results [Glo84] that the optimization problem (1) has a unique solution $u \in H$.

Let $\mathcal{T}_j$ be a consistent partition of Ω in triangles with minimal diameter $h_j = \mathcal{O}(2^{-j})$. The interior nodes and edges of $\mathcal{T}_j$ are denoted by $\mathcal{N}_j$ and $\mathcal{E}_j$, respectively. The finite element space $\mathcal{S}_j \subset H$ contains all continuous functions $v \in H$ which are linear on each triangle $t \in \mathcal{T}_j$. $\mathcal{S}_j$ is spanned by the nodal basis $\Lambda_j = \{\lambda_p^{(j)} \mid p \in \mathcal{N}_j\}$. Replacing H by the finite dimensional approximation $\mathcal{S}_j$ and the functional ϕ by its $\mathcal{S}_j$–interpolate ϕ_j,

$$\phi_j(v) = \sum_{p \in \mathcal{N}_j} \Phi(v(p)) \int_\Omega \lambda_p^{(j)}(x) \, dx, \quad v \in \mathcal{S}_j, \tag{5}$$

we obtain the *discrete optimization problem*

$$u_j \in \mathcal{S}_j : \quad \mathcal{J}(u_j) + \phi_j(u_j) \leq \mathcal{J}(v) + \phi_j(v), \quad v \in \mathcal{S}_j. \tag{6}$$

Observe that the discrete energy $\mathcal{J} + \phi_j$ is finite and continuous on the non–empty, closed, convex set $\mathcal{K}_j = \{v \in \mathcal{S}_j \mid v(p) \in K, \ p \in \mathcal{N}_j\} \subset \mathcal{S}_j$. It is easily seen that the discrete functional ϕ_j still is convex, lower semi–continuous, and proper so that the discrete problem (6) has a unique solution $u_j \in \mathcal{S}_j$. The convergence of the discretization (6) follows from well–known general results [Glo84].

3 MULTILEVEL RELAXATIONS

Assume that $\mathcal{T}_j$ is resulting from j refinements of a given, intentionally coarse triangulation $\mathcal{T}_0$ of Ω. In this way, we obtain a sequence of triangulations $\mathcal{T}_0, \ldots, \mathcal{T}_j$ and of corresponding nested finite element spaces $\mathcal{S}_0 \subset \ldots \subset \mathcal{S}_j$. To avoid additional technicalities, we assume for the moment that each triangulation is uniformly refined, i.e. that each triangle $t \in \mathcal{T}_{k-1}$ is subdivided into four congruent subtriangles to obtain the next triangulation $\mathcal{T}_k$.

Collecting the nodal basis functions $\Lambda_k = \{\lambda_{p_i}^{(k)} \mid i = 1, \ldots, n_k\}$ from all refinement levels, we define the multilevel nodal basis $\Lambda_\mathcal{S}$,

$$\Lambda_\mathcal{S} = \left(\lambda_{p_1}^{(j)}, \ldots, \lambda_{p_{n_j}}^{(j)}, \lambda_{p_1}^{(j-1)}, \ldots, \lambda_{p_{n_{j-1}}}^{(j-1)}, \ldots, \lambda_{p_1}^{(0)}, \ldots, \lambda_{p_{n_0}}^{(0)} \right)$$

which is ordered from fine to coarse. We frequently write $\Lambda_\mathcal{S} = (\lambda_1, \ldots, \lambda_m)$ with $m = n_j + \ldots + n_0$. In the special case of an elliptic selfadjoint problem (i.e. $\phi \equiv 0$) one step of a classical multigrid V–cycle with Gauss-Seidel smoother can be regarded as the successive optimization of the energy functional $\mathcal{J}$ in the direction of the multilevel nodal basis functions $\lambda_l \in \Lambda_\mathcal{S}$ (cf. e.g. [McC92, Xu92, Yse93]). We will use a straightforward extension of this *multilevel relaxation* as the starting point for the construction of monotone multigrid methods for the non–smooth optimization problem (6). To be precise, let us introduce the splitting

$$\mathcal{S}_j = \sum_{l=1}^{m} V_l, \quad V_l = \mathrm{span}\{\lambda_l\}, \quad l = 1, \ldots, m. \tag{7}$$

For a given ν–th iterate $u_j^\nu \in \mathcal{K}_j$ (i.e. with finite energy) one step of a *nonlinear* multilevel relaxation then reads as follows.

Algorithm 3.1 (Nonlinear Multilevel Relaxation)
initialize: $w_0 := u_j^\nu$
for $l = 1$ step 1 until m do

$$\bar{v}_l \in V_l : \quad \mathcal{J}(w_{l-1} + \bar{v}_l) + \phi_j(w_{l-1} + \bar{v}_l) \leq$$
$$\leq \mathcal{J}(w_{l-1} + v) + \phi_j(w_{l-1} + v), \quad v \in V_l \tag{8}$$

$$w_l := w_{l-1} + \omega_l \bar{v}_l, \quad \omega_l \in [0, 1]$$

new iterate: $u_j^{\nu+1} := w_m$

Observe that we have introduced certain damping parameters ω_l which will be useful later on. Assuming

$$\omega_l = 1, \quad l = 1, \ldots, n_j, \tag{9}$$

the leading *fine grid corrections* in direction of $\lambda_l \in \Lambda_j$ can be regarded as one step of the well–known single grid relaxation [Glo84]. The corresponding iteration operator is denoted by $\mathcal{M}_j$ and $\bar{u}_j^\nu := \mathcal{M}_j(u_j^\nu)$ is called smoothed iterate. Note that we have $\bar{u}_j^0 \in \mathcal{K}_j$ for an arbitrary initial iterate $u_j^0 \in \mathcal{S}_j$. The subsequent *coarse grid* corrections of the smoothed iterate $\bar{u}_j^\nu$ in the directions $\lambda_l \in \Lambda_{\mathcal{S}} \setminus \Lambda_j$ are intended to reduce the low frequency contributions of the error.

The proof of the following theorem is based on the global convergence of the leading single grid relaxation [Glo84] and on the monotone decrease of the energy $\mathcal{J}(w_l) + \phi_j(w_l)$ of the intermediate iterates.

Theorem 3.1 *For any initial iterate $u_j^0 \in \mathcal{S}_j$ and any sequence of damping parameters with the property (9) the sequence of iterates $(u_j^\nu)_{\nu \geq 0}$ produced by Algorithm 3.1 converges to the solution u_j of the discrete problem (6).*

In the special case $\phi_j \equiv 0$ Algorithm 3.1 can be implemented as a V–cycle: Representing the bilinear form $a(\cdot, \cdot)$ on the coarse grid spaces $\mathcal{S}_k$ by their values on Λ_k, one can update the residual and evaluate the local corrections without visiting the fine grid. This provides optimal numerical complexity, i.e. $\mathcal{O}(n_j)$ operations, for each iteration step. To find a related implementation for the nonlinear case, we will now consider the local subproblems for the local corrections

$$\bar{v}_l = \bar{z}_l \lambda_l \in V_l$$

in more detail. Using subdifferential calculus [ET76], (8) can be rewritten as the following *scalar inclusion* for the unknown coefficient $\bar{z}_l \in I\!R$

$$0 \in a(\lambda_l, \lambda_l)\bar{z}_l - (\ell(\lambda_l) - a(w_{l-1}, \lambda_l)) + \partial\phi_j(w_{l-1} + \bar{z}_l\lambda_l)(\lambda_l). \tag{10}$$

Observe that $\partial\phi_j(w_{l-1} + z\lambda_l)(\lambda_l)$ is a piecewise linear function in z because the scalar function Φ which generates ϕ_j is piecewise quadratic. Hence, after some tedious calculations, the fine grid corrections in direction of $\lambda_l \in \Lambda_j$ are available in closed form [Korara, Koron].

Let us consider the coarse grid correction in direction of some fixed $\lambda_l \in \Lambda_k$, $k < j$. It is clear that $\bar{z}_l$ can not be computed without evaluating the intermediate iterate w_{l-1} at all nodes $p \in \operatorname{int} \operatorname{supp} \lambda_l$, because the subdifferential $\partial \phi_j(w_{l-1} + z\lambda_l)(\lambda_l)$ is *nonlinear* with respect to the argument $w_{l-1} + z\lambda_l$. This leads to (at least) one additional prolongation for each local coarse grid correction. As a consequence, the number of operations for one complete iteration step is no longer linearly bounded but grows like $\mathcal{O}(n_j \log(n_j))$.

To preserve the optimal numerical complexity of the classical V–cycle, we will now approximate the exact coarse grid corrections $\bar{v}_l$ by a *local linearization* of the subproblems (10) in a neighborhood of the smoothed iterate $\bar{u}_j^\nu$. For this reason, we define the *discrete phases* $\mathcal{N}_j^i(\bar{u}_j^\nu) \subset \mathcal{N}_j$ of $\bar{u}_j^\nu$ by

$$\mathcal{N}_j^i(\bar{u}_j^\nu) = \{p \in \mathcal{N}_j \mid \bar{u}_j^\nu(p) \in (\theta_i, \theta_{i+1})\}, \quad , i = 0, \ldots, N.$$

At the remaining *critical nodes* $\mathcal{N}_j^\bullet(\bar{u}_j^\nu)$,

$$\mathcal{N}_j^\bullet(\bar{u}_j^\nu) = \mathcal{N}_j \setminus \bigcup_{i=0}^N \mathcal{N}_j^i(\bar{u}_j^\nu), \tag{11}$$

$\bar{u}_j^\nu$ has values in the set $\{\theta_1, \ldots, \theta_N\}$ of transition points.

Now the key observation is that $\phi_j(w)$ *is a quadratic functional as long as the discrete phases of w remain invariant.* Such a neighborhood of $\bar{u}_j^\nu$ is given by

$$\mathcal{K}_{\bar{u}_j^\nu} = \{w \in \mathcal{S}_j \mid \underline{\varphi}_{\bar{u}_j^\nu}(p) \le w(p) \le \overline{\varphi}_{\bar{u}_j^\nu}(p), \ p \in \mathcal{N}_j\},$$

where the obstacles $\underline{\varphi}_{\bar{u}_j^\nu}, \overline{\varphi}_{\bar{u}_j^\nu} \in \mathcal{S}_j$ are defined by

$$\begin{aligned} \underline{\varphi}_{\bar{u}_j^\nu}(p) = \theta_i, \ \ \overline{\varphi}_{\bar{u}_j^\nu}(p) = \theta_{i+1}, &\quad \text{if} \quad p \in \mathcal{N}_j^i(\bar{u}_j^\nu), \\ \underline{\varphi}_{\bar{u}_j^\nu}(p) = \overline{\varphi}_{\bar{u}_j^\nu}(p) = \bar{u}_j^\nu(p), &\quad \text{if} \quad p \in \mathcal{N}_j^\bullet(\bar{u}_j^\nu). \end{aligned} \tag{12}$$

Recall that the closed convex subset $\mathcal{K}_{\bar{u}_j^\nu} \in \mathcal{S}_j$ is *fixed by the fine grid correction.*

By construction, the functional ϕ_j on $\mathcal{K}_{\bar{u}_j^\nu}$ can be rewritten as

$$\phi_j(w) = \tfrac{1}{2} b_{\bar{u}_j^\nu}(w, w) - f_{\bar{u}_j^\nu}(w) + \text{const.}, \quad w \in \mathcal{K}_{\bar{u}_j^\nu}, \tag{13}$$

with the symmetric positive semidefinite bilinear form $b_{\bar{u}_j^\nu}(v, w)$,

$$b_{\bar{u}_j^\nu}(v, w) = \sum_{i=0}^N \sum_{p \in \mathcal{N}_j^i(\bar{u}_j^\nu)} b_i v(p) w(p) \int_\Omega \lambda_p^{(j)}(x) \, dx, \tag{14}$$

and the linear functional $f_{\bar{u}_j^\nu}(v)$,

$$f_{\bar{u}_j^\nu}(v) = \sum_{i=0}^N \sum_{p \in \mathcal{N}_j^i(\bar{u}_j^\nu)} f_i v(p) \int_\Omega \lambda_p^{(j)}(x) \, dx. \tag{15}$$

To take advantage of the simple representation of bilinear forms and linear operators on the coarse spaces $\mathcal{S}_k$, $k < j$, we want to constrain the local corrections in such a way that all the intermediate iterates w_l remain in $\mathcal{K}_{\bar{u}_j^\nu}$. Equivalently, *the coarse grid corrections must not cause a change of phase.* Hence, the local subproblems (8) in Algorithm 3.1 are replaced by the quadratic obstacle problems

$$\bar{v}_l \in \mathcal{D}_l^* : \quad \mathcal{J}(w_{l-1} + v_l^*) + \phi_j(w_{l-1} + v_l^*) \le$$
$$\le \mathcal{J}(w_{l-1} + v) + \phi_j(w_{l-1} + v), \quad v \in \mathcal{D}_l^*, \tag{16}$$

with constraints $\mathcal{D}_l^* = \{v \in V_l \mid w_{l-1} + v \in \mathcal{K}_{\bar{u}_j^\nu}\} \subset V_l$. In the light of (13), the energy functional on $\mathcal{D}_l^*$ has the representation

$$\mathcal{J}(w_{l-1} + v) + \phi_j(w_{l-1} + v) = \tfrac{1}{2} a_{\bar{u}_j^\nu}(v, v) - r_{\bar{u}_j^\nu}(w_{l-1})(v) + \text{const.}$$

where we have set $r_{\bar{u}_j^\nu}(w_{l-1}) = \ell_{\bar{u}_j^\nu} - a_{\bar{u}_j^\nu}(w_{l-1}, \cdot)$ and

$$a_{\bar{u}_j^\nu}(\cdot, \cdot) = a(\cdot, \cdot) + b_{\bar{u}_j}(\cdot, \cdot), \quad \ell_{\bar{u}_j^\nu} = \ell + f_{\bar{u}_j^\nu}. \tag{17}$$

The set $\mathcal{D}_l^*$ clearly contains all $v \in V_l$ satisfying

$$\underline{\varphi}_{\bar{u}_j^\nu} - w_{l-1} \le v \le \overline{\varphi}_{\bar{u}_j^\nu} - w_{l-1}. \tag{18}$$

Hence, we still have to evaluate the intermediate iterate $w_{l-1} \in \mathcal{S}_j$ to check whether some v is contained in $\mathcal{D}_l^*$ or not. For this reason, we approximate $\mathcal{D}_l^*$ by replacing the fine grid defect obstacles $w_{l-1} - \underline{\varphi}_{\bar{u}_j^\nu}$, $w_{l-1} - \overline{\varphi}_{\bar{u}_j^\nu}$ appearing in (18) by coarse grid approximations $\underline{\psi}_l$, $\overline{\psi}_l \in V_l$. To make sure that the resulting subset $\mathcal{D}_l = \{v \in V_l \mid \underline{\psi}_l \le v \le \overline{\psi}_l\} \subset V_l$ satisfies the condition $0 \in \mathcal{D}_l \subset \mathcal{D}_l^*$, we require

$$\underline{\varphi}_{\bar{u}_j^\nu} - w_{l-1} \le \underline{\psi}_l \le 0 \le \overline{\psi}_l \le \overline{\varphi}_{\bar{u}_j^\nu} - w_{l-1}. \tag{19}$$

Let us postpone the construction of such *monotone approximations* $\underline{\psi}_l$, $\overline{\psi}_l$ to the next section. We now summarize one complete step of our linearized multilevel relaxation.

Algorithm 3.2 (Linearized Multilevel Relaxation)
   fine grid smoothing: $\bar{u}_j^\nu := \mathcal{M}_j(u_j^\nu)$
   local linearization: $a_{\bar{u}_j^\nu} := a + b_{\bar{u}_j^\nu}$, $\ell_{\bar{u}_j^\nu} := \ell + f_{\bar{u}_j^\nu}$
   coarse grid correction:
        initialization: $w_{n_j} := \bar{u}_j^\nu$
        for $l = n_j + 1$ *step 1 until* m *do*
        update $\mathcal{D}_l$

$$v_l \in \mathcal{D}_l : \quad \tfrac{1}{2} a_{\bar{u}_j^\nu}(v_l, v_l) - r_{\bar{u}_j^\nu}(w_{l-1})(v_l) \le$$
$$\tfrac{1}{2} a_{\bar{u}_j^\nu}(v, v) - r_{\bar{u}_j^\nu}(w_{l-1})(v), \quad v \in \mathcal{D}_l \tag{20}$$

        $w_l := w_{l-1} + v_l$
   new iterate: $u_j^{\nu+1} := w_m$

It is the main result of this section that local linearization can be regarded as local damping.

Lemma 3.1 *For a given intermediate iterate $w_{l-1} \in \mathcal{S}_j$ the local corrections $\bar{v}_l$ and v_l resulting from the subproblems (8) and (20), respectively, are related by*

$$v_l = \omega_l \bar{v}_l \tag{21}$$

with some $\omega_l \in [0, 1]$.

Proof If $\bar{v}_l \in \mathcal{D}_l$, then the inclusion $\mathcal{D}_l \subset \mathcal{D}_l^*$ yields $\bar{v}_l = v_l$. As $\bar{v}_l \in V_l$ and $\underline{\psi}_l, \overline{\psi}_l \in V_l$, we only have to consider the remaining cases $\bar{v}_l < \underline{\psi}_l$ and $\overline{\psi}_l < \bar{v}_l$. In the first case, (19) gives $\bar{v}_l < v_l = \underline{\psi}_l \leq 0$. The second case can be treated in a similar way. $\square$

Lemma 3.1 implies that Algorithm 3.2 is a special case of Algorithm 3.1. In particular, it is globally convergent. By keeping the local coarse grid corrections v_l in $\mathcal{D}_l$, the damping parameters ω_l are *implicitly* selected in such a way that the local linearization (13) remains valid. A similar approach can be used, if the functional ϕ_j is not piecewise linear but piecewise smooth. This will be the subject of a forthcoming paper.

4 STANDARD MONOTONE MULTIGRID METHODS

To complete the construction of a monotone multigrid method, we now derive suitable local obstacles $\underline{\psi}_l$ and $\overline{\psi}_l$, $l = n_j + 1, \ldots, m$. For symmetry reasons, it is sufficient to consider only the upper obstacles $\overline{\psi}_l$. The construction relies on suitable successive restrictions of the initial defect obstacle $\overline{\varphi}_{\bar{u}_j^\nu} - \bar{u}_j^\nu$. To identify the supporting points and the levels of $\lambda_l \in \Lambda_{\mathcal{S}}$, we will use the notation $\lambda_{l_{ik}} = \lambda_{p_i}^{(k)}$, $i = 1, \ldots, n_k$, $k = 0, \ldots, j$. Then the correction $v^{(k)} = v_{p_1}^{(k)} + \ldots + v_{p_{n_k}}^{(k)}$ is the sum of all local corrections $v_{l_{ik}} = v_{p_i}^{(k)}$ in direction of the basis functions $\lambda_{l_{ik}} = \lambda_{p_i}^{(k)}$ on level k. The following lemma is easily proved by induction.

Lemma 4.1 *Assume that the mappings $R_{k+1}^k : \mathcal{S}_{k+1} \to \mathcal{S}_k$, $k = j - 1, \ldots, 0$, are monotone in the sense that*

$$0 \leq R_{k+1}^k v(p) \leq v(p), \quad p \in \mathcal{N}_{k+1}, \tag{22}$$

holds for all non–negative $v \in \mathcal{S}_{k+1}$. Then, for a given smoothed iterate $\bar{u}_j^\nu$ and the initial defect obstacle $\overline{\psi}^{(j)} = \overline{\varphi}_{\bar{u}_j^\nu} - \bar{u}_j^\nu \geq 0$, the recursive restriction

$$\overline{\psi}^{(k)} = R_{k+1}^k(\overline{\psi}^{(k+1)} - v^{(k+1)}), \quad k = j - 1, \ldots, 0, \tag{23}$$

provides local upper obstacles $\overline{\psi}_l \in V_l$ with the property (19) by the definition

$$\overline{\psi}_{l_{ik}} = \overline{\psi}^{(k)}(p_i)\lambda_{p_i}^{(k)}, \quad i = 1, \ldots, n_k. \tag{24}$$

As we are interested in multigrid convergence rates, we want to exclude the trivial choice $R_{k+1}^k \equiv 0$ which would bring back the single grid relaxation. Hence, we will now derive monotone restrictions R_{k+1}^k satisfying

$$\min\{v(q) \mid q \in \mathcal{N}_{k+1} \cap \text{int supp } \lambda_p^{(k)}\} \leq R_{k+1}^k v(p), \quad p \in \mathcal{N}_k, \tag{25}$$

for all non–negative $v \in \mathcal{S}_{k+1}$, instead of the weaker lower estimate in (22). It will turn out later on that such *quasioptimal* monotone restrictions provide asymptotic multigrid convergence rates. Let us select a certain order of the edges $\mathcal{E}_k = \{e_1, \ldots, e_s\}$ with midpoints $p_e \in \mathcal{N}_{k+1}$, $e \in \mathcal{E}_k$. Then the restriction operator $R_{k+1}^k : \mathcal{S}_{k+1} \to \mathcal{S}_k$ is defined by

$$R_{k+1}^k v = I_{\mathcal{S}_k} \circ R_{e_s} \circ \ldots \circ R_{e_1} v, \quad v \in \mathcal{S}_{k+1}. \tag{26}$$

Here $I_{\mathcal{S}_k}$ denotes the $\mathcal{S}_k$–interpolation and the operators $R_e : \mathcal{S}_{k+1} \to \mathcal{S}_{k+1}$, $e \in \mathcal{E}_k$, are of the form

$$R_e v = v + v_1 \lambda_{p_1}^{(k+1)} + v_2 \lambda_{p_2}^{(k+1)}, \quad v \in \mathcal{S}_{k+1}, \tag{27}$$

with $p_1, p_2 \in \mathcal{N}_k$ denoting the vertices of $e = (p_1, p_2) \in \mathcal{E}_k$. The scalars $v_1, v_2 \in \mathbb{R}$ in (27) are chosen such that

$$R_e v(p) \le v(p), \quad p = p_1, p_e, p_2.$$

In particular, we set $v_1 = 0$, if $v(p_1) \le v(p_e)$ or $v(p_1) + v(p_2) \le 2v(p_e)$. In the remaining case, v_1 is determined by

$$v_1 = \begin{cases} 2v(p_e) - v(p_1) - v(p_2), & \text{if} \quad v(p_2) \le v(p_e) \le v(p_1), \\ v(p_e) - v(p_1), & \text{if} \quad v(p_e) \le v(p), \ p = p_1, p_2. \end{cases}$$

The value of v_2 is obtained in a symmetrical way.

The following proposition can be checked by elementary considerations.

Proposition 4.1 *For any fixed enumeration of $\mathcal{E}_k$ the definition (26) provides a quasioptimal upper restriction operator R_{k+1}^k in the sense of (25).*

We will now formulate Algorithm 3.2 as a multigrid V–cycle. For this reason, we rewrite the computation of the correction $v^{(k)}$ from all local subproblems (20) on a fixed level k as one step of a *projected Gauss–Seidel–method*. The corresponding iteration operator for a bilinear form $a = a(\cdot, \cdot)$, a right hand side r, and obstacles $\underline{\psi}, \overline{\psi}$ is denoted by $\bar{\mathcal{M}}_k(a, r, \underline{\psi}, \overline{\psi}) : \mathcal{S}_k \to \mathcal{S}_k$. Recall that $\mathcal{M}_j : \mathcal{S}_j \to \mathcal{S}_j$ stands for the *nonlinear single grid relaxation*. Lower and upper monotone restrictions will be denoted by $\underline{R}_{k+1}^k$ and $\overline{R}_{k+1}^k$, respectively.

Algorithm 4.1 (Standard Monotone Multigrid Method)
    fine grid smoothing: $\bar{u}_j^\nu := \mathcal{M}_j(u_j^\nu)$
    local linearization: $a_{\bar{u}_j^\nu} := a + b_{\bar{u}_j^\nu}, \quad \ell_{\bar{u}_j^\nu} := \ell + f_{\bar{u}_j^\nu}$
    coarse grid correction:
        initialize:
            bilinear form and residual: $a^{(j)} := a_{\bar{u}_j^\nu}, \quad r^{(j)} := \ell_{\bar{u}_j^\nu} - a_{\bar{u}_j^\nu}(\bar{u}_j^\nu, \cdot)$
            defect obstacles: $\underline{\psi}^{(j)} := \underline{\varphi}_{\bar{u}_j^\nu} - \bar{u}_j^\nu, \quad \overline{\psi}^{(j)} := \overline{\varphi}_{\bar{u}_j^\nu} - \bar{u}_j^\nu$
            global correction: $v_j^\nu := 0$
        for $k = j - 1$ *step* -1 *until* 0 *do*
        canonical restrictions: $a^{(k)} := a^{(k+1)}|_{\mathcal{S}_k \times \mathcal{S}_k}, \quad r^{(k)} := r^{(k+1)}|_{\mathcal{S}_k}$
        quasioptimal restrictions: $\underline{\psi}^{(k)} := \underline{R}_{k+1}^k \underline{\psi}^{(k+1)}, \quad \overline{\psi}^{(k)} := \overline{R}_{k+1}^k \overline{\psi}^{(k+1)}$
        coarse grid smoothing: $v^{(k)} := \bar{\mathcal{M}}_k(a^{(k)}, r^{(k)}, \underline{\psi}^{(k)}, \overline{\psi}^{(k)})(0)$

$$update:$$
$$residual: \ r^{(k)} := r^{(k)} - a^{(k)}(v^{(k)}, \cdot)$$
$$defect \ obstacles: \ \underline{\psi}^{(k)} := \underline{\psi}^{(k)} - v^{(k)} \ , \quad \overline{\psi}^{(k)} := \overline{\psi}^{(k)} - v^{(k)}$$
$$for \ k = 0 \ step \ 1 \ until \ j - 1 \ do$$
$$canonical \ interpolation: \ v_j^\nu := v_j^\nu + v^{(k)}$$
$$new \ iterate: \ u_j^{\nu+1} := \bar{u}_j^\nu + v_j^\nu$$

Note that Algorithm 4.1 contains a slightly improved variant of Mandels method [Man84] for linear complementary problems as a special case. See [Kor94] for details.

Assume that the discrete problem (6) is non–degenerate in the sense that

$$p \in \mathcal{N}_j^\bullet(u_j) \Rightarrow \ell(\lambda_p^{(j)}) - a(u_j, \lambda_p^{(j)}) \in \ \text{int} \ \partial\phi_j(u_j)(\lambda_p^{(j)}). \tag{28}$$

Then it can be shown that the discrete phases of the iterates u_j^ν coincide with the discrete phases of the exact finite element solution u_j, as soon as ν is large enough. As a consequence, Algorithm 3.1 without damping (i.e. $\omega_l \equiv 1$, $l = 1, \ldots, m$) is asymptotically reducing to a *linear multigrid method* for a *reduced linear problem* on a corresponding reduced subspace of $\mathcal{S}_j$. It is important to notice that the linearized Algorithm 4.1 is asymptotically reducing to the *same* linear multigrid method, if *quasioptimal* monotone restrictions (cf. (25)) are used. Hence, we can derive asymptotic estimates of the convergence rates of the standard monotone multigrid method by using recent results on related *linear* subspace correction methods [GO95, KY94, Osw94].

Theorem 4.1 *The standard monotone multigrid method is globally convergent.*

Assume that the discrete problem (6) is non–degenerate in the sense of (28). Then the phases of the iterates $(u_j^\nu)_{\nu \geq 0}$ converge to the phases of u_j and the error estimate

$$\|u_j - u_j^{\nu+1}\| \leq (1 - c(j+1)^{-4})\|u_j - u_j^\nu\| \tag{29}$$

holds, if ν is large enough. The positive constant $c < 1$ depends only on the ellipticity of $a(\cdot, \cdot)$, on the maximal coefficient b_i, $i = 0, \ldots, N$, of Φ, and on the initial triangulation $\mathcal{T}_0$.

We emphasize that the estimate (29) describes the worst case. Absolutely no regularity assumptions on the continuous or discrete free boundary enter the constant c. In addition, we have considered the most simple variant of standard monotone multigrid methods. By repeating the (approximate) optimization in the direction of the basis functions $\lambda_p^{(k)}$ on each level $k = j, \cdots, 0$ in reversed order, we obtain a *standard monotone multigrid method with symmetric smoother*. For this variant, we get a $\mathcal{O}(j^2(\log j)^2)$ estimate. We can further improve this bound by imposing regularity conditions on $a(\cdot, \cdot)$ (providing $\mathcal{O}(j^2)$) or by using L^2–like projections instead of modified interpolation operators. In contrast to (29) the latter estimates also hold in the case of more than two space dimensions. However, we then need a certain regularity of the critical set $\mathcal{N}_j^\bullet(u_j)$. A detailed discussion can be found in [KY94, Osw94].

Let us now consider non–uniform refinement. In this situation, the canonical ordering of the multilevel nodal basis $\Lambda_\mathcal{S}$ would contradict our requirement that each multilevel relaxation should start with a fine grid relaxation step. Of course, one could rearrange $\Lambda_{\mathcal{S}}$ in a suitable way. For the implementation in an existing multigrid code it might be simpler to use the search directions $\Lambda = (\Lambda_j, \Lambda_{\mathcal{S}})$ instead of $\Lambda_{\mathcal{S}}$ or, equivalently, to start with a complete fine grid relaxation and then linearize all corrections in direction of $\lambda_l \in \Lambda_{\mathcal{S}}$. Both of these algorithms have the convergence properties stated in Theorem 4.1. The second algorithm will be used in our numerical experiment reported below.

5   TRUNCATED MONOTONE MULTIGRID METHODS

The standard multigrid method relies on the condition that the coarse grid correction must not change the phases of the smoothed iterate $\bar{u}_j^\nu$. In particular, it must not change the values of $\bar{u}_j^\nu$ at the critical nodes $p \in \mathcal{N}_j^\bullet(\bar{u}_j^\nu)$. Hence, all $\lambda_l \in \Lambda_{\mathcal{S}} \setminus \Lambda_j$ with the property

$$\text{int supp } \lambda_l \cap \mathcal{N}_j^\bullet(\bar{u}_j^\nu) \neq \emptyset \tag{30}$$

must not contribute to the coarse grid correction of the standard multigrid method. This leads to a poor representation of the low frequency parts of the error. To improve the convergence rates by improved coarse grid transport, we will now modify all $\lambda_l \in \Lambda_{\mathcal{S}} \setminus \Lambda_j$ with the property (30) according to the actual guess of the free boundary. Again, it is sufficient to consider only uniform refinement. The non–uniform case can be treated in the same way as described above. We define the modified basis functions

$$\tilde{\lambda}_p^{(k)} = T_{j,k}^\nu \lambda_p^{(k)}, \quad p \in \mathcal{N}_k, \tag{31}$$

by using the truncation operators $T_{j,k}^\nu = I_{\mathcal{S}_j^\nu} \circ \ldots \circ I_{\mathcal{S}_k^\nu}$, $k = 0, \ldots, j$. Here $I_{\mathcal{S}_k^\nu} : \mathcal{S}_j \to \mathcal{S}_k^\nu$ denotes the $\mathcal{S}_k^\nu$–interpolation and the spaces $\mathcal{S}_k^\nu \subset \mathcal{S}_k$,

$$\mathcal{S}_k^\nu = \{v \in \mathcal{S}_k \mid v(p) = 0, \ p \in \mathcal{N}_k^\nu\} \subset \mathcal{S}_k, \tag{32}$$

are the reduced subspaces of functions vanishing on $\mathcal{N}_k^\nu = \mathcal{N}_k \cap \mathcal{N}_j^\bullet(\bar{u}_j^\nu)$, $k = 0, \ldots, j$.

Replacing the multilevel nodal basis $\Lambda_{\mathcal{S}}$ by the actual truncation $\tilde{\Lambda}_{\mathcal{S}}^\nu$,

$$\tilde{\Lambda}_{\mathcal{S}}^\nu = \left(\lambda_{p_1}^{(j)}, \ldots, \lambda_{p_{n_j}}^{(j)}, \tilde{\lambda}_{p_1}^{(j-1)}, \ldots, \tilde{\lambda}_{p_{n_{j-1}}}^{(j-1)}, \ldots, \tilde{\lambda}_{p_1}^{(0)}, \ldots, \tilde{\lambda}_{p_{n_0}}^{(0)}\right), \quad \nu \geq 0,$$

we can now derive a globally convergent *truncated monotone multigrid method* by the same reasoning as described in the previous section. The resulting algorithm can be implemented as a variant of the standard monotone multigrid method. More precisely, in the neighborhood of the critical nodes $p \in \mathcal{N}_j^\bullet(\bar{u}_j^\nu)$ (cf. (11)) the restrictions and prolongations appearing in Algorithm 4.1 have to be modified as follows:

Modifications of Algorithm 4.1 (Truncated Monotone Multigrid Method)
modified restrictions of the bilinear form and of the residual:
    treat all entries from the actual critical nodes $\mathcal{N}_j^\bullet(\bar{u}_j^\nu)$ as zero
modified quasioptimal restrictions of the upper (lower) defect obstacle:
    treat all entries from the actual critical nodes $\mathcal{N}_j^\bullet(\bar{u}_j^\nu)$ as ∞ $(-\infty)$
modified prolongations of the corrections:
    prolongate zero to all critical nodes

Again, we can derive asymptotic estimates of the convergence rates by analysing the resulting *reduced linear multigrid method* for the corresponding reduced linear problem.

Related algorithms have been recently considered in [BX94b, BX94a, ?, KY94]. It is easily checked that the underlying splitting of $\mathcal{S}_j$ is now generated by a larger number of subspaces as compared to the standard case. Hence, we can hope for improved asymptotic convergence rates of the truncated multigrid method. This is supported by the numerical results reported below. However, the theoretical analysis suffers from the fact that there is no strengthened Cauchy–Schwarz inequality for the spans of truncated basis functions $\tilde{\lambda}_p^{(k)} \notin \mathcal{S}_k$. Without any additional regularity this leads to even more pessimistic estimates than for the standard case.

Theorem 5.1 *The truncated monotone multigrid method is globally convergent.*

Assume that the discrete problem (6) is non–degenerate in the sense of (28). Then the phases of the iterates $(u_j^\nu)_{\nu \geq 0}$ converge to the phases of u_j and the error estimate

$$\|u_j - u_j^{\nu+1}\| \leq (1 - c(j+1)^{-6})\|u_j - u_j^\nu\| \tag{33}$$

holds, if ν is large enough. The positive constant $c < 1$ depends only on the ellipticity of $a(\cdot,\cdot)$, on the maximal coefficient b_i, $i = 1,\ldots,N$, of Φ, and on the initial triangulation $\mathcal{T}_0$.

As in the standard case, we can derive various improvements of the worst–case result (33). For example, we get a $\mathcal{O}(j^3)$ estimate, if symmetric smoothers are used.

6   NUMERICAL EXPERIMENTS

We will now illustrate the numerical performance of monotone multigrid methods in the framework of an adaptive algorithm. In this case, the underlying hierarchy of triangulations is resulting from the adaptive refinement of an initial triangulation $\mathcal{T}_0$. The adaptive refinement strategy and stopping criteria for the iterative solver on each refinement level are based on a posteriori estimates of the *approximation error* $\|u - \tilde{u}_j\|$ and of the *algebraic error* $\|u_j - \tilde{u}_j\|$ of some given $\tilde{u}_j \in \mathcal{S}_j$. A detailed description is contained in [Koron].

We consider the following model problem involving a jump discontinuity of Stefan type together with an upper obstacle. We choose the bilinear form $a(\cdot,\cdot)$ and the functional ℓ according to

$$a(v,w) = \int_\Omega (\partial_1 v \partial_1 w + \partial_2 v \partial_2 w)\,dx, \quad \ell(v) = \int_\Omega f\,v\,dx$$

with a peak source

$$f(x_1, x_2) = 3000 x_1 x_2 (x_1 - 1)(x_2 - 1) exp\left(-10(0.5 - x_1)^2 (0.5 - x_2)^2\right),$$

and $\Omega = (0,1) \times (0,1)$. The scalar function Φ defined in (4) is given by the parameters $N = 1$, $\theta_0 = -\infty$, $\theta_1 = 0.5$, $\theta_2 = 0.75$, and $b_0 = 400$, $f_0 = 200$, $c_0 = 50$, $b_1 = 0$, $f_1 = -100$, $c_1 = -50$.

The initial triangulation $\mathcal{T}_0$ is obtained by subdividing Ω in four congruent triangles. We now apply the adaptive algorithm as described in [Koron], using the truncated monotone multigrid method as iterative solver. On each refinement level j the discrete problem is solved up to an (estimated) accuracy of 0.5% in order to obtain the approximate finite element solution $\tilde{u}_j \in \mathcal{S}_j$. The whole adaptive algorithm stops as soon as the (estimated) approximation error $\|u - \tilde{u}_j\|$ is less than 5%.

This final accuracy is reached after 8 adaptive refinement steps, providing the triangulation $\mathcal{T}_8$ together with the approximate solution $\tilde{u}_8$ as depicted in Figure 1. Observe the occurrence of a "mushy" region where $\tilde{u}_8 \equiv \theta_1$ and of a contact zone where $\tilde{u}_8 \equiv \theta_2$. Both are reflected by the adaptively refined mesh.

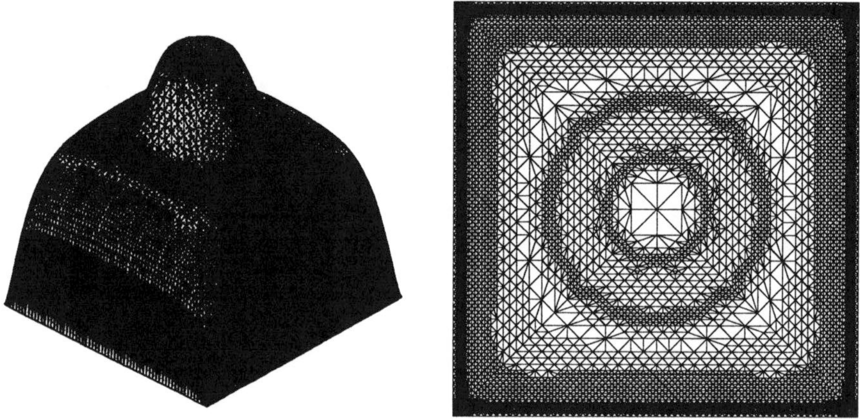

Figure 1 Final Approximation $\tilde{u}_8$ and Final Triangulation $\mathcal{T}_8$

The complete approximation history is given in Table 1. Recall that the refinement depth is the maximal number of successive refinements. The effectivity index is the ratio of the a posteriori estimation and of a sufficiently accurate approximation of the exact error (cf. e.g. [BEKar, Koron]).

Table 1 Approximation History

Level	Depth	Nodes	Iterations	est. Approx. Error	Effectivity
0	0	1	2	7.6 %	0.22
1	1	5	2	16.8 %	1.21
2	2	25	2	14.7 %	1.24
3	3	77	2	13.0 %	1.42
4	4	277	3	10.4 %	1.69
5	5	733	3	8.33 %	1.90
6	6	2937	2	6.2 %	2.06
7	7	4413	2	5.6 %	2.06
8	7	7249	2	4.9 %	1.99

From the moderate number of iterations on each refinement level, it can be hardly perceived that we are dealing with a nonlinear problem. Only the severe underestimation of the error on the initial level indicates that it may be dangerous to start an adaptive algorithm from such a coarse mesh.

In order to compare the convergence properties of the standard monotone multigrid method (STDKH) and of the truncated version (TRCKH), we now consider the iterative solution of the discrete problem on the final triangulation $\mathcal{T}_8$. Starting with the initial iterate $u_8^0 = 0$, we obtain the algebraic errors $\|u_8 - u_8^\nu\|$, $\nu = 0, \ldots, 20$, as shown in Figure 2.

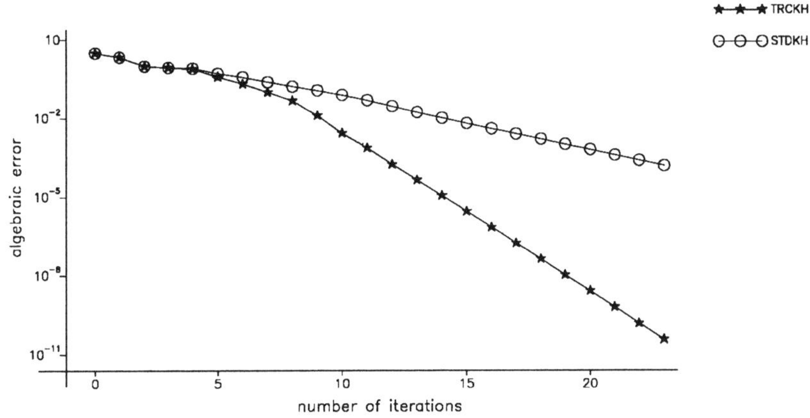

Figure 2   Iteration History: Initial Iterate $u_j^0 = 0$

The overall convergence behavior can be divided into a *transient* phase, dominated by the search for the (discrete) free boundary, and an *asymptotic* phase, corresponding to the iterative solution of the corresponding reduced linear problem. As compared to the standard method STDKH, the truncated version TRCKH exhibits a tremendous improvement of the asymptotic convergence rates, giving a numerical justification of the truncation of nodal basis functions. Note that the transient convergence properties remain basically the same. Replacing the artificial initial iterate zero by the interpolation from the previous level, the transient phase is eliminated from the convergence history. This explains the moderate number of iterations occuring in Table 1.

To study the convergence properties with increasing refinement level j, we consider the *asymptotic efficiency rates* $\rho_j = \sqrt[\nu_0]{\delta_j^{\nu_0}/\delta_j^0}$ where δ_j^ν denotes the algebraic error after ν iteration steps. We choose ν_0 such that $\delta_j^{\nu_0} < 10.^{-12}$. Starting with the interpolated initial iterate, the asymptotic efficiency rates for the standard and for the truncated multigrid method seem to saturate at the values 0.7 and 0.4, respectively. This is better than predicted by the theoretical results. Even for the "bad" initial iterate zero, we observed uniform *global bounds* of the convergence rates. A theoretical verification of these experimental results will be the subject of future research.

Acknowledgment   *The author thanks B. Erdmann from the Konrad–Zuse–Zentrum in Berlin for computational assistance.*

References

[BEK93] Bornemann F., Erdmann B., and Kornhuber R. (1993) Adaptive multilevel methods in three space dimensions. *J. Numer. Meth. Engrg.* 36: 3187–3203.

[BEKar] Bornemann F., Erdmann B., and Kornhuber R. (1996, to appear) A posteriori error estimates for elliptic problems in two and three space dimensions. *SIAM J. Numer. Anal.* 33.

[BX94a] Bank R. and Xu J. (1994) An algorithm for coarsening unstructured meshes. *Preprint* .

[BX94b] Bank R. and Xu J. (1994) The hierarchical basis multigrid method and incomplete LU decomposition. In Keyes and Xu [KX94], pages 163–174.

[Cra88] Crank J. (1988) *Free and Moving Boundary Problems.* Oxford University Press, Oxford.

[DL72] Duvaut G. and Lions J. (1972) *Les inéquations en mécanique et en physique.* Dunaud, Paris.

[ET76] Ekeland I. and Temam R. (1976) *Convex Analysis and Variational Problems.* North–Holland, Amsterdam.

[Glo84] Glowinski R. (1984) *Numerical Methods for Nonlinear Variational Problems.* Springer–Verlag, New York.

[GO95] Griebel M. and Oswald P. (1995) On the abstract theory of additive and multiplicative Schwarz algorithms. *Numer. Math.* 70: 163–180.

[Hac85] Hackbusch W. (1985) *Multi-Grid Methods and Applications.* Springer–Verlag, Berlin.

[HK94] Hoppe R. and Kornhuber R. (1994) Adaptive multilevel–methods for obstacle problems. *SIAM J. Numer. Anal.* 31(2): 301–323.

[Hop93] Hoppe R. (1993) A globally convergent multigrid algorithm for moving boundary problems of two-phase Stefan type. *IMA J. Numer. Anal.* 13: 235–253.

[HR89] Hackbusch W. and Reusken A. (1989) Analysis of a dampened nonlinear multilevel method. *Numer. Math.* 55: 225–246.

[HS90] Hoffmann K.-H. and Sprekels J. (eds) (1990) *Free Boundary Value Problems.* Birkhäuser, Basel.

[Kor94] Kornhuber R. (1994) Monotone multigrid methods for elliptic variational inequalities I. *Numer. Math.* 69: 167 – 184.

[Korara] Kornhuber R. (to appear) Monotone multigrid methods for elliptic variational inequalities II. *Numer. Math.*

[Koron] Kornhuber R. (to appear) A posteriori error estimates for elliptic variational inequalities. *Computers & Mathematics* .

[Korarb] Kornhuber R. (in preparation) *Adaptive Monotone Multigrid Methods for Nonlinear Variational Problems.*

[KX94] Keyes D. and Xu J. (eds) (1994) *Proceedings of the 7th International Conference on Domain Decomposition Methods 1993.* AMS, Providence.

[KY94] Kornhuber R. and Yserentant H. (1994) Multilevel methods for elliptic problems on domains not resolved by the coarse grid. In Keyes and Xu [KX94], pages 49–60.

[Man84] Mandel J. (1984) A multilevel iterative method for symmetric, positive definite linear complementarity problems. *Appl. Math. Optimization* 11: 77–95.

[McC92] McCormick S. (1992) *Multilevel Projection Methods for Partial Differential Equations*. SIAM, Philadalphia.

[Osw94] Oswald P. (1994) Stable subspace splittings for Sobolev spaces and domain decomposition algorithms. In Keyes and Xu [KX94], pages 87–98.

[Xu92] Xu J. (1992) Iterative methods by space decomposition and subspace correction. *SIAM Review* 34: 581–613.

[Yse93] Yserentant H. (1993) Old and new convergence proofs for multigrid methods. *Acta Numerica*, pages 285–326.

Domain Decomposition and Multilevel Techniques for Preconditioning Operators

SERGEI V. NEPOMNYASCHIKH[*]

1 INTRODUCTION

In recent years, domain decomposition methods have been used extensively to efficiently solve boundary value problems for partial differential equations in complex shape domains [4, 13, 16]. On the other hand, multilevel techniques on hierarchical data structures also have developed into an effective tool for the construction and analysis of fast solvers [2, 5, 15, 17]. But the direct realization of multilevel techniques on a parallel computer system for the global problem in the original domain involves difficult communication problems. In this paper, we present and analyze a combination of these two approaches: domain decomposition and multilevel decomposition on hierarchical structures to design optimal preconditioning operators.

Let $\Omega \subset R^2$ be a polygon. In the domain Ω we consider the boundary value problem

$$
\begin{cases}
-\displaystyle\sum_{i,j=1}^{2} \frac{\partial}{\partial x_i}\left(a_{ij}(x)\frac{\partial u}{\partial x_j}\right) + a_0(x)u &= f(x), & x \in \Omega, \\[2ex]
u(x) &= 0, & x \in \Gamma_0, \\[2ex]
\dfrac{\partial u}{\partial n_a} + \sigma(x)u &= 0, & x \in \Gamma_1.
\end{cases}
\tag{1.1}
$$

[*]Computing Center, Siberian Branch of Russian Academy of Sciences, Novosibirsk, 63090, Russia, e-mail: svnep@comcen.nsk.su

Domain Decomposition Methods in Sciences and Engineering, edited by R. Glowinski *et al.*

where

$$\frac{\partial u}{\partial n_a} = \sum_{i,j=1}^{2} a_{ij}(x)\frac{\partial u}{\partial x_j}\cos(n, x_i)$$

is the conormal derivative, n denotes the outward normal to Γ, and Γ_0 is a union of a finite number of curvilinear segments, $\Gamma = \Gamma_0 \cup \Gamma_1, \Gamma_0 = \bar{\Gamma}_0$. Here $\bar{\Gamma}_0$ denotes the closure of Γ_0.

By $H^1(\Omega, \Gamma_0)$ we denote the subspace of the Sobolev space $H^1(\Omega)$

$$H^1(\Omega, \Gamma_0) = \left\{ v \in H^1(\Omega) \,|\, v(x) = 0, \ x \in \Gamma_0 \right\}.$$

We introduce the bilinear form $a(u, v)$ and the linear functional $l(v)$:

$$a(u, v) = \int_\Omega \left(\sum_{i,j=1}^{2} a_{ij}(x)\frac{\partial u}{\partial x_j}\frac{\partial v}{\partial x_i} + a_0(x)uv \right) dx + \int_{\Gamma_1} \sigma(x)uvdx,$$

$$l(v) = \int_\Omega f(x)vdx.$$

Let us suppose that the operator coefficients and the right-hand side of the problem (1.1) are such that the bilinear form $a(u, v)$ is symmetric, elliptic, and continuous on $H^1(\Omega, \Gamma_0) \times H^1(\Omega, \Gamma_0)$, i.e.

$$a(u, v) = a(v, u) \quad \forall u, v \in H^1(\Omega, \Gamma_0),$$

$$\alpha_0\|u\|^2_{H^1(\Omega)} \le a(u, u) \le \alpha_1\|u\|^2_{H^1(\Omega)}, \quad \forall u \in H^1(\Omega, \Gamma_0)$$

and the linear functional $l(v)$ is continuous on $H^1(\Omega, \Gamma_0)$:

$$|l(u)| \le \alpha\|u\|_{H^1(\Omega)}, \quad \forall u \in H^1(\Omega, \Gamma_0).$$

The generalized solution $u \in H^1(\Omega, \Gamma_0)$ of (1.1) is, by definition, a solution to the projection problem [1]

$$u \in H^1(\Omega, \Gamma_0) \,:\, a(u, v) = l(v), \quad \forall v \in H^1(\Omega, \Gamma_0). \tag{1.2}$$

We know that under these assumptions for $a(u, v)$ and $l(v)$ there exists a unique solution of (1.2).

Let Ω be a union of n nonoverlapping subdomains Ω_i,

$$\bar{\Omega} = \bigcup_{i=1}^{n} \bar{\Omega}_i, \quad \Omega_i \cap \Omega_j = \emptyset, \ i \ne j,$$

where Ω_i are polygons with diameters on the order of H. Let us consider a coarse grid triangulation of Ω

$$\Omega_0^h = \bigcup_{i=1}^{n} \Omega_{0,i}^h, \quad \Omega_{0,i}^h = \bigcup_{l=1}^{M_i^{(0)}} \bar{\tau}_{i,l}^{(0)},$$

$$\text{diam}\,(\tau_i^{(0)}) = 0(H)$$

and we refine $\Omega_{0,i}^h$ several times. This results in a sequence of nested triangulations

$$\Omega_{i,0}^h, \Omega_{i,1}^h, \cdots, \Omega_{i,J}^h$$

such that

$$\bar{\Omega}_{i,k}^h = \bigcup_{l=1}^{M_i^{(k)}} \bar{\tau}_{i,l}^{(k)}, k = 0, 1, \cdots, J,$$

where the triangles $\tau_{i,l}^{(k+1)}$ are generated by subdividing triangles $\tau_{i,l}^{(k)}$ into four congruent subtriangles by connecting the midpoints of the edges.

Introduce the spaces

$$W_{i,0} \subset W_{i,1} \subset \cdots \subset W_{i,J} \;\; = \;\; H_h(\Omega_i),$$

$$V_{i,0} \subset V_{i,1} \subset \cdots \subset V_{i,J} \;\; = \;\; H_h(\Gamma_i), \tag{1.3}$$

$$\Gamma_i = \partial\Omega_i, \quad i = 1, 2, \cdots, n.$$

Here the space $W_{i,k}$ consists of real-valued functions which are continuous on Ω and linear on the triangles in $\Omega_{i,k}^h$. The space $V_{i,k}$ is the space of traces on Γ_i of functions from $W_{i,k}$:

$$V_{i,k} = \left\{ \varphi^h \mid \varphi^h = u^h|_{\Gamma_i}, \text{ with } u^h \in W_{i,k} \right\}.$$

We define the space $H_h(\Omega)$ of real continuous functions which are linear on each triangle of Ω^h and vanish at Γ_0.

Let us consider the projection problem

$$u^h \in H_h(\Omega) : \; a(u^h, v^h) = l(v^h), \quad \forall v^h \in H_h(\Omega) \tag{1.4}$$

which is an approximation of the problem (1.2).

Each function $u^h \in H_h(\Omega)$ is put in correspondence with a real column vector $u \in R^N$ whose components are values of the function u^h at the corresponding nodes of the triangulation Ω^h. Then (1.4) is equivalent to the system of mesh equations

$$Au \;=\; f,$$

$$(Au, v) \;=\; a(u^h, v^h), \quad \forall u^h, v^h \in H_h(\Omega), \tag{1.5}$$

$$(f, v) \;=\; l(v^h), \quad \forall v^h \in H_h(\Omega),$$

where u^h and v^h are the respective interpolations of vectors u and v; (f, v) is the Euclidean scalar product in R^N.

The goal of this work is to construct a symmetric positive definite preconditioning operator B for (1.5) so as to satisfy the inequalities

$$c_1(Bu, u) \le (Au, u) \le c_2(bu, u) \tag{1.6}$$

where the positive constants c_1 and c_2 are independent of h and H; the multiplication of a vector by B^{-1} should be easy to implement.

Using a combination of Additive Schwarz and Fictitious Space Methods, optimal preconditioning operators have been constructed in [11, 12, 13] for the case of arbitrary (unstructured) grids. However, that construction involves explicit extension operators whose implementation for three dimensional problems is optimal from the arithmetic cost and the condition number points of view but difficult for practical realization. The main goal of this work is to construct, using the hierarchical structure (1.3), a robust optimal preconditioning operator. One of the crucial points in [11, 12, 13] and this paper is the use of non–exact solvers in subdomains and explicit extension operators.

It means, to construct optimal preconditioning operators, we can design norm preserving operators of functions given at Γ_i into Ω_i with the optimal arithmetic cost (a number of arithmetic operations should be proportional to a number degrees of freedom) and then, instead of exact solvers in subdomains, we can use any spectrally equivalent preconditioning operators. Optimal extension operators have been presented in [8, 9, 11] for unstructured grids and robust explicit extension operators on hierarchical data structures in [5, 14].

The paper is organized as follows. In Section 2, using Additive Schwarz Method, we describe general construction of a preconditioning operator with local multilevel preconditioning operators. In Section 3, we present an optimal multilevel extension of grid functions from boundaries subdomains into inside subdomains. In Section 4, we propose an optimal interface preconditioning operator at the boundaries of the subdomains which involves a multilevel decomposition and corresponding explicit extension operators at interfaces.

2   DOMAIN DECOMPOSITION – ADDITIVE SCHWARZ-METHOD

To design the preconditioning operator for system (1.5), we use the additive Schwarz-Method [7] and employ the main idea of the construction of preconditioners from [13] for the hierarchical grids. Denote by $\overset{\circ}{H}_h(\Omega_i)$ the subspace of $H_h(\Omega_i)$

$$\overset{\circ}{H}_h(\Omega_i) = \left\{ u^h \in H_h(\Omega_i) \mid u^h(x) = 0, \quad x \in \Gamma_i \right\}$$

and define the local preconditioning operators B_i such that

$$B_i : \overset{\circ}{H}_h(\Omega_i) \to \overset{\circ}{H}_h(\Omega_i),$$

$$c_3 \|u^h\|^2_{H^1(\Omega_i)} \le (B_i u, u) \le c_4 \|u^h\|^2_{H^1(\Omega_i)} \quad \forall u^h \in \overset{\circ}{H}_h(\Omega_i),$$

where c_3, c_4 are independent of h and H. We hereafter use the same notation for an operator and its matrix representation. For instance, to define B_i, we can use the so-called BPX–preconditioners [3]. To do it, denote by $\{f_l^{(k)}\}$ nodal basis functions from the k–th level and define

$$B_i^{-1} u^h = \sum_{k=0}^{J} \sum_{f_l^{(k)} \in \overset{\circ}{H}_h(\Omega_i)} (u^h, f_l^{(k)})_{L_2(\Omega_i)} f_l^{(k)}. \tag{2.1}$$

Let us assume that we can define the extension operators t_i

$$t_i \; : \; V_{i,J} \longrightarrow W_{i,J}$$

such that

$$t\varphi^h \;\; = \;\; u^h,$$

$$u^h(x) \;\; = \;\; \varphi^h(x), \;\; x \in \Gamma_i, \tag{2.2}$$

$$\|t_i\varphi^h\|_{H^1(\Omega_i)} \le c_5\|\varphi^h\|_{H^{1/2}(\Gamma_i)} \quad \forall \varphi^h \in V_{i,J},$$

with c_5 independent of h and H. Here $\|\varphi^h\|_{H^{1/2}(\Gamma_i)}$ is the norm [10] in the Sobolev space $H^{1/2}(\Gamma_i)$

$$\|\varphi^h\|^2_{H^{1/2}(\Gamma_i)} = H \int\limits_{\Gamma_i} (\varphi^h(x))^2 dx + \int\limits_{\Gamma_i}\int\limits_{\Gamma_i} \frac{(\varphi^h(x) - \varphi^h(y))^2}{|x-y|^2} dx dy.$$

Then, we can define the extension operator t

$$t: \; H_h(S) \to H_h(\Omega),$$

where $H_h(S)$ is the space of traces of functions from $H_h(\Omega)$ at S

$$S = \bigcup_{i=1}^{n} \Gamma_i$$

and for any $\varphi^h \in H_h(S)$

$$t\varphi^h \;\; = \;\; u^h,$$

$$u^h(x) \;\; = \;\; \varphi^h(x), \quad x \in S,$$

$$\|t\varphi^h\|_{H^1(\Omega)} \;\; \le \;\; c_5\|\varphi^h\|_{H^{1/2}(S)}.$$

Here

$$\|\varphi^h\|^2_{H^{1/2}(S)} = \sum_{i=1}^{n} \|\varphi^h\|^2_{H^{1/2}(\Gamma_i)}.$$

The operator t_i from (2.2) is constructed in Section 3.

Let $\sum$ satisfies the following inequalities

$$c_6\|\varphi^h\|^2_{H^{1/2}(S)} \le (\textstyle\sum \varphi, \varphi) \le c_7\|\varphi^h\|^2_{H^{1/2}(S)} \quad \forall \varphi^h \in H_h(S), \tag{2.3}$$

where c_6, c_7 are independent of h and H. Then, according to [11], we can define the preconditioning operator B as follows

$$B^{-1} = \begin{bmatrix} 0 & & & \\ & B_1^{-1} & & \\ & & \ddots & \\ & & & B_n^{-1} \end{bmatrix} + t\sum{}^{-1} t^*. \tag{2.4}$$

Here 0 is the null-matrix which corresponds to nodes of the triangulation Ω^h at S and B_i is from (2.1).

The following theorem holds

Theorem 2.1 *If operator B is obtained from (2.4), then the constants c_1, c_2 in (1.6) are independent of h and H.*

3   MULTILEVEL EXPLICIT EXTENSION OPERATORS

The main goal of this section is to construct a robust operator t_i from (2.2). In this section, we omit the subscript i.

To design the extension operator

$$t \; : \; V_J \to W_J,$$

we follow to [5, 14]. Denote by $\varphi_i^{(k)}, i = 1, 2, \cdots, N_k$, the nodal basis of V_k and denote by $\Phi_i^{(k)}$ the one-dimensional subspace spanned by this function $\varphi_i^{(k)}$. Define

$$Q_i^{(k)} \; : \; L_2(\Gamma) \to \Phi_i^{(k)}$$

the L_2 orthogonal projection from $L_2(\Gamma)$ onto $\Phi_i^{(k)}$ and denote

$$\tilde{Q}_k = \sum_{i=1}^{N_k} Q_i^{(k)}, \quad k = 0, 1, \cdots, J - 1.$$

For $k = J$ we define $\tilde{Q}_J$ as the L_2 orthogonal projection from $L_2(\Gamma)$ onto V_j. The following lemmas hold [14].

Lemma 3.1 *There exists a positive constant c_8, independent of h and H, such that for any $\varphi^h \in V_J$ we have*

$$\|\varphi_0^h\|_{H^{1/2}(\Gamma)}^2 + \frac{1}{H}\|\varphi_1^h\|_{L_2(\Gamma)}^2 + |\varphi_1^h|_{H^{1/2}(\Gamma)}^2 \leq c_8\|\varphi^h\|_{H^{1/2}(\Gamma)}^2,$$

where

$$\varphi_0^h = \tilde{Q}_0\varphi^h, \quad \varphi_1^h = \varphi^h - \varphi^h. \tag{3.1}$$

Here

$$|\varphi^h|_{H^{1/2}(\Gamma)}^2 = \int_\Gamma \int_\Gamma \frac{(\varphi^h(x) - \varphi^h(y))^2}{|x - y|^2}dxdy.$$

Lemma 3.2 *There exists a positive constant c_9, independent of h and H, such that*

$$\|\varphi_0^h\|^2 + \frac{1}{H}\left(\|\tilde{Q}_0\varphi_1^h\|_{L_2(\Gamma)}^2 + \sum_{k=1}^{J} 2^k\|(\tilde{Q}_k - \tilde{Q}_{k-1})\varphi_1^h\|_{L_2(\Gamma)}^2\right) \leq c_9\|\varphi^h\|_{H^{1/2}(\Gamma)}^2,$$

where φ_0^h, φ_1^h are defined by (3.1).

The construction of operator t is based on the decomposition from Lemma 3.2. Denote by $x_i^{(k)}, i = 1, 2, \cdots, L_k$, the nodes of the triangulation Ω_k^h (we assume that nodes $x_i^{(k)}$ are enumerated first on Γ and then inside Ω) and define the extension operator t in the following way. For any $\varphi^h \in V_J$ set

$$
\begin{aligned}
\psi_0^h &= \tilde{Q}_0 \varphi^h, \\
\psi_k^h &= (\tilde{Q}_k - \tilde{Q}_{k-1})\varphi^h, \quad k = 1, 2, \cdots, J.
\end{aligned}
\tag{3.2}
$$

Then

$$
\varphi^h = \psi_0^h + \psi_1^h + \cdots + \psi_J^h.
$$

Define the extension $u_k^h \in W_k$ as follows

$$
\begin{aligned}
u_0^h(x_i^{(0)}) &= \begin{cases} \psi_0^h(x_i^{(0)}), & x_i^{(0)} \in \Gamma, \\ \bar{\psi}, & x_i^{(0)} \in \Gamma, \end{cases} \\[2mm]
u_k^h(x_i^{(k)}) &= \begin{cases} \psi_k^h(x_i^{(k)}), & x_i^{(k)} \in \Gamma, \\ 0, & x_i^{(k)} \notin \Gamma, \end{cases}
\end{aligned}
\tag{3.3}
$$

$$
k = 1, 2, \cdots, J.
$$

Here $\bar{\psi}$ is, for instance, the meanvalue of the function ψ_0^h on Γ, namely

$$
\bar{\psi} = \frac{1}{N_0} \sum_{i=1}^{N_0} \psi_0^h(x_i^{(0)}).
$$

Define

$$
t\varphi^h = u^h \equiv u_0^h + u_1^h + \cdots + u_J^h.
\tag{3.4}
$$

Remark 3.1 *We can use the L_2 orthogonal projections from $L_2(\Gamma)$ onto V_k instead of $\tilde{Q}_k, k = 0, 1, \cdots, J-1$. But in this case the cost of the decomposition (3.2) is expensive (especially for three dimensional problems).*

Theorem 3.1 *There exists a positive constant c_{10}, independent of h and H, such that*

$$
\|t\varphi^h\|_{H^1(\Omega)} \leq c_{10} \|\varphi^h\|_{H^{1/2}(\Gamma)} \quad \forall \varphi^h \in V_J.
$$

Here the operator t is from (3.2)–(3.4).

Remark 3.2 *It is obvious that*

$$
Q^{(k)}\varphi^h = \frac{(\varphi^h, \varphi_i^{(k)})_{L_2(\Gamma)}}{(\varphi_i^k, \varphi_i^k)_{L_2(\Gamma)}} \varphi_i^{(k)}
$$

and the cost of the action of t and t^* is proportional to the number of nodes of the grid domain.

4 INTERFACE PRECONDITIONING OPERATORS

In this section, we construct an optimal interface preconditioner in the space $H_h(S)$ which satisfies (2.3). To do it, we use the idea of Additive Schwarz Method at interface S from [13]. Let S be a union of K nonoverlapping edges E_i of the triangulation Ω_0^h

$$S = \bigcup_{j=1}^{K} \bar{E}_j, \quad E_j \cap E_i = \emptyset, \quad i \neq j.$$

Split $H_h(S)$ into a vector sum of subspaces

$$H_h(S) = U_0 + U_1 + \cdots + U_k, \tag{4.1}$$

where U_0 is the coarse space which consists of continuous functions linear on the edges E_j, $j = 1, 2, \ldots, K$, and U_j, $j = 1, 2, \ldots, K$, correspond to E_j and are defined below.
Denote by

$$\mathring{U}_j = \{\varphi^h \in H_h(S) \mid \varphi^h(x) = 0, \ x \notin E_j\},$$
$$\tilde{U}_j^{(k)} = V_k\big|_{E_j}, \quad k = 0, 1, \ldots, J.$$

For any edge E_j we define the explicit extension operator τ_j

$$\tau_j \ : \ \tilde{U}_j^{(J)} \to H_h(S)$$

as follows. Denote by $\varphi_{j,i}^{(k)}$, $i = 1, 2, \ldots, I_j^{(k)}$, the nodal basis of $\tilde{U}_j^{(k)}$ (the functions $\varphi_{j,i}^{(k)}$ differ from the functions $\varphi_i^{(k)}$ from Section 3 only at the end points of E_j) and denote by $\Phi_{j,i}^{(k)}$ the one-dimensional subspace spanned by this function $\varphi_{j,i}^{(k)}$. Denote by

$$Q_{j,i}^{(k)} \ : \ L_2(E_j) \to \Phi_{j,i}^{(k)}$$

corresponding L_2 orthogonal projection. Set

$$\tilde{Q}_j^{(k)} = \sum_{i=1}^{I_j^{(k)}} Q_{j,i}^{(k)}, \quad k = 0, 1, \cdots, J - 1,$$

and define $\tilde{Q}_J^{(k)}$ as the L_2 orthgonal oprojection from $L_2(E_j)$ onto $\tilde{U}_j^{(J)}$. Now we can define the extension operator τ_j according to (3.2)–(3.4). For any $\varphi^h \in \tilde{U}_j^{(J)}$ set

$$\begin{aligned}
\psi_0^h &= \tilde{Q}_j^{(0)}\varphi^h, \\
\psi_k^h &= (\tilde{Q}_j^{(k)} - \tilde{Q}_j^{(k-1)})\varphi^h, \quad k = 1, 2, \cdots, J,
\end{aligned} \tag{4.2}$$

and

$$u_k^h = \begin{cases} \psi_k^h(x_i^{(k)}), & x_i^{(k)} \in E_j \\ 0, & x_i^{(k)} \notin E_j, \quad k = 0, 1, \cdots, J, \end{cases} \tag{4.3}$$

$$\tau_j \varphi^h = u_0^h + u_1^h + \cdots + u_J^h.$$

Define

$$U_j = \mathring{U}_j + \tau_j \tilde{U}_j.$$

Then from Theorem 3.1 and [13] we have the following theorem concerning decomposition (4.1).

Theorem 4.1 *There exists a positive constant c_{11}, dependent of h and H, such that for any function $\varphi^h \in H_h(S)$ there exist $\varphi_j^h \in U_j$, $j = 0, 1, \ldots, K$, such that*

$$\varphi_0^h + \varphi_1^h + \cdots + \varphi_k^h = \varphi^h,$$

$$\|\varphi_0^h\|_{H^{1/2}(S)}^2 + \|\varphi_1^h\|_{H^{1/2}(S)}^2 + \cdots + \|\varphi_K^h\|_{H^{1/2}(S)}^2 \le c_{11}\|\varphi^h\|_{H^{1/2}(S)}^2$$

Let the operator $\sum_0$ generates an equivalent norm in U_0, namely

$$c_{12}\|\varphi^h\|_{H^{1/2}(S)}^2 \le (\textstyle\sum_0 \varphi, \varphi) \le c_{13}\|\varphi^h\|_{H^{1/2}(S)}^2 \quad \forall \varphi^h \in U_0, \tag{4.4}$$

where c_{12}, c_{13} independent of h and H. Define local preconditioners for U_j, $j = 1, 2, \ldots, K$. Denote by $\mathring{\sum}_j$ and $\widetilde{\sum}_j$ the BPX-like preconditioners in the spaces $\mathring{U}_j$ and $\tilde{U}_j$, respectivly

$$\mathring{\sum}_i{}^{-1} \varphi^h = \sum_{k=0}^{J} \sum_{\sup \varphi_{j,i}^{(k)} \subset E_j} (\varphi^h, \varphi_{j,i}^{(k)})_{L_2(E_j)} \varphi_{j,i}^{(k)} \quad \forall \varphi^h \in \mathring{U}_j,$$

$$\widetilde{\sum}_i{}^{-1} \varphi^h = \sum_{k=0}^{J} \sum_{\sup \varphi_{j,i}^{(k)} \cap E_j \neq \emptyset} (\varphi^h, \varphi_{j,i}^{(k)})_{L_2(E_j)} \varphi_{j,i}^{(k)} \quad \forall \varphi^h \in \tilde{U}_j.$$

Then, define the interface preconditioning operator $\sum$ in the following way

$$\sum{}^{-1} = \sum_0^+ + \sum_{j=0}^{K} \left(\mathring{\sum}_j{}^{-1} + \tau_j \widetilde{\sum}_j{}^{-1} \tau_j^* \right). \tag{4.5}$$

Here $\sum_0^+$ is a pseudo-inverse of $\sum_0$ from (4.4), τ_j is obtained from (4.2), (4.3), and we extend operator $\mathring{\sum}_j{}^{-1}$ by zero outside E_j. The following theorem holds.

Theorem 4.2 *If operator $\sum$ is defined from (4.5) then the constants c_6, c_7 from (2.3) are independent of h and H.*

Remark 4.1 *The method suggested in this paper can be generalized evidently for three dimensional problems.*

Remark 4.2 *Using combinations of the presented techniques and techniques from [10], effective preconditoning operators for elliptic problems with discontinuous coefficients can be constructed.*

References

[1] J.-P. Aubin. *Approximation of elliptic boundary-value problems*. Wiley–Interscience, New York, London, Sydney, Toronto, 1972.

[2] J. H. Bramble. *Multigrid Methods*. Research Notes in Mathematics Series. Pitman, Boston-London-Melbourne, 1993.

[3] J. H. Bramble, J. E. Pasciak, and J. Xu. Parallel multilevel preconditioners. *Mathematics of Computation*, 55(191):1–22, 1990.

[4] M. Dryja and O. B. Widlund. Multilevel additive methods for elliptic finite element problems. In W. Hackbusch, editor, *Parallel Algorithms for Partial Differential Equations*, pages 58–69, Braunschweig, 1991. Vieweg–Verlag. Proc. of the Sixth GAMM–Seminar, Kiel, January 19–21, 1990.

[5] G. Haase, U. Langer, A. Meyer, and S. V. Nepomnyaschikh. Hierarchical extension and local multigrid methods in domain decomposition preconditioners. *East–West J. Numer. Math.*, 2(3):173–193, 1994.

[6] W. Hackbusch. *Multi–Grid Methods and Applications*, volume 4 of *Springer Series in Computational Mathematics*. Springer–Verlag, Berlin, 1985.

[7] A. M. Matsokin and S. V. Nepomnyaschikh. A Schwarz alternating method in a subspace. *Soviet Mathematics*, 29(10):78–84, 1985.

[8] A. M. Matsokin and S. V. Nepomnyaschikh. Norms in the space of traces of mesh functions. *Sov. J. Numer. Anal. Math. Modelling*, 3:199–216, 1988.

[9] S. V. Nepomnyaschikh. *Domain decomposition and Schwarz methods in a subspace for the approximate solution of elliptic boundary value problems*. PhD thesis, Computing Center of the Siberian Branch of the USSR Academy of Sciences, Novosibirsk, 1986.

[10] S. V. Nepomnyaschikh. Domain decomposition methods for elliptic problems with discontinuous coefficients. In R. Glowinski, Y. A. Kuznetsov, G. A. Meurant, and J. Periaux, editors, *Domain decomposition methods for partial differential equations*, pages 242–251, Philadelphia, 1991. SIAM. Proceedings of the 4th International Symposium, Moscow, 1990.

[11] S. V. Nepomnyaschikh. Method of Splitting into Subspaces for Solving Elliptic Boundary Value Problems in Complex–form Domains. *Sov. J. Numer. Anal. Math. Modelling*, 6:151–168, 1991.

[12] S. V. Nepomnyaschikh. Mesh theorems on traces, normalization of function traces and their inversion. *Sov. J. Numer. Anal. Math. Modelling*, 6:1–25, 1991.

[13] S. V. Nepomnyaschikh. Decomposition and Fictitious Domains Methods for Elliptic Boundary Value Problems. In T. F. Chan, D. E. Keyes, G. A. Meurant, T. S. Scroggs, and R. G. Voigt, editors, *5th Conference on Domain Decomposition Methods for PDE*, pages 62–72, Philadelphia, 1992. SIAM.

[14] S. V. Nepomnyaschikh. Optimal multilevel extension operators. Preprint SPC 95_3, Technische Universität Chemnitz–Zwickau, Fakultät für Mathematik, 1995.

[15] P. Oswald. *Multilevel Finite Element Approximation: Theory and Applications.* Teubner Skripten zur Numerik. B. G. Teubner Stuttgart, 1994.

[16] P. Le Tallec. Domain Decomposition Methods In Computational Mechanics. *Computational Mechanics Advances (North Holland)*, 1(2):121–220, Feb. 1994.

[17] J. Xu. Iterative methods by space decomposition and subspace correction. *SIAM Review*, 34:581–613, 1992.

Cascadic Multigrid Methods

FOLKMAR A. BORNEMANN AND PETER DEUFLHARD

1   Introduction

In this contribution we consider linear scalar elliptic problems on general domains with space dimension $d \leq 3$. For the numerical solution of such problems with finite elements, *multigrid methods* are a both popular and efficient choice, cf. Hackbusch [Hac85]. Such methods work on a sequence of grid levels $j = 0, 1, \dots \ell$, where in our notation $j = 0$ denotes the coarse grid level and $j > 0$ the refinement levels. For the case of *adaptive* grids, the local multigrid method suggested by [BPWX91] turned out to be the method of choice. In the adaptive setting it is favorable to consider *nested iterations*, wherein the computed solution on the previous level serves as starting point for the iteration on the new level. A typical distinction of different types of nested multigrid methods is made by the number p of correction cycles on each level: W-cycles are characterized by $p = 2$, V-cycles by $p = 1$.

For the first time the case $p = 0$, i.e., performing *no* coarse grid correction at all, was seriously considered by Deuflhard [Deu94]. He pointed out that the use of an a posteriori algorithmic control of this kind of nested iteration in combination with the conjugate gradient method gave reasonable results in practice. He called the method *cascadic conjugate gradient method*. As a distinctive feature this method performs *more iterations on coarser levels* so as to obtain less iterations on finer levels. Shaidurov [Sha94] as well as Bornemann and Deuflhard [BD96] proved accuracy and optimality of this approach with respect to the *energy norm*. The latter two authors could also show, that the conjugate gradient method can be replaced by any smoothing iteration like the traditional candidates SSOR or damped Jacobi. Because of this fact, they called this kind of iteration *cascadic multigrid methods*.

Domain Decomposition Methods in Sciences and Engineering, edited by R. Glowinski *et al.*
© 1997 John Wiley & Sons, Ltd.

In order to convey the basic structure, we give a schematic comparison of the cascadic multigrid method with the nested V-cycle multigrid method:

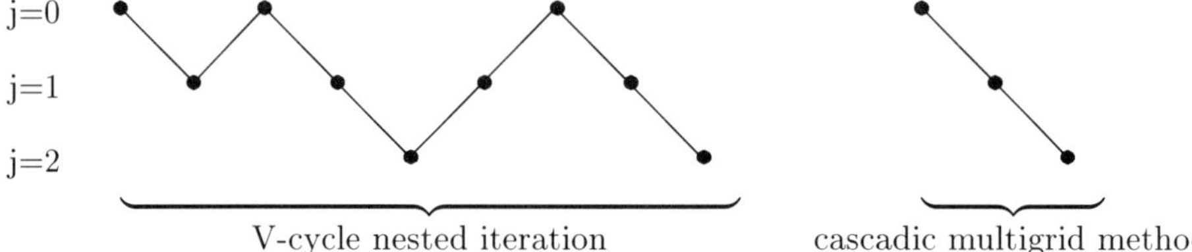

The purpose of the present contribution is to survey the main methods of the proof, the main results for the case of adaptive triangulations and to include some numerical comparisons with nested multigrid.

2   Analysis for general smoothers

In this section, we analyze the cascadic multigrid method with respect to accuracy and computational complexity using a general smoother as iterative method on each discretization level.

Let $\Omega \subset \mathbf{R}^d$ be a polygonal Lipschitz domain. We consider an elliptic Dirichlet problem on Ω in the weak formulation:

$$u \in H_0^1(\Omega) : \qquad a(u, v) = (f, v)_{L^2} \quad \forall v \in H_0^1(\Omega).$$

For the sake of a clear notation we only consider the H^2-regular case, i.e.,

$$\|u\|_{H^2} \le c \|f\|_{L^2} \qquad \forall f \in L^2(\Omega).$$

The induced energy-norm will be denoted by

$$\|u\|_a^2 = a(u, u) \qquad \forall u \in H_0^1(\Omega).$$

Given a nested family of triangulations $(\mathcal{T}_j)_{j=1}^{\ell}$, the spaces of linear finite elements are

$$X_j = \{u \in C(\bar{\Omega}) : u|_T \in P_1(T) \quad \forall T \in \mathcal{T}_j, \quad u|_{\partial\Omega} = 0\},$$

where $P_1(T)$ denotes the linear functions on the triangle T. We have

$$X_0 \subset X_1 \subset \ldots \subset X_\ell \subset H_0^1(\Omega).$$

The finite element approximations are given by

$$u_j \in X_j : \qquad a(u_j, v_j) = (f, v_j)_{L^2} \qquad \forall v_j \in X_j.$$

In this section we consider *quasi-uniform* triangulations with meshsize parameter

$$\frac{1}{c} 2^{-j} \le h_j = \max_{T \in \mathcal{T}_j} \operatorname{diam} T \le c\, 2^{-j}.$$

For ease of notation, we will use the symbol c for any positive constant, that only depends on the bilinear form $a(\cdot,\cdot)$, on Ω and the shape regularity as well as the quasi-uniformity of the triangulations. All other dependencies will be stated explicitly.

Denoting the *basic iterative procedure* on each level by $\mathcal{I}$, the cascadic multigrid method can be written as:

$$
\begin{aligned}
&\text{(i)} \quad u_0^* = u_0 \\
&\text{(ii)} \quad j = 1, \ldots, \ell: \quad u_j^* = \mathcal{I}_{j,m_j} u_{j-1}^*.
\end{aligned}
\tag{1}
$$

Here $\mathcal{I}_{j,m_j}$ denotes m_j steps of the basic iteration applied on level j.

We call a cascadic multigrid method *optimal* for level ℓ (with respect to the energy norm), if we obtain *both accuracy*

$$
\|u_\ell - u_\ell^*\|_a \approx \|u - u_\ell\|_a,
$$

which means that the iteration error is comparable to the approximation error, *and multigrid complexity*

$$
\text{amount of work} \ = O(n_\ell),
$$

where $n_\ell = \dim X_\ell$.

We consider the following type of basic iterations for the finite-element problem on level j started with $u_j^0 \in X_j$:

$$
\|u_j - \mathcal{I}_{j,m_j} u_j^0\|_a \le \|\mathcal{S}_{j,m_j}(u_j - u_j^0)\|_a
$$

with a *linear* mapping $\mathcal{S}_{j,m_j} : X_j \to X_j$ for the error propagation. We call the basic iteration an *energy reducing smoother*, if it obeys the *smoothing properties*

$$
\begin{aligned}
&\text{(i)} \quad \|\mathcal{S}_{j,m_j} v_j\|_a \le c\, \frac{h_j^{-1}}{m_j^\gamma} \|v_j\|_{L^2} \qquad \forall v_j \in X_j, \\
&\text{(ii)} \quad \|\mathcal{S}_{j,m_j} v_j\|_a \le \|v_j\|_a
\end{aligned}
\tag{2}
$$

with a parameter $0 < \gamma \le 1$. As is shown in [Hac85] the symmetric Gauß-Seidel, the SSOR and the damped Jacobi iteration are smoothers in the sense of (2) with parameter $\gamma = 1/2$, for which the operator $\mathcal{S}_{j,m_j}$ is simply defined by

$$
u_j - \mathcal{I}_{j,m_j} u_j^0 = \mathcal{S}_{j,m_j}(u_j - u_j^0).
$$

This simple construction does not work for the conjugate gradient method because of its well-known *nonlinearity*. However, the conjugate gradient method can be put into our framework by means of a nontrivial construction as given by the authors in [BD96] which yields $\gamma = 1$.

The smoothing property (2(i)) resembles an *inverse* inequality which is exactly accompanied by the following *approximation property*

$$
\|u_j - u_{j-1}\|_{L^2} \le c\, h_j \|u_j - u_{j-1}\|_a \qquad j = 1, \ldots, \ell,
\tag{3}
$$

which can be proved by the standard Aubin-Nitsche duality argument.

The smoothing property and the approximation property are now the main building blocks of the central convergence estimate for the cascadic multigrid method (1).

Theorem 1 *The error of the cascadic multigrid method with a smoother as basic iteration can be estimated by*

$$\|u_\ell - u_\ell^*\|_a \le c \sum_{j=1}^{\ell} \frac{1}{m_j^\gamma} \|u_j - u_{j-1}\|_a \le c \sum_{j=1}^{\ell} \frac{h_j}{m_j^\gamma} \|f\|_{L^2}.$$

Proof. For $j = 1, \ldots, \ell$ we get by the linearity of $\mathcal{S}_{j,m_j}$

$$\begin{aligned}
\|u_j - u_j^*\|_a &= \|u_j - \mathcal{I}_{j,m_j} u_{j-1}^*\|_a \le \|\mathcal{S}_{j,m_j}(u_j - u_{j-1}^*)\|_a \\
&\le \|\mathcal{S}_{j,m_j}(u_j - u_{j-1})\|_a + \|\mathcal{S}_{j,m_j}(u_{j-1} - u_{j-1}^*)\|_a.
\end{aligned}$$

The first term can be estimated by the smoothing property (2(i)) and the approximation property (3):

$$\|\mathcal{S}_{j,m_j}(u_j - u_{j-1})\|_a \le c \frac{h_j^{-1}}{m_j^\gamma} \|u_j - u_{j-1}\|_{L^2} \le c \frac{1}{m_j^\gamma} \|u_j - u_{j-1}\|_a.$$

If we estimate the second term by property (2(ii)) of a smoother, we thus get

$$\|u_j - u_j^*\|_a \le \frac{c}{m_j^\gamma} \|u_j - u_{j-1}\|_a + \|u_{j-1} - u_{j-1}^*\|_a. \tag{4}$$

The start $u_0^* = u_0$ and induction yields the assertion. $\square$

Since we have

$$2^{\ell-j} h_\ell/c \le h_j \le c\, 2^{\ell-j} h_\ell$$

Theorem 1 leads us to consider sequences $m_1, \ldots, m_\ell$ of the kind

$$m_j = \lceil \beta^{\ell-j} m_\ell \rceil, \tag{5}$$

for some fixed $\beta > 0$. Now, a simple evaluation of geometric sums gives conditions for accuracy and optimality:

Lemma 2 *Let the number m_j of iterations on level j be given by (5). The cascadic multigrid method yields the error*

$$\|u_\ell - u_\ell^*\|_a \le c \cdot \frac{1}{1 - (2/\beta^\gamma)} \cdot \frac{h_\ell}{m_\ell^\gamma} \|f\|_{L^2}, \qquad \text{for} \quad \beta > 2^{1/\gamma},$$

and a computational cost proportional to

$$\sum_{j=1}^{\ell} m_j n_j \le c \cdot \frac{1}{1 - \beta/2^d} \cdot m_\ell n_\ell, \qquad \text{for} \quad \beta < 2^d.$$

This Lemma shows that the two goals accuracy and multigrid complexity are not contradicting each other as long as

$$\gamma > 1/d.$$

As shown in [BD96] either the accuracy or the complexity has to deteriorate logarithmically for $\gamma = 1/d$.

Summarizing, we have proved that a plain symmetric Gauß-Seidel, SSOR or damped Jacobi iteration as basic iteration is optimal for $d = 3$, whereas the conjugate gradient method is optimal for $d \ge 2$.

3 Adaptive cascadic multigrid methods

In this section, we develop an a posteriori control for the number m_j of iteration on each level. This will be done for adaptively chosen triangulations, thus dropping the assumption of quasi-uniformity.

Using a *diagonally preconditioned* stiffness matrix the authors show in [BD96] for the smoothing iterations considered

$$\|\mathcal{S}_{j,m_j} v_j\|_a \leq \frac{c}{m_j^\gamma} \left(\sum_{T \in \mathcal{T}_j} h_T^{-2} \|v_j\|_{L^2(T)}^2 \right)^{1/2}.$$

In order to compensate the hidden local inverse inequality we make a local and a global approximation assumption:

$$\begin{aligned}
&\text{(i)} \quad h_T^{-2}\|u_j - u_{j-1}\|_{L^2(T)}^2 \leq c\|u_j - u_{j-1}\|_{H^1(T)}^2, \qquad \forall T \in \mathcal{T}_j \\
&\text{(ii)} \quad \|u - u_j\| \leq c\,n_j^{-1/d}\|f\|_{L^2},
\end{aligned} \tag{6}$$

which are heuristically justified for *adaptive* triangulations. Note that quasi-uniform triangulations do not satisfy assumption (ii) for problems which are not H^2-regular.

The proof of Theorem 1 gives the estimate

$$\|u_\ell - u_\ell^*\|_a \leq c \sum_{j=1}^{\ell} \frac{1}{m_j^\gamma}\|u_j - u_{j-1}\|_a \leq c \sum_{j=1}^{\ell} \frac{1}{m_j^\gamma\, n_j^{1/d}}\|f\|_{L^2}.$$

We can now extend the accuracy and optimality result Lemma 2 to the adaptive case. Here we have to use additionally, that the sequence of number of unknowns belongs to a geometric progression:

$$n_j < \sigma_0 n_j \leq n_{j+1} \leq \sigma_1 n_j \qquad j = 0, 1, \ldots.$$

With the choice of the iteration numbers m_j on level j as

$$m_j = \left\lceil m_\ell \left(\frac{n_\ell}{n_j} \right)^{(d+1)/2d\gamma} \right\rceil, \tag{7}$$

we get for $d > 1$ under assumption (6) the final error

$$\|u_\ell - u_\ell^*\|_a \leq \frac{c}{m_\ell^\gamma\, n_\ell^{1/d}}\|f\|_{L^2}$$

and for $\gamma = 1$ the complexity

$$\sum_{j=1}^{\ell} m_j\, n_j \leq c\, m_\ell\, n_\ell.$$

However, at the intermediate level j we do *not* know the number n_ℓ of nodal points at the final level ℓ, which means that so far our iteration is not yet implementable.

To make it implementable, we *define* the final level ℓ as the first level on which the approximation error is below some user given tolerance TOL. Hence assumption (6) gives us the relation

$$\frac{\|u - u_j\|_a}{\text{TOL}} \approx \left(\frac{n_\ell}{n_j}\right)^{1/d}, \tag{8}$$

which leads us to replace (7) by

$$m_j = \left\lceil m_\ell \left(\frac{\|u - u_j\|_a}{\text{TOL}}\right)^{(d+1)/2\gamma} \right\rceil. \tag{9}$$

This algorithm is closest to the a priori choice of the parameters m_j. However, in an actual computation, the basic iteration can be accurate enough much earlier than stated by theory. Therefore, we now go back to the crucial recursion (4), i.e.,

$$\|u_j - u_j^*\|_a \leq \frac{c}{m_j^\gamma}\|u - u_{j-1}\|_a + \|u_{j-1} - u_{j-1}^*\|_a,$$

which we simply turn into a termination criterion for the basic iteration by inserting (9). We thus end up with the termination criterion

$$\|u_j - u_j^*\|_a \leq \rho \left(\frac{\text{TOL}}{\|u - u_j\|_a}\right)^{(d+1)/2} \|u - u_{j-1}\|_a + \|u_{j-1} - u_{j-1}^*\|_a, \tag{10}$$

where $0 < \rho \leq 1$ is some safety factor. Note that the smoothing parameter γ dropped out since we stress the accuracy aspect of our analysis for the adaptive control. Herein the approximation error $\|u - u_j\|_a$ is not known, but can be replaced by the estimate

$$\|u - u_j\|_a \approx \|u - u_{j-1}\|_a \left(\frac{n_{j-1}}{n_j}\right)^{1/d} \approx \epsilon_{j-1} \left(\frac{n_{j-1}}{n_j}\right)^{1/d},$$

where ϵ_{j-1} denotes some estimate of the discretization error on the previous level, which is certainly provided by any adaptive algorithm; cf. [DLY89, BEK96]. If we replace the iteration errors $\|u_j - u_j^*\|_a$, $\|u_{j-1} - u_{j-1}^*\|_a$ by appropriate estimates δ_j, δ_{j-1}, the design of an implementable control strategy for the adaptive cascadic multigrid method is completed:

$$\delta_j \leq \rho \left(\frac{\text{TOL}}{\epsilon_{j-1}} \left(\frac{n_j}{n_{j-1}}\right)^{1/d}\right)^{(d+1)/2} \epsilon_{j-1} + \delta_{j-1}. \tag{11}$$

4   A numerical experiment

The above designed adaptive cascadic multigrid method has been implemented in four different variants regarding to the choice of the underlying basic iteration:

- **CCG:** the conjugate gradient method, i.e., no coarse grid correction (cascadic conjugate gradient method).

- **CSGS:** the symmetric Gauß-Seidel iteration, i.e., again no coarse grid correction.
- **BPX:** the conjugate gradient method with the so-called BPX multilevel preconditioner due to Bramble, Pasciak and Xu [BPX90, Xu92, BY93]. This implementation is closest to the original "cascade principle" of Deuflhard, Leinen and Yserentant [DLY89, BEK93].
- **V-cycle:** a multigrid V-cycle using a Jacobi smoother, local — i.e., on new nodal points and neighbors only — in the case of locally refined triangulations. This implementation can be viewed as a robust and efficient *automatic choice* of the number m_j of V-cycles on level j for the usual *nested multigrid method*.

The variants were tested using the elliptic problem

$$-\Delta u = 0, \qquad u|_{\Gamma_1} = 10^3, \quad u|_{\Gamma_2} = 0, \quad \partial_n u|_{\Gamma_3} = 0$$

on a domain Ω which is a unit square with slit

$$\Omega = \{x \in \mathbf{R}^2 : |x|_\infty \leq 1\} \cap \{x \in \mathbf{R}^2 : |x_2| \geq 0.03 x_1\}.$$

The boundary pieces are

$$\Gamma_1 = \{x \in \Omega : x_1 = 1, x_2 \geq 0.03\}, \quad \Gamma_2 = \{x \in \Omega : x_1 = 1, x_2 \leq -0.03\},$$

and $\Gamma_3 = \partial\Omega \setminus (\Gamma_1 \cup \Gamma_2)$.

The accuracy parameter was set to a relative $\mathrm{TOL}_\mathrm{rel} = 2.24 \cdot 10^{-2}$, i.e., $\mathrm{TOL} = \|u\|_a \cdot \mathrm{TOL}_\mathrm{rel} = 412.52 \cdot \mathrm{TOL}_\mathrm{rel}$. Throughout the safety factor was set to $\rho = 0.4$. These choices yield 15 refinement levels with a final triangulation of roughly 4400 nodal points, which allows one to estimate the *discretization error* using two further *uniform* refinements:

$$\frac{\|u - u_{15}\|_a}{\|u\|_a} \approx 2 \cdot 10^{-2}, \qquad \frac{\|u - u_{15}\|_{L^2}}{\|u\|_{L^2}} \approx 1 \cdot 10^{-4},$$

where $\|u\|_a = 412.52$ and $\|u\|_{L^2} = 1121.36$.

Table 1   Algebraic errors of the variants for $\mathrm{TOL}_\mathrm{rel} = 2.24 \cdot 10^{-2}$

variant	$\dfrac{\|u_{15} - u_{15}^*\|_a}{\|u_{15}\|_a}$	$\dfrac{\|u_{15} - u_{15}^*\|_{L^2}}{\|u_{15}\|_{L^2}}$	$\sum \dfrac{\mathrm{CPU\text{-}time}}{n_{15}}$
CCG	$6 \cdot 10^{-2}$	$1 \cdot 10^{-2}$	2.14 ms
CSGS	$6 \cdot 10^{-2}$	$1 \cdot 10^{-2}$	2.70 ms
BPX	$4 \cdot 10^{-3}$	$2 \cdot 10^{-4}$	3.53 ms
V-cycle (adap.)	$2 \cdot 10^{-3}$	$1 \cdot 10^{-4}$	4.27 ms

Table 1 shows the behavior of the (purely) *algebraic error* for the different variants. In *energy norm* they are comparable to the required accuracy. However, the variants without coarse grid corrections are slightly less accurate by a factor of three, whereas the nested multigrid like variants BPX and V-cycle stay nicely below the discretization error.

With respect to the L^2-*norm* only the variants which include in some sense a coarse grid correction (BPX and V-cycle) give satisfactory results. This fact points out that the given termination criterion (11) can be viewed as a robust tool to control the number of iterations in nested multigrid methods. In terms of Schwarz methods [Xu92] one observes that the *additive* variant BPX is by a factor of two less accurate than the *multiplicative* V-cycle, but slightly faster.

Remark. The reader should note, that the term *optimality* has been used in this paper with respect to the *energy norm*. As Table 1 already indicates things are totally different for the L^2- or L^∞-norm. By means of simple examples one can *prove* that the cascadic multigrid method *cannot* be optimal with respect to the L^2-norm. There is a rather precise theoretical understanding of this phenomenon, which will be subject of a forthcoming publication [BK95].

REFERENCES

[BD96] Bornemann F. A. and Deuflhard P. (1996) The cascadic multigrid method for elliptic problems. *Numer. Math.* to appear.

[BEK93] Bornemann F. A., Erdmann B., and Kornhuber R. (1993) Adaptive multilevel methods in three space dimensions. *Inter. J. Numer. Meth. Engrg.* 36: 3187–3202.

[BEK96] Bornemann F. A., Erdmann B., and Kornhuber R. (1996) A posteriori error estimates for elliptic problems in two and three space dimensions. *SIAM J. Numer. Anal.* to appear.

[BK95] Bornemann F. A. and Krause R. (1995) manuscript in preparation.

[BY93] Bornemann F. A. and Yserentant H. (1993) A basic norm equivalence for the theory of multilevel methods. *Numer. Math.* 64: 455–476.

[BPWX91] Bramble J. H., Pasciak J. E., Wang J., and Xu J. (1991) Convergence estimates for multigrid algorithms without regularity assumptions. *Math. Comp.* 57: 23–45.

[BPX90] Bramble J. H., Pasciak J. E., and Xu J. (1990) Parallel multilevel preconditioners. *Math. Comp.* 55: 1–22.

[Deu94] Deuflhard P. (1994) Cascadic conjugate gradient methods for elliptic partial differential equations. Algorithm and numerical results. In Keyes D. and Xu J. (eds) *Proceedings of the 7th International Conference on Domain Decomposition Methods 1993*. AMS, Providence.

[DLY89] Deuflhard P., Leinen P., and Yserentant H. (1989) Concepts of an adaptive hierarchical finite element code. *IMPACT Comput. Sci. Engrg.* 1: 3–35.

[Hac85] Hackbusch W. (1985) *Multi-Grid Methods and Applications.* Springer-Verlag, Berlin, Heidelberg, New York.

[Sha94] Shaidurov V. V. (1994) *Some estimates of the rate of convergence for the cascadic conjugate-gradient method.* Otto-von-Guericke-Universität, Magdeburg. Preprint.

[Xu92] Xu J. (1992) Iterative methods by space decomposition and subspace correction. *SIAM Review* 34: 581–613.

Domain Decomposition Methods and Multilevel Preconditioners for Nonconforming and Mixed Methods for Partial Differential Problems

ZHANGXIN CHEN AND RICHARD E. EWING

ABSTRACT

In this paper we describe recent development of domain decomposition methods and multilevel preconditioners for solution of the discrete systems which arises from the application of nonconforming and mixed finite element methods to partial differential problems.

1. INTRODUCTION

A lot of numerical methods for solving partial differential problems have been developed in past years. Among the most popular and often used methods are Galerkin finite element methods, finite volume methods, and mixed finite element methods. The mixed methods have certain advantages over the Galerkin methods and the finite volume methods [3]. In particular, they preserve mass element by element, and produce a direct and accurate approximation for the vector unknown of the differential problems, which is the variable of primary interest in many applications such as the velocity field in the flow equation in highly heterogeneous media [16] and the electric field in the potential equation in semiconductor devices [5, 7, 8].

Due to their saddle point property, it is known that the linear systems arising from the mixed methods are generally harder to solve than those arising from comparable Galerkin methods. In particular, there has been little theory for constructing good preconditioners and developing efficient domain decomposition methods for solving the systems of algebraic equations arising from the mixed methods. An alternate approach was suggested by means of a nonmixed formulation. Namely,

Z. Chen, Department of Mathematics, Box 156, Southern Methodist University, Dallas, Texas 75275–0156, USA, e-mail: `zchen@golem.math.smu.edu`. R. Ewing, Institute for Scientific Computation, Texas A&M University, College Station, Texas 77843–3404, USA, e-mail: `ewing@ewing.tamu.edu`. Partly supported by the Department of Energy under contract DE–ACOS–840R21400.

Domain Decomposition Methods in Sciences and Engineering, edited by R. Glowinski *et al.*
© 1997 John Wiley & Sons, Ltd.

it has been shown that the mixed finite element methods are equivalent to a modification of nonconforming Galerkin methods [**1, 2, 4, 18**]. The nonconforming methods yield a symmetric and positive definite problem, which can be more easily solved using domain decomposition methods.

The purpose of this paper is to describe recent development of domain decomposition methods and multilevel preconditioners for solution of the discrete systems which arises from the application of nonconforming finite element methods to partial differential problems. Then, based on the equivalence mentioned above between the mixed and nonconforming methods, we discuss the domain decomposition methods and multilevel preconditioners for the mixed methods for the partial differential problems. In the next section we consider the domain decomposition methods, and then in the third section we present the multilevel preconditioners. In the final section we discuss the equivalence between the mixed and nonconforming finite element methods.

2. DOMAIN DECOMPOSITION METHODS

Let Ω be a bounded domain in $\mathbb{R}^d$, and let N_h be a nonconforming finite element space associated with a triangulation $\mathcal{E}_h$ of Ω, where h is the discretization parameter. We are concerned with the approximate problem: Find $u_h \in N_h$ such that

$$(2.1) \qquad a_h(u_h, v) = (f, v), \quad \forall v \in N_h,$$

where the bilinear form $a_h(\cdot, \cdot)$ is assumed to be symmetric and positive definite, $(\cdot, \cdot)$ denotes the $L^2(\Omega)$ inner product (for simplicity), and $f \in L^2(\Omega)$. It is well known that the linear system arising from (2.1) is not well conditioned. The aim of this section is to develop an additive Schwarz algorithm for (2.1). For this, let $\{\Omega_j\}_{j=1}^J$ be an overlapping domain decomposition of Ω with the overlap parameter δ. The decomposition is assumed to align with the boundary $\partial\Omega$. Associated with each Ω_j, let N_h^j be a nonconforming finite element space of the same form as N_h, whose elements have support in Ω_j. The finite element space N_h is assumed to be represented as a sum of $J + 1$ subspaces:

$$(2.2) \qquad N_h = N_h^0 + N_h^1 + \ldots + N_h^J,$$

where N_h^0 is a so-called coarse space. We now define the operators $\Pi_j : N_h \to N_h^j$, $j = 0, 1, \ldots, J$, by

$$(2.3) \qquad a_h(\Pi_i v, w) = a_h(v, w), \quad \forall w \in N_h^j,$$

and the operator $\Pi : N_h \to N_h$ by

$$(2.4) \qquad \Pi = \sum_{j=0}^J \Pi_j.$$

In (2.3) we could use so-called inexact solvers; the analysis is the same. We now define a Schwarz algorithm for (2.1).

Additive algorithm. The additive Schwarz algorithm for (2.1) is given by

$$(2.5) \qquad \Pi u_h = f_h, \quad f_h = \sum_{j=0}^J f_j,$$

where $f_j \in N_h^j$ satisfies

$$a_h(f_j, v) = (f, v), \quad \forall v \in N_h^j, \ j = 0, 1, \ldots, J.$$

Note that (2.1) and (2.5) have the same solution. The following abstract convergence result for bounds on the condition number of Π can be found in [**17**]. Assume that

(A1) There is a constant $\mathcal{C}$ such that every $v \in N_h$ can be represented by $v = \sum_{j=0}^{J} v_j$ with $v_j \in N_h^j$ satisfying

$$\sum_{j=0}^{J} a_h(v_j, v_j) \leq \mathcal{C} a_h(v, v).$$

(A2) Let $\kappa = (\kappa_{ij})$ be a symmetric matrix with $\kappa_{ij} \geq 0$ satisfying

$$|a_h(v_i, v_j)| \leq \kappa_{ij} a_h(v_i, v_i)^{1/2} a_h(v_j, v_j)^{1/2}, \quad \forall v_i \in N_h^i, \ v_j \in N_h^j, \ i, j = 1, \ldots, J.$$

Then we have

$$\lambda_{\min}(\Pi) \geq \mathcal{C}^{-1} \text{ and } \lambda_{\max}(\Pi) \leq \rho(\kappa) + 1,$$

where $\rho(\kappa)$ is the spectral radius of κ.

We now apply the above theory to analyze a couple of examples.

Example 2.1. Let N_h be the P_1 nonconforming finite element space of the second order elliptic problem in $\mathbb{R}^2$. That is,

$$N_h = \{v \in L^2(\Omega) : v|_E \in P_1(E), \ \forall E \in \mathcal{E}_h; \ v \text{ is continuous at the midpoints}$$

of interior sides and vanishes at the midpoints of sides on $\partial\Omega\},$

where $\mathcal{E}_h$ is a partition of Ω into triangles. We need to assume a structure to our family of partitions. In the first step, let $\mathcal{E}_H$ be a coarse triangulation of Ω into nonoverlapping triangular substructures Ω_j', $j = 1, \ldots, J$. Then, in the second step we refine $\mathcal{E}_H$ into triangles to have a triangulation $\mathcal{E}_h$, $h < H$. Finally, let $\{\Omega_j\}_{j=1}^{J}$ be an overlapping domain decomposition of Ω by extending Ω_j' with the overlap parameter δ defined by

$$\delta = \min\{\text{dist}(\partial\Omega_j \setminus \partial\Omega, \partial\Omega_j' \setminus \partial\Omega), j = 1, \ldots, J\}.$$

Finally, define the coarse space

$$(2.6) \qquad N_h^0 = \{v \in N_h : v = R_h\varphi, \ \varphi \in U_H\},$$

where R_h is the nodal interpolation operator into N_h and U_H is the P_1-conforming space associated with $\mathcal{E}_H$. With these choices, it follows from the above abstract result [**15**] that the condition number $c(\Pi)$ of Π can be estimated by

$$(2.7) \qquad c(\Pi) = O(1 + H/\delta).$$

Example 2.2. Let N_h be the rectangular nonconforming finite element space of the second order elliptic problem in $\mathbb{R}^2$. Namely,

$$N_h = \Big\{ \xi : \xi|_E = a_E^1 + a_E^2 x + a_E^3 y + a_E^4(x^2 - y^2), \ a_E^i \in \mathbb{R}, \ \forall E \in \mathcal{E}_h;$$

$$\text{if } E_1 \text{ and } E_2 \text{ share an edge } e, \text{ then } \int_e \xi|_{\partial E_1} \, ds = \int_e \xi|_{\partial E_2} \, ds;$$

$$\text{and } \int_{\partial E \cap \partial\Omega} \xi|_{\partial\Omega} \, ds = 0 \Big\},$$

where $\mathcal{E}_h$ is a partition of Ω into rectangles. The overlapping domain decomposition $\{\Omega_j\}_{j=1}^J$ and spaces $\{N_h^j\}_{j=1}^J$ can be similarly defined as in Example 2.1. Finally, the coarse space is defined as follows. Let $\widehat{\mathcal{E}}_h$ be the triangulation of Ω into triangles obtained by connecting the two opposite vertices of the rectangles in $\mathcal{E}_h$. Associated with $\widehat{\mathcal{E}}_h$, let $\widehat{N}_h$ be the P_1 nonconforming finite element space in Example 2.1. Then we define the operator $\widehat{\mathcal{I}}_h : N_h \to \widehat{N}_h$ by the relation: If $v \in N_h$ and e is an edge of a triangle in $\widehat{N}_h$, then $\widehat{\mathcal{I}}_h v \in \widehat{N}_h$ is defined by

$$\frac{1}{|e|}(\widehat{\mathcal{I}}_h v, 1)_e = \frac{1}{|e|}(v, 1)_e.$$

Now we have

$$N_h^0 = \{v \in N_h : v = R_h \varphi, \ \varphi \in \widehat{N}_h^0\},$$

where $\widehat{N}_h^0$ is given as in (2.6), and R_h is the interpolation operator into N_h. The result (2.7) can be shown here by means of the operator $\widehat{\mathcal{I}}_h$ introduced above.

Other choices for N_h^0 in Examples 2.1 and 2.2 can be made [**12**]. It follows from (2.7) that if we use a generous overlapping, then the condition number of Π is uniformly bounded. With the same technique, two-level multiplicative algorithms and multilevel Schwarz algorithms can be developed [**12, 21**]. Moreover, exploiting the equivalence between the nonconforming and mixed methods [**6, 10, 12, 13**] (also see the fourth section), the Schwarz algorithms can be applied to the mixed methods [**3**]. Fourth order problems can be also analyzed using the present theory.

3. MULTILEVEL PRECONDITIONERS

Let

$$N_0 \to N_1 \to \cdots \to N_J$$

be a sequence of nonconforming finite element spaces associated with a increasing sequence of triangulations of Ω such that there is an intergrid transfer operator between two adjacent spaces:

$$I_j : N_{j-1} \to N_j, \quad j = 1, \ldots, J.$$

We are again concerned with the approximate problem: Find $u_J \in N_J$ such that

$$(3.1) \qquad\qquad a_J(u_J, v) = (f, v), \quad \forall v \in N_J,$$

where $a_J(\cdot, \cdot)$ is assumed to be equivalent to a discrete energy scalar product $(\cdot, \cdot)_{\mathcal{E}}$ on N_J:

$$(3.2) \qquad\qquad a_J(v, v) \approx \|v\|_{\mathcal{E}}^2, \quad \forall v \in N_J.$$

We define the operators $R_j : N_j \to N_J$ by

$$R_j = I_J I_{J-1} \cdots I_{j+1}, \quad j = 0, \ldots, J-1.$$

Let $\{b_j(\cdot, \cdot)\}_{j=0}^J$ be a family of bilinear forms defined by

$$b_j(v, w) = \alpha_j(v, w), \quad \forall v, w \in N_j,$$

where the positive constants α_j are determined by the (inverse) inequalities

$$\|v\|_{\mathcal{E}}^2 \leq \alpha_j \|v\|^2, \quad \forall v \in N_j, j = 0, \ldots, J.$$

Finally, introduce the operators $T^j : N_J \to N_j$ by

$$b_j(T^j v, w) = a_J(v, R_j w), \quad \forall w \in N_j,$$

and the elements $f^j \in N_j$ by

$$b_j(f^j, w) = (f, R_j w), \quad \forall w \in N_j.$$

Now it can be seen that the problem (3.1) is equivalent to

(3.3)
$$\mathcal{P}_J u_J = f_J,$$

where

$$\mathcal{P}_J = \sum_{j=0}^{J} R_j T^j, \quad f_J = \sum_{j=0}^{J} R_j f^j.$$

Assume that (3.2) and the following condition are satisfied:

$$b_j(v - I_j P_{j-1} v, v - I_j P_{j-1} v) \le C_J \|v - I_j P_{j-1} v\|_{\mathcal{E}}^2, \quad \forall v \in N_j,$$

where the operators $P_{j-1} : N_j \to N_{j-1}$ are given by

$$(P_{j-1} v, w)_{\mathcal{E}} = (v, w)_{\mathcal{E}}, \quad \forall v \in N_{j-1}, j = 1, \dots, J.$$

Then we have the following abstract result on the condition number of $\mathcal{P}_J$ [14, 19]:

(3.4)
$$C_* \max_{j=0,\dots,J} \tau_j \le c(\mathcal{P}_J) \le C^* C_J \sum_{j=0}^{J} \tau_j,$$

where the constants C_* and C^* are independent of j and

$$\tau_j = \sup\{\|R_j v\|_{\mathcal{E}}^2 / \|v\|_{\mathcal{E}}^2, \ v \in N_j, \ v \ne 0\}.$$

We now again consider the two examples presented in §2. Let h_0 and $\mathcal{E}_{h_0} = \mathcal{E}_0$ be given. For each integer $0 < j \le J$, let $h_j = 2^{-j} h_0$ and $\mathcal{E}_{h_j} = \mathcal{E}_j$ be constructed by connecting the midpoints of the edges of the triangle or rectangle in $\mathcal{E}_{j-1}$. Associated with each $\mathcal{E}_j$, let N_j be the nonconforming space defined as in Example 2.1 or Example 2.2. The intergrid transfer operators $I_j : N_{j-1} \to N_j$ are defined in the usual way by nodal or edge value averaging procedures [6]. Then, using the result (3.4), we have for both examples:

$$c(\mathcal{P}_J) = O(J).$$

Finally, thanks to the equivalence between the nonconforming and mixed finite element methods (see the next section), the theory presented in this section applies to the mixed methods [11], where algebraic multilevel preconditioners for both nonconforming and mixed methods are also constructed. Fourth order differential problems can be also analyzed using the present theory. Due to the page limitation, numerical results are not shown here; please refer to [11, 12, 14, 19] for the numerical results on the domain decomposition methods and multilevel preconditioners.

4. MIXED FINITE ELEMENT METHODS

We concentrate on the model problem

(4.1)
$$\begin{aligned} -\nabla \cdot (a\nabla u) &= f \quad &\text{in } \Omega, \\ u &= 0 \quad &\text{on } \partial\Omega. \end{aligned}$$

The Raviart-Thomas space [20] over triangles is given by

$$\Lambda_h = \left\{ v \in (L^2(\Omega))^2 : v|_E = \left(a_E^1 + a_E^2 x, \ a_E^3 + a_E^2 y \right), \ a_E^i \in \mathbb{R}, \ E \in \mathcal{E}_h \right\},$$
$$W_h = \left\{ w \in L^2(\Omega) : w|_E \text{ is constant for all } E \in \mathcal{E}_h \right\},$$
$$L_h = \left\{ \mu \in L^2(\partial\mathcal{E}_h) : \mu|_e \text{ is constant, } e \in \partial\mathcal{E}_h; \ \mu|_e = 0, \ e \subset \partial\Omega \right\},$$

where $\partial\mathcal{E}_h$ denotes the set of all interior edges. Then the hybrid form of the mixed method for (4.1) is to seek $(\sigma_h, u_h, \lambda_h) \in \Lambda_h \times W_h \times L_h$ such that

$$
\begin{aligned}
\sum_{E\in\mathcal{E}_h}(\nabla \cdot \sigma_h, w)_E &= (f,w), & \forall\, w \in W_h, \\
(4.2) \quad (\alpha\sigma_h, v) - \sum_{E\in\mathcal{E}_h}\left[(u_h, \nabla \cdot v)_E - (\lambda_h, v \cdot \nu_E)_{\partial E}\right] &= 0, & \forall\, v \in \Lambda_h, \\
\sum_{E\in\mathcal{E}_h}(\sigma_h \cdot \nu_E, \mu)_{\partial E} &= 0, & \forall\, \mu \in L_h,
\end{aligned}
$$

where ν_E denotes the unit outer normal to E and $\alpha = a^{-1}$. The solution σ_h is introduced to approximate the vector field

$$\sigma = -a\nabla u,$$

which is the variable of primary interest in many applications. Since σ lies in the space

$$H(\text{div}; \Omega) = \left\{ v \in (L^2(\Omega))^2 : \nabla \cdot v \in L^2(\Omega) \right\},$$

and we do not require that Λ_h be a subspace of $H(\text{div}; \Omega)$, the last equation in (4.2) is used to enforce that the normal components of σ_h are continuous across the interior edges in $\partial\mathcal{E}_h$, so in fact $\sigma_h \in H(\text{div}; \Omega)$.

There is no continuity requirement on the spaces Λ_h and W_h, so σ_h and u_h can be locally (element-by-element) eliminated from (4.2). In fact, from [6], (4.2) can be algebraically condensed to the symmetric, positive definite system for the Lagrange multiplier λ_h:

$$(4.3) \qquad\qquad\qquad\qquad M_h\lambda_h = F_h,$$

where the contributions of the triangle E to the stiffness matrix M_h and the right-hand side F_h are

$$(4.4) \qquad\qquad m_{ij}^E = \frac{\overline{\nu}_E^i \cdot \overline{\nu}_E^j}{(\alpha, 1)_E}, \quad F_i^E = -\frac{(\alpha J_E^f, \overline{\nu}_E^i)_E}{(\alpha, 1)_E} + (J_E^f, \nu_E^i)_{e_E^i},$$

where ν_E^i denotes the outer unit normal to the edge e_E^i, $\overline{\nu}_E^i = |e_E^i|\nu_E^i$, $|e_E^i|$ is the length of e_E^i, $J_E^f = (f,1)_E(x,y)/(2|E|)$, and $|E|$ denotes the area of E.

Let P_h denote the $L^2(\Omega)$ projection operator onto W_h, $\alpha_h = P_h\alpha$, and $f_h = P_h f$. Also, set

$$\tilde{f}_h|_E = \frac{f_h}{2}\left(3 - \frac{\alpha}{\alpha_h}\right)|_E,$$

and

$$\tilde{a}_h(\psi, \varphi) = \sum_{E\in\mathcal{E}_h}(\alpha_h^{-1}\nabla\psi, \nabla\varphi)_E.$$

Then as shown in [6], the system (4.3) corresponds to the system arising from the triangular nonconforming finite element method: Find $\psi_h \in N_h$ such that

$$(4.5) \qquad\qquad \tilde{a}_h(\psi_h, \varphi) = (\tilde{f}_h, \varphi), \quad \forall\varphi \in N_h,$$

where N_h is defined as in Example 2.1. Hence the previous methods can be used to solve (4.3), i.e., the mixed method (4.2). It should be also noted that the natural degrees of freedom, i.e., the values at the midpoint of edges of L_h and N_h are the same.

After the computation of λ_h, σ_h and u_h (if they are needed) can be recovered as follows. Set $\sigma_h|_E = (a_E + b_E x, c_E + b_E y)$ and $\overline{f}_E = f_h|_E$. Then it follows from [6] that

$$b_E = \frac{\overline{f}_E}{2},$$

$$a_E = -\frac{1}{(\alpha,1)_E}\left(\sum_{i=1}^{3}|e_E^i|\nu_E^{i(1)}\lambda_h|_{e_E^i} + \frac{\overline{f}_E}{2}(\alpha,x)_E\right),$$

$$c_E = -\frac{1}{(\alpha,1)_E}\left(\sum_{i=1}^{3}|e_E^i|\nu_E^{i(2)}\lambda_h|_{e_E^i} + \frac{\overline{f}_E}{2}(\alpha,y)_E\right),$$

and

$$u_h|_E = \frac{1}{2|E|}\left((\alpha\sigma_h,(x,y))_E + \sum_{i=1}^{3}\lambda_h|_{e_E^i}((x,y),\nu_E^i)_{e_E^i}\right), \quad E \in \mathcal{E}_h.$$

We now consider a modified version of the mixed method (4.2) in which the coefficient α is projected into the space W_h [9]: Find $(\sigma_h, u_h, \lambda_h) \in \Lambda_h \times W_h \times L_h$ such that

$$\sum_{E\in\mathcal{E}_h}(\nabla\cdot\sigma_h, w)_E = (f, w), \qquad \forall\, w \in W_h,$$

$$(\alpha_h\sigma_h, v) - \sum_{E\in\mathcal{E}_h}[(u_h, \nabla\cdot v)_E - (\lambda_h, v\cdot\nu_E)_{\partial E}] = 0, \qquad \forall\, v \in \Lambda_h,$$

$$\sum_{E\in\mathcal{E}_h}(\sigma_h\cdot\nu_E, \mu)_{\partial E} = 0, \qquad \forall\, \mu \in L_h.$$

Associated with this projected formulation, the linear system has the form in place of (4.4):

$$(4.6) \qquad m_{ij}^E = \frac{\overline{\nu}_E^i\cdot\overline{\nu}_E^j}{(\alpha,1)_E}, \quad F_i^E = -\frac{(J_E^f, \overline{\nu}_E^i)_E}{|E|} + (J_E^f, \nu_E^i)_{e_E^i}, \quad E \in \mathcal{E}_h.$$

The corresponding nonconforming system becomes: Find $\psi_h \in N_h$ such that

$$(4.7) \qquad \tilde{a}_h(\psi_h, \varphi) = (f_h, \varphi), \quad \forall\varphi \in N_h.$$

The present systems in (4.6) and (4.7) are simpler than the corresponding systems in (4.4) and (4.5). The advantage of the projected mixed formulation over the usual one is more obvious for the mixed finite element method over rectangles [6]. The three-dimensional case can be considered similarly.

References

[1] T. Arbogast and Zhangxin Chen, On the implementation of mixed methods as nonconforming methods for second order elliptic problems, *Math. Comp.* (1995) **64**, 943–972.

[2] D. N. Arnold and F. Brezzi, Mixed and nonconforming finite element methods: implementation, postprocessing and error estimates, *RAIRO Modél. Math. Anal. Numér.* (1985) **19**, 7–32.

[3] F. Brezzi and M. Fortin, *Mixed and Hybrid Finite Element Methods*, Springer-Verlag, New York, 1991.

[4] Zhangxin Chen, Analysis of mixed methods using conforming and nonconforming finite element methods, *RAIRO Modèl. Math. Anal. Numer.* (1993) **27**, 9–34.

[5] Zhangxin Chen, Finite element analysis of the one-dimensional full-drift diffusion model, *SIAM J. Numer. Anal.* (1995) **32**, 455–483.

[6] Zhangxin Chen, Equivalence between and multigrid algorithms for mixed methods for second order elliptic problems, *East-West J. Numer. Math.*, in press.

[7] Zhangxin Chen and B. Cockburn, Error estimates for a finite element method for the drift diffusion semiconductor device equations, *SIAM J. Numer. Anal.* (1994) **31**, 1062–1089.

[8] Zhangxin Chen and B. Cockburn, Analysis of a finite element method for the drift diffusion semiconductor device equations: the multidimensional case, *Numer. Math.* (1995) **71**, 1–28

[9] Zhangxin Chen and J. Douglas, Jr., Approximation of coefficients in hybrid and mixed methods for nonlinear parabolic problems, *Mat. Aplic. Comp.* (1991) **10**, 137–160.

[10] Zhangxin Chen and R. E. Ewing, Recent development of multigrid algorithms for mixed and nonconforming methods for second order elliptic problems, In: *Proceedings of Eighth Copper Mountain Conference on Multigrid Methods*, Copper Mountain, Colorado, April 2–7, 1995, to appear.

[11] Zhangxin Chen, R. E. Ewing, Y. Kuznetsov, R. Lazarov, and S. Maliassov, Multilevel preconditioners for mixed methods for second order elliptic problems, IMA Preprint Series #1269, 1994.

[12] Zhangxin Chen, R. Ewing, and R. Lazarov, Domain decomposition algorithms for mixed methods for second order elliptic problems, *Math. Comp.* (1996) **65**, to appear.

[13] Zhangxin Chen, D. Y. Kwak, and Y. J. Yon, Multigrid algorithms for nonconforming and mixed methods for symmetric and nonsymmetric problems, IMA Preprint Series #1277, 1994.

[14] Zhangxin Chen and P. Oswald, Simple V-cycles for the nonconforming Q1 element, Preprint, 1995.

[15] L. Cowsar, Domain decomposition methods for nonconforming finite element spaces of Lagrange type, In: *Proceedings of the Sixth Copper Mountain Conference on Multigrid Methods*, N. Melson et al., eds., NASA Conference Publication 3224, 1993, pp. 93–109.

[16] J. Douglas, R. E. Ewing and M. Wheeler, The approximation of the pressure by a mixed method in the simulation of miscible displacement, *RAIRO Anal. Numér.* (1983) **17**, 17–33.

[17] M. Dryja and O. Widlund, Domain decomposition algorithms with small overlap, *SIAM J. Sci. Comput.* (1994) **15**, 604–620.

[18] L. D. Marini, An inexpensive method for the evaluation of the solution of the lowest order Raviart–Thomas mixed method, *SIAM J. Numer. Anal.* (1985) **22**, 493–496.

[19] P. Oswald, Intergrid transfer operators and multilevel preconditioners for nonconforming discretizations, Preprint, 1995.

[20] P. Raviart and J. Thomas, A mixed finite element method for second order elliptic problems, In: *Mathematical Aspects of the FEM*, Lecture Notes in Mathematics, **606**, Springer-Verlag, Berlin & New York, 1977, pp. 292–315.

[21] P. Vassilevski and J. Wang, An application of the abstract multilevel theory to nonconforming finite element methods, *SIAM J. Numer. Anal.* (1995) **32**, 235–248.

Compensation Method of an Optimal-order Wilson Nonconforming Multigrid

Liming Ma [1] and Qianshun Chang[2]

1   Introduction

It is well known that the nonconforming finite elements are widely used in solving elliptic boundary value problems, because they have fewer degrees of freedom, simpler basis functions and better convergence behavior. As usual, we apply them with two methods. One is the so called tolerance method, and the other is the penalty method. But they both have some disadvantages: The convergence order is lower than for conforming elements with the same degree of piecewise polynomial interpolation if the former method is used (cf.[1] and [2]). Using the second method, the convergence order is only half of that of conforming elements with the same degree of piecewise polynomial (see[3]). To increase the convergence order, the compensation method was introduced in [4]. Through applying this method, the same accuracy order as for conforming elements with the same degree of piecewise piecewise polynomial can be obtained.

Additionally, the nonconforming multigrid method is also a very efficient method for solving the elliptic boundary value problem. Its characteristic feature is its fast convergence. Moreover, one can obtain an acceptable approximation of the discrete problem at an expense of computational work proportional to the number of unknowns. It is not only the complexity which is optimal, also the constants of proportionality are so small that other methods can hardly surpass the multigrid efficiency.

In the present paper, we will give a new method by combining above two methods

[1] Graduate School, Academia Sinica, Beijing 100039, China
[2] Institute of Applied Mathematics,Academia Sinica,Beijing 100080,China

Domain Decomposition Methods in Sciences and Engineering, edited by R. Glowinski *et al.*

for the Wilson element, such that the accuracy order using Wilson nonconforming element is almost the same as for a quadratic conforming element.

The paper is organized as follows. We will begin with a discussion of the compensation method for the Wilson element. The intergrid transfer operator is defined and its properties are discussed in section 3. In section 4, the $k-$level iteration and nested iteration are given. The last section contains the error estimate for the new method.

2   Compensation for the Wilson Element

For simplicity, the Poisson's equation is considered:

$$\begin{cases} -\Delta u = f \text{ in } \Omega \\ u = 0 \text{ on } \partial\Omega, \end{cases} \tag{2.1}$$

where $f \in L^2(\Omega)$, and Ω is a rectangular domain. The variational form of the problem (2.1) is: Find $u \in H_0^1(\Omega)$, such that

$$a(u, v) = (f, v), \forall v \in H_0^1(\Omega), \tag{2.2}$$

where

$$a(u, v) = \int_\Omega \nabla u \nabla v dx,$$

$$(f, v) = \int_\Omega f v dx.$$

For $h_k (k \geq 1)$ in a null sequence, let τ_k be a subdivision of $\bar{\Omega}$ into rectangles P; τ_{k+1} can be obtained by connecting the midpoints of the edges of $P \in \tau_k$. Then $h_k = 2h_{k+1}$. In addition, we assume that $\tau_k (k \geq 1)$ satisfy the regular and uniformly conditions, namely, there exist two constants σ, γ, which are independent of P, such that

$$\bar{h}_{k.P} \leq \sigma h_{k,P}, \quad h_k \leq \gamma \bar{h}_{k,P}, \quad \forall P \in \tau_k,$$

where $\bar{h}_{k,P}$ and $h_{k,P}$ denote the diameters of P, i.e. the length of the longest sides of P, and of the smallest sides of P, respectively. $h_k = \max_{P \in \tau_k} \bar{h}_{k.P}$.

Let V_k be the Wilson finite element space associated with τ_k. For every $v \in V_k$, it has the following properties:
(i) $v|_P$ is quadratic polynomial;
(ii) v is continuous at the vertices and vanishes at the vertices along $\partial\Omega$;
(iii) The remainning two degrees of freedom are the mean values of second derivatives $\frac{\partial^2 v}{\partial x_i^2} (i = 1, 2)$ for each $P \in \tau_k$.

If $u, v \in V_k$, set

$$a_k(u, v) = \sum_{P \in \tau_k} \int_P \nabla u \nabla v dx,$$

$$b_k(u, v) = \sum_{P \in \tau_k} \int_{\partial P} \{\bar{u}_n[v] + \bar{v}_n[u] + \frac{\theta}{h}[u][v]\} ds,$$

and

$$c_k(u, v) = a_k(u, v) + b_k(u, v),$$

where

$$[u] = u^+ - u^-,$$
$$u_n = \frac{\partial u}{\partial n},$$
$$\bar{u}_n = \frac{1}{2}(u_n^+ + u_n^-),$$

and n is the normal direction of ∂P pointing in the "+" direction. The positive constant θ, which is independent of h_k, will be determined later.

Now, a new discrete variational problem is given as follows.

Find $u_k \in V_k$, such that

$$c_k(u_k, v) = (f, v), \quad \forall v \in V_k. \tag{2.3}$$

Obviously, if V_k is conforming, then $b_k(u, v) = 0$, i.e. the problem (2.3) becomes a standard conforming finite problem.

For the tolerance method, the approximation solution u_k^* is obtained by solving the problem: Find $u_k^* \in V_k$, such that

$$a_k(u_k^*, v) = (f, v), \quad \forall v \in V_k. \tag{2.4}$$

From [1], the following error estimates hold.

$$\|u - u_k^*\|_k \leq C h_k |u|_{H^2(\Omega)}, \tag{2.5}$$

$$\|u - u_k^*\|_{L^2(\Omega)} \leq C h_k^2 |u|_{H^2(\Omega)}. \tag{2.6}$$

We can easily get the following lemmas.

Lemma 2.1 *If $v \in V_k$, then*

$$\int_{\partial p} u^2 ds \leq C h_k^{-1} |u|_{o.p}^2$$

holds (see [4]).

Lemma 2.2 *If $|v|_{1.h_k}^2 = \sum_{P \in \tau_k} |v|_{1.P}^2$, then $|v|_{1.h_k}$ is a norm on V_k.*

Theorem 2.1 *The bilinear form $c_k(u, v)$ is not only uniformly V_h elliptic, but also bounded. So the problem (2.3) has a unique solution.*

We omit the proof.

Theorem 2.2 *If u_k is the solution of problem (2.3) and $u \in H^3(\Omega)$, then there exists a constant C independent of h_k, such that*

$$|u - u_k|_{1.h_k} \leq C h_k^2 |u|_{3.\Omega},$$

$$|u - u_k|_{L^2(\Omega)} \leq C h_k^3 |u|_{3.\Omega}.$$

Proof. We can prove this theorem by using Green's formula on each element P in τ_k and a duality argument.

3   The Intergrid Transfer Operator and its Properties

The intergrid transfer operator I_{k-1}^k is defined as follows. For any $v \in V_{k-1}$, I_{k-1}^k satisfies following conditions:

(i) If A is a vertex of element $P \in \tau_k$ inside Ω and a vertex of a element in τ_{k-1}, then

$$(I_{k-1}^k v)(A) := v(A).$$

(ii) If A is a common vertex of two elements P_1 and P_2 in τ_{k-1}, then

$$(I_{k-1}^k v)(A) := \frac{1}{2}[v|_{P_1}(A) + v|_{P_2}(A)].$$

If A is in the interior of a rectangle in τ, then

$$(I_{k-1}^k v)(A) := v(A).$$

(iii) $I_{k-1}^k v = 0$ at the vertices along $\partial\Omega$.

(iv) The remaining rest degrees of freedom are the mean values of the second derivatives of v on $P \in \tau_k$.

We define a mesh-dependent energy norm by

$$\|v\|_k := \sqrt{c_k(v, v)}.$$

From Theorem 2.1, we have the following lemma.

Lemma 3.1 *The norm $|v|_{1.h_k}$ is equivalent to the energy norm $\|v\|_k$.*

Now, we give two properties of the operator I_{k-1}^k; proofs can be found in [7].

Property A. *There exists a constant C, independent of h_k, such that*

$$\|I_{k-1}^k v\|_{L^2(\Omega)} \leq C\|v\|_{L^2(\Omega)} , \forall v \in V_{k-1}.$$

Property B. *There exists a constant C, independent of h_k, such that*

$$\|I_{k-1}^k v\|_k \leq C\|v\|_{k-1}, \forall v \in V_{k-1}.$$

Assume that $0 \leq \lambda_1 \leq \lambda_2 \cdots \leq \lambda_{N_k}$ and $\phi_1, \phi_2, \ldots, \phi_{N_k}$ are the eigenvalues and eigenfunctions of $c_k(\cdot, \cdot)$ with respect to $(\cdot, \cdot)$, respectively. From the inverse inequality, there exists a constant C^*, such that

$$\lambda_{n_k} \leq C^* h_k^{-2}. \tag{3.1}$$

If $v = \sum_{i=1}^{i=N_k} c_i \phi_i$, then the discrete norm $\|\|v\|\|_{s.k}$ can be defined as follows.

$$\|\|v\|\|_{s,k}^2 := \sum_{i=1}^{i=N_k} \lambda_i^s c_i^2, s \in R.$$

Obviously, $\|\|v\|\|_{0,k} = \|v\|_{L^2(\Omega)}, \|\|v\|\|_{1.k} = \|v\|_k.$

4 The k-Level Iteration and the Nested Iteration

If z_0 is an initial value of the solution; then the approximation $MG(k, z_0, G)$ can be obtained by solving following problem: Find $z \in V_k$, such that

$$c_k(z, v) = G(v), \ \forall v \in V_k, G \in V_k',$$

where V_k' denote the conjugate space of V_k and $c_k(u, v)$ is defined as in section 2.

If $k = 1$, then $z_1 := MG(1, z_0, G)$ is the solution using a direct method. For $k \geq 1$, z_m are obtained by solving the equation:

$$(z_i - z_{i-1}, v) = \Lambda_k^{-1}(G(v) - c_k(z_{i-1}, v)), \ \forall v \in V_k,$$

where $1 \leq i \leq m$, $\Lambda_k \leq C^* h_k^{-2}$ (see (3.1)), and m is an integer to be determined later. In addition, q_p is obtained by a $(k-1)$-level iteration p times $(p = 2, 3)$, namely,

$$q_0 = 0,$$

$$q_i = MG(k-1, q_{i-1}, \bar{G}), \ 1 \leq i \leq p,$$

where

$$\bar{G}(v) \ = \ G(I_{k-1}^k v) - c_k(z_m, I_{k-1}^k v)$$
$$= \ c_k(z - z_m, I_{k-1}^k v), \ \forall v \in V_{k-1}.$$

The approximate solution is $MG(k, z_0, G) := z_m + I_{k-1}^k q_p$.

The full multigrid method is defined as follows. Let $\hat{u}_1$ is the solution obtained by using a direct method. The approximation $\hat{u}_k (k \geq 2)$ is obtained, recursively, by

$$u_0^j = I_{j-1}^j \hat{u}_{j-1},$$

$$u_l^j = MG(j, u_{l-1}^j, G), \ 1 \leq l \leq r, \ G(v) = \int_G fv dx_1 dx_2,$$

$$\hat{u}_j = u_r^j.$$

Here r is a positive integer to be determined later.

5 The Error Estimate

In this section, we will prove that the convergence order of the multigrid method with compensation is almost the same as that of the conforming quadratic element.

Theorem 5.1 *If the number of smoothing steps m is large enough, then the k-level iteration is a contraction for the energy norm.*

The proof is given in [7].

Theorem 5.2 *If $\hat{u}_k$ is the approximation using the full multigrid algorithm and r is large enough, then exists a constant C, independent of k, such that*

$$\|u - \hat{u}_k\|_{L^2} \leq Ch_k^2(h_k|u|_{3,\Omega} + |u|_{2,\Omega}),$$

$$\|u - \hat{u}_k\|_k \le Ch_k(h_k|u|_{3.\Omega} + |u|_{2.\Omega}),$$

where we assume that $u \in H^3(\Omega) \cap H_0^1(\Omega)$ is the solution of problem (2.2).
Proof: We only prove the first inequality because the proof of the second inequality is simpler.

Let $\bar{\pi}_{k-1}$ be the Lagrange interpolant operator. Then,

$$I_{k-1}^k(\bar{\pi}_{k-1}u) = \bar{\pi}_{k-1}u,$$
$$\|u - \bar{\pi}_{k-1}u\|_{L^2(\Omega)} \le Ch_k^2|u|_{2.\Omega}.$$

Thus,

$$\|u_k - \hat{u}_k\|_{L^2(\Omega)}$$
$$\le \gamma^r[\|u_k - u\|_{L^2(\Omega)} + \|u - \pi_{k-1}^-u\|_{L^2(\Omega)} + \|I_{k-1}^k(\pi_{k-1}^-u - \hat{u}_{k-1}\|_{L^2(\Omega)}]$$
$$\le \frac{C\gamma^r}{1 - 2C\gamma^r}(h_k^3|u|_{3.\Omega} + h_k^2|u|_{2.\Omega}) .$$

Choosing γ such that $1 - 2C\gamma > 0$, we find

$$\|u_k - \hat{u}_k\|_{L^2(\Omega)} \le Ch_k^2(|u|_{2.\Omega} + h_k|u|_{3.\Omega}).$$

The first inequality follows by using a triangle inequality.

REFERENCES

[1] Ciarlet P. G. (1978) *The Finite Element Method For Elliptic Problem*, North-Holland, Amsterdam.

[2] Stummel F. (1979) The generalized patch test. *SIAM J. Numer. Anal.*, 16:449-471.

[3] Babuska I. and Zlamal M. (1973) Nonconforming element in the finite element method with penalty. *SIAM J. NUmer. Anal.*,10: 863-875.

[4] Zhan Chongxi (1987) Nonconforming finite elements with compensation for plate bending problems. *J. Compute. Math.*, 5: 238-248.

[5] Shi Zhongci (1986) A remark on the optimal order of convergence of Wilson nonconforming element. *J.Comput.Math.*, 2:159-163.

[6] Agmon S.(1965) *Lectures on Elliptic Boundary Value Problems*, D.Van Nostrand Co, Int.

[7] Yu Xijun (1993) A multigrid method of nonconforming Wilson finite element. *Math. Numer. Sinica* 2: 346-351(in Chinese).

CONVERGENCE ESTIMATES FOR MULTIGRID ALGORITHMS WITH KACZMARZ SMOOTHING

KAB SEOK KANG AND DO YOUNG KWAK*

Abstract. We prove some estimates for the convergence of multigrid algorithms with Kaczmarz smoothing applied to symmetric positive definite problems. To estimate the convergence of multigrid algorithms, we assume some properties of smoothing procedure which are satisfied by Kaczmarz, Gauss-Seidel, and Jacobi smoothing and are weaker than earlier assumptions. We use product formula for multigrid algorithm to show the V-cycle convergence factor is $\delta = 1 - \frac{1}{C(j-1)}$. The theory is presented in an abstract setting which can be applied to finite element multigrid. Also, numerical example is provided.

1. Introduction. Many authors presented various convergence analyses of multigrid methods which are often based on certain assumptions concerning the smoothing process. These assumptions are sometimes verified for specific examples in [2, 5, 6]. In this paper, we provide a weaker assumption under which multigrid algorithms are shown to converge.

Assumptions concerning the smoothing process which given in [6] are satisfied by point, line and block Jacobi and point, line and block Gauss-Seidel smoothing but not by Kaczmarz smoothing, for which the constant C_R the assumption (C.1) in [6] grows with $1/h^2$. Instead, the latter satisfies a weaker condition introduced in this paper.

The outline of the remainder of this paper is as follows. Section 2 describes the basic multigrid algorithm in an abstract setting and gives some of the conditions on the smoothers. In §3, we prove multigrid convergence under the weaker assumption. Kaczmarz smoothing procedures are described and analyzed in §4. Some numerical experiments showing the Kacmarz smoother satisfies our weak assumption are reported there. Finally, in §5, we discuss the finite element multigrid applications.

2. The Multigrid Algorithms. We assume that there is a sequence of nested finite-dimensional inner product spaces $\mathcal{M}_1 \subset \mathcal{M}_2 \subset \ldots \subset \mathcal{M}_j$ with $(\cdot,\cdot)_k$. In addition, we assume that there are symmetric positive definite operators $A_k : \mathcal{M}_k \to \mathcal{M}_k$ for $k = 1,\ldots,j$. We denote $A(\cdot,\cdot) = (A_j\cdot,\cdot)_j$. The multigrid algorithm is an iterative procedures for the solution of the problem on $\mathcal{M}_j$, i.e., given $f \in \mathcal{M}_j$ find $u \in \mathcal{M}_j$ satisfying

$$
(1) \qquad A_j u = f.
$$

We define the projectors $P^0_{k-1} : \mathcal{M}_k \to \mathcal{M}_{k-1}$ and $P_{k-1} : \mathcal{M}_k \to \mathcal{M}_{k-1}$ by, for all $\phi \in \mathcal{M}_{k-1}$, $(P^0_{k-1}v, \phi)_{k-1} = (v, \phi)_k$ and $A(P_{k-1}v, \phi) = A(v, \phi)$, respectively. '$T$' and '$*$' will denote adjoint with respect to $(\cdot,\cdot)_k$ and $A(\cdot,\cdot)$, respectively.

Also, we require a sequence of linear smoothing operators $R_k : \mathcal{M}_k \to \mathcal{M}_k$ for $k = 2,\ldots,j$. We shall always take $R_1 = A_1^{-1}$.

We set

$$
R_k^{(l)} = \begin{cases} R_k & \text{if } l \text{ is odd,} \\ R_k^T & \text{if } l \text{ is even} \end{cases}
$$

and set $K_k = I - R_k A_k$.

* Department of Mathematicss, Korea Advanced Institute of Science and Technology, Taejon, Korea 305–701, email: `dykwak@math.kaist.ac.kr, kks002@math.kaist.ac.kr`.

Domain Decomposition Methods in Sciences and Engineering, edited by R. Glowinski *et al.*
© 1997 John Wiley & Sons, Ltd.

We next define a multigrid process for iteratively computing the solution of (1).

Algorithm Set $B_1^s = A_1^{-1}$. Assume that B_{k-1}^s has been defined and define $B_k^s g$ for $g \in \mathcal{M}_k$ as follows;

1. Set $v^0 = 0$ and $q^0 = 0$.
2. Define v^i for $i = 1, 2, \ldots, m(k)$ by $v^i = v^{i-1} + R_k^{(i+m(k))}(g - A_k v^{i-1})$.
3. Define $w^{m(k)} = v^{m(k)} + q^p$, where q^i, for $i = 1, \ldots, p$, is defined by

$$q^i = q^{i-1} + B_{k-1}^s[P_{k-1}^0(g - A_k v^{m(k)}) - A_{k-1} q^{i-1}].$$

4. Define w^i for $i = m(k)+1, \ldots, 2m(k)$ by $w^i = w^{i-1} + R_k^{(i+m(k))}(g - A_k w^{i-1})$.
5. Set $B_k^s g = w^{2m(k)}$.

In the above algorithm, by defining $B_k^n g = w^{m(k)}$, we get nonsymmetric multigrid algorithm B_k^n. From the above algorithm, fundamental recurrence relations for the nonsymmetric and the symmetric multigrid algorithm are

$$
\begin{aligned}
I - B_k^n A_k &= [(I - P_{k-1}) + (I - B_{k-1}^n A_{k-1})^p P_{k-1}]\bar{K}_k^{(m(k))}, \\
I - B_k^s A_k &= (\bar{K}_k^{(m(k))})^*[(I - P_{k-1}) + (I - B_{k-1}^s A_{k-1})^p P_{k-1}]\bar{K}_k^{(m(k))}
\end{aligned}
$$

on $\mathcal{M}_k$ where

$$
\bar{K}_k^{(m(k))} = \begin{cases} (K_k^* K_k)^{m(k)/2} & \text{if } m(k) \text{ is even,} \\ K_k(K_k^* K_k)^{(m(k)-1)/2} & \text{if } m(k) \text{ is odd.} \end{cases}
$$

To estimate the convergence of multigrid algorithm, we need some conditions concerning the smoothing operators. The conditions which were often assumed by many authors ([1–7]) are

(C.1) There is a constant C_R which does not depend on k such that the smoothing procedure satisfies

$$\lambda_k^{-1}\|u\|_k^2 \leq C_R(\bar{R}_k u, u)_k \qquad \text{for all } u \in \mathcal{M}_k.$$

Here, $\bar{R}_k$ is either $(I - K_k^* K_k)A_k^{-1}$ or $(I - K_k K_k^*)A_k^{-1}$. λ_k is the largest eigenvalue of A_k.

(C.2) Let $T_k = R_k A_k$. There is a constant $\theta < 2$ not depending on k satisfying

$$A(T_k v, T_k v) \leq \theta A(T_k v, v).$$

But (C.1) is not satisfied by some smoothers, say Kaczmarz smoothing. Thus we modify (C.1) as follows

(SM.1) There is a constant C_R which does not depend on k such that the smoothing procedure satisfies

$$\lambda_k^{-2}(A_k u, u)_k \leq C_R(\bar{R}_k u, u)_k \qquad \text{for all } u \in \mathcal{M}_k.$$

3. Convergence Estimates for Multigrid Algorithms. To estimate the convergence of multigrid algorithm, we need some properties concerning the operator A_k and the subspaces. These are as follows:

(O.1) There exists a sequence of linear operators $Q_k : \mathcal{M}_j \to \mathcal{M}_k$ for $k = 1, \ldots, j$, with $Q_j = I$ satisfying the following properties. There are constants C_1 and C_2 not depending on k for which

$$(2) \quad \begin{aligned} (A_k^{-1}(Q_k - Q_{k-1})u, (Q_k - Q_{k-1})u)_k &\leq C_1 \lambda_k^{-2}(A_k u, u)_k \quad \text{for } k = 2, \ldots, j, \\ A(Q_k u, Q_k u) &\leq C_2 A(u, u) \quad \text{for } k = 1, \ldots, j-1. \end{aligned}$$

THEOREM 1. *Assume that (O.1) hold. Let R_k hold (SM.1) and (C.2). Let B_j^s be defined by symmtric multigrid algorithm with $p = 1$. Then*

$$(3) \qquad A((I - B_j^s)v, v) \leq \delta_j A(v, v) \quad \text{for all } v \in \mathcal{M}_j$$

hold with $\delta_j = 1 - \frac{1}{C(j-1)}$ where $C = [(1 + C_2^{1/2})(2\theta/(2 - \theta))^{1/2} + (C_R C_1)^{1/2}]^2$.
PROOF. We observe that

$$I - B_j^s A_j = (I - B_j^n A_j)^*(I - B_j^n A_j)$$

and for $T_k = (I - (\bar{K}_k^{(m(k))})^*)P_k$, that

$$(I - B_j^n A_j)^* = (I - T_j)(I - T_{j-1}) \cdots (I - T_1).$$

To use a product analysis, we set $E_0 = I$ and $E_k = (I - T_k)E_{k-1}$. We get

$$\begin{aligned} A(u, u) - A(E_j u, E_j u) &= \sum_{k=1}^{j}[A(E_{k-1}u, E_{k-1}u) - A(E_k u, E_k u)] \\ &= \sum_{k=1}^{j} A((2I - T_k)E_{k-1}u, T_k E_{k-1}u). \end{aligned}$$

Note that $I - B_j^n A_j = E_j^*$ and hence the inequality (3) will follow if we show that

$$(4) \qquad A(u, u) \leq C(j - 1) \sum_{k=1}^{j} A((2I - T_k)E_{k-1}u, T_k E_{k-1}u).$$

From the fact that $Q_j = I$, we get

$$\begin{aligned} (5) \qquad A(u, u) &= \sum_{k=2}^{j} A(E_{k-1}u, (Q_k - Q_{k-1})u) + A(u, Q_1 u) \\ &\quad + \sum_{k=2}^{j} A((I - E_{k-1})u, (Q_k - Q_{k-1})u). \end{aligned}$$

For the first sum on the right-hand side (5), from (O.1) and (SM.1), we see that

$$\sum_{k=2}^{j} A(E_{k-1}u, (Q_k - Q_{k-1})u) = \sum_{k=2}^{j}(A_k^{-1}A_k^2 P_k E_{k-1}u, (Q_k - Q_{k-1})u)_k$$

$$\leq \sum_{k=2}^{j} A(A_k P_k E_{k-1}u, A_k P_k E_{k-1}u)^{1/2} \cdot (A_k^{-1}(Q_k - Q_{k-1})u, (Q_k - Q_{k-1})u)_k^{1/2}$$

$$\leq (C_R C_1)^{1/2} A^{1/2}(u, u) \sum_{k=2}^{j} A^{1/2}((I - K_k^* K_k)P_k E_{k-1}u, P_k E_{k-1}u)$$

$$\leq (C_R C_1(j - 1))^{1/2} A^{1/2}(u, u) \left(\sum_{k=2}^{j} A((I - K_k^* K_k)P_k E_{k-1}u, P_k E_{k-1}u) \right)^{1/2}$$

The remainder of proof is the same with the proof of Theorem 4.3 in [6]. $\square$

4. Smoothing Procedures in Multigrid Algorithms. We define the multicative smoother(Gauss-Seidel smoother) by the following algorithm.

Algorithm 4.1. Let $f \in \mathcal{M}_k$, we define $R_k f \in \mathcal{M}_k$ as follows;

1. Set $v_0 = 0$.
2. Define v_i for $i = 1, \ldots, l$ by $v_i = v_{i-1} + A_{k,i}^{-1} Q_k^i (f - A_k v_{i-1})$ where $A_{k,i}$ is an i-th diagonal element of A_k and Q_k^i is a projection onto span$\{e_i\}$ with respect to $(\cdot, \cdot)_k$.
3. Set $R_k f = v_l$.

Let P_k^i is the projection onto span$\{e_i\}$ with respet to $(A_k \cdot, \cdot)_k$. It immediately follows from the identity $A_{k,i} P_k^i = Q_k^i A_k$ that $K_k = (I - P_k^l) \cdots (I - P_k^1)$. It was shown that Gauss-Seidel smoother satisfies (C.1)([6]).

Theorem 2 Let R_k be a smoother which satisfy (C.1). Then R_k satisfy (SM.1).
PROOF. Let $u = A_k w$, then (C.1) becomes

$$\lambda_k^{-1}(A_k w, A_k w)_k \leq C_R[(A_k w, w)_k - (A_k K_k w, K_k w)_k].$$

From the fact, for all $w \in \mathcal{M}_k$, $(A_k w, w)_k \leq \lambda_k (w, w)_k$, (SM.1) is satisfied. $\square$

Kaczmarz smoother is defined by the following algorithm.

Algorithm 4.2 Let $f \in \mathcal{M}_k$. We define $R_k f \in \mathcal{M}_k$ as follows:

1. Set $v_0 = 0$.
2. Define v_i for $i = 1, \ldots, l$ by $v_i = v_i - \frac{a_i}{a_i^T a_i}(a_i^T v_{i-1} - f_i)$ where $a_i^T = i^{\text{th}}$ row of A_k.
3. Set $R_k f = v_l$.

From the above algorithm, we know that

$$(6) \qquad K_k = I - R_k A_k = (I - B_k^l) \cdots (I - B_k^1) \text{ where } B_k^i = \frac{a_i a_i^T}{a_i^T a_i} = \frac{A_k^T Q_k^i A_k}{(A_k A_k^T)_{ii}}.$$

Let S_k^i be the projection with respect to $((A_k A_k^T)\cdot, \cdot)_k$. Then we get $(A_k A_k^T)_{ii} S_k^i = Q_k^i (A_k A_k^T)$. From $B_k^i = A_k^T S_k^i A_k^{-T}$, (6) becomes $K_k = A_k^T (I - S_k^l) \cdots (I - S_k^1) A_k^{-T}$. Above presentation reflects that Kaczmarz iteration can be regarded as a Gauss-Seidel iteration applied to $A_k A_k^T u = f$. In fact, from Gauss-Seidel iteration, we know that $(I - S_k^l) \cdots (I - S_k^1) = I - (D + L)^{-1}(A_k A_k^T)$ where $A_k A_k^T = D + L + L^T$. Therefore $K_k = I - A_k^T (D + L)^{-1} A_k$.

Numerical verification of (SM.1) and (C.2). We consider an elliptic partial differential equation of the form

$$(7) \qquad \begin{aligned} -\sum_{i,j=1}^2 \frac{\partial}{\partial x_i}\left(a_{ij}(x)\frac{\partial v}{\partial x_j}(x)\right) &= f(x) \quad \text{in } \Omega, \\ v(x) &= 0 \quad \text{on } \partial\Omega \end{aligned}$$

where Ω is a unit square and $(a_{ij})_{i,j=1}^2$ is a symmetric positive matrix.

To obtain A_k, we discretize Ω by quasi-uniform triangular element with $h_k = 2^{-k}$ and define $\mathcal{M}_k$ to be the set of piecewise linear functions which vanish on $\partial\Omega$.

First, for $(a_{ij}) = I$, we calculate C_R's in (C.1) and (SM.1) of damped Jacobi with $\omega = 0.8$, Gauss-Seidel, and Kaczmarz smoothing.

h	Jacobi		Gauss-Seidel		Kaczmarz	
	(C.1)	(SM.1)	(C.1)	(SM.1)	(C.1)	(SM.1)
1/8	1.409707	1.409707	1.118052	1.118052	2.281606	1.349255
1/16	1.519271	1.519271	1.123504	1.123504	8.349859	1.371804
1/32	1.551332	1.551332	1.124665	1.124665	32.667680	1.376693
1/64	1.559570	1.559776	1.124900	1.124900	129.937300	1.377789

Table I

Table I show that Kaczmarz smoother does not satisfy (C.1).

Next, we calculate C_R in (SM.1) and θ in (C.2) of Kaczmarz smoother for other (a_{ij})'s.

h	I		$p(x)I$		$q(x)$	
	C_R	θ	C_R	θ	C_R	θ
1/8	1.349255	1.496815	1.321122	1.455358	1.322647	1.460811
1/16	1.371804	1.516820	1.342990	1.479987	1.341607	1.483321
1/32	1.376693	1.522054	1.356463	1.495506	1.353792	1.497318
1/64	1.377789	1.523371	1.364844	1.505585	1.361640	1.506282

Table II

Here $p(x) = e^{0.2x+0.7y}$ and $q(x) = \mathrm{diag}(e^{0.2x+0.3y}, x+0.5)$.

5. Finite Elements Applications . We shall consider the problem of approximating the solution v of (7). The form A corresponding to the above operator is given by

$$A(v, w) = \sum_{i,j=1}^{n} \int_{\Omega} a_{ij} \frac{\partial v}{\partial x_i} \frac{\partial w}{\partial x_j} dx \qquad \text{for all } v, w \in \mathcal{M}_k.$$

Clearly, $U \in H_0^1(\Omega)$ is the solution of

$$A(U, \theta) = (F, \theta) \qquad \text{for all } \theta \in H_0^1(\Omega).$$

By positive definiteness of $\{a_{ij}\}$, $\|\cdot\|_A = A^{1/2}(\cdot, \cdot)$ is a norm on $H_0^1(\Omega)$ and this norm is equivalence to $\|\cdot\|_1$ which denote $H^1(\Omega)$-norm.

We let τ_k^i be a sequence of quasi-uniform triangulations of size h_k for $k = 1, \ldots, j$. We define $\mathcal{M}_k$ as in the previous section. Since Ω is polygonal, the subspaces are nested.

Let $\{y_k^i\}$ be the collection of nodes corresponding to the triangulation for $\mathcal{M}_k$. Let $(u, v)_k = h_k^2 \sum_i u(y_k^i) v(y_k^i)$. Note that the quasi-uniformity of the triangulations implies that the norm $\|\cdot\|_k$ is equivalent to the L^2 norm on the subspace $\mathcal{M}_k$. The operator A_k, $k = 1, \ldots, j$, are then defined by, for all $\phi \in \mathcal{M}_k$, $(A_k v, \phi)_k = A(v, \phi)$.

Let Q_k denote the $L^2(\Omega)$ projection onto $\mathcal{M}_k$. We know that, since the triangulations are quasi-uniform and inverse property, for all $v \in H_0^1(\Omega)$,

$$(8) \qquad \|(I - Q_k)v\| \leq ch_k\|v\|_1 \qquad \text{and} \qquad \|Q_k v\|_1 \leq C\|v\|_1.$$

From (8) and definition, we get

$$(9) \quad
\begin{aligned}
&(A_k^{-1}(Q_k - Q_{k-1})v, (Q_k - Q_{k-1})v)_k \\
&= \sup_{u \in \mathcal{M}_k} \frac{((Q_k - Q_{k-1})v, u)^2}{(A_k u, u)_k} = \sup_{u \in \mathcal{M}_k} \frac{((Q_k - Q_{k-1})v, (Q_k - Q_{k-1})u)^2}{(A_k u, u)_k} \\
&\leq \sup_{u \in \mathcal{M}_k} \frac{\|(Q_k - Q_{k-1})v\|^2 \|(Q_k - Q_{k-1})u\|^2}{(A_k u, u)_k} \leq \frac{ch_k^2\|v\|_1^2 ch_k^2\|u\|_1^2}{\|u\|_1^2} \\
&\leq Ch_k^4(A_k v, v)_k \leq C\lambda_k^{-2}(A_k v, v)_k
\end{aligned}$$

since $\|(Q_k - Q_{k-1})v\| \le \|(I - Q_k)v\| + \|(I - Q_{k-1})v\|$ and $\lambda_k = O(h_k^{-2})$. Combining (8) and (9) shows that (O.1).

REFERENCES

[1] R.E. Bank and C.C. Douglas, *Sharp estimates for multigrid rates of convergence with general smoothing and acceleration*, SIAM J. Numer. Anal. **22** (1985), 617–633.

[2] R.E. Bank and T. Dupont, *An optimal order process for solving finite element equations*, Math. Comp. **36** (1981), 35–51.

[3] D. Braess and W. Hackbush, *A new convergence proof for the multigrid method including the V-cycle*, SIAM J. Numer. Anal. **20** (1983), 967–975.

[4] J.H. Bramble, D.Y. Kwak, and J.E. Pasciak, *Uniform convergence of multigrid V-cycle iterations for indefinite and nonsymmetric problems*, SIAM J. Numer. Anal. **31** (1994).

[5] J.H. Bramble and J. E. Pasciak, *New convergence estimates for multigrid algorithms* Math. Comp. **49** (1987), 311–329.

[6] J.H. Bramble and J. E. Pasciak, *The analysis of smoothers for multigrid algorithms* Math. Comp. **58** (1992), 467–488.

[7] J.H. Bramble and J.E. Pasciak, *New estimates for multigrid algorithms including the V-cycle*, Math. Comp. **60** (1994), 447–471.

[8] J.H. Bramble, J. E. Pasciak, J. Wang, and J. Xu, *Convergence estimates for multigrid algorithms without regularity assumptions* Math. Comp. **56** (1991), 23–45.

[9] W. Hackbush, *Multi-grid methods and applications*, Springer-Verlag, NewYork, 1985.

[10] S. McCormick (Ed.) *Multigrid methods*, SIAM, Philadelphia, PA, 1987.

Two–level Method with Coarse Space Size Independent Convergence

PETR VANĚK [1], RADEK TEZAUR [1], MARIAN BREZINA [1], and JITKA KŘÍŽKOVÁ [2]

Dedicated to Honza.

1 INTRODUCTION

The basic disadvantage of the standard two-level method is the strong dependence of its convergence rate on the size of the coarse-level problem. In order to obtain the optimal convergence result, one is limited to using a coarse space which is only a few times smaller than the size of the fine-level one. Consequently, the asymptotic cost of the resulting method is the same as in the case of using a coarse-level solver for the original problem. Today's two-level domain decomposition methods typically offer an improvement by yielding a rate of convergence which depends on the ratio of fine and coarse level only polylogarithmically ([BPS86], [BPS89], [DSW94], [Man93], [FR91], [MT]). However, these methods require the use of local subdomain solvers for which straightforward application of iterative methods is problematic, while the usual application of direct solvers is expensive.

We suggest a method diminishing significantly these difficulties. Following the

[1] Center for Comp. Mathematics, University of Colorado at Denver, Denver, CO 80217; pvanek@tiger.cudenver.edu, rtezaur@tiger.cudenver.edu, mbrezina@tiger.cudenver.edu
[2] UWB, Americká 42, 30614 Plzeň, Czech Republic; krizkova@kirke.zcu.cz

Domain Decomposition Methods in Sciences and Engineering, edited by R. Glowinski *et al.*

unpublished technical report [VK95], we develop a simple abstract framework based on the concept of smoothed aggregation introduced in [VMB] with aggregates derived from the system of nonoverlapping subdomains. We show that the smoothing of the coarse-space by an appropriate polynomial of degree about N_{es}^{-d} (symbol d denotes the dimension of the problem to be solved and N_{es} is the characteristic number of degrees of freedom per subdomain) can assure the coarse-space size independent convergence. The associated cost is significantly smaller than that of the local solvers in the case of standard domain decomposition. Moreover, it decreases as d increases.

Because of the page limit, we only apply the abstract framework to for the case of scalar equation with jumps in coefficients. More general problems and numerical experiments will be treated in [TVB].

2 ABSTRACT FRAMEWORK

We are interested in a numerical solution of a system of linear algebraic equations

$$Ax = b \tag{1}$$

with a symmetric positive definite $n \times n$ matrix A. Let $P : \mathbb{R}^m \to \mathbb{R}^n$, $m << n$ be a linear injective tentative prolongator and $S \in [\mathbb{R}^n]$ a symmetric smoother commuting with A. Let us set

$$A_S = S^2 A, \quad S' = I - \frac{\omega}{\rho} A_S, \quad \omega \in (0, 2), \quad \rho = \rho(A_S). \tag{2}$$

Furthermore, let $\mathbf{x} \leftarrow \mathcal{S}_S(\mathbf{x}, \mathbf{b})$ and $\mathbf{x} \leftarrow \mathcal{S}_{S'}(\mathbf{x}, \mathbf{b})$ be relaxation methods consistent with (1) such that their linear parts are matrices S and S'. Our algorithm is a standard variational two-level method with a smoothed prolongator SP, a pre-smoother $\mathcal{S}_S$ and a post-smoother $\mathcal{S}_{S'}$.

Algorithm 1 *Given the initial approximation* $\mathbf{x}$,

   repeat

     1. $\mathbf{x} \leftarrow \mathcal{S}_S(\mathbf{x}, \mathbf{b})$,
     2. solve $(P^T A_S P)\mathbf{v} = P^T S(A\mathbf{x} - \mathbf{b})$,
     3. $\mathbf{x} \leftarrow \mathbf{x} - SP\mathbf{v}$,
     4. $\mathbf{x} \leftarrow \mathcal{S}_{S'}(\mathbf{x}, \mathbf{b})$

   until convergence;
     5. Post process $\mathbf{x} \leftarrow \mathcal{S}_S(\mathbf{x}, \mathbf{b})$.

In the following, we will prove that, for a suitable S and P, steps 1–4 of the algorithm ensure convergence independent of the dimension ratio n/m in the A_S−norm. The postprocessing step 5 of the algorithm enables us to prove the same result in the energy norm of the original problem (1). The main disadvantage of the convergence estimate in A_S−norm is its indirect coarse-space dependence as for a smaller coarse-space we need a more powerfull smoother S to get the optimal convergence result.

Assumption 2 *Let the smoother S be a symmetric matrix that commutes with A, and $\rho(S) \leq 1$. We assume that the tentative prolongator P satisfies the weak approximation property in the following form: For every $\mathbf{u} \in \mathbb{R}^n$, there exists $\mathbf{v} \in \mathbb{R}^m$ such that*

$$\|\mathbf{u} - P\mathbf{v}\| \leq C_1 C_D(m, n)\rho^{-1/2}(A)\|\mathbf{u}\|_A. \tag{3}$$

For the prolongator smoother S, we require

$$\rho(S^2 A) \leq C_2^2 C_D^{-2}(m, n)\rho(A), \tag{4}$$

where $C_D(m, n), C_1, C_2 > 0$, and C_1, C_2 do not depend on m and n.

Remark 3 *For second order problems, we typically have $C_D(m, n) = H/h$ (the ratio of local meshsizes on the coarse and the fine level). In Section 3 we construct S as a suitable polynomial in A for which $\rho(S^2 A) \approx N^{-2}\rho(A)$, where N denotes the degree of S. In order to satisfy Assumption 2, we need $N \approx H/h$. This choice yields a coarse level matrix $(SP)^T A(SP)$ with a number of nonzero entries per row uniformly bounded with respect to H/h. Detailed arguments will be given in Section 3.*

The following theorem shows that, under Assumption 2, the convergence rate of Algorithm 1 is independent of dimensions m and n of the coarse and fine spaces.

Theorem 4 *Let $\mathbf{e}_i$ denote the error after i iterations given by steps 1–4 of Algorithm 1 , and $\mathbf{e}_i^S = S\mathbf{e}_0$ the error smoothed by step 5. Then, it holds that*

$$\|\mathbf{e}_{i+1}\|_{A_S}^2 \leq (1 - C_3)\|\mathbf{e}_i\|_{A_S}^2, \quad \text{and} \quad \|\mathbf{e}_i^S\|_A^2 \leq (1 - C_3)^i\|\mathbf{e}_0\|_A^2, \tag{5}$$

where $C_3 = \frac{(C_1 C_2)^{-2}\omega(2-\omega)}{1+(C_1 C_2)^{-2}\omega(2-\omega)} > 0$. Here ω is the damping parameter from (2).

Proof. Since $\rho(S) \leq 1$ and $\mathbf{e}_i^S = S\mathbf{e}_i$, we have $\|\mathbf{e}_i^S\|_A = \|\mathbf{e}_i\|_{A_S}$ and $\|\mathbf{e}_0\|_{A_S} \leq \|\mathbf{e}_0\|_A$. Therefore, the second inequality in (5) follows from the first one.

It is obvious that the linear part of the steps 1–4 is given by

$$S'[I - SP(P^T A_S P)^{-1}P^T SA]S = S'S[I - P(P^T A_S P)^{-1}P^T A_S].$$

Thus, the method can be viewed as a standard two-level method for solving a problem with matrix A_S (in place of A) and prolongator P (in place of SP).

Since $I - P(P^T A_S P)^{-1}P^T A_S$ is the A_S–orthogonal projection onto A_S–orthogonal complement of Range(P) (i.e., projection onto Ker$(P^T A_S)$), $S'S = SS'$, $\rho(S) \leq 1$, and $\rho(S') \leq 1$, we have

$$\|S'S[I - P(P^T A_S P)^{-1}P^T A_S]\|_{A_S}^2 \leq \sup_{\mathbf{x} \in \text{Ker}\,(P^T A_S)} \min\left\{\frac{\|S\mathbf{x}\|_{A_S}^2}{\|\mathbf{x}\|_{A_S}^2}, \frac{\|S'\mathbf{x}\|_{A_S}^2}{\|\mathbf{x}\|_{A_S}^2}\right\}. \tag{6}$$

In the rest of the proof, we will show that at least one of the expressions in the minimum above is bounded by $1 - C_3$ for any $\mathbf{x} \in \text{Ker}\,(P^T A_S)$.

We first express $\|S'\mathbf{x}\|_{A_S}/\|\mathbf{x}\|_{A_S}$ in terms of $\|S\mathbf{x}\|_{A_S}/\|\mathbf{x}\|_{A_S}$. It is easy to see that

$$\frac{\|A_S\mathbf{x}\|^2}{\|\mathbf{x}\|_{A_S}^2} \geq C_E\rho(A_S) \quad \text{implies} \quad \frac{\|S'\mathbf{x}\|_{A_S}^2}{\|\mathbf{x}\|_{A_S}^2} \leq 1 - C_E\omega(2 - \omega). \tag{7}$$

Let us recall that $A_S = AS^2$, and A and S commute. Hence, we can write $\|S^2\mathbf{x}\|_A^2 = \|S\mathbf{x}\|_{A_S}^2$, and

$$\frac{\|A_S\mathbf{x}\|^2}{\|\mathbf{x}\|_{A_S}^2} = \frac{\|A_S\mathbf{x}\|^2}{\|S^2\mathbf{x}\|_A^2}\frac{\|S^2\mathbf{x}\|_A^2}{\|\mathbf{x}\|_{A_S}^2} = \frac{\|AS^2\mathbf{x}\|^2}{\|S^2\mathbf{x}\|_A^2}\frac{\|S\mathbf{x}\|_{A_S}^2}{\|\mathbf{x}\|_{A_S}^2}. \tag{8}$$

Now, consider $\mathbf{x} \in \mathrm{Ker}\,(P^T A_S) = \mathrm{Ker}\,(P^T A S^2)$. Then, setting $\mathbf{u} = S^2\mathbf{x}$, we have $\mathbf{u} \in \mathrm{Ker}\,(P^T A) = \mathrm{Range}\,(P)^{\perp_A}$. From the weak approximation condition (3), we estimate the ratio $\frac{\|AS^2\mathbf{x}\|^2}{\|S^2\mathbf{x}\|_A^2}$ using the standard orthogonality argument: For $\mathbf{v} \in \mathbb{R}^m$ from (3), we obtain

$$\|\mathbf{u}\|_A^2 = (A\mathbf{u}, \mathbf{u}) = (A\mathbf{u}, \mathbf{u} - P\mathbf{v}) \leq \|A\mathbf{u}\|\,\|\mathbf{u} - P\mathbf{v}\| \leq C_1 C_D(m,n)\rho^{1/2}(A)\|A\mathbf{u}\|\,\|\mathbf{u}\|_A.$$

Therefore,

$$\frac{\|A\mathbf{u}\|}{\|\mathbf{u}\|_A} \geq C_1^{-1}C_D^{-1}(m,n)\rho^{1/2}(A).$$

Substituting this estimate into (8) and using Assumption 2, we get

$$\frac{\|A_S\mathbf{x}\|^2}{\|\mathbf{x}\|_{A_S}^2} \geq C_1^{-2}C_D^{-2}(m,n)\rho(A)\frac{\|S\mathbf{x}\|_{A_S}^2}{\|\mathbf{x}\|_{A_S}^2} \geq (C_1 C_2)^{-2}\rho(A_S)\frac{\|S\mathbf{x}\|_{A_S}^2}{\|\mathbf{x}\|_{A_S}^2}. \tag{9}$$

Thus, by (7)

$$\frac{\|S'\mathbf{x}\|_{A_S}^2}{\|\mathbf{x}\|_{A_S}^2} \leq [1 - \frac{\|S\mathbf{x}\|_{A_S}^2}{\|\mathbf{x}\|_{A_S}^2}(C_1 C_2)^{-2}\omega(2 - \omega)].$$

Since $\|S\mathbf{x}\|_{A_S}^2/\|\mathbf{x}\|_{A_S}^2 \leq 1$, we may finally write using (6)

$$\|S'S[I - P(P^T A_S P)^{-1}P^T A_S]\|_{A_S}^2 \leq \sup_{\alpha \in [0,1]} \min\left\{\alpha, 1 - \alpha(C_1 C_2)^{-2}\omega(2 - \omega)\right\}.$$

The expression on the right hand side is bounded by $\frac{1}{1+(C_1 C_2)^{-2}\omega(2-\omega)}$ which completes the proof. $\square$

3   EXAMPLE OF A TENTATIVE COARSE SPACE AND PROLONGATOR SMOOTHER

Let $\Omega \subset \mathbb{R}^2$ be a Lipschitz domain and $\mathcal{T}$ be a shape-regular (locally quasiuniform) finite element mesh on Ω. Let $V_{\mathcal{T}}$ be $P1$ or $Q1$ finite element space associated with the mesh $\mathcal{T}$ with zero Dirichlet boundary conditions imposed at some nodes of $\mathcal{T} \cap \partial\Omega$. Note that, for the purpose of numerical solution of the discretized problem, we do not need any assumptions on the form or measure of the part of the boundary with Dirichlet conditions imposed. For simplicity, we assume that the finite element basis functions φ_i are scaled so that $\|\varphi_i\|_{L^\infty} = 1$. We consider the following elliptic model problem: Find $u \in V_{\mathcal{T}}$ such that

$$A_\Omega(u,v) = (f,v)_{L^2(\Omega)} \quad \forall v \in V_{\mathcal{T}}, \quad A_\Omega(u,v) = \sum_{i=1}^{2}\int_\Omega a(x)\frac{\partial u(x)}{\partial x_i}\frac{\partial v(x)}{\partial x_i}dx. \tag{10}$$

We allow large variation of $a(x)$ between the subdomains; more detailed assumptions on $a(x)$ will be specified below.

Finite element discretization of (10) on V_T leads to a system of linear algebraic equations with a symmetric positive definite matrix A_T. In order to accomodate discontinuities of the coefficient $a(x)$, we will solve the system with a diagonally scaled matrix $A = D^{-1/2}A_T D^{-1/2}$ instead, where $D = \mathrm{diag}(A_T)$.

For the sake of construction of the tentative prolongator P we need a system $\{\Omega_i\}_{i=1}^m$ of closed disjoint subdomains of Ω such that each subdomain Ω_i is a simply connected closure of an aggregate of elements. We assume that each node of the underlying finite element mesh belongs to exactly one of the subdomains and that there is a layer one element wide between two neighboring subdomains. Further, we assume that the family of subdomains $\{\Omega_i\}$ satisfies the following properties:

Assumption 5

(i) We assume that there is about the same number of elements in each subdomain Ω_i. Let us denote the characteristic number of elements per subdomain by N_{es} and set $\hat{h} = N_{es}^{-1/2}$. We require that subdomain Ω_i can be mapped onto a reference subdomain $\hat{\Omega} = [0,1] \times [0,1]$ by a one-to-one locally Lipschitz mapping G_i:

$$\|\partial G_i(x)\| \le C\frac{\hat{h}}{h(x)}, \quad \|\partial G_i^{-1}(\hat{x})\| \le \frac{h(x)}{c\hat{h}}, \quad \forall \hat{x} = G_i(x), x \in \Omega_i, \qquad (11)$$

where $h(x)$ is the local meshsize in the neighborhood of x and $c, C > 0$ are constants uniform with respect to i. Symbol $\|\cdot\|$ denotes a matrix operator norm.

(ii) The coefficient $a(x)$ is allowed only a modest variation within each subdomain in the sense that $a(x) \approx a_i > 0, \quad \forall x \in \Omega_i$.

(iii) If Ω_i and Ω_j are adjacent subdomains (there exists $T \in T$ so that $\partial T \cap \partial \Omega_i \ne \emptyset$ and $\partial T \cap \partial \Omega_j \ne \emptyset$) and $a_i >> a_j$, the jump in $a(x)$ occurs along $\partial \Omega_i$. In other words, the discontinuity is located on the boundary of the subdomain with the larger value of $a(x)$.

Conditions (11) imply that an element $T \subset \Omega_i$ of size $h(x)$ is mapped by G_i onto $G_i(T)$ of size about $\hat{h}$. Thus, G_i maps locally quasi-uniform mesh on Ω_i onto a quasi-uniform mesh of meshize $\hat{h}$, and subdomains Ω_i are reasonable aggregates consisting of about $N_{es} = \hat{h}^{-2}$ elements. If the mesh T is quasi-uniform ($h(x) \approx h$), then $h(x)/\hat{h}$ can be viewed as the characteristic size of Ω_i. Note that, if a subdomain decomposing algorithm uses only the adjacency of elements or nodes and generates shape regular subdomains in the case of quasiuniform mesh, then for locally quasi-uniform meshes it can be expected to generate subdomains satisfying (11).

The purpose of assumption *(iii)* is to ensure that the basis function φ_j associated with a node $v_j \in \Omega_i$, satisfies

$$a(\varphi_j, \varphi_j) \approx a_i. \qquad (12)$$

If *(iii)* were not satisfied, (12) could be violated for the basis functions corresponding to the nodes on $\partial \Omega_j$ adjacent to a subdomain Ω_i with $a_i >> a_j$.

The tentative prolongator based on scaled aggregations is defined as follows.

Algorithm 6

(i) Set $P_{ij} = \begin{cases} 1, & \textit{if the node } v_i \textit{ belongs to subdomain } \Omega_i, \\ 0, & \textit{otherwise.} \end{cases}$

(ii) Set $P \leftarrow D^{1/2}P$, *where* $D = \mathrm{diag}(A_{\mathcal{T}})$.

For each subdomain, we introduce an index set F_i of all unconstrained (with no Dirichlet boundary conditions imposed) degrees of freedom associated with Ω_i. Let $\Pi : \mathbb{R}^n \to V_{\mathcal{T}}$ denotes the finite element interpolator given by $\Pi\mathbf{x} = \sum_{j=1}^n x_j\varphi_j$, the local interpolator $\Pi_i\mathbf{x} = \sum_{j \in F_i} x_j\varphi_j$, and discrete $l^2(F_i)$−norm $\|\mathbf{x}\|_{l^2(F_i)}^2 = \sum_{j \in F_i} x_j^2, \mathbf{x} \in \mathbb{R}^n$.

Let Ω_i' be a subset of Ω_i consisting of all elements $T \subset \Omega_i$ such that all degrees of freedom on T are unconstrained. On each subdomain we define a linear mapping $Q_i : \mathbb{R}^n \to \mathbb{R}^n$ (acting on the degrees of freedom of F_i only) to be the $l^2(F_i)$-orthogonal projection onto the one-dimesional space of vectors spanned by $\mathbf{c} \in \mathbb{R}^n$ such that $c_j = 1$ for $j \in F_i$, zero elsewhere.

Lemma 1 (Discrete scaled Poincaré-Friedrichs inequality) *For every* $\mathbf{u} \in \mathbb{R}^n$, *it holds that*

$$\|\mathbf{u} - Q_i\mathbf{u}\|_{l^2(F_i)} \leq CN_{es}^{1/2}|\Pi_i\mathbf{u}|_{H^1(\Omega_i)}, \tag{13}$$

the constant $C > 0$ *depends on the constants from* (11) *and on the aspect ratios of elements in* Ω_i *only.*

Proof. Consider a transformed function $\hat{u} = u \circ G_i^{-1}$ i.e. $\hat{u}(\hat{x}) = u(G_i^{-1}(\hat{x})), \hat{x} \in \hat{\Omega}$. Let us define a weighted L^2−norm by $\|u\|_{L_h^2} = \|u(x)/h(x)\|_{L^2}$. Owing to (11), H^1−seminorm scales uniformly, i.e. $|u|_{H^1(\Omega_i)} \approx |\hat{u}|_{H^1(\hat{\Omega})}$. It can be easily seen that

$$\|u\|_{L_h^2(\Omega_i)} \approx \hat{h}^{-1}\|\hat{u}\|_{L^2(\hat{\Omega})} \quad \text{and} \quad \|\Pi_i\mathbf{x}\|_{L_h^2(\Omega')} \approx \|\mathbf{x}\|_{l^2(F_i)}.$$

Let $\mathbf{c} \in \mathbb{R}^n$ be the vector given by $c_j = 1$ for $j \in F_i$, zeroes elsewhere (as in the definition of Q_i). Then, by the equivalence of $l^2(F_i)$ and $L_h^2(\Omega_i')$ for finite element functions , $\Omega_i' \subset \Omega_i$, and the scaling above, we have for $\alpha \in \mathbb{R}$

$$\begin{aligned} \|\mathbf{u} - \alpha\mathbf{c}\|_{l^2(F_i)} &\approx \|\Pi_i\mathbf{u} - \alpha\|_{L_h^2(\Omega_i')} \tag{14} \\ \leq \|\Pi_i\mathbf{u} - \alpha\|_{L_h^2(\Omega_i)} &\approx \hat{h}^{-1}\|(\Pi_i\mathbf{u} - \alpha) \circ G_i^{-1}\|_{L^2(\hat{\Omega})}, \end{aligned}$$

Using the definition of Q_i, inequality (14), Poincaré-Friedrichs inequality on $\hat{\Omega}$ and the uniform scaling of H^1 seminorm, we obtain

$$\begin{aligned} \|\mathbf{u} - Q_i\mathbf{u}\|_{l^2(F_i)} &= \inf_{\alpha \in \mathbb{R}^1}\|\mathbf{u} - \alpha\mathbf{c}\|_{l^2(F_i)} \leq C\hat{h}^{-1}\inf_{\alpha \in \mathbb{R}^1}\|(\Pi_i\mathbf{u}) \circ G_i^{-1} - \alpha\|_{L^2(\hat{\Omega})} \\ &\leq C\hat{h}^{-1}|(\Pi_i\mathbf{u}) \circ G_i^{-1}|_{H^1(\hat{\Omega})} \leq C\hat{h}^{-1}|\Pi_i\mathbf{u}|_{H^1(\Omega_i)}, \end{aligned}$$

concluding the proof. $\square$

Now we are ready to prove the weak approximation property (3).

Lemma 2 (Weak approximation property) *Under Assumption 5, the inequality* (3) *is satisfied with* $C_D(m,n) = N_{es}^{1/2}$ *and* C_1 *that depends only on constants from* (11) *and aspect ratios of elements.*

Proof. We set $Q = D^{1/2}(\sum_{i=1}^{m} Q_i)D^{-1/2}$. Let $\mathbf{u} \in \mathbb{R}^n$, $\mathbf{x} = D^{1/2}\mathbf{u}$. Then, using (12),

$$\|\mathbf{u}\|_A^2 = \|\mathbf{x}\|_{A_T}^2 = A_\Omega(\Pi\mathbf{x}, \Pi\mathbf{x}) \geq \sum_{i=1}^{m} A_{\Omega_i}(\Pi_i\mathbf{x}, \Pi_i\mathbf{x}) \geq C\sum_{i=1}^{m} a_i|\Pi_i\mathbf{x}|_{H^1(\Omega_i)},$$

$$\|\mathbf{u} - Q\mathbf{u}\|^2 = \|D^{1/2}(I - \sum_{i=1}^{m} Q_i)\mathbf{x}\|^2 \leq C\sum_{i=1}^{m} a_i\|\mathbf{x} - Q_i\mathbf{x}\|_{l^2(F_i)}^2.$$

From here, setting $P\mathbf{v} = Q\mathbf{u}$ and using Lemma 1 and $\rho(A) \leq C$, the statement follows.
$\square$

In the rest of this section we will discuss the choice of the prolongator smoother S. Let $\bar{\rho}$ be the estimate of $\rho(A)$ satisfying

$$\rho(A) \leq \hat{\rho} \leq C_\rho \rho(A). \tag{15}$$

For any integer $i \geq 0$, we define $\hat{\rho}_i = \frac{\hat{\rho}}{9^i}$, $A_0 = A$ and

$$S_i = \prod_{j=0}^{i-1} W_j, \quad W_j = I - \frac{4}{3}\hat{\rho}_j^{-1}A_j, \quad A_j = W_{j-1}^2 A_{j-1}. \tag{16}$$

It is easy to see that $\deg(S_i) \leq \frac{3}{2}3^i$. We choose the prolongator smoother $S = S_K$ for K such that

$$\deg(S_{K+1}) \geq qN_{es}^{1/2} \geq \deg(S_K), \tag{17}$$

where $q \in (0, 1]$ is a given parameter.

Theorem 7 *Let the tentative prolongator P be given by Algorithm 6 with the system of subdomains $\{\Omega_i\}_{i=1}^m$ satisfying Assumption 5. Let the prolongator smoother S be defined by (15)–(17). Then, the statement of Theorem 4 is valid with the constant C_3 independent of the meshsize, coefficients a_i, constant N_{es}, and boundary conditions. Moreover, the coarse-level matrix and the smoothed prolongator SP have a uniformly bounded number of nonzero entries per row.*

Proof. Due to Lemma 2, the approximation property (3) is satisfied with $C(m, n) = N_{es}^{1/2}$. Let us show that (4) holds with the same $C(m, n)$. From definition (16), we have $S^2A = A_{K+1}$. By induction, we can prove $\rho(A_i) \leq \hat{\rho}_i$: For $i = 0$, the inequality holds by (15); assume it holds for $j \leq i$. Then, by (16)

$$\rho(A_{i+1}) = \max_{t \in \sigma(A_i)} (1 - \frac{4}{3}\hat{\rho}_i^{-1}t)^2 t \leq \max_{t \in [0, \hat{\rho}_i]} (1 - \frac{4}{3}\hat{\rho}_i^{-1}t)^2 t \leq \hat{\rho}_{i+1}.$$

Hence, $\rho(A_K) \leq 9^{-K}\hat{\rho}$. Considering that, by (17), $K \approx \log_3 N_{es}^{1/2}$, we get (4). The optimal convergence result now follows from Theorem 4.

Let us show that the number of nonzero entries per row of the coarse-level matrix $A_c = (SP)^T A(SP)$ is bounded uniformly with respect to N_{es}. It is easy to see that $[A_c]_{ij}$ can be nonzero only if $\text{supp}(\Pi SP\mathbf{e}^i) \cap \text{supp}(\Pi SP\mathbf{e}^j) \neq \emptyset$, where $\mathbf{e}^i$ is the i-th canonical basis vector of $\mathbb{R}^m$. Clearly, $\text{supp}(\Pi P\mathbf{e}^i)$ is the domain Ω_i with one belt of surrounding elements added. Bounded overlaps of such supports are obvious. The smoother S adds at most qN_{es} strips of elements. Consequently, each support has a nonempty intersection with only a bounded number of other supports. $\square$

Theorem 8 *Let the assumptions of Theorem 7 be fulfilled and the Choleski factorization be used to solve the coarse-level problem. Then, the optimal number of elements per subdomain is $N_{es} \approx n^{2/5}$ and the system (1) can be solved to the level of truncation error in $O(n^{1.2})$ operations.*

Proof. We only consider the components of the algorithm which cost more than $O(n)$ operations. During the setup, such procedures involve evaluation of SP ($O(N_{es}^{1/2}n)$ operations) and Choleski factorization of the coarse-level matrix, which costs $O(m^2) = O(n^2/N_{es})$ operations. As Theorem 7 assures the optimal convergence result, we only have to perform $O(1)$ iterations. Nonscalable procedures during each iteration are the smoothing ($O(N_{es}^{1/2}n)$ operations) and the back substitution ($O(m^{1.5}) = O((n/N_{es}^{1/2})^{1.5})$). The statement follows by trivial manipulations. $\square$

REFERENCES

[BPS86] Bramble J. H., Pasciak J. E., and Schatz A. H. (1986) The construction of preconditioners for elliptic problems by substructuring, I. *Math. Comp.* 47: 103–134.

[BPS89] Bramble J. H., Pasciak J. E., and Schatz A. H. (1989) The construction of preconditioners for elliptic problems by substructuring, IV. *Math. Comp.* 53: 1–24.

[DSW94] Dryja M., Smith B. F., and Widlund O. B. (December 1994) Schwarz analysis of iterative substructuring algorithms for elliptic problems in three dimensions. *SIAM J. Numer. Anal.* 31(6): 1662–1694.

[FR91] Farhat C. and Roux F. X. (1991) A method of finite element tearing and interconnecting and its parallel solution algorithm. *Int. J. Numer. Meth. Engng.* 32.

[Man93] Mandel J. (1993) Balancing domain decomposition. *Comm. in Numerical Methods in Engrg.* 9: 233–241.

[MT] Mandel J. and Tezaur R. Convergence of a substructuring method with Lagrange multipliers. To appear in Numer. Math.

[TVB] Tezaur R., Vaněk P., and Brezina M. Two-level method for solids. In preparation.

[VK95] Vaněk P. and Křížková J. (1995) Two–level method on unstructured meshes with convergence rate independent of the coarse-space size. Technical Report 35, Center of Computational Mathematics/UCD.

[VMB] Vaněk P., Mandel J., and Brezina M. Algebraic multigrid by smoothed aggregation for second and fourth order elliptic problems. To appear in Computing.

A Multigrid Method for Nonlinear Parabolic Problems

Xijun Yu

1   Introduction

The finite element methods for solving nonlinear parabolic problems have been studied by many authors; see, e.g., Douglas and Dupont [5], Wheeler [4], Luskin [3]. These atthors have proposed various ways of solving the problems numerically and they have established optimal order convergence rates of methods, such as the linearized methods, the predictor-corrector methods, the extrapolation methods, the alternating direction methods and different iterative methods [2]. Multigrid methods for solving parabolic problems have been studied by some authors; see Hachbusch [14-15], Bank and Dupont [12], Brandt and Greenwald [16] as well as Yu [13]. But these methods are given mainly for linear parabolic equations. For nonlinear parabolic problems Hachbusch and Brandt in [14], [15], [16] have given multigrid methods by using integral differential equations and the frozen-τ technique.

In this paper, we present a multigrid procedure for two-dimensional nonlinear parabolic problems. The method is an extension of our earlier algorithm given in [13] for linear parabolic problems. The iterative methods for solving the system of nonlinear algebraic equations are avoided because the unknown function $U_k^{n+\theta}$ in the nonlinear coefficient $a(x, U_k^{n+\theta})$ and the right term $f(x, t, U_k^{n+\theta})$ in the system of nonlinear algebraic equations is replaced by $I_k U_{k-1}^{n+\theta}$ in the multigrid procedure, where I_k denotes a intergrid transfer operator, θ a weight function and $U_{k-1}^{n+\theta}$ the solutions of the equation on level $k - 1$. We analyze the convergence of our algorithm and the computational cost for N time steps. The computational cost is asymptotically $O(NN_k)$ where N_k is the dimension of the discrete finite element space and N is

[1] Laboratory of Computational Physics, Institute of Applied Physics and Computational
   Mathematics P.O.Box 8009, Beijing 100088, China

Domain Decomposition Methods in Sciences and Engineering, edited by R. Glowinski *et al.*

the number of time steps. In addition, the methods can be applied to more general nonlinear parabolic problems.

2　Notations and preliminaries

We consider an initial value problem of the following nonlinear parabolic equation:

$$
\begin{cases}
\dfrac{\partial u}{\partial t} = \nabla(a(x,u)\nabla u) + f(x,t,u), & (x,t) \in \Omega \times [0,T], \\
u(x,t) = 0, & (x,t) \in \partial\Omega \times [0,T], \\
u(x,0) = u_0(x), & x \in \partial\Omega,
\end{cases}
\tag{2.1}
$$

where $\Omega \subset R^2$ is a convex polygonal domain, ∇ is a gradient operator with respect to $x = (x_1, x_2)$. Assume that the nonlinear coefficient $a(x,p)$ satisfies the condition: There are constants K_0, $K_1 > 0$ such that

$$
0 < K_0 \le a(x,u) \le K_1, \ \forall(x,p) \in \bar{\Omega} \times R^1.
\tag{2.2}
$$

$a(x,p)$ and $f(x,t,p)$ satisfy uniform Lipschitz conditions with respect to p, i.e., there is a constant $L > 0$ such that

$$
\begin{aligned}
|a(x,p_1) - a(x,p_2)| &\le L|p_1 - p_2|, \ \forall(x,p) \in \bar{\Omega} \times R^1, \\
|f(x,t,p_1) - f(x,t,p_2)| &\le L|p_1 - p_2|, \ \forall(x,t,p) \in \bar{\Omega} \times [0,T] \times R^1.
\end{aligned}
\tag{2.3}
$$

The variational form of problem (2.1) is : Find a continuously differentiable mapping $u(t) = u(x,t) : [0,T] \to H_0^1(\Omega)$ such that

$$
\begin{cases}
(\dfrac{\partial u}{\partial t}, v) + a(u; u, v) = (f(u), v), \ \forall v \in H_0^1(\Omega), \\
(u(x,0), v) = (u_0(x), v),
\end{cases}
\tag{2.4}
$$

where $a(u,v) = \int_\Omega a(x,u)\nabla u \nabla v dx$, $(f(u),v) = \int_\Omega f(x,t,u)v dx$. Assume that the solution of problem (2.4) exists and is unique, and that the solution is smooth enough for finite element analysis.

Let Γ_1 be an initial mesh partition of domain Ω (a triangulation or quadrilateral partition). Γ_k ($k \ge 1$) is a partition obtained by connecting the midpoints of the edges of elements in Γ_{k-1}. Then $\Omega = \cup_{\tau \in \Gamma_k}$ and $h_k = \frac{1}{2}h_{k-1}$ where $h_k = \max_{\tau \in \Gamma_k} h_\tau$.

Let $\mathcal{M}_k(k \ge 1)$ be a finite element space of piecewise linear or quadratic functions associated with the decompositions Γ_k ($k \ge 1$). Then $\mathcal{M}_{k-1} \subset \mathcal{M}_k \subset H_0^1(\Omega)$.

Let $\Delta t > 0$ be a time step size, $t_n = n\Delta t$ ($n = 1, 2, \cdots, N$), $N = [\frac{T}{\Delta t}]$. Let

$$
\begin{aligned}
t_{n+\theta} &= \frac{1}{2}(1+\theta)t_{n+1} + \frac{1}{2}(1-\theta)t_n, \ U^n = U(x,t_n), \\
U^{n+\theta} &= \frac{1}{2}(1+\theta)U^{n+1} + \frac{1}{2}(1-\theta)U^n, \\
f(U^{n+\theta}) &= f(x,t_{n+\theta}, U^{n+\theta}),
\end{aligned}
$$

where $\theta \in [0,1]$.

The finite element method for solving the variational problem (2.4) is: Find $\{U^j\}_{j=1}^N : \bar{J} \to \mathcal{M}_k$ such that

$$\left\{ \begin{array}{l} (\dfrac{U^{n+1} - U^n}{\Delta t}, v) + a(U^{n+\theta}; U^{n+\theta}, v) = (f(U^{n+\theta}), v), \\[2mm] (u(x,0), v) = (u_0(x), v), \quad \forall v \in \mathcal{M}_k. \end{array} \right. \tag{2.5}$$

(2.5) is the Crank-Nicolson scheme for $\theta = 0$. (2.5) is the fully implicit scheme for $\theta = 1$. For $\forall \theta \in [0, 1]$, obviously, (2.5) is a system of nonlinear algebraic equations for each time $t_j = j\Delta t$.

3   Time-Dependent Full Multigrid Method

Let I_k be an intergrid transfer operator, $I_k : \mathcal{M}_{k-1} \to \mathcal{M}_k$. I_k is defined as the piecewise linear function or as the average of values of the neighboring nodal points. Let I_k^t be the conjugate operator of I_k or the restriction operator, $I_k^t : \mathcal{M}_k \to \mathcal{M}_{k-1}$ which satisfies

$$(I_k^t u_k, v_{k-1}) = (u_k, I_k v_{k-1}), \quad \forall u_k \in \mathcal{M}_k, \ v_{k-1} \in \mathcal{M}_{k-1}. \tag{3.1}$$

By the nested property of the finite element spaces, there exists a matrix $B_k = [b_{ij}]_{N_{k-1} \times N_k}$ such that $I_k = B_k^T$, $I_k^t = B_k^{[13]}$.

If the solutions U_{k-1}^{n+1} and U_{k-1}^n on level $k - 1$ as well as U_k^n on level k are known, then we obtain a system of linearized algebraic equations as follows:

$$\left\{ \begin{array}{l} (\dfrac{U_k^{n+1} - U_k^n}{\Delta t}, v) + a(I_k U_{k-1}^{n+\theta}; U_k^{n+\theta}, v) = (f(I_k U_{k-1}^{n+\theta}), v), \\[2mm] (u(x,0), v) = (u_0(x), v), \quad \forall v \in \mathcal{M}_k. \end{array} \right. \tag{3.2}$$

In the following, we will give the time-dependent k level algorithm for solving the system of linear algebraic equations (3.2). Assume that the solutions U_{k-1}^{n+1} and U_{k-1}^n on level $k - 1$ and U_k^n on level k are known. Then an initial approximate value of the solution at $(n + 1)$th step time on the k level is taken as:

$$U_{k,0}^{n+1} = U_k^n + I_k(U_{k-1}^{n+1} - U_{k-1}^n). \tag{3.3}$$

1) **Pre-smoothing**: Perform ν_1 time smoothing iterations on level k:

$$U_{k,\nu_1}^{n+1} = S_k^{\nu_1} U_{k,0}^{n+1} \tag{3.4}$$

where S_k is a smoothing operator, such as the Jacobi, Gauss-Seidel and the preconditioned conjugate gradient iteration.

2) **Coarse grid correction**: The coarse grid equation is that $\forall v \in \mathcal{M}_{k-1}$,

$$\begin{aligned} &(\frac{\hat{U}_{k-1}^{n+1} - U_{k-1}^n}{\Delta t}, v) + a(U_{k-1}^{n+\theta}; \hat{U}_{k-1}^{n+\theta}, v) = (f(U_{k-1}^{n+\theta}), v) + [(f(I_k U_{k-1}^{n+\theta}), I_k v) \\[2mm] &- (\frac{U_{k,\nu_1}^{n+1} - U_k^n}{\Delta t}, I_k v) - a(I_k U_{k-1}^{n+1}; \frac{1}{2}(1 + \theta)U_{k,\nu_1}^{n+1} + \frac{1}{2}(1 - \theta)U_k^n, I_k v)], \end{aligned} \tag{3.5}$$

where

$$\hat{U}_{k-1}^{n+\theta} = \frac{1}{2}(1+\theta)\hat{U}_{k-1}^{n+1} + \frac{1}{2}(1-\theta)U_{k-1}^{n}.$$

Let $\hat{U}_{k-1,p}^{n+1}$ be the solution of (3.5) obtained by using p time iterations and $\hat{U}_{k-1,0}^{n+1} = U_{k-1}^{n+1}$ as the initial value. Then, the corrected value U_{k,ν_1+1}^{n+1} of the iterative solution of (3.4) on level $k-1$ is defined as:

$$U_{k,\nu_1+1}^{n+1} = U_{k,\nu_1}^{n+1} + I_k(\hat{U}_{k-1,p}^{n+1} - U_{k-1}^{n+1}). \tag{3.6}$$

3)**Post-smoothing**: Perform ν_2 time smoothing iterations on level k:

$$U_{k,\nu_1+\nu_2+1}^{n+1} = S_k^{\nu_2} U_{k,\nu_1+1}^{n+1}. \tag{3.7}$$

Thus we obtain an approximate solution of the equation (3.2) at the $(n+1)'s$ time step on level k as

$$U_k^{n+1} = U_{k,\nu_1+\nu_2+1}^{n+1}.$$

The full multigrid scheme is defined as a recursive process over the mesh level k. If we carry out the multigrid operation for each time step n, we get a time-dependent full multigrid method.

The k level algorithm depends on the solution U_{k-1}^{n+1}, U_{k-1}^{n} and U_k^{n}. Therefore the full multigrid iterative procedure depends on the solution U_k^0 for $k = 1, 2, \cdots$ and U_1^n for $n = 1, 2, \cdots, N$.

The approximate solutions $U_k^0 (k = 1, 2, \cdots)$ are determined by the following scheme.

1) For $k = 1$, $U_1^0 = \bar{U}_1^0$ is obtained by exactly solving equation (3.8).

2) For $k > 1$, U_k^0 is obtained by using $I_k U_{k-1}^0$ as the initial value of the multigrid iterations. The exact solution $\bar{U}_k^0 (k = 1, 2, \cdots)$ satisfies the equation:

$$(\bar{U}_k^0, v) + a(u_0; \bar{U}_k^0, v) = (f(u_0(x)), v), \ \forall v \in \mathcal{M}_k. \tag{3.8}$$

The solution U_1^n $(n = 1, 2, \cdots, N)$ for the different θ values will be considered in the following two situations.

1) When $\theta \neq 0$, U_1^{n+1} is obtained by solving the following linear equation:

$$(\frac{U_1^{n+1} - U_1^n}{\Delta t}, v) + a(U_1^n; U_1^{n+\theta}, v) = (f(U_1^n), v), \ \forall v \in \mathcal{M}_1, \tag{3.9}$$

for $n = 0, 1, 2, \cdots, N-1$.

2) When $\theta = 0$, U_1^1 is obtained by applying the predictor and corrector twice. Let U_1^* be a solution of the following predictor equation,

$$(\frac{U_1^* - U_1^0}{\Delta t}, v) + a(U_1^0; (U_1^* + U_1^0)/2, v) = (f(U_1^0), v), \ \forall v \in \mathcal{M}_1. \tag{3.10}$$

Here $U_1^{*\frac{1}{2}} = (U_1^* + U_1^0)/2$, and U_1^{**} the solution of the following corrector equation,

$$(\frac{U_1^{**} - U_1^0}{\Delta t}, v) + a(U_1^{*\frac{1}{2}}; (U_1^{**} + U_1^0)/2, v) = (f(U_1^{*\frac{1}{2}}), v), \ \forall v \in \mathcal{M}_1. \tag{3.11}$$

Set $U_1^{**\frac{1}{2}} = (U_1^{**} + U_1^0)/2$. Then U_1^1 is obtained by the equation:

$$(\frac{U_1^1 - U_1^0}{\Delta t}, v) + a(U_1^{**\frac{1}{2}}; U_1^{\frac{1}{2}}, v) = (f(U_1^{**\frac{1}{2}}), v), \ \forall v \in \mathcal{M}_1. \tag{3.12}$$

The solution $U_1^{n+1} (n = 1, 2, \cdots, N-1)$ is obtained by applying the modified Crank-Nicolson method.

$$(\frac{U_1^{n+1} - U_1^n}{\Delta t}, v) + a(EU_1^n; U_1^{n+\frac{1}{2}}, v) = (f(EU_1^n), v), \ \forall v \in \mathcal{M}_1, \tag{3.13}$$

where $EU_1^n = \frac{3}{2}U_1^n - \frac{1}{2}U_1^{n-1}$.

4   Convergence Analysis

Let u be the solution of (2.1) which satisfies

$$u \in L^\infty(H^3), \quad \frac{\partial u}{\partial t} \in L^2(H^1) \cap L^\infty(H^2), \quad \frac{\partial^2 u}{\partial t^2} \in L^\infty(H^1), \quad \frac{\partial^3 u}{\partial t^3} \in L^2(L^2) \cap L^1(H^1). \tag{4.1}$$

Then under the conditions (2.2) and (2.3), the finite element solution of (2.5) has the following error estimate; see [3-5].

Lemma 1. *Let u be the solution of (2.4). Let $\tilde{U}_k^n (n \geq 1)$ and $\tilde{U}_k^0$ be the solutions of (2.5) and (3.8), respectively. Then for $\theta \in [0,1]$, there are constants $c^*, \tau_0 > 0$ independent of $h_k, \{\tilde{U}_k^n\}$ and Δt such that for $\Delta t \leq \tau_0$, we have*

$$\|u - \tilde{U}_k^n\|_{L^2} + h_k\|u - \tilde{U}_k^n\|_{H_0^1} \leq \begin{cases} c^*(h_k^2 + \Delta t^2), \ \theta = 0 \\ c^*(h_k^2 + \Delta t), \ \theta \neq 0 \end{cases}. \tag{4.2}$$

We will now prove that the finite element solution of the discrete equation (3.2) still satisfies (4.2).

Lemma 2. *Assume that we have obtained the finite element solutions $\bar{U}_{k-1}^{n+1}$, $\bar{U}_{k-1}^n$ on level $k-1$ and $\bar{U}_k^n$ on level k and let $\bar{U}_k^{n+1}$ be the finite element solution of (3.2), and let $\tilde{U}_k^{n+1}$ be the finite element solution of (2.5) on level k. Then, for $\theta \in [0,1]$, $\Delta t \sim O(h_k^2)$, there are constants $c^*, \tau_0 > 0$, independent of $h_k, \{\tilde{U}_k^n\}, \{\bar{U}_k^n\}$ and Δt, such that for $\Delta t \leq \tau_0$, we have*

$$\|\tilde{U}_k^n - \bar{U}_k^n\|_{L^2} + h_k\|\tilde{U}_k^n - \bar{U}_k^n\|_{H_0^1} \leq \begin{cases} c^*(h_k^2 + \Delta t^2), \ \theta = 0 \\ c^*(h_k^2 + \Delta t), \ \theta \neq 0 \end{cases}. \tag{4.3}$$

Applying Lemma 1, Lemma 2, and the triangle inequality, we obtain the following convergence result for the finite element solution of the equation (3.2).

Theorem 1. *Let u be the solution of (2.4) and satisfy conditions (2.2), (2.3) and (4.1). Let $\bar{U}_k^n (n \geq 2)$ be the solution of (3.2) and let $\bar{U}_1^n, \bar{U}_k^0$ be the solutions of (3.9)-(3.13) and (3.8), respectively. Then for $\theta \in [0,1]$ and $\Delta t \sim O(h_k^2)$, there are the constants $c^*, \tau_0 > 0$ independent of $h_k, \{\bar{U}_k^n\}$ and Δt such that for $\Delta t \leq \tau_0$, we have*

$$\|u - \bar{U}_k^n\|_{L^2} + h_k\|u - \bar{U}_k^n\|_{H_0^1} \leq \begin{cases} c^*(h_k^2 + \Delta t^2), \ \theta = 0 \\ c^*(h_k^2 + \Delta t), \ \theta \neq 0 \end{cases} \tag{4.4}$$

Lemma 3. *Assume that u satisfies conditions (2.2) (2.3) and (4.1). Let $\bar{U}_{k-1}^{n+1}$ be the solution of (3.2) on level $k-1$ and $\hat{U}_{k-1}^{n+1}$ be the solutions of (3.5). Then for $\theta \in [0,1]$, and $\Delta t \sim O(h_k^2)$, we have*

$$\|\hat{U}_{k-1}^{n+1} - \bar{U}_{k-1}^{n+1}\|_{L^2} + h_k\|\hat{U}_{k-1}^{n+1} - \bar{U}_{k-1}^{n+1}\|_{H_0^1}$$
$$\leq R_{k-1} + c[\|\bar{U}_k^{n+1} - U_{k,\nu_1}^{n+1}\|_{L^2} + \Delta t\|\nabla(\bar{U}_k^{n+1} - U_{k,\nu_1}^{n+1}\|_{L^2}]^{\frac{1}{2}}, \tag{4.5}$$

where

$$R_{k-1} = \begin{cases} c^*(h_{k-1}^2 + \Delta t^2), & \theta = 0, \\ c^*(h_{k-1}^2 + \Delta t), & \theta \neq 0. \end{cases}$$

The constants c^*, c depend on $K_0, K_1, L, \|\nabla u\|_{L^\infty(L^\infty)}$.

Let $\hat{U}_{k-1,p}^{n+1}$ be an approximate solution of equation (3.5) obtained by p smoothing iterations. Then there exists a constant $0 < \gamma < 1$ such that

$$\|\hat{U}_{k-1}^{n+1} - \hat{U}_{k-1,p}^{n+1}\|_{L^2}^2 + \frac{1}{2}(1+\theta)\Delta t K_0\|\nabla(\hat{U}_{k-1}^{n+1} - \hat{U}_{k-1,p}^{n+1})\|_{L^2}^2$$
$$\leq \gamma^p[\|\hat{U}_{k-1}^{n+1} - \bar{U}_{k-1}^{n+1}\|_{L^2}^2 + \frac{1}{2}(1+\theta)\Delta t K_1\|\nabla(\hat{U}_{k-1}^{n+1} - \bar{U}_{k-1}^{n+1})\|_{L^2}^2]. \tag{4.6}$$

Therefore, the error of the coarse corrective solution of (3.6) satisfies the inequality

$$\|\bar{U}_k^{n+1} - U_{k,\nu_1+1}^{n+1}\|_{L^2}^2 + \frac{1}{2}(1+\theta)\Delta t K_0\|\nabla(\bar{U}_k^{n+1} - U_{k,\nu_1+1}^{n+1})\|_{L^2}^2$$
$$\leq cI_1 + (1+\gamma^p)R_{k-1}^2 \tag{4.7}$$

where

$$R_{k-1} = \begin{cases} c^*(h_{k-1}^2 + \Delta t^2), & \theta = 0, \\ c^*(h_{k-1}^2 + \Delta t), & \theta \neq 0, \end{cases}$$

and

$$I_1 = \|\bar{U}_k^{n+1} - U_{k,\nu_1}^{n+1}\|_{L^2}^2 + \frac{1}{2}(1+\theta)\Delta t K_0\|\nabla(\bar{U}_k^{n+1} - U_{k,\nu_1}^{n+1})\|_{L^2}^2.$$

Inequality (4.7) shows that the error of the coarse corrective solution is bounded by the error of the solution of (3.2) and the error of the finite element solution of (3.5).

The smoothing iterative method of (3.2) satisfies the estimate:

$$\|\bar{U}_k^{n+1} - U_{k,\nu_1}^{n+1}\|_{L^2}^2 + \frac{1}{2}(1+\theta)\Delta t K_0\|\nabla(\bar{U}_k^{n+1} - U_{k,\nu_1}^{n+1})\|_{L^2}^2$$
$$\leq \rho(S_k^{\nu_1})[\|\bar{U}_k^{n+1} - U_{k,0}^{n+1}\|_{L^2}^2 + \frac{1}{2}(1+\theta)\Delta t K_1\|\nabla(\bar{U}_k^{n+1} - U_{k,0}^{n+1})\|_{L^2}^2]. \tag{4.8}$$

Thus by (4.7) and (4.8), the k level algorithm defined in (3.3)-(3.7) satisfies:

Theorem 2. *Let $\bar{U}_k^{n+1}$ be the exact solution of (3.2) and let $\bar{U}_{k,\nu_1+\nu_2+1}^{n+1}$ be the iterative solution of the k level algorithm for (3.2). If there exists a constant $0 < \gamma < 1$ such that (4.6) holds for the level $k-1$, then for $\nu_1 + \nu_2$ large enough, we have*

$$\|\bar{U}_k^{n+1} - U_{k,\nu_1+\nu_2+1}^{n+1}\|_{L^2}^2 + \frac{1}{2}(1+\theta)K_0\Delta t\|\nabla(\bar{U}_k^{n+1} - U_{k,\nu_1+\nu_2+1}^{n+1})\|_{L^2}^2$$
$$\leq \gamma R_{k-1}^2 + \gamma[\|\bar{U}_k^{n+1} - U_{k,0}^{n+1}\|_{L^2}^2 + \frac{1}{2}(1+\theta)K_0\Delta t\|\nabla(\bar{U}_k^{n+1} - U_{k,0}^{n+1})\|_{L^2}^2 \tag{4.9}$$

Theorem 3. *Let u be the solution of (2.4) and satisfy conditions (2.2), (2.3) and (4.1). Let $U^n_{k,\nu_1+\nu_2+1}$ be the k level iterative solution of (3.3)-(3.7) Then there are constants $c^*, \tau_0 > 0$ independent of h_k and Δt such that if $\Delta t \sim O(h_k^2)$ and $\Delta t \le \tau_0$,*

$$\|u - U^{n+1}_{k,\nu_1+\nu_2+1}\|_{L^2} + h_k\|u - U^{n+1}_{k,\nu_1+\nu_2+1}\|_{H^1_0} \le R_k. \tag{4.10}$$

Theorem 4. *Assume that conditions (2.2), (2.3) and (4.1) hold. Then the approximate solution defined by multigrid algorithm satisfies the inequality:*

$$\|u(t_{n+1}) - U^{n+1}_k\|_{L^2} + h_k\|u(t_{n+1}) - U^{n+1}_k\|_{H^1_0} \le R_k \tag{4.11}$$

where the constant c^* is independent of $h_k, \Delta t$ and $\{U^n_k\}$.

REFERENCES

[1] Douglas J. Jr., Dupont T. and Ewing R. E. (1979) Incomplete iteration for time-stepping a Galerkin method for a quasilinear parabolic problem, *SIAM. J. Numer. Anal.* 16: 503-522.

[2] Douglas J. Jr. (1979) *Effective time-stepping methods for the numerical solution of nonlinear parabolic problems*, the Mathematics of Finite Elements and Applications III, 1978 (Whiteman J. R. ed.), Academic Press, New york, 289-304.

[3] Luskin M. (1979) A Galerkin method for nonlinear parabolic equations with nonlinear boundary conditions. *SIAM. J. Numer. Anal.* 16: 284-299.

[4] Wheeler M. F. (1973) A priori L^2 error estimates for Galerkin approximations to parabolic partial differential equations. *SIAM. J. Numer. Anal.* 10:723-759.

[5] Douglas J. Jr. and Doupont T. (1970) Galerkin methods for parabolic equations. *SIAM. J. Numer. Anal.* 7:575-626.

[6] Missirlis N. M. and Evans D. J. (1981) On the convergence of some generalized preconditioned iterative methods. *SIAM. J. Numer. Anal.* 18: 591-596.

[7] Axelsson O. (1974) On preconditioning and convergence acceleration in sparse matrix problems. *CERN European Organization for Nuclear Research,* Geneva.

[8] Douglas J. Jr. (1976) *Preconditioned conjugate gradient iteration applied to Galerkin methods for a mildly nonlinear Dirichlet problem*, Sparse Matrix Computations, Academic Press, Inc., New york, 333-348.

[9] Johson O. G., Micchelli C. A. and Paul G. (1983) Polynomial preconditioners for conjugate gradient calculations. *SIAM. J. Numer. Anal.* 20: 362-376.

[10] Hayes L. T. (1981) Galerkin alternating-direction methods for nonrectangular regions using patch approximations. *SIAM. J. Numer. Anal.* 18:627-643.

[11] Bramble J. H., Ewing R. E. and Li Gang (1989) Alternating direction multistep methods for parabolic problems – iterative stabilization. *SIAM. J. Numer. Anal.* 26:904-919.

[12] Bank R. E. and Dupont T. (1981) An optimal order process for solving finite element equations. *Math. Comp.* 36: 35-51.

[13] Yu Xijun, A parabolic multigrid method, (submitted).

[14] Hackbusch W. (1981) Fast numerical solution of time-periodic parabolic problems by a multigrid method, *SIAM. J. Sci. Stat. Comput.* 2: 198-206.

[15] —- Numerical solution of linear and nonlinear parabolic control problems.

[16] Brandt A. and Greenwald J. (1991) parabolic multigrid revisited, *Intern. Series of Numer. Math.* 98: 143-154.

[17] Ciarlet P. G. (1978) *The finite element method for elliptic problems*, Netherlands.

Part III Domain Decomposition and Parallel Computing

On Convergence of the Parallel Schwarz Algorithm with Pseudo-Boundary and the Parallel Multisplitting Iterative Method

Dawei Chang

1 Introduction

Suppose that we are given a linear system

$$Ax = b \tag{1}$$

where $A \in R^{n \times n}$ is a nonsingular matrix, $x, b \in R^n$ are vectors. In order to compute the solution of (1) iteratively, O'Leary and White propose multisplitting methods in [6] which are based on several splittings of the matrix A. More precisely, in [6] a multisplitting of A is defined as a collection of triples (M_k, N_k, E_k), $k = 1, 2, \cdots, K$, such that for all k, M_k, N_k, E_k are $n \times n$ matrices, each M_k is nonsingular, $A = M_k - N_k$, and E_k is a diagonal matrix with nonnegative entries satisfying $\sum_{k=1}^{K} E_k = I$. The corresponding multisplitting method to solve (1) is given by the iteration

$$x^{m+1} = \sum_{k=1}^{K} E_k y^{m,k}, \quad m = 0, 1, \cdots \tag{2}$$

where

$$M_k y^{m,k} = N_k x^m + b, \quad k = 1, 2, \cdots, K.$$

This multisplitting method has a natural parallelism, since the calculations of $y^{m,k}$ for various k are independent and may therefore be performed in parallel. Moreover, the i-th component of $y^{m,k}$ need not be computed if the corresponding diagonal entry of E_k is zero. This may result in considerable savings of computational time. Convergence results for method (2) were first given in [6]. Later, Neumann and Plemmons [5] obtained more qualitative results for one of cases considered in [6].

---

[1] Department of Mathematics, Shaanxi Normal University, Xi'an, 710062, P.R.China

Domain Decomposition Methods in Sciences and Engineering, edited by R. Glowinski *et al.*

2   Parallel Multisplitting TOR Method

Suppose that A is a nonsingular $n \times n$ matrix, for $k = 1, 2, \cdots, K$, L_k, F_k, U_k, E_k are $n \times n$ matrices, L_k and F_k are strictly lower triangular matrix satisfying
(1) $A = D - L_k - F_k - U_k$, where $D = \text{diag}(A)$ is an $n \times n$ and are diagonal matrix and nonsingular, and each U_k is zero-diagonal matrix.
(2) $\sum_{k=1}^{K} E_k = I$ ($n \times n$-identity matrix), where each E_k is diagonal matrix and $E_k \geq 0$.

Then the collection of triples $(D - L_k - F_k, U_k, E_k)$ $(k = 1, 2, \cdots, K)$ is called a multisplitting of A.

For real numbers ω, α and β, we define the following function G_k: $R^n \longrightarrow R^n$, for $k = 1, 2, \cdots, K$

$$G_k(x) = [D - (\alpha L_k + \beta F_k)]^{-1}\{[(1-\omega)D + (\omega - \alpha)L_k + (\omega - \beta)F_k + \omega U_k]x + \omega b\}$$

Multisplitting TOR (MTOR) Method

For any starting vector $x^0 \in R^n$

$$x^{m+1} = \sum_{k=1}^{k=K} E_k G_k(x^m) \qquad m = 0, 1, 2, \cdots$$

until convergence.

Now we define the matrix

$$T_{MTOR}(\omega, \alpha, \beta) = \sum_{k=1}^{K} E_k [D - \alpha L_k - \beta F_k]^{-1}[(1-\omega)D + (\omega - \alpha)L_k + (\omega - \beta)F_k + \omega U_k] \tag{3}$$

and the vector

$$g_{MTOR}(\omega, \alpha, \beta) = \sum_{k=1}^{K} E_k [D - \alpha L_k - \beta F_k]^{-1}\omega b.$$

Then from the multisplitting TOR (MTOR) Method, we get

$$x^{m+1} = T_{MTOR}(\omega, \alpha, \beta)x^m + g_{MTOR}(\omega, \alpha, \beta) \ , \quad m = 0, 1, 2, \cdots. \tag{4}$$

For the MTOR method, corresponding to particular choices of the parameter set (ω, α, β) to be $(1, 0, 0)$, $(1, 1, 1)$, $(\omega, 0, 0)$, (ω, ω, ω) and (ω, γ, γ), it naturally reduces to parallel multisplitting Jacobi (MP), Gauss-Seidel (MGS), JOR (MJOR), SOR (MSOR) and AOR (MAOR) method, where MSOR method is the relaxed parallel multisplitting method in [2]; MAOR method is the parallel multisplitting AOR algorithm in [9]. Thus the MTOR-method is a improvement and an generalization algorithm of [2] and [9]. Hence, a general series of parallel multisplitting method for solving the system of linear equation (1) is formed, which makes the new method more flexible and applicable.

3   Convergence of the MTOR Method

We first need to introduce several known concepts and useful lemmas.

A vector $x \in R^n$ is called nonnegative (positive), denoted $x \geq 0$ $(x > 0)$ if $x_i \geq 0$ $(x_i > 0)$ holds for all components of $x = (x_1, x_2, \cdots, x_n)^T$.

Similarly, a matrix A is called nonnegative, if all of its entries are nonnegative.

For two matrices we write $A \geq B$, when $A - B \geq 0$, and for two vectors $x \geq y (x > y)$, when $x - y \geq 0 (x - y > 0)$. Given a matrix $A = (a_{ij})$, we define its absolute value by $|A| = (|a_{ij}|)$. It follows that $|A| \geq 0$ and that $|AB| \leq |A||B|$ for any two matrices A and B

For any matrix $A = (a_{ij})$, such that $a_{ij} \leq 0$ for $i \neq j$ and $A^{-1} \geq 0$, A is called a M-matrix (see [8]).

For any matrix $A = (a_{ij}) \in R^{n \times n}$, we define its comparison matrix $\langle A \rangle = (\langle \alpha_{ij} \rangle)$ by

$$\langle \alpha_{ij} \rangle = \left\{ \begin{array}{lll} |a_{ij}|, & \text{if} & i = j \\ -|a_{ij}|, & \text{if} & i \neq j \end{array} \right.$$

A matrix A is called H-matrix if its comparison matrix $< A >$ is an M-matrix. Now we introduce several useful lemmas.

Lemma 1 [2] *Let A be an H-matrix, $D = diag(A)$, and $A = D - B$, then*
(1) A is nonsingular.
(2) $|A^{-1}| \leq \langle A \rangle^{-1}$
(3) $|D|$ is nonsingular and $\rho(|D|^{-1}|B|) < 1$.

Lemma 2 [8] *Suppose A, B satisfy $|A| \leq B$, then $\rho(A) \leq \rho(B)$.*

Lemma 3 [8] *Suppose that A is a nonnegative irreducible matrix. Then the spectral radius $\rho(A)$ of A is an eigenvalue of A and the eigenvector x corresponding to $\rho(A)$ satisfies $x > 0$.*

Theorem 1 *Suppose that A is an H-matrix, with a multisplitting*

$$(D - L_k - F_k, U_k, E_k), \ k = 1, 2, \cdots, K$$

such that

$$\langle A \rangle = |D| - |L_k| - |F_k| - U_k| = |D| - |B|$$

where $D = diag(A)$ is $n \times n$, diagonal and nonsingular, each L_k and F_k is a strictly lower triangular matrix, each U_k is a zero-diagonal matrix. Then MTOR method (5) converges for any starting vector $x^0 \in R^n$ provided that the parameters ω, α, β satisfy

$$0 \leq \alpha \, , \ \beta \leq \omega \, , \ 0 < \omega < \frac{2}{1 + \rho(|D|^{-1}|B|)}. \tag{5}$$

Proof. Since $\rho(T_{MTOR}(\omega, \alpha, \beta)) \leq \rho(|T_{MTOR}(\omega, \alpha, \beta)|)$ by Lemma 2, where $T_{MTOR}(\omega, \alpha, \beta)$ is the iteration matrix given by (3), we only need to show that $\rho(|T_{MTOR}(\omega, \alpha, \beta)|) < 1$.

As A is an H-matrix, D is a diagonal matrix, L_k and F_k are strictly lower triangular matrices, we easily see that $D - \alpha L_k - \beta F_k$ are H-matrices for $k = 1, 2, \cdots, K$. Using the result (2) of Lemma 1 and the definition of comparison matrix, we get

$$|(D - \alpha L_k - \beta F_k)^{-1}| \leq \langle D - \alpha L_k - \beta F_k \rangle^{-1} = |D| - \alpha|L_k| - \beta|F_k|.$$

First let the inequalities $0 \leq \alpha \leq \omega$, $0 \leq \beta \leq \omega$, $0 < \omega \leq 1$ hold.

For $k = 1, 2, \cdots, K$, we define the matrices

$$M_k = |D| - \alpha|L_k| - \beta|F_k|, \tag{6}$$

and

$$N_k^1 = (1 - \omega)|D| + (\omega - \alpha)|L_k| + (\omega - \beta)|F_k| + \omega|U_k|. \tag{7}$$

From (6), (7), we obtain

$$N_k^1 = M_k - \omega|D| - \omega|B| = M_k - \omega(|D| - |B|). \tag{8}$$

We take absolute values of both sides of (3) and obtain

$$|T_{MTOR}(\omega, \alpha, \beta)| \leq \sum_{k=1}^{K} E_k M_k^{-1} N_k^1 I - \omega \sum_{k=1}^{K} E_k M_k^{-1} |D|(I - |D|^{-1}|B|). \tag{9}$$

Let $e = [1, 1, \cdots, 1]^T \in R^n$. Since $|D|^{-1}|B|$ is nonnegative, the matrix $J_\epsilon = |D|^{-1}|B| + \epsilon e e^T$ has only positive entries and is irreducible for any $\epsilon > 0$. By Lemma 3, we known that $\rho(J_\epsilon)$ is an eigenvalue of J_ϵ and the corresponding eigenvector $x_\epsilon \geq 0$ satisfying

$$J_\epsilon x_\epsilon = (|D|^{-1}|B| + \epsilon e e^T)x_\epsilon = \rho(J_\epsilon)x_\epsilon.$$

Moreover, since $0 < \omega \leq 1$, we have

$$1 - \omega + \omega\rho(|D|^{-1}|B|) < 1.$$

By the continuity of the spectral radius, we also get

$$1 - \omega + \omega\rho(J_\epsilon) < 1 \tag{10}$$

if $\epsilon > 0$ is sufficient small.

By (9), we have

$$|T_{MTOR}(\omega, \alpha, \beta)| \leq I - \omega \sum_{k=1}^{K} E_k M_k^{-1}|D|[I - (|D|^{-1}|B| + \epsilon e e^T)]$$

$$= I - \omega \sum_{k=1}^{K} E_k M_k^{-1}|D|(I - J_\epsilon) \tag{11}$$

and by multiplying by x_ϵ,

$$|T_{MTOR}(\omega, \alpha, \beta)|x_\epsilon \leq x_\epsilon - \omega \sum_{k=1}^{K} E_k M_k^{-1}|D|(1 - \rho(J_\epsilon))x_\epsilon. \tag{12}$$

From the definition of M_k, the M_k are H-matrices. By Lemma 1, we get

$$M_k \leq |D| \ , \ M_k^{-1} \geq |D|^{-1}.$$

By (10) and (12), we have

$$|T_{MTOR}(\omega,\gamma,\bar{\gamma})|x_\epsilon \le x_\epsilon - \omega\sum_{k=1}^{K}E_k|D|^{-1}|D|(I-\rho(J_\epsilon))x_\epsilon$$

$$= (1-\omega+\omega\rho(J_\epsilon))x_\epsilon < x_\epsilon. \tag{13}$$

By exercise 2 of [8], p.48,

$$\rho(|T_{MTOR}(\omega,\alpha,\beta)|) < 1$$

holds.

Next let the inequalities $1 < \alpha \le \omega$, $1 < \beta \le \omega$, $1 < \omega < 2/(1+\rho(|D|^{-1}|B|))$ hold. We define matrices

$$N_k^2 = (\omega-1)|D| + (\omega-\alpha)|L_k| + (\omega-\beta|F_k| + \omega|U_k|. \tag{14}$$

From (6) and (14), then

$$N_k^2 = M_k - [(2-\omega)|D| - \omega|B|]. \tag{15}$$

We take absolute values of both sides of (3) and have

$$|T_{MTOR}(\omega,\gamma,\bar{\gamma})| \le \sum_{k=1}^{K}E_kM_k^{-1}N_k^2 \le I - \sum_{k=1}^{K}E_kM_k^{-1}|D|[(2-\omega)I-\omega|D|^{-1}|B|]. \tag{16}$$

As in the previous proof, let $e = [1,1,\cdots,1]^T \in R^n$ and let $x_\epsilon > 0$ denote the vector satisfying $J_\epsilon = (J+\epsilon ee^T)x_\epsilon = \rho(J_\epsilon)x_\epsilon$, where $\epsilon > 0$ is sufficiently small such that $\omega - 1 + \omega\rho(J_\epsilon) < 1$, since $1 < \omega < 2/(1+\rho(|D|^{-1}|B|))$.

From (16) we get

$$|T_{MTOR}(\omega,\alpha,\beta)| \le I - \sum_{k=1}^{K}E_kM_k^{-1}|D|[(2-\omega)I-\omega J_\epsilon] \tag{17}$$

and multiplying by x_ϵ, then

$$\begin{aligned}|T_{MTOR}(\omega,\alpha,\beta)|x_\epsilon &\le x_\epsilon - \sum_{k=1}^{K}E_k|D|^{-1}|D|[2-\omega-\omega\rho(J_\epsilon)]x_\epsilon\\ &= x_\epsilon - [2-\omega-\omega\rho(J_\epsilon)]x_\epsilon = [\omega-1+\omega\rho(J_\epsilon)]x_\epsilon\\ &< x_\epsilon.\end{aligned}$$

Thus $\rho(|T_{MTOR}(\omega,\alpha,\beta)|) < 1$ follows again by exercise of [8], p.48.

Under the assumption of the theorem, this completes the proof.

Theorem 1 implies the following Corollaries

Corollary 1 *Under the conditions of Theorem 1, the MSOR method converges to the unique solution $x^* \in R^n$ of the system of weakly nonlinear equations (1) for any starting vector $x^0 \in R^n$ provided that the parameter ω satisfies*

$$0 < \omega < \frac{2}{1+\rho(|D|^{-1}|B|)}. \tag{18}$$

Corollary 2 *Under the conditions of Theorem 1, the MAOR method converges to the unique solution $x^* \in R^n$ of the system of weakly nonlinear equations (1) for any starting vector $x^0 \in R^n$ provided that the parameter ω satisfies*

$$0 \leq \gamma \leq \omega, \, 0 < \omega < \frac{2}{1 + \rho(|D|^{-1}|B|)}. \tag{19}$$

4 Block MTOR (BMTOR) Method

By splitting the number set $\{1, 2, \cdots, n\}$ into K nonempty subset J_k $(k = 1, 2, \cdots, K)$, i.e.

$$J_k \subset \{1, 2, \cdots, n\}, \; \cup_{k=1}^{K} J_k = \{1, 2, \cdots, n\} \; k = 1, 2, \cdots, K,$$

we define the splitting matrices corresponding to the nonsingular matrix $A \in R^{n \times n}$ as follows:

$$D = \mathrm{diag}(A), \quad D \text{ is nonsingular}$$

$$L_k = (l_{ij}^k) \; l_{ij}^k = \begin{cases} -a_{ij}^k, 1 \leq j < [i/2] \; i, j \in J_k \\ 0, \quad \text{otherwise} \end{cases} \tag{20}$$

$$F_k = (f_{ij}^k) \; f_{ij}^k = \begin{cases} -a_{ij}^k, \; [i/2] \leq j < i \; i, j \in J_k \\ 0, \quad \text{otherwise} \end{cases} \tag{21}$$

$$U_k = (u_{ij}^k) \; u_{ij}^k = \begin{cases} 0 \; , \; j = i \\ -(a_{ij} + l_{ij}^k + f_{ij}^k) \; , \quad \text{otherwise} \end{cases} \tag{22}$$

with

$$A = D - L_k - F_k - U_k, \, k = 1, 2, \cdots, K.$$

Here $[a]$ is used to denote the integer part of a positive number a. The nonnegative diagonal matrices E_k $(k = 1, 2, \cdots, K)$ are introduced with $e_i^k \geq 0$ for $i \in J_k$, $e_i^k = 0$ for $i \notin J_k$, and $\sum_{k=1}^{K} E_k = I$ (identity matrix).

With these matrices, a block multisplitting of the matrix A results and denoted by

$$(D - L_k - F_k, U_k, E_k), \; k = 1, 2, \cdots, K.$$

Now we construct the block MTOR (BMTOR) method for solving the system of linear equations (1) as follows:

BMTOR method

For any starting vector $x^0 \in R^n$, for $m = 0, 1, 2, \cdots$, until convergence

$$x^{m+1} = \sum_{k=1}^{K} E_k x^{m,k}$$

where

$$a_{ii}x_i^{m,k} - \alpha \sum_{1 \leq j \leq [i/2]} l_{ij}^k x_j^{m,k} - \beta \sum_{[i/2] \leq j \leq n} f_{ij}^k x_j^{m,k}$$

$$= (1 - \omega)a_{ii}x_i^m + (\omega - \alpha) \sum_{1 \leq j \leq [i/2]} l_{ij}^k x_j^m$$

$$+ (\omega - \beta) \sum_{[i/2] \leq j \leq n} f_{ij}^k x_j^m + \omega \sum_{i \neq j} u_{ij}^k x_j^m + \omega b_i \ , \ i \in J_k$$

$$x_i^{m+1} = \sum_{k=1}^{K} e_i^k x_i^{m,k} \ , \quad i = 1, 2, \cdots, n.$$

Here $\alpha, \beta \geq 0$ are relaxation factors and $\omega > 0$ is an acceleration parameter.

The BMTOR method is a block MTOR method for numerically solving the system of linear equation (1) in synchronous parallel environments. For different k, the lower dimensional systems of equations (whose dimensions equal the number of elements included in the J_k) corresponding to the k-th splitting can be solved on the k-th processor of a multiprocessor system. A convergence theorem of the BMTOR method can be obtained in a similar way as for the MTOR method, so we will not demonstrate it here in detail.

REFERENCES

[1] Berman A. and Plemmom R. J. (1979) *Nonnegative Matrices in the Mathematical Sciences*, Academic, New York.

[2] Frommer A. and Mager G. (1898) Convergence of relaxed parallel multisplitting method. *Linear Algebra Appl.*, 119:141-152.

[3] Hadjidimos A. (1978) Accelerated overrelaxation method. *Math. Comp.*, 32:149-157.

[4] Kuang J. and Ji J. (1988) A surrey of AOR and TOR methods. *J. Comp. Appl. Math.*, 24:3-12.

[5] Neumann M. and Plemmons P. J. (1987) Convergence of parallel multisplitting iterative method for M-matrix. *Linear Algebra Appl.*, 88/89:559-573.

[6] O'Leary D. P. and White R. E. (1985) Multi-splitting of matrix and parallel solution of linear systems, *SIAM J. Algebra Disc. Method*, 6:630-640.

[7] Ortega J. M. (1988) *Introduction to Parallel and Vector Solution of Linear Systems*, Plenum Press, New York.

[8] Varga R. S. (1962) *Matrix Iterative Analysis*, Prentice-Hall, Englewood Cliffs, N. J.

[9] Wang Deren (1991) On the convergence of parallel multisplitting AOR algorithm, *Linear Algebra Appl.*, 154-156:473-486.

[10] White R. E. (1989) Multisplitting with deferent weighting schemes, *SIAM J. Matrix Anal. Appl*, 10:481-493.

[11] White R. E. (1986) Parallel algorithm for nonlinear problems, *SIAM J. Algebra Disc. Method*, 7:137-149.

[12]   Young D. M. (1971) *Iterative Solution of Large Linear Systems,* Academic
Press, New York.

Estimates of Convergence Rate of Parallel Multisplitting Iterative Methods with Their Applications

Tongxiang Gu

1 Introduction

Consider large-scale systems of linear algebraic equations

$$Ax = b \tag{1}$$

where $A \in R^{n \times n}$ and $b \in R^n$. O'Leary and White [8] proposed a parallel multisplitting iterative method (the PMI-method) for solving (1) in 1985. Through multisplitting of A

$$A = M_l - N_l; \quad \text{with} \quad \det(M_l) \neq 0; \quad l = 1, 2, \cdots, k, \tag{2}$$

they constructed the iterative procedures

$$y_l^m = M_l^{-1} N_l x^m + M_l^{-1} b; \qquad l = 1, 2, \cdots, k \tag{3}$$

for (1). By introducing weighting matrices E_l for $l = 1, 2, \cdots, k$ with

$$0 \le E_l \le I; \qquad \sum_{l=1}^{k} E_l = I, \tag{4}$$

where I is the $n \times n$ identity matrix, they combined (3) and obtained the PMI-method:

$$x^{m+1} = \sum_{l=1}^{k} E_l y_l^m; \qquad m = 0, 1, 2, \cdots \tag{5}$$

---

[1] Department of Mathematics, Henan Normal University, Xinxiang, Henan, 453002, P.R.China

Domain Decomposition Methods in Sciences and Engineering, edited by R. Glowinski *et al.*
© 1997 John Wiley & Sons, Ltd.

The triple (M_l, N_l, E_l) $l = 1, 2, \cdots, k$ is called a multisplitting of the matrix A and we can rewrite (5) in the equivalent form as:

$$x^{m+1} = Hx^m + Gb \tag{6}$$

where $H = \sum_{l=1}^{k} E_l M_l^{-1} N_l; G = \sum_{l=1}^{k} E_l M_l^{-1}.$

Many authors have presented different schemes based on different multisplittings of the coefficient matrix A, for example, PMI-GS and PMI-SGS [7], PMI-SOR [2], PMI-AOR [10], PMI-GSOR and PMI-GAOR [3]. The convergence of these methods were proved under different conditions. However, the methods used to prove the convergence in [2], [3], [7], [10] were inconvenient and varied. The author knows of new previous estimates of the convergence rate of PMI-method. So it is necessary to simplify and unify the proof of the convergence of PMI-method and to estimate the convergence rate of PMI-methods simply and practically.

2　Estimates of the Convergence Rate of PMI-methods

It is well known that the estimate of asymptotic convergence rate $R(H)$ of iterative method is equivalent to the estimate of the spectral radius $\rho(H)$ of the iterative matrix H, because $R(H) = -\log(\rho(H))$. We will denote M_l, N_l, E_l by (m_{ij}^l), (n_{ij}^l), (e_{ij}^l), respectively, and omit the index $l = 1, 2, \cdots, k$. We denote $\sum_{l=1}^{k}, \sum_{j=1}^{n}, \sum_{\substack{j=1 \\ j \neq i}}^{n}, \max_{1 \leq l \leq k}, \max_{1 \leq i \leq n}$, $\min_{1 \leq l \leq k}, \min_{1 \leq i \leq n}$ by $\sum_l, \sum_j, \sum_{j \neq i}, \max_l, \max_i, \min_l, \min_i$, respectively.

Theorem 1. *Let A be nonsingular and (M_l, N_l, E_l) be a multisplitting of A. If M_l is SDD (strictly diagonally dominant), then*

$$\rho(H) \leq \parallel H \parallel_\infty \leq \max_l \{\max_i \{\sum_j \frac{|n_{ij}^l|}{|m_{ii}^l| - \sum_{j \neq i} |m_{ij}^l|}\}\}. \tag{7}$$

Proof. It is well known that $\rho(H) \leq \parallel H \parallel_\infty$, so we need only to prove the right inequality.

Let $n_l = (n_1^l, n_2^l, \cdots, n_n^l)^T$ be an n-vector and $M_l^{-1} n_l = x_l := (x_1^l, x_2^l, \cdots, x_n^l)^T$. Thus,

$$\sum_l E_l M_l^{-1} n_l = \sum_l E_l x_l = \left(\sum_l e_{11}^l x_1^l, \sum_l e_{22}^l x_2^l, \cdots, \sum_l e_{nn}^l x_n^l \right)^T.$$

Since $\sum_l e_{ii}^l = 1$ for $i = 1, 2, \cdots, n$, we have

$$\left\parallel \sum_l E_l M_l^{-1} n_l \right\parallel_\infty = \max_i \{| \sum_l e_{ii}^l x_i^l |\} \leq \max_l \{\max_i \{|x_i^l|\}\} := |x_{i_0}^{l_0}|.$$

Consider the i_0-th equation of $M_{l_0} x_{l_0} = n_{l_0}$.

$$|m_{i_0 i_0}^{l_0}| \cdot |x_{i_0}^{l_0}| = \left| n_{i_0}^{l_0} - \sum_{j \neq i} m_{i_0 j}^{l_0} x_{i_0}^{j} \right| \leq |n_{i_0}^{l_0}| + |x_{i_0}^{l_0}| \cdot \sum_{j \neq i} |m_{i_0 j}^{l_0}|$$

and this implies

$$|x_{i_0}^{l_0}| \leq \frac{|n_{i_0}^{l_0}|}{|m_{i_0 i_0}^{l_0}| - \sum_{j \neq i_0} |m_{i_0 j}^{l_0}|} \leq \max_l \{ \max_i \{ \frac{|n_i^l|}{|m_{ii}^l| - | \sum_{j \neq i} m_{ij}^l|} \} \}.$$

Hence

$$\left\| \sum_l E_l M_l^{-1} n_l \right\|_\infty \leq \max_l \{ \max_i \{ \frac{|n_{ij}^l|}{|m_{ii}^l| - | \sum_{j \neq i} m_{ij}^l|} \} \}.$$

M_l is SDD. Let $D_l = \text{diag}(M_l)$, $C_l = D_l - M_l$, then $\langle M_l \rangle = |D_l| - |C_l|$, $|D_l^{-1} C_l| \leq |D_l|^{-1}|C_l|$ (in fact, the equality holds). We have

$$\rho(D_l^{-1} C_l) \leq \rho(|D_l|^{-1}|C_l|) < 1.$$

Hence M_l is invertible and

$$|M_l^{-1}| = |(D_l - C_l)^{-1}| = |\sum_{j=0}^{\infty} (D_l^{-1} C_l)^j D_l^{-1}|$$

$$\leq \sum_{j=0}^{\infty} (|D_l|^{-1}|C_l|)^j |D_l|^{-1} = (|D_l| - |C_l|)^{-1} = \langle M_l \rangle^{-1}.$$

Furthermore

$$|M_l^{-1} N_l| \leq \langle M_l \rangle^{-1} |N_l|; \quad |H| \leq \sum_l E_l |M_l^{-1} N_l| \leq \sum_l E_l \langle M_l \rangle^{-1} |N_l| := F.$$

Let $N_l = (N_1^l, N_2^l, \cdots, N_n^l)$, where N_i^l is the i-th column vector of N_l. Then

$$\begin{aligned}
\rho(H) \quad & \leq \| H \|_\infty \leq \| F \|_\infty = \max_i \{ \sum_j (\sum_l E_l \langle M_l \rangle^{-1} |N_l|)_{ij} \} \\
& = \max_i \{ \sum_j (\sum_l E_l \langle M_l \rangle^{-1} |N_1^l|, \sum_l E_l \langle M_l \rangle^{-1} |N_2^l|, \cdots, \sum_l E_l \langle M_l \rangle^{-1} |N_n^l|)_{ij} \} \\
& = \max_i \{ \sum_j (E_l \langle M_l \rangle^{-1} \sum_l |N_j^l|)_i \} = \| \sum_l E_l \langle M_l \rangle^{-1} \sum_j |N_j^l| \|_\infty \\
& \leq \max_l \{ \max_i \{ \sum_j \frac{(|N_j^l|)_i}{|m_{ii}^l| - \sum_{j \neq i} |m_{ij}^l|} \} \} = \max_l \{ \max_i \{ \sum_j \frac{|n_{ij}^l|}{|m_{ii}^l| - \sum_{j \neq i} |m_{ij}^l|} \} \}.
\end{aligned}$$

This proves the theorem.

Remark. We can see from the above proof that (7) holds for $M_l \in R^{n \times n}, N_l \in R^{n \times m}$ under the assumption that M_l is SDD. We can make the same remark on the following theorem and corollaries 1 and 2.

Theorem 2. *Under the assumptions of theorem 1 and assumming that M_l is a L-matrix, $N_l \geq 0$, it follows*

$$\min_i\{\sum_j(\sum_l E_l M_l^{-1} N_l)_{ij}\} \geq \min_l\{\min_i\{\sum_j \frac{|n_{ij}^l|}{|m_{ii}^l| - \sum_{j\neq i} |m_{ij}^l|}\}\}. \tag{8}$$

The proof of this theorem is analogous to that of theorem 1.

Corollary 1. *Under the assumptions of theorem 2, we have*

$$\min_l\{\min_i\{\sum_j \frac{n_{ij}^l}{\sum_j m_{ij}^l}\}\} \leq \min_i\{\sum_j(\sum_l E_l M_l^{-1} N_l)_{ij}\} \leq \rho(H) \leq \| H \|_\infty$$

$$= \max_i\{\sum_j(\sum_l E_l M_l^{-1} N_l)_{ij}\} \leq \max_l\{\max_i\{\sum_j \frac{n_{ij}^l}{\sum_j m_{ij}^l}\}\}. \tag{9}$$

In particular, if $\sum_j n_{ij}^l / \sum_j m_{ij}^l \equiv k\,(constant)$, $\rho(H) = \| H \|_\infty = k$. If $\sum_j n_{ij}^l / \sum_j m_{ij}^l \not\equiv$ constant and H is irreducible, the two middle inequalities are strict.
Proof. The first part is obtained from theorem 1 and 2, while the second part can be obtained from theorem 9 in §1.3 of [6].

Corollary 2 ([4, 5]). *Under the assumptions of theorem 2, we have*

$$\min_i\{\sum_j \frac{n_{ij}}{\sum_j m_{ij}}\} \leq \min_i\{\sum_j(M^{-1}N)_{ij}\} \leq \rho(M^{-1}N) \leq \| M^{-1}N \|_\infty$$

$$= \max_i\{\sum_j(M^{-1}N)_{ij}\} \leq \max_i\{\sum_j \frac{n_{ij}}{\sum_j m_{ij}}\}. \tag{10}$$

Proof. We obtain (10) from corollary 1 taking $k = 1$ and $M_1 = M, N_1 = N$. We can get a further direct result from theorem 1.

Corollary 3. *If M_l, N_l satisfy*

$$|m_{ii}^l| \geq \sum_{j\neq i} |m_{ij}^l| + \sum_j |n_{ij}^l| \qquad i = 1, 2, \cdots, n \tag{11}$$

and

$$|m_{ii}^l| > \sum_{j\neq i} |m_{ij}^l|; \qquad\qquad i = 1, 2, \cdots, n, \tag{12}$$

then

$$\rho(H) \leq \| H \|_\infty \leq 1. \tag{13}$$

The second inequality in (13) will be strict when the inequality (11) is strict. The PMI-method is then convergent.

3 Numerical Examples

Consider the systems of linear equations (1), where

$$A = \begin{bmatrix} 5 & -2 & -2 \\ -4 & 10 & -4 \\ -2 & -2 & 5 \end{bmatrix}; \quad b = \begin{bmatrix} 1 \\ 1 \\ 1 \end{bmatrix}.$$

We split A into several two-splittings as follows

$$(a) \qquad M_1 = \begin{bmatrix} 5 & -1 & -1 \\ -2 & 10 & -2 \\ -1 & -1 & 5 \end{bmatrix}; \quad M_2 = \begin{bmatrix} 5 & -2 & 0 \\ -4 & 10 & 0 \\ 0 & -2 & 5 \end{bmatrix};$$

$$N_1 = M_1 - A; \quad N_2 = M_2 - A; \quad E_1 = \mathrm{diag}(0,0,1); \quad E_2 = \mathrm{diag}(1,1,0);$$

(b) Take M_1, M_2, N_1, N_2 as in (a) and take $E_1 = \mathrm{diag}(1,0,0); E_2 = \mathrm{diag}(0,1,1)$;
(c) Take M_1, N_1, E_1, E_2 as in (a) and take $M_2 = \mathrm{diag}(5,10,5); N_2 = M_2 - A$.
  It follos from Corollary 2 that

$$\begin{aligned} \rho(H) &= 2/3 && \text{for case (a) and case (b),} \\ 2/3 &< \rho(H) < 4/5 && \text{for case (c).} \end{aligned}$$

In fact, we get from practical computing that

$$\begin{aligned} (a) \quad & \det(\lambda I - H) = \lambda(3\lambda - 2)(9\lambda + 5)/27 && \text{and } \rho(H) = 2/3, \\ (b) \quad & \det(\lambda I - H) = (3\lambda - 2)(6\lambda + 1)(18\lambda + 5)/972 && \text{and } \rho(H) = 2/3, \\ (c) \quad & \det(\lambda I - H) = (5\lambda + 2)(15\lambda^2 - 6\lambda - 4)/75 && \text{and} \\ & 2/3 < \rho(H) = (6 + \sqrt{276})/30 \approx 22.61/30 < 4/5 \quad . \end{aligned}$$

4 Applications

In this section, we give some convergence and divergence theorems of relaxed PMI-methods by using the estimates established in §2. In order to do this, the concepts of optimally scaled matrix and its several properties introduced in [5] are useful.

Lemma 1 ([5]). *Let $A = (a_{ij})$ bea irreducible matrix with nonzero diagonal entries, $D = diag(A), B = D - A$. Then there exists a diagonal matrix $Q = diag(q_1, q_2, \cdots, q_n)$ with positive diagonal entries such that $\widetilde{A} = (\widetilde{a}_{ij}) = AQ$,*

$$\sum_{j \neq i} |\widetilde{a}_{ij}| / |\widetilde{a}_{ii}| = \rho(|D|^{-1}|B|) \qquad i = 1, 2, \cdots, n \tag{14}$$

here $\widetilde{A}$ is unique except for a constant factor. Furthermore, for an arbitrary $\overline{A} = (\overline{a}_{ij}) = \widetilde{A} P$ where $P = diag(p_1, p_2, \cdots, p_n)$ with $0 < p_i \neq constant$ for $i = 1, 2, \cdots, n$, we have

$$\min_i \{ \sum_{j \neq i} |\overline{a}_{ij}| / |\overline{a}_{ii}| \} \leq \rho(|D|^{-1}|B|) \leq \max_i \{ \sum_{j \neq i} |\overline{a}_{ij}| / |\overline{a}_{ii}| \}. \tag{15}$$

We call $\widetilde{A}$ the optimally scaled matrix of A. Several properties can be established for it.

Property 1 ([5]). Under the assumptions of lemma 1, we have

$$\rho(\mid\widetilde{D}\mid^{-1}\mid\widetilde{B}\mid) = \|\mid\widetilde{D}\mid^{-1}\mid\widetilde{B}\mid\|_\infty = \rho(|D|^{-1}|B|)$$

where $\widetilde{D} = \operatorname{diag}(\widetilde{A})$, $\widetilde{B} = \widetilde{D} - \widetilde{A}$.

Property 2 ([5]). Under the assumptions of lemma 1, the following four properties are equivalent.

(a) $\rho(|D|^{-1}|B|) < 1$, $\qquad\qquad\qquad$ (b) $\rho(\mid\widetilde{D}\mid^{-1}\mid\widetilde{B}\mid) < 1$,

(c) A is H-matrix, i. e., $\langle A\rangle$ is M-matrix, (d) $\widetilde{A}$ is H-matrix, i.e., $\langle\widetilde{A}\rangle$ is M-matrix .

And if one of the above holds, $\widetilde{A}$ is SDD, i.e., $\mid\widetilde{a}_{ii}\mid > \sum_{j\neq i}\mid\widetilde{a}_{ij}\mid$.

Property 3 ([5]). Under the assumptions of lemma 1, if A and $\widetilde{A}$ have matrix splittings $A = M - N; \widetilde{A} = \widetilde{M} - \widetilde{N}$ where $\widetilde{M} = MQ, \widetilde{N} = NQ, M^{-1}$ exists. Here Q is given in lemma 1. Then

$$\rho(M^{-1}N) = \rho(\widetilde{M}^{-1}\widetilde{N}). \tag{16}$$

Let $(D - L_l, U_l, E_l)$ be a multisplitting of A, where $D = \operatorname{diag}(A)$, L_l is strictly lower triangular. Let $R = \operatorname{diag}(r_1, r_2, \cdots, r_n)$, $\Omega = \operatorname{diag}(\omega_1, \omega_2, \cdots, \omega_n)$ be relatation matrices, where $r_i \geq 0, \omega_i > 0$ for $i = 1, 2, \cdots, n$ and let $\omega > 0, r \geq 0$. We can then write the iterative matrices of PMI-SOR [2], PMI-GSOR [3], PMI-AOR [10] and PMI-GAOR [3] as follows:

$$\mathcal{L}_\omega(A) = \sum_l E_l(D - \omega L_l)^{-1}[(1 - \omega)D + \omega U_l],$$

$$\mathcal{L}_\Omega(A) = \sum_l E_l(D - \Omega L_l)^{-1}[(I - \Omega)D + \Omega U_l],$$

$$\mathcal{L}_{r,\omega}(A) = \sum_l E_l(D - rL_l)^{-1}[(1 - \omega)D + (\omega - r)L_l + \omega U_l],$$

$$\mathcal{L}_{R,\Omega}(A) = \sum_l E_l(D - RL_l)^{-1}[(I - \Omega)D + (\Omega - R)L_l + \Omega U_l].$$

If we let $D = \operatorname{diag}(A), B = D - A$, then the Jacobi iterative matrix is $J(A) = D^{-1}B$. We denote the intervals $\left(0, \frac{2}{1+\rho(J(A))}\right)$ and $\left[0, \frac{2}{1+\rho(J(A))}\right)$ as I_A and $\widetilde{I}_A$, respectively. We then have the following theorem.

Theorem 3. *Let A be an H-matrix with $\langle A\rangle = D - |L_l| - |U_l|$. Then*
(a) $\rho(\mathcal{L}_\omega(A)) < 1, \forall\omega \in I_A$.
(b) $\rho(\mathcal{L}_\Omega(A)) < 1, \forall\omega_i \in I_A(\forall i)$.
(c) $\rho(\mathcal{L}_{r,\omega}(A)) < 1 , \forall r \leq \omega, r \in\widetilde{I}_A, \omega \in I_A$.
(d) $\rho(\mathcal{L}_{R,\Omega}(A)) < 1 , \forall r_i \leq \omega_i, r_i \in\widetilde{I}_A, \omega_i \in I_A(\forall i)$.
Proof. Since $(a), (b), (c)$ are special cases of (d), we only give a proof of (d). We

first assume that A is irreducible. Hence, by theorem 1, we have

$$\rho(\mathcal{L}_{R,\Omega}(A)) = \rho(\mathcal{L}_{R,\Omega}(\widetilde{A})) \leq \parallel \mathcal{L}_{R,\Omega}(\widetilde{A}) \parallel_\infty$$

$$\leq \max_l\{\max_i\{\frac{|1-\omega_i||\widetilde{a}_{ii}|+(\omega_i-r_i)\sum_{j<i}|l_{ij}^l|++\omega_i\sum_j|u_{ij}^l|}{\widetilde{a}_{ii}-r_i\sum_{j<i}|l_{ij}^l|}\}\} \tag{17}$$

$$= \max_l\{\max_i\{\frac{|1-\omega_i|+\omega_i\rho(|J(A)|)-r_i\sum_{j<i}|l_{ij}^l|/|\widetilde{a}_{ii}|}{1-r_i\sum_{j<i}|l_{ij}^l|/|\widetilde{a}_{ii}|}\}\}.$$

When $\omega_i \in I_A$ for $i = 1, 2, \cdots, n$, we obtain $|1 - \omega_i| + \omega_i\rho(|J(A)|) < 1$. When $a, b, c > 0, a < b$, we have $(a - c)/(b - c) < a/b$. Hence, when $r_i \leq \omega_i, r_i \in \widetilde{I}_A, \omega_i \in I_A(\forall i)$, we get from (17) that

$$\rho(\mathcal{L}_{R,\Omega}(\widetilde{A})) = \rho(\mathcal{L}_{R,\Omega}(A)) < |1 - \omega_i| + \omega_i\rho(|J(A)|) < 1.$$

If A is reducible, we can change some zero entries of A into sufficient a small positive number $\epsilon > 0$ such that A change into A_ϵ and A_ϵ is irreducible. We can work with A_ϵ as above. Finally, we can show that theorem 3 holds when A is reducible by taking $\epsilon \to 0$ and using the continuity of the spectral radius of the entries of the matrix.

Remark. Theorem 3 shows that PMI-SOR, PMI-GSOR, PMI-AOR and PMI-GAOR are convergent if the parameters are in the intervals given in theorem 3. This is in keeping with results given in [2], [3], [10]. We unify the proof of convergence of these relaxed PMI-methods. Using theorem 1, we can get more general results than those given in [2], [3], [10] and theorem 3.

Theorem 4. *If A is SDD, $\langle A \rangle = |D| - |L_l| - |U_l|$ and $\omega > 0, \omega_i > 0, 0 \leq r \leq \omega, 0 \leq r_i \leq \omega_i$ for $i = 1, 2, \cdots, n$ are all smaller than $2|a_{ii}|/\sum_j|a_{ij}|$, then all relaxed PMI-methods in theorem 3 are convergent.*

Theorem 5. *If A is an irreducible L-matrix but not a M-matrix, then*
(a) $\rho(\mathcal{L}_\omega(A)) \geq 1$ for sufficiently small ω .
(b) $\rho(\mathcal{L}_\Omega(A)) \geq 1$ for sufficiently small $\omega_i(\forall i)$.
(c) $\rho(\mathcal{L}_{r,\omega}(A)) \geq 1$ for sufficiently small r, ω with $0 \leq r \leq \omega$.
(d) $\rho(\mathcal{L}_{R,\Omega}(A)) \geq 1$ for suitable small r_i, ω_i with $0 \leq r_i \leq \omega_i(\forall i)$.

Proof. We also only show that (d) holds. First, we choose r_i sufficiently small such that $D - RL_l$ is SDD. Then we have from theorem 2 and property 3, that

$$\rho(\mathcal{L}_{R,\Omega}(A)) = \rho(\mathcal{L}_{R,\Omega}(\widetilde{A}))$$

$$\geq \min_l\{\min_i\{\frac{[1 - \omega_i] + \omega_i\rho(|J(A)|) - r_i\sum_{j<i}|l_{ij}^l|/|\widetilde{a}_{ii}|}{1 - r_i\sum_{j<i}|l_{ij}^l|/|\widetilde{a}_{ii}|}\}\} \tag{18}$$

and from property 2, we get that the right hand of inequality (18) is not smaller than unit for any l and i. So (d) holds.

REFERENCES

[1] Berman A. and Plemmons R. J. (1979) *Nonnegative Matrices in the Mathematical and Science*, Academic Press, New York.

[2] Frommer F. and Mayer G. (1989) Convergence of relaxed parallel multisplitting methods. *Lin. Alg. Appl.* 119:141-152.

[3] Gu Tongxiang and Wang Nengchao (1992) A class of multisplitting iterative methods. *Proceedings of Third National Conference on Parallel Algorithm of China,* Huazhong Uni. Sci. Tech. Press, Wuhan, China, 186-190.

[4] Hu Jiagan (1982) Estimates of $\| B^{-1}A \|$ and their applications. *Numerica Mathematica Sinica,* 4(3):272-282.

[5] Hu Jiagan (1983) Scaling transformation and convergence of splitting of matrix,*Numerica Mathematica Sinica,* 5(1):72-78.

[6] Hu Jiagan (1991) *Iterative Methods for Solving Linear Systems of Algebraic Equations*, Science Press, Beijing, China.

[7] Neumann M. and Plemmons R. J. (1987) Convergence of parallel multisplitting iterative methods for M-matrices.*Lin. Alg. Appl.* 88/89:559-573.

[8] O'Leary D. P. and White R. E. (1985) Multi-splitting of matrices and parallel solution of linear systems. *SIAM J. Alg. Disc. Meth.* 6(4):630-640.

[9] Varga R. S. (1962) *Matrix Iterative Analysis*, Prentice-Hall, Englewood Cliffs, N. J.

[10] Wang Deren (1991) On the convergence of parallel multisplitting AOR algorithm. *Lin. Alg. Appl.* 154/156:473-486.

[11] Young D. M. (1971) *Iterative Solution of Large Linear Systems*, Academic Press, New York .

Splitting Extrapolation Method for Solving Multidimensional Problems in Parallel

C. B. Liem, T. M. Shih and T. Lu

ABSTRACT. The splitting extrapolation method is an important method in numerical solution of multidimensional problems. Two types of splitting extrapolation algorithms with their applications in solving partial differential equations are disscussed. Numerical experiments show that the method is superior than the Richardson extrapolation method in the sense of parallelism, computational complexity and computer storage needed.

INTRODUCTION

Many mathematical models of scientific and engineering problems are described by partial differential equations. Despite the significant progress made in digital computers over the last two decades, the solution of high dimensional problems with complicated domains still remains difficult. It is due to the fact that the computational complexity and computer storage required increase exponentially with respect to the dimension. In order to overcome this difficulty, it is necessary to develop parallel algorithms with high accuracy.

In recent years, promising progress has been made in parallel computational methods. The following three types of methods have a common characteristic: large scale multidimensional problems are subdivided into smaller problems.

(a) Domain decomposition methods, including multilevel methods and the fast adaptive composite grid methods [4];
(b) Sparse grid combination techniques [7]; and
(c) Splitting extrapolation methods [1, 2, 3, 5, 6].

The splitting extrapolation, which are also called the multivariate Richardson extrapolation, was first established in 1983 by Q. Lin and T. Lu [1]. It is an ideal method for dealing with the so called "dimensional effect" arised in solving multidimensional problems. The most recent development of the method can be found in the monograph [8]. A comprehensive review on splitting extrapolation methods and sparse grid combination techniques can be found in [7].

THE PRINCIPLE OF SPLITTING EXTRAPOLATION

In order to find the approximate solution of a continuous problem, one first choose a suitable grid parameter h and an appropriate discretization scheme, so as

C. B. Liem and T. M. Shih, Department of Applied Mathematics, The Hong Kong Polytechnic University, Kowloon, Hong Kong. E-mail: MACBLIEM@POLYU.EDU.HK
T. Lu, Chengdu institute of Computer Applications, Academia Sinica, Chengdu, Sichuen, China
Supported by the University Grant Council of Hong Kong under Grant No. 354/018

Domain Decomposition Methods in Sciences and Engineering, edited by R. Glowinski *et al.*

to convert the continuous problem into a set of algebraic equations and then obtain the numerical solution $u(h)$. The accuracy of $u(h)$ depends on h. For instance, a finite difference scheme for a s-dimensional problem can naturally include s independent grid parameters, i.e., $h = (h_1, h_2, \cdots, h_s)$. However, independent grid parameters can be chosen according to the scale as well as the geometry of the domain. The number of these parameters can even be larger than the dimension s.

For many continuous problems, it can be shown that under certain assumptions, there is an asymptotic expansion of the error between the numerical solution $u(h)$ and the exact solution u:

$$(1) \qquad u(h) = u + \sum_{1 \leq |\alpha| \leq m} C_\alpha h^{2\alpha} + O\left(h_0^{2m+1}\right),$$

where $h = (h_1, h_2, \cdots, h_s)$, $\alpha = (\alpha_1, \alpha_2, \cdots, \alpha_s)$, $|\alpha| = \alpha_1 + \alpha_2 + \cdots + \alpha_s$, $h^{2\alpha} = h_1^{2\alpha_1} \cdots h_s^{2\alpha_s}$ and $h_0 = \max_{1 \leq i \leq s} h_i$, here $h_1, h_2, \cdots, h_s$ are independent grid parameters. If $h_1 = h_2 = \cdots = h_s$, (1) is the classical Richardson asymptotic expansion. Obviously, (1) indicates that the error of $u(h)$ is $O\left(h_0^2\right)$. It is expected that, based on (1), a cheaper and more accurate solution can be obtained by the method of splitting extrapolation, i.e., instead of taking a global refinement in all directions as suggested by the classical Richardson extrapolation, one needs only to carry out some unidirectional refinements. In 1990, we proposed two types of unidirectional refinements [3].

Type 1. Given an initial grid parameter $h = (h_1, \cdots, h_s)$, we choose successively the refined grid parameters $\frac{h}{2^\beta} = (\frac{h_1}{2^{\beta_1}}, \cdots, \frac{h_s}{2^{\beta_s}})$, $0 \leq |\beta| \leq m$, and obtain the corresponding approximate solution $u\left(\frac{h}{2^\beta}\right)$.

Type 2. Choose successively the refined grid parameter $\frac{h}{(1+\beta)} = (\frac{h_1}{1+\beta_1}, \cdots, \frac{h_s}{1+\beta_s})$, $0 \leq |\beta| \leq m$, and the corresponding approximate solution is denoted by $u\left(\frac{h}{1+\beta}\right)$.

$u\left(\frac{h}{2^\beta}\right)$ (or $u\left(\frac{h}{(1+\beta)}\right)$), $0 \leq |\beta| \leq m$, can be evaluated in parallel. Furthermore, by using the extrapolation coefficients $\{a_\beta \text{ or } \tilde{a}_\beta, 0 \leq |\beta| \leq m\}$, we can obtain the following approximations with m splits.

Type 1.

$$(2) \qquad u_m(h) = \sum_{0 \leq |\beta| \leq m} a_\beta u\left(\frac{h}{2^\beta}\right),$$

and

Type 2.

$$(3) \qquad \tilde{u}_m(h) = \sum_{0 \leq |\beta| \leq m} \tilde{a}_\beta u\left(\frac{h}{2^\beta}\right).$$

Both $u_m(h)$ and $\tilde{u}_m(h)$ are of order $2m+1$.

It is known that the extrapolation coefficients a_β satisfy the following equations

$$
(4) \qquad
\begin{cases}
\displaystyle\sum_{0\le|\beta|\le m} a_\beta = 1 \\[2ex]
\displaystyle\sum_{0\le|\beta|\le m} \frac{a_\beta}{2^{2(\beta,\alpha)}} = 0, \quad 0 \le |\alpha| \le m
\end{cases} .
$$

Similar relations also hold for $\tilde{a}_\beta$.

The exact extrapolation coefficients can be computed in advance. In the following, for a two-dimensional case and $0 \le |\beta| \le 3$, terms in the asymptotic expansion (1), the corresponding approximate solutions and the extrapolation coefficients a_β are arranged in triangular patterns:

$$
\begin{array}{cccc}
1 & h_x^2 & h_x^4 & h_x^6 \quad \cdots \\[1ex]
h_y^2 & h_x^2 h_y^2 & h_x^4 h_y^2 \quad \cdots \\[1ex]
h_y^4 & h_x^2 h_y^4 \quad \cdots \\[1ex]
h_y^6 \quad \cdots \\[1ex]
\cdots
\end{array}
$$

$$
\begin{array}{cccc}
u\left(h_x, h_y\right) & u\left(\tfrac{h_x}{2}, h_y\right) & u\left(\tfrac{h_x}{4}, h_y\right) & u\left(\tfrac{h_x}{8}, h_y\right) \quad \cdots \\[2ex]
u\left(h_x, \tfrac{h_y}{2}\right) & u\left(\tfrac{h_x}{2}, \tfrac{h_y}{2}\right) & u\left(\tfrac{h_x}{4}, \tfrac{h_y}{2}\right) & \cdots \\[2ex]
u\left(h_x, \tfrac{h_y}{4}\right) & u\left(\tfrac{h_x}{2}, \tfrac{h_y}{4}\right) & \cdots \\[2ex]
u\left(h_x, \tfrac{h_y}{8}\right) & \cdots \\[1ex]
\cdots
\end{array}
$$

$$
\begin{array}{cccc}
-\dfrac{97}{567} & \dfrac{148}{135} & -\dfrac{64}{27} & \dfrac{4096}{2835} \quad \cdots \\[2ex]
\dfrac{148}{135} & -\dfrac{80}{27} & \dfrac{256}{135} \quad \cdots \\[2ex]
-\dfrac{64}{27} & \dfrac{256}{135} \quad \cdots \\[2ex]
\dfrac{4096}{2835} \quad \cdots \\[1ex]
\cdots
\end{array}
$$

As a special case, when the number of split $m = 1$, the extrapolation coefficients for a s-dimensional problem are:

$$
a_{(0,\cdots,0)} = -\frac{(4s-3)}{3}, \quad a_{(1,0,\cdots,0)} = \cdots = a_{(0,\cdots,0,1)} = \frac{4}{3},
$$

i.e., there are only two different values and $\tilde{a}_\beta = a_\beta$.

Some of the coefficients a_β and $\tilde{a}_\beta$, $0 \le |\beta| \le m$, can be found in [4] and [5].

MULTIVARIATE ASYMPTOTIC ERROR EXPANSION OF THE NUMERICAL SOLUTION TO PDE

Finite Difference Methods

Consider the following semilinear elliptic equation:

$$(5) \qquad \begin{cases} \Delta u = f\left(x, u\right), & \text{in } \Omega = \left(0, 1\right)^s, \\ \quad u = 0, & \text{on } \partial\Omega. \end{cases}$$

where $f'_u\left(x, u\right) \ge 0, h = \left(h_1, h_2, \cdots, h_s\right)$, and $h_i = \frac{1}{N_i}$, for $i = 1, 2, \cdots, s$.

Using the central difference scheme, we have the following difference equation:

$$\Delta^h u^h = \sum_{i=1}^s h_i^{-2} \left\{ u^h\left(x_1, x_2, \cdots, x_i - h_i, \cdots, x_s\right) - 2u^h\left(x_1, \cdots, x_i, \cdots, x_s\right) \right.$$

$$(6) \qquad \left. + u^h\left(x_1, \cdots, x_i + h_i, \cdots, x_s\right) \right\} = f\left(x, u^h\right), \ x = \left(x_1, \cdots, x_s\right) \in \Omega^h,$$

$$u^h = 0, x \in \partial\Omega^h,$$

where $\Omega^h = \left\{x = \left(x_1, \cdots, x_s\right) : x_i = jh_i, 1 \le j \le N_i, 1 \le i \le s\right\}$.

Theorem 1. If $u \in C^{7+\sigma}\left(\Omega\right), 0 < \sigma \le 1$, then $\exists w_\beta \in C^{5+\sigma - 2|\beta|}\left(\Omega\right), 1 \le |\beta| \le 2$, such that

$$u^h - u + \sum_{1 \le |\beta| \le 2} w_\beta h^{2\beta} = O\left(h_0^{5+\sigma}\right),$$

where $h_0 = \max_{1 \le i \le s} h_i$.

Remark: If Ω is a smooth domain, the above theorem will still hold if quadratic interpolation polynomials are applied at irregular points.

Finite Element Methods

One of the earliest work on splitting extrapolation of finite element methods was published by Q. Lin and T. Lu in 1983. A recent development is the monograph by Q. Lin and Q. Zhu (1994). If $s = 2$, using bilinear elements on rectangular grids, one can prove the following theorem for equation (5):

Theorem 2. If $\Im^h$ is a regular rectangular subdivision of Ω, and $u \in W_q^3\left(\Omega\right) \cap H_0^1\left(\Omega\right) \cap \left(\Pi_{e \in \Im^h} W_q^4\left(e\right)\right)$, where $1 \le q \le \infty$, then there exist functions w_1 and w_2, independent of $h = \left(h_1, h_2\right)$, such that

$$(7) \qquad \left\| u^h - u^I - h_1^2 w_1^I - h_2^2 w_2^I \right\|_0 \le c h_0^4 \left\| u \right\|_{4,2}' ,$$

and

$$(8) \qquad \left\| u^h - u^I - h_1^2 w_1^I - h_2^2 w_2^I \right\|_\infty \le c h_0^4 \left| \ln h_0 \right| \left\| u \right\|_{4,\infty}'$$

where $h_0 = \max\{h_1, h_2\}$, $\|\cdot\|_0$ denotes the norm in $L^2(\Omega)$, $\|\cdot\|_\infty$ denotes the norm in $L^\infty(\Omega)$, u^I denotes the interpolation of u, and

$$\|u\|_{n,p}' = \left(\sum_{e \in \Im^h} \|u\|_{n,p,e}^p \right)^{\frac{1}{p}}, \quad \text{for } p = 2, \infty.$$

The above theorem shows that one split can be used while under stronger conditions, more splits are allowed.

Similar result remains true for three dimensional case. But a more important case is that under certain conditions, one can choose independent grid parameters according to the size and the geometry of the problem. In general, the more the independent grid parameters, the higher the parallelism and more computer CPU and storage can be saved. In order to explain the idea, consider the following two figures:

In figure 1, there are three independent grid parameters for a two-dimensional problem.

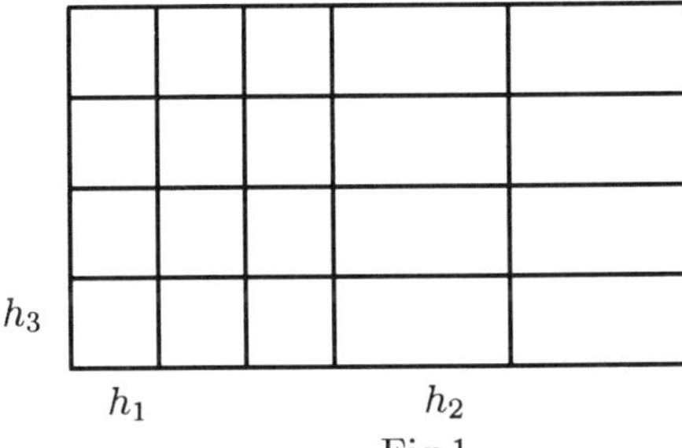

Fig.1

In figure 2, higher accuracy is needed near the point A. Therefore we can use six independent grid sizes where h_2 and h_5 can be chosen to be smaller than others.

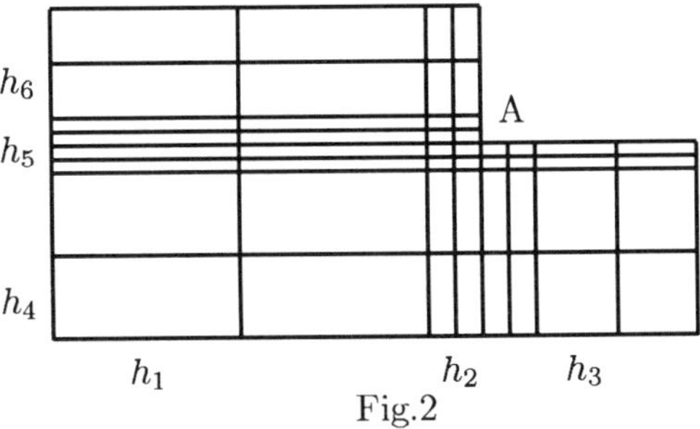

Fig.2

TABLE 1. Comparison of numerical results of three extrapolation methods

No. of split	Richardson extrapolation		Type 1 SEM		Type 2 SEM	
	max error	CPU	max error	CPU	max error	CPU
0	3.91E-4	0.04	3.91E-4	0.04	3.91E-4	0.04
1	2.44E-6	0.71	6.79E-6	0.52	6.79E-6	0.49
2	7.61E-9	13.60	1.90E-7	2.53	1.90E-7	2.09
3	3.23E-11	285.67	5.06E-9	12.98	5.32E-9	7.41
4					1.41E-10	21.94
5					7.89E-12	57.11

TABLE 2. Comparison of computer storage needed by three extrapolation methods

No. of split	Richardson extrapolation	Type 1 SEM	Type 2 SEM
0	162	162	162
1	1,769	339	339
2	16,956	945	945
3	149,063	2,135	2,135
4	1,250,370	4,410	3,430
5	2,048,543	9,051	5,411

NUMERICAL EXPERIMENTS

Example 1: Consider the three dimensional Poisson equation:

$$\begin{cases} \Delta u = f & , \text{ in } \quad \Omega = (0,1)^3 , \\ u = 0 & , \text{ on } \quad \partial\Omega. \end{cases}$$

The exact solution is $u = \prod_{i=1}^{3} \left[x_i \left(1 - x_i\right) \cos\left(\frac{\pi x_i}{2}\right) \right]$. Using the seven-point finite difference scheme with initial grid sizes $h_1 = h_2 = h_3 = \frac{1}{4}$, the results are recorded in Tables 1—3.

TABLE 3. Comparision of the degree of parallelism for three extrapolation methods, N is the processor used

No. of split	Richardson extrapolation		Type 1 SEM		Type 2 SEM	
	N	max CPU	N	max CPU	N	max CPU
1	2	0.33	4	0.13	4	0.13
2	3	12.94	10	0.38	10	0.38
3	4	272.01	20	1.42	20	0.67
4	5	$\cdots$	35	$\cdots$	35	1.44
5	6	$\cdots$	56	$\cdots$	56	2.54

TABLE 4. Maximum relative errors

err0	err1	err2	err3	err4
4.120E-2	4.125E-2	4.125E-2	4.125E-2	4.119E-2

err5	err6	errspl	errRich	
2.518E-2	2.498E-2	3.189E-3	7.535E-3	

TABLE 5. Relative errors at some points

Coordinates	err0	errspl	errRich
$(1/2, 1/2)$	8.55E-3	2.43E-5	3.93E-5
$(3/2, 1/2)$	9.63E-3	1.03E-5	3.05E-5
$(5/2, 1/2)$	5.09E-3	8.80E-5	4.36E-5
$(7/2, 1/2)$	3.84E-3	1.37E-4	1.88E-5
$(8/2, 1/2)$	1.18E-2	6.80E-5	7.23E-5
$(9/2, 1/2)$	2.44E-2	7.85E-4	3.56E-4

Example 2: Consider the two dimensional Poisson equation:

$$\begin{cases} \Delta u = f & , \quad \text{in} \quad \Omega = (0,5) \times (0,1), \\ u = 0 & , \quad \text{on} \quad \partial\Omega. \end{cases}$$

The exact solution is $u(x,y) = xy(1-y)\left(1 - \frac{x}{5}\right)e^{xy}$. Using rectangular elements and choose six independent grid sizes h_i $(i = 1, 2, \cdots, 6)$ in five subdomains Ω_i $(i = 1, 2, \cdots, 5)$.(Fig.3). Let $u_0 = u(h_1, h_2, \cdots, h_6)$, $u_i = u\left(h_1, \cdots, \frac{h_i}{2}, \cdots, h_6\right)$, denote the maximum relative error of u_i $(i = 0, 1, \cdots, 6)$, of the splitting extrapolation solution and of Richardson extrapolation by erri, errspl and errRich respectively. The results are shown in Table 4 and Table 5.

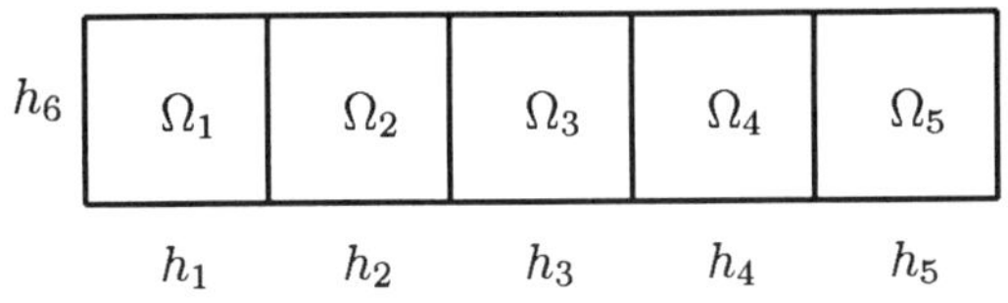

h_6 | Ω_1 | Ω_2 | Ω_3 | Ω_4 | Ω_5

h_1 h_2 h_3 h_4 h_5

Fig.3

CONCLUSION

If there exists the asymptotic expansion (1), the numerical accuracy of the splitting extrapolation and the Richardson extrapolation methods are comparable, while the former is superior in the sense of parallelism, computational complexity and computer storage needed.

REFERENCES

(1) T. Lu, T. M. Shih and C. B. Liem, The Splitting Extrapolation Method, World Scientific Co. Singapore, 1995.

(2) Q. Lin, T. Lu, Splitting extrapolation for multidimensional problems, J. Comp. Math., 1(1983), 45-51.

(3) T. Lu, T. M. Shih and C. B. Liem, An analysis of splitting extrapolation for multidimensional problems, System Sci. & Math. 3(1990),261-272.

(4) T. Lu, T. M. Shih and C. B. Liem, Domain Decomposition Methods-New Numerical Techniques for Solving PDE, Science Press, Beijing, 1993, (in Chinese).

(5) T. M. Shih, C. B. Liem and T. Lu, Coefficients of the splitting extrapolation method and their applications in multidimensional problems, Research Report IMS-60, Inst. of Math. Sci., Academia Sinica, Chengdu, 1994.

(6) U. Rüde, Extrapolation and related technique for solving elliptic equations, Bericht I9135, Inst. für Informatik, TU München, 1991.

(7) U. Rüde, Multilevel, extrapolation and sparse grid method, in Bericht I9319, Inst. für Informatik, TU München, 1993.

(8) Q. Lin and Q. D. Zhu, Pretreatment and Posttreatment Theory of Finite Elements, Shanghai Science Press, 1994, (in Chinese).

Implementation of Non-overlapping Domain Decomposition Methods on Parallel Computer ADENA

ATSUSHI SUZUKI[1]

This paper discusses parallel efficiency of non-overlapping Domain Decomposition methods(DDM) for elliptic problem with the results of the implementation on parallel computer ADENA, and describes DDM solver for Stokes problem.

1 NON-OVERLAPPING METHOD FOR ELLIPTIC PROBLEM

In the Non-overlapping Domain Decomposition method, appropriate boundary condition on inner-boundary need to be found to get a solution satisfying original problem. There are two methods differed in boundary data. One method requires continuity of Dirichlet data on inner-boundary[GDP84]. Neumann data on the interface is obtained by preconditioned Conjugate Gradient solver. We call '*Neumann type*' method in this report. The other method requires continuity of Neumann data on inner-boundary [BW86]. This method is well-known as Schur complement method, we call '*Dirichlet type*' method.

In the following section we discuss details of two algorithms for the elliptic problem

[1] Division of Applied Systems Science, Faculty of Engineering, Kyoto University,
Kyoto, 606-01, Japan
e-mail : suzuki@kuamp.kyoto-u.ac.jp

Domain Decomposition Methods in Sciences and Engineering, edited by R. Glowinski *et al.*

in two dimensional space:

$$(E) \begin{cases} -\triangle u = f & \text{in} \quad \Omega \\ u = g & \text{on} \quad \Gamma = \partial\Omega \end{cases}. \tag{1}$$

For the simplicity of representation, we discuss the two sub-domains problem. Domain Ω is divided into two sub-domains $\{\Omega_i\}_{i=1}^2$ and γ_{12} is the interface between the two sub-domains : $\gamma_{12} = \partial\Omega_1 \cap \partial\Omega_2$.

1.1 Interface Problem with Lagrange Multiplier

Non-overlapping DDM is formulated as minimization problem with constraint. Let $\{J_i(v_i)\}_{i=1}^2$ be cost function in sub-domain and $b(\{v_1, v_2\}, q)$ be constraint on the interface, the minimization problem is following:

$$\text{Find} \quad \{u_1, u_2\} \in V \quad \text{such that} \quad J_1(u_1) + J_2(u_2) \leq J_1(v_1) + J_2(v_2) \; \forall\{v_1, v_2\} \in V, \tag{2}$$

$$V := \{\{v_1, v_2\} \in X_1 \times X_2\} \mid b(\{v_1, v_2\}, q) = 0 \; \forall q \in M\}.$$

By use of Lagrangean : $\mathcal{L}(\{v_1, v_2\}, q) := J_1(v_1) + J_2(v_2) + b(\{v_1, v_2\}, q)$, the solution of this minimization problem is equal to the solution of the saddle point problem :

$$\text{Find} \quad (\{u_1, u_2\}, p) \in (X_1 \times X_2) \times M \quad \text{such that}$$

$$\mathcal{L}(\{u_1, u_2\}, q) \leq \mathcal{L}(\{u_1, u_2\}, p) \leq \mathcal{L}(\{v_1, v_2\}, p) \quad \forall(\{v_1, v_2\}, q) \in (X_1 \times X_2) \times M.$$

1.2 Neumann type

In Neumann type, let $X_i := \{v_i \in H^1(\Omega_i) \mid v_i = 0 \text{ on } \partial\Omega_i \setminus \gamma_{12}\}$ and $M := H^{-\frac{1}{2}}(\gamma_{12})$, and cost function and constraint are

$$J_i(v_i) := \frac{1}{2}\int_{\Omega_i} \nabla u_i \nabla v_i dx - \int_{\Omega_i} f_i v_i dx \quad \text{and} \quad b(\{v_1, v_2\}, q) := -\int_{\gamma_{12}} (v_1 - v_2)q d\gamma.$$

The constraint means Dirichlet data on the interface γ_{12} is continuous. The dual problem of equation(2) is a variational problem called 'interface problem' which determine Neumann data on the interface.

$$\text{Find} \quad p \in H^{-\frac{1}{2}}(\gamma_{12}) \quad \text{such that} \quad \alpha(p, q) = F(q) \qquad \forall q \in H^{\frac{1}{2}}(\gamma_{12}) \tag{3}$$

$$\alpha(p, q) = \int_{\gamma_{12}} (Ap)q d\gamma \qquad Ap = u_1(p) - u_2(p) \qquad F(q) = \int_{\gamma_{12}} (w_2 - w_1)q d\gamma \tag{4}$$

$$\begin{cases} -\triangle u_i = 0 & \text{in} \quad \Omega_i \\ u_i = 0 & \text{on} \quad \partial\Omega_i \setminus \gamma_{12} \\ \frac{\partial u_i}{\partial n_i} = (-1)^{i-1}p & \text{on} \quad \gamma_{12} \end{cases} \qquad \begin{cases} -\triangle w_i = f_i & \text{in} \quad \Omega_i \\ w_i = g_i & \text{on} \quad \partial\Omega_i \setminus \gamma_{12} \\ \frac{\partial w_i}{\partial n_i} = 0 & \text{on} \quad \gamma_{12} \end{cases} \tag{5}$$

1.3 Dirichlet type

In Dirichlet type, we consider dualization of the Dirichlet problem and divergence space to deal with Dirichlet data in weak sense. let $X_i := \{v_i \in H(\text{div} ; \Omega_i) \mid \nabla \cdot v_i + f_i = 0\}$

and $M := H^{\frac{1}{2}}(\gamma_{12})$, cost function and constraint are

$$J_i(v_i) := \frac{1}{2}\int_{\Omega_i}|v_i|^2 dx - \int_{\partial\Omega_i\setminus\gamma_{12}} g_i v_i \cdot n_i d\gamma \quad \text{and} \quad b(\{v_1,v_2\},q) := -\int_{\gamma_{12}}(v_1 \cdot n_1 + v_2 \cdot n_2)q d\gamma.$$

The constraint means Neumann data is continuous on γ_{12}. The dual problem of equation(2) is a variational problem called 'interface problem' which determine Dirichlet data on the interface. Usual formulation is used to describe Drichilet problem in sub-domain.

$$\text{Find } p \in H^{\frac{1}{2}}(\gamma_{12}) \text{ such that } \alpha(p,q) = F(q) \quad \forall q \in H^{-\frac{1}{2}}(\gamma_{12}) \tag{6}$$

$$\alpha(p,q) = \int_{\gamma_{12}}(\mathcal{A}p)q d\gamma \quad \mathcal{A}p = \frac{\partial u_1(p)}{\partial n_1} + \frac{\partial u_2(p)}{\partial n_2} \quad F(q) = -\int_{\gamma_{12}}\left(\frac{\partial w_1}{\partial n_1} + \frac{\partial w_2}{\partial n_2}\right)q d\gamma \tag{7}$$

$$\begin{cases} -\triangle u_i = 0 & \text{in} \quad \Omega_i \\ u_i = 0 & \text{on} \quad \partial\Omega_i \setminus \gamma_{12} \\ u_i = p & \text{on} \quad \gamma_{12} \end{cases} \quad \begin{cases} -\triangle w_i = f_i & \text{in} \quad \Omega_i \\ w_i = g_i & \text{on} \quad \partial\Omega_i \setminus \gamma_{12} \\ w_i = 0 & \text{on} \quad \gamma_{12} \end{cases} \tag{8}$$

Calculation of the gap of Neumann data on γ_{12} is done by Gauss-Green's formula exactly,

$$\int_{\gamma_{12}}\left(\frac{\partial u_1}{\partial n_1} + \frac{\partial u_2}{\partial n_2}\right)q d\gamma = \sum_{i=1}^{2}\int_{\Omega_i}\nabla u_i \cdot \nabla \tilde{q}_i dx,$$

where $\tilde{q}$ stands for extension of function q on γ_{12}, in descrete case FEM basis on γ_{12}. This $L^2(\gamma_{12})$ inner product is replaced by $H^1(\gamma_{12})$ inner product as the preconditioner for Conjugate Gradient solver of equation(6).

1.4 *Homogeneous Neumann Boundary Problem*

The Poisson equation with homogeneous Neumann boundary condition must satisfy the compatibility condition, and its solution is unique except ambiguity of a constant.

$$\text{Find } u \in H^1(\Omega) \setminus \mathbb{R} \quad \begin{cases} -\triangle u = f & \text{in } \Omega \\ \frac{\partial u}{\partial n} = 0 & \text{on } \Gamma \end{cases} \quad \int_{\Omega} f dx = 0 \tag{9}$$

Neumann type DDM can solve the problem with balancing procedure to satisfy compatibility condition in each sub-domain, and to adjust the constant over sub-domains. This procedure is described as following.

Initial data of search vector p^0 of CG solver must satisfy the compatibility condition:

$$\begin{cases} -\triangle u_i^0 = f_i & \text{in} \quad \Omega_i \\ \frac{\partial u_i^0}{\partial n_i} = 0 & \text{on} \quad \partial\Omega_i \setminus \gamma_{12} \\ \frac{\partial u_i^0}{\partial n_i} = (-1)^{i-1}p^0 & \text{on} \quad \gamma_{12} \end{cases} \quad \int_{\Omega_i} f_i dx + (-1)^{(i-1)}\int_{\gamma_{12}} p^0 d\gamma = 0. \tag{10}$$

On each step of CG iteration, sub-problems with Neumann boundary data p^n which must satisfy the compatibility condition are solved.

$$\begin{cases} -\triangle u_i^n = 0 & \text{in} \quad \Omega_i \\ \frac{\partial u_i^n}{\partial n_i} = 0 & \text{on} \quad \Gamma_i \\ \frac{\partial u_i^n}{\partial n_i} = (-1)^{i-1}p^n & \text{on} \quad \gamma_{12} \end{cases} \quad \int_{\gamma_{12}} p^n = 0. \tag{11}$$

Because each solution u_i^n has ambiguity of a constant, the constant $\{c_i\}_{i=1}^2$ in each sub-domain should be determined for next step of CG iteration,

$$\int_{\gamma_{12}} (u_1^n - c_1^n) - (u_2^n - c_2^n)d\gamma = 0. \tag{12}$$

If the domain is decomposed into many sub-domains, the matrix with one dimensional kernel have to be solved to determine the constants of sub-domains. Because of the singularity of this matrix, the direct solver is efficient than iterative solvers. This solver must be processed in a single processor to avoid the redundant cost for communication.

2 MATRIX REPRESENTATION OF INTERFACE PROBLEM

In this section, we give a matrix representation of 'interface problem'. From finite element base functions φ_i, matrices A, B are obtained : $(A)_{ij} = a(\varphi_j, \varphi_i)$, $(B)_{ij} = \int_{\gamma_{12}} \varphi_j \varphi_i d\gamma$. Suppose that $u_I^{(i)}$ denotes the values on inner nodes and $u_B^{(i)}$ on the interface γ_{12}. Also matrix $A^{(i)}$ is decomposed into four parts.

2.1 Neumann type

$$\left\{ \begin{array}{rcl} -\triangle u_i &=& 0 \quad \text{in} \quad \Omega_i \\ u_i &=& 0 \quad \text{on} \quad \partial\Omega_i \setminus \gamma_{12} \\ \frac{\partial u_i}{\partial n_i} &=& (-1)^{i-1}p \quad \text{on} \quad \gamma_{12} \end{array} \right. \qquad \left(\begin{array}{cc} A_{II}^{(i)} & A_{IB}^{(i)} \\ A_{IB}^{(i)}{}^{T} & A_{BB}^{(i)} \end{array} \right) \left(\begin{array}{c} u_I^{(i)} \\ u_B^{(i)} \end{array} \right) = \left(\begin{array}{c} 0 \\ (-1)^{i-1}Bp \end{array} \right),$$

By eliminating data in sub-domain $u_I^{(i)}$ matrix representation of 'interface problem' operator is obtained:

$$\mathcal{A}_h = \left((A_{BB}^{(1)} - A_{IB}^{(1)}{}^{T} A_{II}^{(1)}{}^{-1} A_{IB}^{(1)})^{-1} + (A_{BB}^{(2)} - A_{IB}^{(2)}{}^{T} A_{II}^{(2)}{}^{-1} A_{IB}^{(2)})^{-1} \right) B. \tag{13}$$

We use main part of the operator $\mathcal{A}_h$, $(A_{BB}^{(1)}{}^{-1} + A_{BB}^{(2)}{}^{-1})B$ as a preconditioner for CG solver. We use a direct solver (modified Cholesky decomposition solver) for elliptic problem in sub-domain. The reason is that direct solver is faster than iterative solver in small size problem.

2.2 Dirichlet type

$$\left\{ \begin{array}{rcl} -\triangle u_i &=& 0 \quad \text{in} \quad \Omega_i \\ u_i &=& 0 \quad \text{on} \quad \partial\Omega_i \setminus \gamma_{12} \\ u_i &=& p \quad \text{on} \quad \gamma_{12} \end{array} \right. \qquad \left(\begin{array}{cc} A_{II}^{(i)} & A_{IB}^{(i)} \\ 0 & E_{BB} \end{array} \right) \left(\begin{array}{c} u_I^{(i)} \\ u_B^{(i)} \end{array} \right) = \left(\begin{array}{c} 0 \\ p \end{array} \right)$$

Neumann data on the the interface are calculated by data $u_I^{(i)}$ exactly. A matrix representation of 'interface problem' operator is obtained :

$$\mathcal{A}_h = B^{-1} \left((A_{BB}^{(1)} - A_{IB}^{(1)}{}^{T} A_{II}^{(1)}{}^{-1} A_{IB}^{(1)}) + (A_{BB}^{(2)} - A_{IB}^{(2)}{}^{T} A_{II}^{(2)}{}^{-1} A_{IB}^{(2)}) \right). \tag{14}$$

The matrix representation of discrete 'interface problem' is equivalent to Schur complement matrix except for matrix B^{-1}. As in Neumann type, we use $B^{-1}(A_{BB}^{(1)} + A_{BB}^{(2)})$ as a preconditioner for CG solver.

These preconditioner are easily extended to multi-sub-domains problems. In Neumann type method, pieces of the interface are independent each other(shown in left of figure 1), the residual vector of 'interface problem' are calculated in parallel over all pieces of the interface. While, in Dirichlet type method, because of cross points, the interface has complicated structure(shown in right of figure 1). Therefore the residual vector given in $H^1(\gamma)$ inner product are calculated using the iterative method which requires cooperation of all processors for pieces of the interface.

Figure 1   The structure of inner-boundary

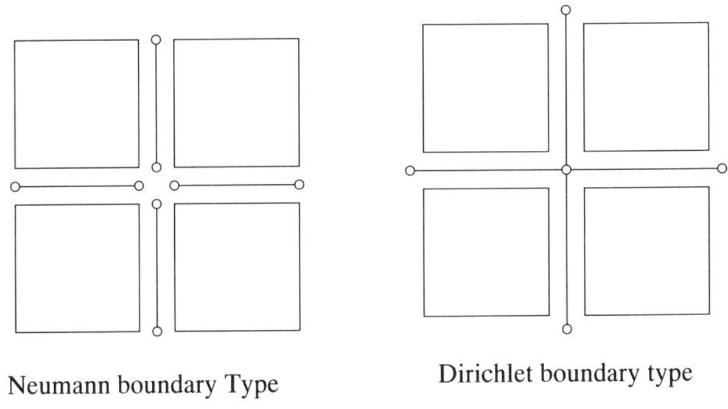

3   IMPLEMENTATION ON PARALLEL COMPUTER

3.1   *ADENA massively parallel computer*

The ADENA is distributed memory system, which has 256 CPUs with 2M bytes local memory. By using '*Alternating Data Edition*' network, all processors can communicate with two times operations of data-transfer. Moreover it is very easy to gather information of all processors to one processor, and also, to broadcast data in one processor to other all processors. This ability is powerful for inner product calculation, over all edges of sub-domains, which is required in the procedure of updating boundary data on interface in the DDM.

3.2   *Results of elliptic problem on ADENA*

The test problem is $u - \triangle u = f$ in $[0,1] \times [0,1]$, and the exact solution is $u = \sin(0.5\pi(x + 0.1))\sin(0.5\pi(y + 0.1))$. Square region contains 480×480 nodes. Table 1 shows the results in two type algorithms. Time for data transfer is less than 0.7% of total time of solver. Dirichlet type solver is faster than Neumann type solver. However in Dirichlet type, as described in section2.2, due to the cross point of boundary interface, time for updating boundary data is about 10% at 256 processors, against 1.7% in Neumann type.

Table 1   Results of elliptic problem on ADENA (480×480 nodes, $\varepsilon = 10^{-10}$)

	Neumann type				Dirichlet type			
# processors	100	144	225	256	100	144	225	256
# iteration	303	353	442	475	165	178	197	203
total time	395.40	277.72	189.57	171.71	199.14	129.28	78.69	68.50
parallel efficiency %	100.0	98.8	92.6	89.9	100.0	106.9	112.4	113.5
boundary data update	2.83	2.60	2.48	2.92	7.11	6.60	7.02	6.96
data transfer	0.77	0.85	1.08	1.08	0.35	0.36	0.41	0.39

time (sec.)

Table 2 shows the results of homogeneous Neumann boundary problem with fixed number of CPU and various size of problem. As viewed in section 1.4, balancing procedure is the bottle neck of the algorthim. From this result, we conclude that the algorithm is efficient if only the condition is satisfied : the number of sub-domain is less than the number of nodes in sub-domain.

Table 2   Homogeneous Neumann boundary problem on ADENA
(256CPUs, $\varepsilon = 10^{-10}$)

problem size	64×64	96×96	128×128	160×160	192×192	224×224	256×256
total time	5.269	6.553	7.842	9.174	10.868	12.879	15.628
direct solver	0.174	0.440	0.903	1.615	2.672	4.121	6.135
balancing	4.850	5.826	6.604	7.184	7.777	8.293	8.963

time (sec.)

3.3   Results on vector parallel computer VPP500

The Fujitsu VPP500 is vector parallel system which has vector processors with cross-bar network. Vector parallel machine can simulate massively parallel machine by replacing parallel loop over sub-domains with vector loop. However, to use vector parallel machine effectively, vector loop should be long enough. The results of Dirichlet type of elliptic problem is shown in table 3. From this results, we need more fast convergence of 'interface problem' and low-cost algorithm for calculation of norms on the interface .

Table 3 Results of elliptic problem with Dirichlet type DDM on VPP500
$(480 \times 480$ nodes, $\varepsilon = 10^{-10})$

# sub-domains		16×16	20×20	30×30	40×40	60×60	80×80
# iterations		223	246	296	337	368	412
1 CPU	total time	6.433	6.123	6.233	7.628	11.963	20.140
	direct solver	5.355	4.660	3.665	3.318	3.428	3.743
	interface	0.798	1.131	2.102	3.699	7.682	15.224
2 CPUs	total time	5.289	4.580	4.516	5.435	7.773	12.758
	direct solver	3.934	2.906	2.088	1.806	1.585	1.936
	interface	1.139	1.424	2.099	3.214	5.648	10.104
4 CPUs	total time	4.477	3.986	—	4.062	5.504	8.424
	direct solver	3.396	2.372	—	1.059	0.882	0.926
	interface	1.211	1.430	—	2.717	4.266	7.049

time (sec.)

4   DOMAIN DECOMPOSITION SOLVER FOR STOKES PROBLEM

We consider a generalized Stokes problem in two dimension :

$$(S) \begin{cases} \alpha u - \nu \triangle u + \nabla p = f & \text{in} \quad \Omega \\ \nabla \cdot u = 0 & \text{in} \quad \Omega \end{cases} \qquad \begin{matrix} u = g \text{ on } \Gamma = \partial\Omega \\ (\text{with } \int_\Gamma g \cdot n d\gamma = 0 \end{matrix} \quad , \qquad (15)$$

where α and ν are positive constants. Generalized Stokes problem is solved by saddle point algorithm which contains two kinds of elliptic problem. One is problem for velocity with Dirichlet boundary condition, the other is problem for pressure with Neumann boundary condition for preconditioner : $(\nu I_h^{-1} + \alpha(-\triangle)_h^{-1})$ to improve convergence of saddle point solver[CC88].

We consider two alternatives of DDM approaches. One approach is to consider generalized Stokes problems in each sub-domain. This approaches needs CG solver for the saddle point problem of each sub-problem, in each iterative procedure of DDM. Therefore this method is not better because of high cost of iterative solver for sub-problems.

The other is to apply DDM only to the elliptic solver in generalized Stokes saddle point solver. In this method we can use the direct solver as elliptic solver, which is effective for small size sub-problems.

We use Dirichlet type DDM for elliptic problem of velocity, and Neumann type DDM for pressure.

The test problem is the kernel of implicit scheme for time dependent cavity flow problem. $\alpha(= 68.28)(\nu(= 0.001715))$ is reciprocally proportional to the time step(Reynolds number, respectively). We use $P1$ iso $P2/P1$ mesh for velocity/pressure. Table4 shows computation time by second in various numbers of pressure nodes. In small size problems, time for preconditioner of pressure consumes almost time of the solver.

Table 4   Result of generalized Stokes problem on ADENA (256 CPUs, $\varepsilon = 10^{-10}$)

# pressure node	64×64	96×96	128×128	160×160	192×192	224×224	256×256
solver total	174.80	215.51	278.44	361.73	468.03	598.71	797.24
velocity	16.72	23.45	43.78	78.58	127.48	201.67	307.33
Neumann prob.	145.99	174.82	209.27	246.93	289.72	328.94	398.06
mass matrix	4.71	7.18	11.67	17.84	26.28	35.69	49.00

time (sec.)

5   CONCLUSION

We implemented the DDM for elliptic problem on massively parallel computer ADENA. Numerical results show that Dirichlet type is more fast and stable than Neumann type. Another implementation on vector parallel computer shows that we need much faster algorithm which can deal huge number of sub-domains. Application of elliptic DDM to the generalized Stokes problem is suitable for parallel processing. To get more high performance, we need to develop better preconditioner for Stokes solver suitable for DDM.

Acknowledgment. The author wihes to thank Professor T. Nogi for his help in this study.

References

[BF91] Brezzi F. and Fortin M. (1991) *Mixed and Hybrid Finite Element Methods.* Springer-Verlag.

[BW86] Bjørstad P. E. and Widlund O. B. (December 1986) Iterative methods for the solution of elliptic problems on regions partitioned into substructures. *SIAM J. Numer. Anal.* 23(6): 1097–1120.

[CC88] Cahouet J. and Chabard J. P. (1988) Some fast 3d finite element solvers for the generalized stokes problem. *Int. J. Numer. Methods in Fluids.* 8: 869–895.

[GDP84] Glowinski R., Dinh Q. V., and Périaux J. (1984) *Domain Decomposition for Elliptic Problems*, volume 5 of *Finite Elements in Fluids*, chapter 3, pages 45–106. John Wiley & Sons Limited.

[Su94] Suzuki A. (1994(to appear)) Implementation of domain decomposition methods on parallel computer ADENART. In Ecer A., Périaux J., and Satofuka N. (eds) *Proceedings of Parallel CFD '94.* Elsevier Science.

A Multi-Parameter Parallel Algorithm for Local Higher Accuracy Approximation

Aihui Zhou

1 Introduction

One may employ mesh refinement techniques if higher accuracy approximation is required. It is recognized that the global uniform refinement on the whole domain leads to simple and usually vectorizable algorithms but wastes time and memory, while local refinement on subdomains minimizes the size of the discrete problem and improves the accuracy locally but leads to a lower accuracy in the whole domain.

It has been shown recently that a combination of such discrete solutions related to refinement on subdomains can yield an approximation of higher accuracy and the procedure can be done in parallel using multi-processor computers if the exact solution is globally smooth (see [8, 9]). This technique is based on a so-called multi-parameter error resolution. The crucial point of this approach is to choose certain independent mesh parameters: According to its geometry, the domain is divided into some subdomains and covered with different meshes so that the number of independent mesh parameters, say p, is as large as possible, and an approximation of higher accuracy can be computed $(p + 1)$ processors in parallel.

We shall prove here that the parallel algorithm based on the multi-parameter error resolution can produce an approximation of higher accuracy even if the exact solution is only locally smooth.

---

[1] Institute of Systems Science, Academia Sinica, Beijing 100080, China

Domain Decomposition Methods in Sciences and Engineering, edited by R. Glowinski *et al.*

2   A Parallel Algorithm

Consider the finite element solution of the following Dirichlet boundary value problem,

$$
\begin{cases}
-\Delta u & = f, \ \text{in } \Omega, \\
u & = 0, \ \ \text{on } \partial\Omega,
\end{cases}
\tag{2.1}
$$

where $\Omega \subset R^2$ is a polygonal convex domain. Divide Ω into subdomains $T = \{\Omega_j : j = 1, 2, \cdots, m\}$ so that T is a quasi-uniform partition of Ω and each Ω_j is a triangle or a parallelogram. On each Ω_j a uniform mesh is imposed, and a quasi-uniform global partition of Ω is formed. Denote the mesh size(s) on Ω_j by $h_{j,1}$ if Ω_j is a trangle, and by $h_{j,1}$ and $h_{j,2}$, the mesh sizes along the directions of the edges of Ω_j, if Ω_j is a parallelogram $(j = 1, \cdots, m)$. Among these mesh parameters, some are independent, say $h_1, \cdots, h_p$. It can be proved that the number of independent parameters can be equal to or greater than 2 even that $p >> 2$. Let $h = max\{h_j : j = 1, \cdots, p\}$, T^h be the partition on Ω and let S^h be the conforming finite element space on T^h consisting of functions which are linear or bilinear on each triangular element or each parallelogram element in T^h. The interpolated finite element approximation corresponding to $h_1, \cdots, h_p$ is denoted by $u(h_1, \cdots, h_p)$. Set $u_0 = u(h_1, \cdots, h_p)$ and $u_j = u(h_1, \cdots, h_{j-1}, h_j/2, h_{j+1}, \cdots, h_p)$. Then a parallel algorithm for approximations of higher accuracy is given by

Algorithm.

Step 1. Compute $u_j (0 \le j \le p)$ in parallel.

Step 2. Set $u^c = (4\sum_{j=1}^{p} u_j - (4p - 3)u_0)/3$.

It is proved later that the composite numerical solution u^c is a higher accuracy approximation to the exact solution u of (2.1) in $D \subset\subset D^* \subset \Omega$ if u is smooth on D^*.

3   Multi-Parameter Error Resolution

In the following, $L^q(\Omega)(1 \le q < \infty), H^s(\Omega), H_0^s(\Omega)$ and $W^{s,q}(\Omega)(s = 1, 2, \cdots)$ are the usual Lebesgue and Sobolev spaces respectively. Set $S_0^h = H_0^1(\Omega) \cap S^h$.

The Galerkin projection $R_h u \in S_0^h$ of the solution u of (2.1) is determined by

$$
\int_\Omega \nabla R_h u \, \nabla v = \int_\Omega fv, \ \ \forall v \in S_0^h.
\tag{3.1}
$$

To discuss the multi-parameter error resolution, we shall compare $R_h u$ with the Lagrange's interpolation $i_h u$ of u. It is known that

$$
\int_\Omega \nabla (R_h u - i_h u) \, \nabla v = \int_\Omega \nabla (u - i_h u) \, \nabla v, \ \ \forall v \in S_0^h.
\tag{3.2}
$$

By the Euler-MacLaurin formula and the integral identity ([3, 5]), the following result is obtained.

$1, \cdots, p, i = 1, 2)$ *of 3rd, 2nd and 1st order, respectively, such that*

$$\int_{\Omega} \nabla(u - i_h u) \, \nabla v = \sum_{j=1}^{p} h_j^2 \left(\int_{\Omega} L_j u N_{j,1} v + \int_{\Gamma} M_j u N_{j,2} v \right)$$

$$+ \begin{cases} O(h^3)|v|_{1,1}, & if \, u \in H_0^1(\Omega) \cap C^4(\Omega), \\ O(h^2)\|u\|_{3,s,\Omega}|v|_{1,t,\Omega}, & if \, u \in H_0^1(\Omega) \cap W^{3,s}(\Omega)(s \geq 1), \end{cases} \tag{3.3}$$

where $t = s/(s-1)$ *and* $\Gamma = \cup_{1 \leq i,j \leq m}(\partial\Omega_i \cap \partial\Omega_j)$.

For any fixed $z \in \Omega$, let G_z, $G_z^h \in S_0^h$ be the Green function and discrete Green function, respectively. Then from Lemma A 1 and A 2 in [1], one can obtain the following result

Proposition 2. *If* T^h *is a quasi-uniform partition, then*

(i) for any $z \in \Omega, 1 \leq t < 2$,

$$\|G_z - G_z^h\|_{1,t,\Omega} \leq ch^{1-2/s} \, |\ln h \, |^{1/2} \, . \tag{3.4}$$

(ii) for any $z \in D \subset\subset D^* \subset \Omega$,

$$h^{-1}\|G_z - G_z^h\|_{1,\infty,\Omega\backslash D^*} + \|G_z\|_{1,\infty,\Omega\backslash D^*} + \|G_z^h\|_{1,\infty,\Omega\backslash D^*} \leq c. \tag{3.5}$$

Thus, we obtain the following relation between the Galerkin projection and the interpolation.

Theorem 1. *If* $u \in H_0^1(\Omega) \cap W^{3,s}(\Omega)(s > 2)$, *then there exists* $\{w_1(u), \cdots, w_p(u)\} \subset H_0^1(\Omega) \cap C(\Omega)$ *such that the following multi-parameter error resolution formula*

$$R_h u = i_h u + \sum_{j=1}^{p} w_j(u)h_j^2 + o(h^2) \tag{3.6}$$

holds on $C(\Omega_0)$, *where* $\Omega_0 \subset\subset \Omega \backslash \Gamma_{00}$ *with* $\Gamma_{00} \equiv$ *the set of all interior macro-vertices of* T^h.

Proof. From propositions 1 and 2,

$$\int_{\Omega} \nabla(u - i_h u) \, \nabla G_z^h = \sum_{j=1}^{p} h_j^2 \left(\int_{\Omega} L_j u N_{j,1} G_z + \int_{\Gamma} M_j u N_{j,2} G_z \right)$$

$$+ \sum_{j=1}^{p} h_j^2 \left(\int_{\Omega} L_j u N_{j,1}(G_z^h - G_z) + \int_{\Gamma} M_j u N_{j,2}(G_z^h - G_z) \right) + O(h^2)|G_z^h|_{1,t,\Omega}$$

$$= \sum_{j=1}^{p} w_j(u)h_j^2 + O(h^2)\|u\|_{3,s,\Omega}, \tag{3.7}$$

where $w_j(u) = \int_{\Omega} L_j u N_{j,1} G_z + \int_{\Gamma} M_j u N_{j,2} G_z.$

Combining (3.2) and (3.7), we obtain

$$R_h u = i_h u + \sum_{j=1}^{p} w_j(u) h_j^2 + O(h^2) \|u\|_{3,s,\Omega}. \tag{3.8}$$

Similarly, we also obtain the following result (cf [8, 9])

$$R_h u = i_h u + \sum_{j=1}^{p} w_j(u) h_j^2 + O(h^4 |\ln h|) \tag{3.9}$$

on $C(\Omega_0)$ with $\Omega_0 \subset\subset \Omega \setminus \Gamma_{00}$ if the exact soultion is smooth enough, e.g., $u \in H_0^1(\Omega) \cap C^4(\Omega)$.

It is easy to see that a linear functional F_h defined by

$$F_h(u) = h^{-2}(R_h u - i_h u - \sum_{j=1}^{p} w_j(u) h_j^2) \tag{3.10}$$

satisfies:

$$| F_h(u) | \le c|u|_{3,s,\Omega}, \quad \forall u \in H_0^1(\Omega) \cap W^{3,s}(\Omega). \tag{3.11}$$

Thus, combining (3.9) and (3.11), we obtain the theorem by a functional argument.

Theorem 2. *If $f \in W^{1,r}(\Omega)(r > 1)$ and $u \in H_0^1(\Omega) \cap W^{3,s}(D^*)(s > 2)$, then there exists $\{w_1(u), \cdots, w_p(u)\} \subset H_0^1(\Omega) \cap C(\Omega)$ such that the following multi-parameter error resoltuion formula*

$$R_h u = i_h u + \sum_{j=1}^{p} w_j(u) h_j^2 + o(h^2) \tag{3.12}$$

holds on $C(D)$, where $D \subset\subset D^* \setminus \Gamma_{00} \subset \Omega$.

Proof. First of all, we have $u \in H_0^1(\Omega) \cap W^{3,r^*}(\Omega)$ for some $r^* > 1$(see Grisvard [2]). Let $\omega \in C_0^\infty(\Omega)$ satisfty $\omega \equiv 1$ on D_1 and $supp\omega \subset\subset D^*$, where $D \subset\subset D_1 \subset\subset D^*$. Define $u_1 = \omega u$ and $u_2 = u - u_1$. Then $u_1 \in H_0^1(\Omega) \cap W^{3,s}(\Omega)$ and $u_2 \in H_0^1(\Omega) \cap W^{3,r^*}(\Omega)$.

Let $z \in D$, and let l_h be a linear functional defined by $l_h(u) = F_h(u_2)$. Then from proposition 2

$$| l_h(u) | \le ch^{-2} \sum_{j=1}^{p} h_j^2 \left(\int_\Omega L_j u_2 N_{j,1}(G_z^h - G_z) + \int_\Gamma M_j u_2 N_{j,2}(G_z^h - G_z) \right)$$

$$+ c|u_2|_{3,1,\Omega} \|G_z^h\|_{1,\infty,\Omega \setminus D_1}$$

$$\le c|u|_{3,1,\Omega}. \tag{3.13}$$

Thus, theorem 1 and a functional argument yield $l_h(u) \longrightarrow 0$ as $h \longrightarrow 0$, i.e.

$$h^{-2}(R_h u_2 - i_h u_2 - \sum_{j=1}^{p} w_j(u_2) h_j^2) = o(1). \tag{3.14}$$

On the other hand, by theorem 1, we have

$$h^{-2}(R_h u_1 - i_h u_1 - \sum_{j=1}^{p} w_j(u_1) h_j^2) = o(1). \tag{3.15}$$

Combining (3.14) and (3.15), we complete the proof.

Remark. Using the argument in [3, 5], there exists an interpolation operator I_h so that

$$I_h R_h u = i_h u + \sum_{j=1}^{p} w_j h_j^2 + o(h^2) \tag{3.16}$$

holds on $C(D)$. $I_h R_h u$ is determined by the parameters $h_1, \cdots, h_p$ and is denoted by $u(h_1, \cdots, h_p)$. Thus, a so-called multi-parameter splitting extrapolation technique leads to

$$u^c \equiv (4 \sum_{j=1}^{p} u_j - (4p-3)u_0)/3 = u + o(h^2), \tag{3.17}$$

which holds on $C(D)$, where $D \subset\subset D^* \setminus \Gamma_{00} \subset \Omega$.

4   Other Partitions with Multi-Parameter Resolution.

It is pointed out that similar results can be expected for other kinds of partitions (cf. [8]): Divide Ω into several convex quadrilateral $T = \{\Omega_1, \cdots, \Omega_m\}$ such that T is quasi-uniform. Let Φ_i:

$$x_1(\xi, \eta) = a_{i,1}(1-\xi)(1-\eta) + a_{i,2}\xi(1-\eta) + a_{i,3}\xi\eta + a_{i,4}(1-\xi)\eta,$$

$$x_2(\xi, \eta) = b_{i,1}(1-\xi)(1-\eta) + b_{i,2}\xi(1-\eta) + b_{i,3}\xi\eta + b_{i,4}(1-\xi)\eta$$

be the bilinear coordinate transformations from the unit square $[0,1]^2$ to $\Omega_i (i = 1, \cdots, m)$, where $(a_{i,j}, b_{i,j})(j = 1, 2, 3, 4)$ are the 4 vertices of the Ω_i. Under the mapping, a line parallel to $\xi-$ or $\eta-$ axis in $[0,1]^2$ is transformed into the line linking the two equipartition points of a two opposite edges in Ω_i, and globally a quasi-uniform partition T^h on Ω is formed. For a function v defined on Ω_i, let $\hat{v}$ be the function defined on $[0,1]^2$ by

$$\hat{v} = v \circ \Phi_i. \tag{4.1}$$

Conversely, a function $\hat{v}$ defined on $[0,1]^2$ determines a function v on Ω_i satisfying (4.1). Let $[0,1]^2$ be covered by a uniform mesh. Define

$$S_0^h(\Omega) = \{v \in H_0^1(\Omega) : v \circ \Phi_i \text{ is piecewise bilinear on } [0,1]^2, i = 1, 2, \cdots\}, \tag{4.2}$$

$$u = \hat{u} \circ \Phi_i^{-1}, \quad \text{on } \Omega_i,$$

$$i_h u = \hat{i_h \hat{u}} \circ \Phi_i^{-1}, \quad \text{on } \Omega_i,$$

where $\hat{i_h \hat{u}}$ is the piecewise bilinear interpolant of $\hat{u}$ on $[0,1]^2$. $i_h u(x) = u(x)$ holds for the nodal points x of Ω and $S_0^h(\Omega)$ is determined by some parameters, say $h_1, \cdots, h_p$.

By induction, it can be proved that for any polygonal domain and with a proper choice of $\{\Omega_1, \cdots, \Omega_m\}$, p satisfies $p \geq 2$ even $p >> 2$. Given

$$\int_{\Omega_i} \partial_1(u - i_h u)\partial_1 v = \int_{[0,1]^2} p_{12}(\partial_\xi(\hat{u} - \hat{i_h\hat{u}})\partial_\eta\hat{v} + \partial_\eta(\hat{u} - \hat{i_h\hat{u}})\partial_\xi\hat{v})$$

$$+ \int_{[0,1]^2} p_{11}\partial_\xi(\hat{u} - \hat{i_h\hat{u}})\partial_\xi\hat{v} + p_{22}\partial_\eta(\hat{u} - \hat{i_h\hat{u}})\partial_\eta\hat{v},$$

where

$$p_{11} = (\partial_\eta x_2)^2/J, \ \ p_{22} = (\partial_\eta x_2)^2/J,$$

$$p_{12} = -\partial_\eta x_2 \partial_\xi x_2/J, \ \ J = |\partial_\xi x_1 \partial_\eta x_2 - \partial_\eta x_1 \partial_\xi x_2|,$$

if $R_h u$ satisfies (3.1) for $S_0^h(\Omega)$ defined by (4.2). Then there exists an interpolation operator I_h (cf. [3, 5]) and functions w_i such that the following formula

$$I_h R_h u = u + \sum_{i=1}^{p} w_i h_i^2 + o(h^2) \tag{4.3}$$

holds on $C(D)$, where $D \subset\subset D^* \setminus \Gamma_{00} \subset \Omega$.

REFERENCES

[1] Blum H., Lin Q. and Rannacher R. (1986) Asymptotic error expansion and Richardson extrapolation for linear finite elements. *Numer. Math.* 49:11-37.

[2] Grisvard P. (1976) Behavior of the solutions of an elliptic boundery value problem in a polygonal or polyhedral domain, in *Numerical Solution of Partial Differential Equations-III (B. Hubbard, ed.)*, Academic Press, New-York, 207-274.

[3] Lin Q. (1990) *An integral identity and interpolated postprocess in super-convergence*, Research Report 90-07, Inst. of Sys. Sci., Academia Sinica.

[4] Lin Q. and Lü T. (1983) The splitting extrapolation method for multi-dimensional problems. *J. Comp. Math.* 1:45-51.

[5] Lin Q., Yan N. and Zhou A. (1991) A rectangle test for interpolated finite elements. in *Proc. of Sys. Sci. & Sys. Engrg.*, Great Wall Culture Publish Co., Hong Kong, 217-229.

[6] Lin Q. and Zhu Q. D. (1986) Local asymptotic expansion and extrapolation for finite elements. *J. Comp. Math.*, 3: 263-265.

[7] Rannacher R. (1987) Extrapolation techniques in the finite element method (A Survey). in *Proc. of the Summer School in Numer. Analysis at Helsinki*.

[8] Zhou A., Liem C. B. and Shih T. M. (1994) A parallel algorithm based on multi-parameter asymptotic error expansion. in *Proc. of Conference on Scientific Comput.*, Hong Kong, March 17-19.

[9] Zhou A., Liem C. B., Shih T. M. and Lü T. (1994) A multi-parameter splitting extrapolation and a parallel algorithm. Research Report IMS-61, Inst. of Math. Sci., Academia Sinica.

Part IV Domain Decomposition Methods for Advection-Diffusion, Transport and Wave Problems

EXACT CONTROLLABILITY AND DOMAIN DECOMPOSITION METHODS WITH NON-MATCHING GRIDS FOR THE COMPUTATION OF SCATTERING WAVES

M.O. BRISTEAU, E.J. DEAN, R. GLOWINSKI, V. KWOK, J. PERIAUX

ABSTRACT. The main goal of this article is to discuss the solution of scattering problems for coated obstacles by a methodology combining exact controllability techniques and domain decomposition algorithms. In this article, domain decomposition techniques are used to split the computational domain into smaller ones and also to take into account the different physical properties taking places in different regions, which may require the coupling of local approximate solutions defined over non-matching finite element meshes; this coupling is easily achieved through a weak formulation of the interface, using Lagrange multipliers. The results of numerical experiments for documented aerospace test cases are presented, showing the good performances of these new methods, particularly for non-convex reflectors.

1. Introduction.

In a recent publication (see [1]), the authors have introduced a novel approach for the computation of scattering waves by coated obstacles. This methodology rests on the combination of:

(i) An *exact controllability* formulation à la *Hilbert Uniqueness Methods* (HUM, [2]), which allows the fast calculation of limit cycles for wave equations.

(ii) *Domain decomposition* methods for non-overlapping subdomains where the interface compatibility conditions are enforced via *Lagrange multipliers*. (For wave equations this method was introduced in Reference [3], to our knowledge.)

In References [1] and [3] the domain decomposition methodology was used to decouple subregions with different physical properties, such as air and the coating material, requiring meshes which do not match at the interface if one wishes to use the same time discretization step in all subregions. In the present article, we go one step further by introducing domain decompositions whose main goal is to further split the computational domain (for distributed parallel computing purposes, for example). The resulting methodology is particularly robust and well-suited to scattering by non-convex reflectors, which is typical in industrial applications.

Domain Decomposition Methods in Sciences and Engineering, edited by R. Glowinski *et al.*

2. The Generalized Helmholtz Equation and its Equivalent Wave Problem.

The *Helmhlotz equation* (Maxwell equations in two dimensions for the T.M. mode) for *heterogeneous media* is given by

$$(2.1) \qquad \epsilon k^2 U + \nabla \cdot \mu^{-1} \nabla U = F \text{ in } \Omega,$$

where ϵ and μ are the *permitivity* and *permeability coefficients*, respectively (we assume that ϵ and μ are positive); Ω is the region of $\mathbb{R}^2$ where the propagation phenomenon is taking place. We suppose that Ω is *bounded* (see Figure 2.1 below) by an artificial boundary Γ, where an *absorbing boundary* condition is prescribed. For simplicity, we specify on Γ

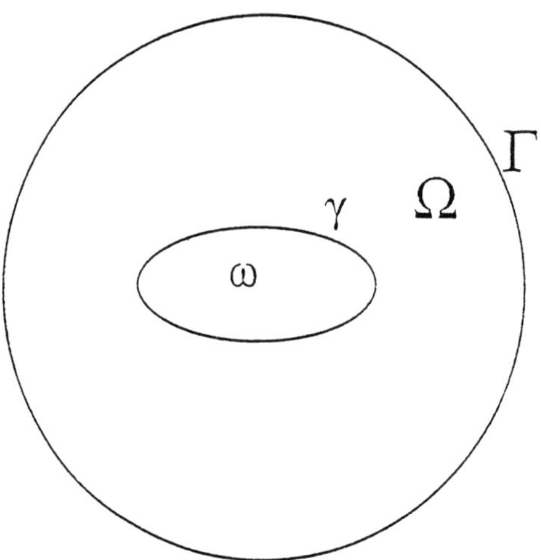

Figure 2.1

$$(2.2) \qquad \frac{\partial U}{\partial n} + \sqrt{\epsilon \mu} ik U = 0,$$

to be completed by a Dirichlet condition on the boundary γ of the reflector ω, namely

$$(2.3) \qquad U = G,$$

where $-G$ is the incident field at γ. All of the above functions are complex valued. The wave number k is given by

$$(2.4) \qquad k = 2\pi/T,$$

T being the time period.

Remark 2.1. if Ω is "filled" with air (or void), $F = 0$ in (2.1) for incident monochromatic waves. The function U represents the scattered field. Condition (2.2) assumes that Γ is located in air (or void).

Following [4], we observe that (2.1)-(2.3) is equivalent to finding T-periodic solutions of

$$(2.5) \qquad \epsilon u_{tt} - \nabla \cdot \mu^{-1} \nabla u = f \text{ in } \Omega \times (0, T)(= Q),$$

$$(2.6) \qquad u = g \text{ on } \gamma \times (0, T)(= \sigma),$$

$$(2.7) \qquad \frac{\partial u}{\partial n} + \sqrt{\epsilon \mu} \frac{\partial u}{\partial t} = 0 \text{ on } \Gamma \times (0, T)(= \Sigma)$$

where $u(x, t) = \text{Re}(U(x)e^{ikt})$. From the T-periodicity, u satisfies

$$(2.8) \qquad u(0) = u(T), \; u_t(0) = u_t(T).$$

3. Exact Controllability Solution Methods.

In order to solve (2.1)-(2.3) or (2.5)-(2.8) by controllability methods inspired from HUM, we observe that solving the above problems is equivalent to finding a pair $\{e_0, e_1\}$ so that

$$(3.1) \qquad u(0) = e_0, u_t(0) = e_1,$$
$$(3.2) \qquad u(T) = e_0, u_t(T) = e_1,$$

with u solution of (2.5)-(2.7). Problem (2.5)-(2.7), (3.1), (3.2) is an *exact controllability problem* which can be solved by methods directly inspired from the J.L. Lions *Hilbert Uniqueness Method* (HUM) (see, e.g., [2], [5], [6] for an introduction to this method and for further applications).

4. Least Squares Formulation of Problem (2.5)-(2.7), (3.1), (3.2).

In order to apply HUM to the solution of problem (2.5)-(2.7), (3.1), (3.2) an appropriate choice for the space containing $\{e_0, e_1\}$ is

$$(4.1) \qquad E = V_g \times L^2(\Omega), \text{ with } V_g = \{\varphi \mid \varphi \in H^1(\Omega), \varphi \mid_\gamma = g(0)\}.$$

A least squares formulation is given by

$$(4.2) \qquad \text{Min}_{\mathbf{v} \in E} J(v)$$

with $\mathbf{v} = \{v_0, v_1\}$ and

$$(4.3) \qquad J(\mathbf{v}) = \frac{1}{2} \int_\Omega [\mu^{-1}|\nabla(y(T) - v_0)|^2 + \epsilon|y_t(T) - v_1|^2] \, dx,$$

where, in (4.3), y is the solution of

$$(4.4) \qquad \epsilon y_{tt} - \nabla \cdot \mu^{-1} \nabla y = f \text{ in } Q,$$

$$(4.5) \qquad y = g \text{ on } \sigma,$$

$$(4.6) \qquad \frac{\partial y}{\partial n} + \sqrt{\epsilon \mu} \frac{\partial y}{\partial t} = 0 \text{ on } \sigma,$$

$$(4.7) \qquad y(0) = v_0, y_t(0) = v_1.$$

The choice of J is quite natural indeed since the total energy of the system is given by

$$\mathcal{E}(t) = \frac{1}{2} \int_\Omega [\epsilon |y_t(t)|^2 + \mu^{-1} |\nabla y(t)|^2] \, dx.$$

The least squares problem (4.2) can be solved by a *conjugate gradient algorithm* operating in space E; such an algorithm is described in Section 6 of [1].

5. Calculation of J'.

In order to solve problem (4.2) by a conjugate gradient algorithm, we need to know the differential J' of J. To compute J' we shall use the *perturbation method* described in Section 5 of [1]. We obtain then

$$(5.1) \qquad \begin{cases} \langle J'(\mathbf{v}), \mathbf{w} \rangle = & \int_\Omega \mu^{-1} \nabla(v_0 - y(T)) \cdot \nabla w_0 \, dx \\ & - \int_\Omega \epsilon p_t(0) w_0 \, dx + \int_\Gamma \sqrt{\frac{\epsilon}{\mu}} p(0) w_0 d\Gamma \\ & + \int_\Omega \epsilon p(0) w_1 \, dx + \int_\Omega \epsilon(v_1 - y_t(T)) w_1 \, dx, \end{cases}$$

for all $\mathbf{v} = \{v_0, v_1\} \in E$ and all $\mathbf{w} = \{w_0, w_1\} \in E$, with $E_0 = V_0 \times L^2(\Omega)$ and

$$V_0 = \{z \mid z \in H^1(\Omega), \ z = 0 \text{ on } \gamma\};$$

in (5.1), the function p is (uniquely) defined by the following adjoint system

$$(5.2) \qquad \epsilon p_{tt} - \nabla \cdot \mu^{-1} \nabla p = 0 \text{ in } Q,$$

$$(5.3) \qquad \frac{\partial p}{\partial n} - \sqrt{\epsilon \mu} \frac{\partial p}{\partial t} = 0 \text{ on } \sigma,$$

$$(5.4) \qquad\qquad p = 0 \text{ on } \sigma,$$

$$(5.5) \qquad\qquad p(T) = y_t(T) - v_1,$$

$$(5.6) \qquad \begin{cases} \int_\Omega \epsilon p_t(T) z \, dx = \int_\Gamma \sqrt{\frac{\epsilon}{\mu}} (y_t(T) - v_1) z \, d\Gamma \\ \quad - \int_\Omega \mu^{-1} \nabla(y(T) - v_0) \cdot \nabla z \, dx, \forall z \in V_0. \end{cases}$$

From the knowledge of J' we can solve problem (4.2) by a conjugate gradient algorithm operating in space $E = V_g \times L_2(\Omega)$. As already mentioned, such an algorithm is described in [1, Section 6]: each iteration requires the solution of *one forward wave equation* (such as (4.4)-(4.7)), of *one backward wave equation* (such as (5.2)-(5.6)), and of an *elliptic problem* in Ω, associated to the bilinear form

$$\{\varphi, \psi\} \to \int_\Omega \nabla\varphi \cdot \nabla\psi \, dx.$$

6. A Finite-Difference/Finite-Element Implementation.

The practical implementation of the controllability method is based upon a *time discretization* by a *centered explicit finite-difference scheme*. This scheme is combined with *piecewise-linear finite-element approximations* for the *space* variables. We use *mass lumping* through numerical integration by the *trapezoidal rule* to obtain a *diagonal mass matrix* for the acceleration terms. The fully discrete scheme has to satisfy a stability condition such as $\Delta t \leq Ch$, where C is independent of h and Δt. To obtain accurate solutions, we need to have h at least *ten times smaller* than the wavelenth; consequently, Δt has to be at least ten times smaller than the period. If we assume that the number of iterations is independent of h and Δt (assumptions supported by numerical experiments), the solution of the Helmholtz equation via the new approach involves a number of operations which for a given value of k is proportional to the number of grid points; for this estimate we do not take into account the time spent at solving the elliptic problems mentioned in Section 5, which in two dimensions is negligible compared to the time required to solve the wave equations; in three dimensions it remains to see if the same conclusion holds.

Remark 6.1. In [6] and [7] one may find the detailed description of the solution method for a controllability problem for the wave equation closely related to the one discussed in this article. The problem in [6] and [7] is space and time discretized by finite-element and finite-difference methods similar to those advocated above.

7. Combining Controllability and Comain Decomposition Methods.

7.1 Generalities.

In the particular case of the scattering of waves by coated materials (see Figure 7.1 for such a situation), the waves propagate faster in the surrounding air (or void) than in the coating. If one wishes to use the same time discretization step Δt everywhere with a *local* CFL number of the order of one, it is necessary to employ local meshes with quite different space discretization steps. This observation leads quite naturally to a *domain decomposition method* associated to the physical heterogeneities of the problem; furthermore, it leads to the use of space discretization meshes which do not match at the interface (see Figure 7.2). The crucial point when combining exact controllability and domain decomposition with non-matching finite-element meshes is to develop efficient solution methods for wave problems such as (2.5)-(2.7), (3.1) when ϵ and μ are piecewise continuous. For simplicity we shall discuss first the solution of (2.5)-(2.7), (3.1) for the physical situation associated to figure 7.1 whose notation has been kept.

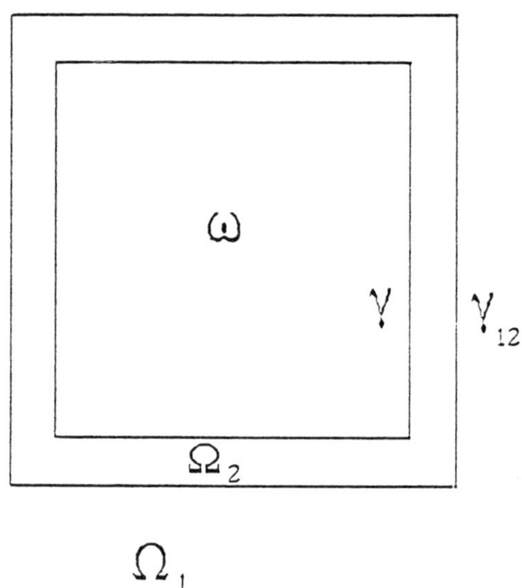

Figure 7.1

7.2 A Domain Decomposition Method for Linear Wave Problems.
The problem to be solved is described by

$$(7.1) \qquad \epsilon_i \frac{\partial^2 u_i}{\partial t^2} - \nabla \cdot (\mu_i^{-1} \nabla u_i) = -\epsilon_i \frac{\partial^2 u^{inc}}{\partial t^2}$$

$$+ \nabla \cdot (\mu_i^{-1} \nabla u^{inc}) \text{ in } Q_i = \Omega_i \times (0, T), \ \forall i = 1, 2,$$

$$(7.2) \qquad u_2 = -u^{inc} \text{ on } \sigma,$$

$$(7.3) \qquad \sum_{i=1}^{2} \mu_i^{-1} \frac{\partial u_i}{\partial n_i} = - \sum_{i=1}^{2} \mu_i^{-1} \frac{\partial u^{inc}}{\partial n_i} \text{ on } \sigma_{12} = \gamma_{12} \times (0, T),$$

$$(7.4) \qquad u_1 = u_2 \text{ on } \sigma_{12},$$

$$(7.5) \qquad \frac{\partial u_1}{\partial n_1} + \sqrt{\epsilon_1 \mu_1} \frac{\partial u_1}{\partial t} = 0 \text{ on } \sigma,$$

$$(7.6) \qquad u_i(0) = e_0 \, |_{\Omega_i}, \frac{\partial u_i}{\partial t}(0) = e_1 \, |_{\Omega_i}, \ \forall i = 1, 2,$$

where, in (7.1)-(7.6), u_i is the local *scattered field* and u^{inc} the *incident field*. In the case of monochromatic plane waves, the incident field has the following complex representation:

$$u^{inc}(x, t) = -e^{ik(t - \sum_{j=1}^{d} a_j x_j)}$$

with

$$x = \{x_j\}_{j=1}^{d} (d = 2 \text{ or } 3)$$

and

$$a = \{a_j\}_{j=1}^{d}$$

satisfying

$$\sum_{j=1}^{d} a_j^2 = 1$$

(we assume that in air (or void) the propagation velocity is $c = 1$).

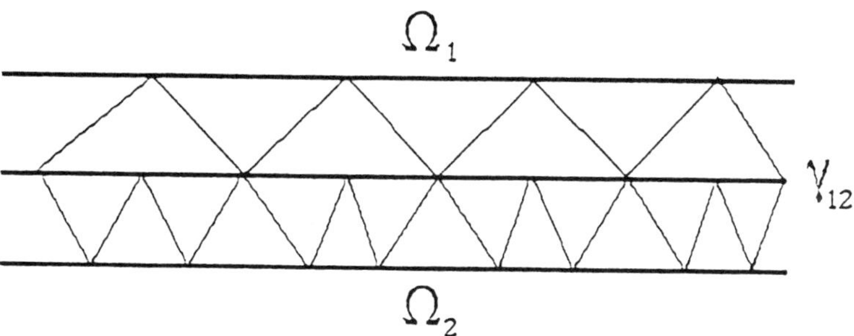

Figure 7.2

Remark 7.1. If medium $i = 1$ is either air or void, then the right-hand side of (7.1) vanishes (incidentally this right-hand side is the function f in (2.5), to be used from now on).

To solve (7.1)-(7.6) by domain decomposition, we observe (following [3]) that this system is equivalent to finding $\{u_1, u_2, \lambda\}$ so that

$$
(7.7) \quad
\begin{cases}
\sum_{i=1}^{2} \int_{\Omega_i} \epsilon_i \frac{\partial^2 u_i}{\partial t^2} v_i \, dx + \sum_{i=1}^{2} \int_{\Omega_i} \mu_i^{-1} \nabla u_i \cdot \nabla v_i \, dx + \int_{\Gamma} \sqrt{\frac{\epsilon_1}{\mu_1}} \frac{\partial u_1}{\partial t} v_1 \, d\Gamma \\[2mm]
\quad = \int_{\gamma_{12}} \lambda(v_2 - v_1) \, d\gamma_{12} - \sum_{i=1}^{2} \int_{\gamma_{12}} \mu_i^{-1} \frac{\partial u^{inc}}{\partial n_i} v_i \, d\gamma_{12} \\[2mm]
\quad + \sum_{i=1}^{2} \int_{\Omega_i} f v_i \, dx, \ \forall \{v_1, v_2\} \in V_1 \times V_2,
\end{cases}
$$

$$
(7.8) \quad \int_{\gamma_{12}} \mu(u_2 - u_1) \, d\gamma_{12} = 0, \ \forall \mu \in L^2(\gamma_{12}),
$$

$$
(7.9) \quad u_2 = g \text{ on } \sigma,
$$

$$
(7.10) \quad u_i(0) = e_0 \mid_{\Omega_i}, \ \frac{\partial u_i}{\partial t}(0) = e_1 \mid_{\Omega_i}, \ \forall i = 1, 2.
$$

In (7.7), V_1 and V_2 are defined by

$$
(7.11) \quad V_1 = H^1(\Omega_1), \ v_2 = \{v_2 \in H^1(\Omega_2), v_2 = 0 \text{ on } \gamma\}.
$$

Remark 7.2. The *Lagrange multiplier* treatment of matching conditions such as (7.3) and (7.4) is not new; it has been used systematically for the solution of elliptic or parabolic problems (see, e.g., [3], [8], and [9] and the references therein).

Remark 7.3. In the particular case of either air/air or coating/coating interfaces, everything which has been written previously still holds (particularly *Remark 7.1*); actually we have now a much simpler situation since for these two cases we can use local meshes which match at the subdomain interfaces. Such a situation will be considered in Section 8.

7.3. Time Discretization and Domain Decomposition.

In this section we shall use, for simplicity, the formalism of the continuous problem. However, the following approximation makes full sense only when V_1, V_2, and $L^2(\gamma_{12})$ have been approximated by well-chosen finite-dimensional spaces; we shall address this issue in Section 7.4. Let $\Delta t(> 0)$ be a *time discretization step*; denoting by u^n an approximation of $u(n\Delta t)$, we approximate (7.7)-(7.10) by

$$
(7.12) \quad
\begin{cases}
\sum_{i=1}^{2} [\int_{\Omega_i} \epsilon_i \frac{u_i^{n+1} + u_i^{n-1} - 2u_i^n}{|\Delta t|^2} v_i \, dx + \int_{\Omega_i} \mu_i^{-1} \nabla u_i^n \cdot \nabla v_i \, dx] \\[2mm]
\quad + \int_{\Gamma} \sqrt{\frac{\epsilon_1}{\mu_1}} \frac{u_1^{n+1} - u_1^{n-1}}{2\Delta t} v_1 \, d\Gamma = \int_{\gamma_{12}} \lambda^n (v_2 - v_1) \, d\gamma_{12} \\[2mm]
\quad - \sum_{i=1}^{2} \int_{\gamma_{12}} \mu_i^{-1} \frac{\partial u^{inc}}{\partial n_i}(n\Delta t) v_i \, d\gamma + \sum_{i=1}^{2} \int_{\Omega_i} f^n v_i \, dx, \ \forall \{v_1, v_2\} \in V_1 \times V_2,
\end{cases}
$$

$$(7.13) \qquad \int_{\gamma_{12}} \mu(u_2^{n+1} - u_1^{n+1})d\gamma_{12} = 0, \ \forall \mu \in L^2(\gamma_{12}), \forall n \geq 0,$$

$$(7.14) \qquad u_2^{n+1} = g^{n+1} \text{ on } \gamma,$$

$$(7.15) \qquad u_i^0 = e_0 \ |_{\Omega_i}, u_i^1 - u_i^{-1} = 2\Delta t e_1 \ |_{\Omega_i}, \ \forall i = 1, 2.$$

Combining (7.13), (7.14) will imply that relation (7.13) also holds for $n = -1$ and $n = -2$; this property is important for accuracy and stability purposes.

Remark 7.4. Scheme (7.12)-(7.15) is a domain decomposition implementation of the well-known *second-order explicit finite-differences scheme* for the wave equation.

7.4 Space/Time Discretization and Domain Decomposition.

In this section, we shall discuss a *finite-element* realization of scheme (7.12)-(7.15) and also the stability of the corresponding fully discrete scheme. We shall also discuss practicalities such as the use of *numerical integration* and the *iterative* or *direct computation* of the discrete multipliers.

7.4.1. A Fully Discrete Domain Decomposition Method.

Let W_1, W_2, and Λ be finite-dimensional subspaces of V_1, V_2 and $L^2(\gamma_{12})$, respectively. We denote by H_{2h} the discrete equivalent of $H^1(\Omega_2)$. We approximate then (7.12)-(7.15) by

$$(7.16) \qquad \left\{ \begin{aligned} &\sum_{i=1}^2 [\int_{\Omega_i} \epsilon_i \frac{u_i^{n+1} + u_i^{n-1} - 2u_i^n}{|\Delta t|^2} v_i \ dx \\ &+ \int_{\Omega_i} \mu_i^{-1} \nabla u_i^n \cdot \nabla v_i dx] + \int_{\Gamma} \sqrt{\frac{\epsilon_1}{\mu_1}} \frac{u_1^{n+1} - u_1^{n-1}}{2\Delta t} v_1 d\Gamma \\ &= \int_{\gamma_{12}} \lambda^n (v_2 - v_1)d\gamma_{12} - \sum_{i=1}^2 \int_{\gamma_{12}} \mu_i^{-1} \frac{\partial u^{inc}}{\partial n_i}(n\Delta t)v_i d\gamma + \sum_{i=1}^2 \int_{\Omega_i} f^n v_i dx, \\ &\forall \{v_1, v_2\} \in W_1 \times W_2; \{u_1^{n+1}, u_2^{n+2}\} \in W_1 \times H_{2h}^1, \end{aligned} \right.$$

$$(7.17) \qquad \int_{\gamma_{12}} \mu(u_2^{n+1} - u_1^{n+1})d\gamma_{12} = 0, \ \forall \mu \in \Lambda, \forall n \geq 0; \lambda^n \in \Lambda,$$

$$(7.18) \qquad u_2^{n+1} = g_h^{n+1} \text{ on } \gamma,$$

$$(7.19) \qquad u_i^0 = e_{0h} \ |_{\Omega_i}, \ u_i^1 - u_i^{-1} = 2\Delta t e_{1h} \ |_{\Omega_i}, \ \forall i = 1, 2.$$

In (7.16)-(7.10), h is a space-discretization parameter and e_{0h}, e_{1h} and h_h are approximations of e_0, e_1, and g, respectively; we suppose that e_{0h} and e_{1h} are continuous at γ_{12} (we have dropped the subscript h in (7.16)-(7.19), where u_1^j, u_2^j, λ^j are approximations of the related functions in (7.12)-(7.15)).

7.4.2. Practical Solutions of Problems (7.16)-(7.18).

Expanding $u_1^{n+1}, u_2^{n+1}, \lambda^n$ on vector bases of W_1, H_{2h}^1, and Λ, we reduce the solution of problem (7.16)-(7.18) to the solution of a linear system of the following (saddle-point) type:

$$(7.20) \qquad \begin{cases} Ax + B^t y = b, \\ Bx = c. \end{cases}$$

In (7.20), A is an $N \times N$ matrix, symmetric and positive definite (possibly diagonal; see [3, Section 4.1]), B is a $M \times N$ matrix (with $M << N$), and b and c belong to $\mathbb{R}^N$ and $\mathbb{R}^M$, respectively. Problem (7.6) has a unique solution in $\mathbb{R}^N \times (\mathbb{R}^M / \ker B^t)$ if and only if c belongs to $R(B)$ (the range of B). From the fact that A is a regular matrix, we can eliminate x from (7.20); we obtain then that y is a solution of the following linear system

$$(7.21) \qquad BA^{-1}B^t y = BA^{-1}b - c.$$

For the particular problem considered here, matrix $BA^{-1}B^t$ is well-conditioned (on $\mathbb{R}^M / \ker B^t$) for small values of Δt; this property strongly suggests a *conjugate gradient* solution for problem (7.21) (and (7.20), consequently). Such an algorithm is described in, e.g., [1, Section 8.4.2], [10], [11].

7.4.3. Further Comments.

The domain decomposition method discussed in the above paragraphs can be generalized easily to *strip domain decompositions* like the one shown in Figure 7.2(a). In the case of *box domain decompositions*, like the one of Figure 7.2(b), crossing points (like C) require special attention as shown in [3, Section 4.3].

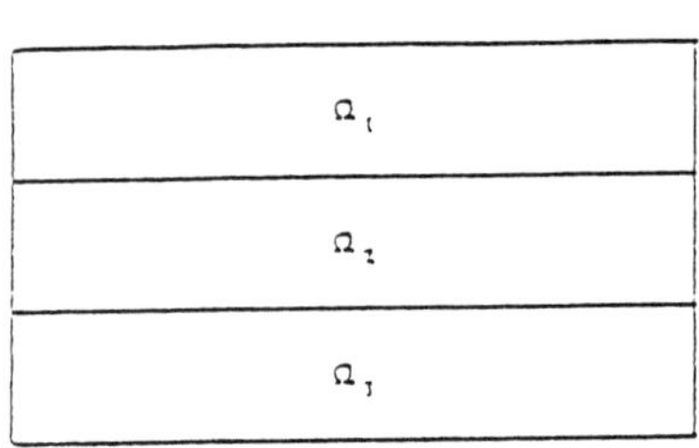

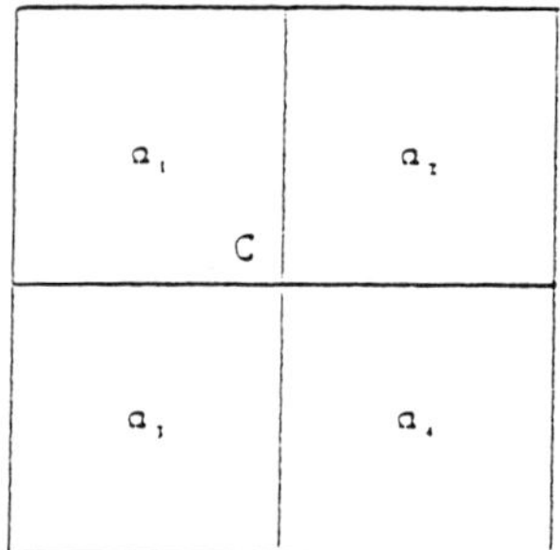

Figure 7.2(a)   Figure 7.2(b)

8. Numerical Experiments.

The above methods have been applied to the solution of various test problems in [1, Section 9] and in [12, Chapter 4]. The test problem that we consider here concerns the scattering of a planar wave by a two-piece reflector consisting of two identical NACA 0012 airfoils whose chords are parallel; the distance between the two airfoils is λ_0 (wavelenth in air); we suppose that the chords are horizontal with the leading edges on the left and that both airfoils are perfect reflectors. This two-piece reflector is illuminated by an incident planar wave coming from the upper left with a 45 degree angle of incidence. We suppose that the frequency of the incident wave is $f = 1.2$ GHz implying that the wavelength in air is $\lambda_0 = .25$m. We suppose that the distance between the airfoils is λ_0, that the chord length is $4\lambda_0$. We suppose next that both airfoils have been coated by a material whose relative permitivity and permeability are 7.4 and 1.4, respectively, implying that waves propagate in the coating 3.2 times slower than in air or void; the thickness of the coating is $\lambda_0/10$ in the normal direction. In order to simulate numerically the scattering of the incident wave by the above reflectors, we embed them in a rectangular domain; the distance between the boundary of the two-piece reflector and the boundary of the embedding domain has been taken equal to $3\lambda_0$. We have chosen as time step $\Delta t = T/75$, where $T = 1/f = .83 \times 10^{-9}$. The space discretization mesh is 4 times finer in the coating regions than in air. We have visualized in Figure 8.1 the four subdomain decomposition which has been used to compute the scattered field:

(i) The two coating regions; each contains 2816 triangles and 1760 vertices; we shall denote by Ω_3 (resp. Ω_4) the upper (resp. lower) one (see lower part of the figure).

(ii) A region denoted by Ω_2, which closely surrounds the coated reflectors and which totally located in air (middle part of the figure); it contains 2312 triangles and 1329 reflectors.

(iii) Finally, the "rectangular" ring (upper part of the figure) which surrounds the airfoils and the three other subdomains; it has been denoted by Ω_1 and contains 21,668 triangles and 11,130 vertices.

We have visualized on Figures 8.2 through 8.5 the contours of the scattered field in $\Omega_1, \Omega_2, \Omega_3, \Omega_4$, respectively. For further details on these numerical experiments see [12].

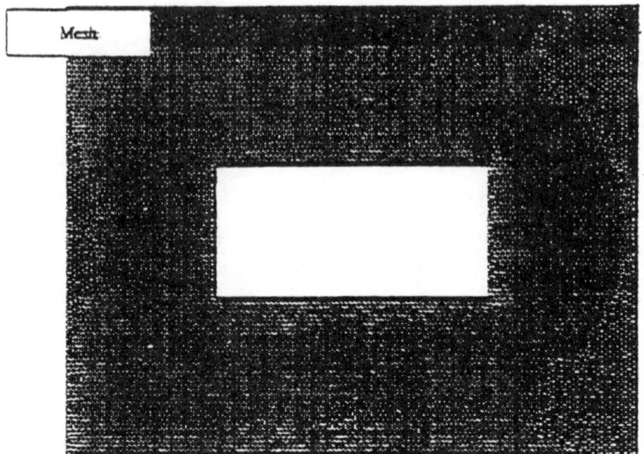

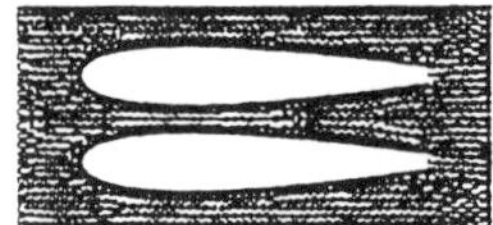

Figure 8.1: The subdomains and their finite element grid.

Figure 8.2

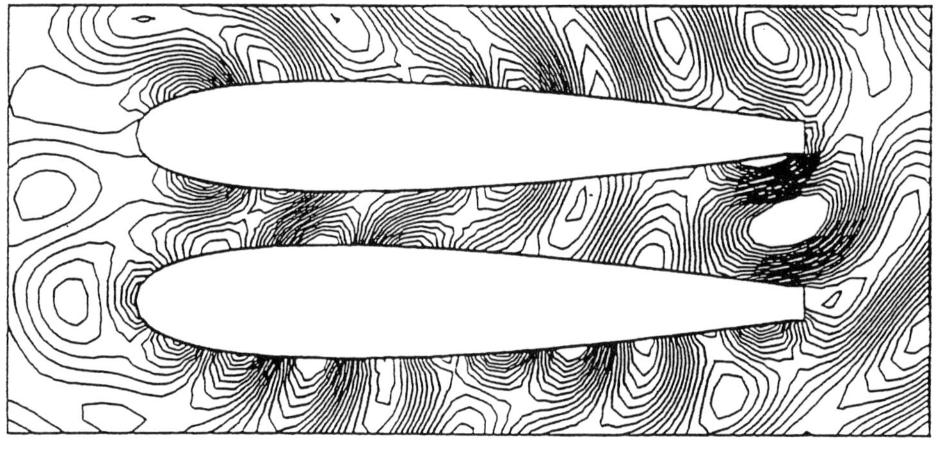

Figure 8.3

Figure 8.4

Figure 8.5

Acknowledgments.

The authors would like to thank A. Bamberger, Ch. de la Foye, F. El Dabaghi, P. Joly, A. Lanusse, P. Le Tallec, J. L. Lions, P. Perrier, A. Priou, H. Steve, and Q. H. Tran for helpful comments and suggestions. The supports of DRET (Contract 93.388), IFP, DARPA (Contract AFOSR-90-0334), NSF (Grants INT 86 1268, DMS 8822522, RII 8917691, DMS 9046924, DMS 9408151) and of the Texas Board of Higher Education (Grants 003652156ARP, 003652146ATP, 003652091ARP) are also acknowledged.

References.

[1] M. O. BRISTEAU, E. J. DEAN, R. GLOWINSKI, V. KWOK, J. PERIAUX, Application of exact controllability to the computation of scattering waves, in *Control Problems in Industry*, I. Lasieka and B. Morton, eds., Birkhäuser, Boston, 1995, pp.17-41.

[2] J. L. LIONS, Exact controllability, stabilization, and perterbation of distributed systems, *SIAM Review*, (1988), **30**, pp.1-68.

[3] E. J. DEAN, R. GLOWINSKI, A domain decomposition method for the wave equation, in *Les Grands Systèmes des Sciences et de la Technologie*, J. Horowitz and J. L. Lions, eds., Masson, Paris, 1993, pp.241-264.

[4] M. O. BRISTEAU, R. GLOWINSKI, J. PERIAUX, Using exact controllability to solve the Helmholtz equation at high wave numbers, Chapter 12 of *Mathematical and Numerical Aspects of Wave Propagation*, R. Kleinman, Th. Angel, D. Colton, F. Santosa, I. Strakgold, eds., SIAM, Philadelphia, 1993, pp.113-127.

[5] R. GLOWINSKI, J. L. LIONS, Exact and approximate controllability for distributed parameter systems (I), *Acta Numerica*, 1994, pp.269-378.

[6] R. GLOWINSKI, J. L. LIONS, Exact and approximate controllability for distributed parameter systems (II), *Acta Numerica*, 1995, pp.159-333.

[7] R. GLOWINSKI, Ensuring well-posedness by analogy: Stokes problem and boundary control for the wave equation, *Journal of Computational Physics*, **103**, (1992), pp.189-221.

[8] R. GLOWINSKI, M. F. WHEELER, Domain Decomposition and Mixed Finite Element Methods for Elliptic Problems, in *Domain Decomposition Methods for Partial Differential Equations*, R. Glowinski, G. H. Golub, G. Meurant, J. Periaux, eds., SIAM, Philadelphia, 1988, pp.144-172.

[9] L. C. COWSAR, E. J. DEAN, R. GLOWINSKI, P. LE TALLEC, C. H. LI, J. PERIAUX, M. F. WHEELER, Decomposition principles and their applications in Scientificy Computing, in *Parallel Processing for Scientific Computing*, J. Dongarra, K. Kennedy, P. Messina, D. G. Sorensen, R. G. Voigt, eds., SIAM, Philadelphia, 1992, pp.213-237.

[10] M. FORTIN, R. GLOWINSKI, *Augmented Lagrangians*, North-Holland, Amsterdam, 1983.

[11] R. GLOWINSKI, P. LE TALLEC, *Augmented Lagrangian and Operator Splitting Methods in Nonlinear Mechanics*, SIAM, Philadelphia, 1989.

[12] V. KWOK, *Méthodes de Controlabilité Exacte et de Décomposition de Domaines*

pour la Résolution Numérique des Equations de l'Electro-Magnétisme en Régime Harmonique dans un Milieu Hétérogène, Ph.D. dissertation, Université P. et M. Curie, Paris, France, July 1995.

Affiliations.

M. O. Bristeau, INRIA, B.16, 78153 Le Chesnay, France. E. J. Dean, University of Houston, Department of Mathematics, Houston, Texas, 77204-3476, USA. R. Glowinski, University of Houston, Department of Mathematics, Houston, Texas, 77204-3476, USA. V. Kwok, Dassault Aviation, 78 Quai Marcel Dassault, 92214 Saint Cloud, France. J. Periaux, Dassault Aviation, 78 Quai Marcel Dassault, 92214 Saint Cloud, France.

Schwarz Domain Decomposition Method with Time Stepping along Characteristic for Convection Diffusion Equations

Hongxing Rui and Danping Yang

1 Introduction

The multiplicative Schwarz domain decomposition method is a powerful iteration method for solving elliptic equations and other stationary problems. A systematic theory has been developed for elliptic finite element problems in the past few years, see [1, 2, 7, 8, 9]. In this paper, we are interested in solving the parabolic convection diffusion problems. We use time-stepping along characteristic method mentioned by Douglas, Russell [6], which is powerful especially for convection-dominated equations, and Galerkin approximation in the space variables. At a fixed time level, the resulting equation is equivalent to an elliptic problem which depends on a time-step parameter Δt. This suggests that we might apply the multiplicative Schwarz domain decomposition method, originally proposed for elliptic equations to the parabolic cases at every time level. The crucial mathematical questions is then to know how the convergence rate depends on the space mesh and the time step parameters. In the present paper, we introduce two kinds of domain decomposition algorithm, give the convergence rate and error estimates which tell us that after iterating only one cycle at every time-level, the global approximate solutions converge to the exact solution.

Other domain decomposition methods for parabolic problem can be found in [2,

[0] This work is supported by China State Major Key Project for Basic Researches and TCTPFT of the State Education Commission
[1] Mathematics Department, Shandong University, Jinan, 250100, P.R.China

5, 8]. In [8], Lions has given a kind of Schwarz alternating algorithm in the case of two subdomains for the heat equation and a convergence result but without an error estimate. In [5] Dawson, Dupont have given a kind of nonoverlapping domain decomposition method for parabolic equations, but since they have used explicit schemes at the intersection points, a stability condition is needed in the convergence analysis. In [2], Cai has given a kind of additive Schwarz algorithms for parabolic convection diffusion equations.

The outline of the paper is as following: In the next section, we will introduce two kinds of multiplicative Schwarz algorithms with time-stepping along characteristic. In Sect. 3, we will give the convergence rate of this algorithm, we also give L^2 error estimates with a fixed number of iterations at every time level, which tell us that when $h, \Delta t$ are sufficiently small, the approximate solution converges after a fixed number of iterations at every time level. Throughout this paper, c and C, with or without subscripts, denote generic, strictly positive constants. Their values may be different at different occurrences, but they are independent of the mesh parameters h and the time step Δt, which will be introduced later.

2   Multiplicative Schwarz Algorithms

Without loss of generality, we consider the following model problem in a bounded polygonal domain $\Omega \subset R^2$

$$\begin{cases} \frac{\partial u}{\partial t} + b \cdot \nabla u - \sum_{i,j=1}^{2} \frac{\partial}{\partial x_j}\left(a_{ij}\frac{\partial u}{\partial x_i}\right) = f, & \text{in} \quad \Omega, \\ u = 0, & \text{on} \quad \partial\Omega, \\ u(x,0) = u^0(x), & \text{in} \quad \Omega. \end{cases} \tag{1}$$

Here $b = (b_1, b_2), b \cdot \nabla u = b_1\frac{\partial u}{\partial x_1} + b_2\frac{\partial u}{\partial x_2}$ and $J = (0, T]$ denotes the time interval. The coefficient satisfy $a_{ij} = a_{ji}$ and there exists a positive constant γ such that

$$\sum_{ij=1}^{2} a_{ij}\xi_i\xi_j \geq \gamma|\xi|^2, \quad \forall \xi = (\xi_1, \xi_2)^{\top} \in R^2. \tag{2}$$

The variational formulation of problem (1) is: For $t \in J$, find $u(t) \in H_0^1(\Omega)$ such that

$$\begin{cases} (\frac{\partial u}{\partial t}, v) + a(u, v) + (b \cdot \nabla u, v) = (f, v), & v \in H_0^1(\Omega), \\ (u(0), v) = (u^0, v) \end{cases} \tag{3}$$

where

$$a(u, v) = \int_\Omega \sum_{ij=1}^{2} a_{ij}\frac{\partial u}{\partial x_i}\frac{\partial v}{\partial x_j} dx. \tag{4}$$

Let Δt denote the time-step, and let $t^n = n\Delta t$ and $u^n = u(t^n)$. For any point $x = (x_1, x_2)$, let

$$\bar{x} = \begin{cases} x - b\Delta t = (x_1 - b_1\Delta t, x_2 - b_2\Delta t), & \text{when} \quad x - b\Delta t \in \Omega \\ 2Y(x - b\Delta t) - X(x - b\Delta t), & \text{when} \quad x - b\Delta t \notin \Omega \end{cases} \tag{5}$$

where $Y(x) \in \partial\Omega$ denotes the point of projection of x, $X(x) \in \Omega$ denotes the symmetric point of x, we also let $\bar{u}^{n-1} = u(\bar{x}, t^{n-1})$. Then

$$\frac{u^n - \bar{u}^{n-1}}{\triangle t} = \frac{\partial u}{\partial t} + b \cdot \nabla u + O(\frac{\partial^2 u}{\partial \tau^2} \triangle t) \tag{6}$$

where τ denotes the unit vector in the characteristic direction of the transport term $(\frac{\partial u}{\partial t} + b \cdot \nabla u)$. Then the form (3) can be changed to

$$(\frac{u^n - \bar{u}^{n-1}}{\triangle t}, v) + a(u, v) = (f^{n-1}, v) + (\rho^n, v), \tag{7}$$

where

$$\rho^n = \frac{u^n - \bar{u}^{n-1}}{\triangle t} - (\frac{\partial u}{\partial t} + b \cdot \nabla u)^n = O(\frac{\partial^2 u}{\partial \tau^2} \triangle t).$$

We now divide Ω into overlapping subdomains $\Omega_1, \Omega_2, \cdots, \Omega_p$ satisfying Condition(A):

Condition(A): For any $x \in \bar{\Omega}$ there exist an open domain D_x and an $i_0 \in \{1, 2, \cdots, p\}$ such that $x \in D_x$ and $D_x \cap \Omega \subset \Omega_{i_0}$.

Extending the elements in $H_0^1(\Omega_j)$ to Ω by zero, we give the semi-discrete multiplicative Schwarz algorithms:

Scheme I: Let $U^0 = u^0$, for $n \geq 1$, we find $U^n \in H_0^1(\Omega)$ by three steps:

1) Set $U_0^n = U^{n-1}$.

2) Find $U_{jp+i}^n (j = 0, 1, \cdots, m-1, i = 1, 2, \cdots, p)$ such that

$$\begin{cases} (\frac{U_{jp+i}^n - \bar{U}^{n-1}}{\triangle t}, v) + a(U_{jp+i}^n, v) = (f^n, v), & v \in H_0^1(\Omega_i), \\ U_{jp+i}^n = U_{jp+i-1}^n, & x \in \Omega \setminus \Omega_i. \end{cases} \tag{8}$$

3) Let $U^n = U_{mp}^n, x \in \Omega$. Here m denotes the iteration time at the time-level in question.

Problem (8) are continuous in the spatial variables; in practical computation we can use an appropriate numerical method to solve it. Next we shall give a kind of multiplicative Schwarz algorithm combined with Galerkin finite element method. Let T_h denote a quasi-uniform triangulation of Ω which is aligned with the above domain decomposition, let h be the mesh parameter. $M_h \subset H_0^1(\Omega)$ denotes a standard finite element space such that

$$\inf_{\varphi \in M_h} (\|u - \varphi\| + h\|u - \varphi\|_1) \leq C\|u\|_{r+1} h^{r+1}, \ u \in H_0^1(\Omega) \cap H^{r+1}(\Omega). \tag{9}$$

Let $T_{i,h}$ denote the restriction of T_h to Ω_i, let $M_h(\Omega_i)$ denote the restriction of M_h to Ω_i, and let $M_h^0(\Omega_i) = M^h(\Omega_i) \cap H_0^1(\Omega_i)$. Set the initial approximation $W_h^0 \in M_h$ satisfying

$$\|W_h^0 - u^0\| \leq Ch^{r+1}. \tag{10}$$

We give the following multiplicative Schwarz algorithm :

Scheme II: For $n \geq 1$ find $W_h^n \in M_h$ in three steps:

1). Let $W_0^n = W_h^{n-1}$

2). Find $W_{jp+i}^n (j = 0, 1, \cdots, m-1, i = 1, 2, \cdots, p)$ satisfying

$$\begin{cases} (\frac{W_{jp+i}^n - \bar{W}^{n-1}}{\triangle t}, v) + a(W_{jp+i}^n, v) = (f^n, v), & v \in M_h^0(\Omega_i), \\ W_{jp+i}^n = W_{jp+i-1}^n, & x \in \Omega \setminus \Omega_i. \end{cases} \tag{11}$$

3) Let $W^n = W^n_{mp}, x \in \Omega$.

It is clear that the solutions of Scheme I and Scheme II are unique.

3   Convergence Analysis and Error Estimates

First we give two lemmas given Condition (A). We use the notations

$$\begin{cases} A(u,v) = (u,v) + \Delta t a(u,v), \forall v \in H_0^1(\Omega), \\ \|u\|_a = (a(u,u))^{\frac{1}{2}}, \\ \|u\|_A = (A(u,u))^{\frac{1}{2}} = (\|u\|_0^2 + \Delta t\|u\|_a^2)^{\frac{1}{2}}. \end{cases} \tag{12}$$

Lemma 3.1 *Suppose that the domain decomposition satisfies Condition (A). Then for $u \in H_0^1(\Omega)$ there exist a decomposition $u = \sum\limits_{i=1}^{p} u_l, u_l \in H_0^1(\Omega_l)$ such that*

$$\sum_{l=1}^{p} \|u_l\|_A^2 \le (1 + C_1\Delta t)\|u\|_A^2, \ \forall u \in H_0^1(\Omega), \tag{13}$$

where C_1 denotes a constant independent of Δt and u.

Lemma 3.2 *Under the condition of Lemma 3.1, there exists a constant C_2 independent of $h, \Delta t$. For $u \in M_h$, there exists a decomposition $u = \sum\limits_{l=1}^{p} u_l, u_l \in M_h^0(\Omega_l)$ such that*

$$\sum_{l=1}^{p} \|u_l\|_A^2 \le (1 + C_2(\Delta t + h))\|u\|_A^2. \tag{14}$$

In order to estimate the convergence rate, we need one of the following stronger but reasonable conditions, which can be easily satisfied.

Condition (B): The subregion $\Omega_i(1 \le i \le p)$ can be divided into four parts:

$$D_j = \sum_{r_{j-1} \le i \le r_j} \Omega_i, \ j = 1, 2, 3, 4, \ r_0 = 0, \ r_4 = p \tag{15}$$

Subdomains in D_j are disjoint and $\{D_1, D_2\}, \{D_3, D_4\}, \{D_1 \cup D_2, D_3 \cup D_4\}$ are domain decompositions of $D_1 \cup D_2, D_3 \cup D_4, \Omega$ respectively, which satisfy Condition (A) for $p = 2$.

Condition (C): The subregion $\Omega_i(1 \le i \le p)$ can be divided into k parts:

$$D_j = \sum_{r_{j-1} \le i \le r_j} \Omega_i, \ j = 1, 2, ..., k, \ r_0 = 0, \ r_k = p \tag{16}$$

such that: (1) $\{\Omega_j, r_{j-1} \le i \le r_j\}$ is a domain decompositions of D_j satisfying condition (A) and for $r_{j-1} + 1 \le i, l \le r_j, \Omega_i \cap \Omega_l = \emptyset$, if $l \ne i - 1, i + 1$; (2) $\{D_j, 1 \le j \le k\}$ is a domain decomposition of Ω satisfying Condition(A) and for $1 \le j, l \le k, \Omega_j \cap \Omega_l = \emptyset$, if $l \ne j - 1, j + 1$.

Remark 1. When Condition (B) holds, the above method can be parallelized by coloring the subdomains and solving in parallel on disjoint subdomains of the same color.

For error estimates of Scheme I, defining the operator $R_i : H_0^1(\Omega) \to H_0^1(\Omega_i)$ such that

$$A(R_i u - u, v) = 0, \forall v \in H_0^1(\Omega_i). \tag{17}$$

Theorem 3.1 *Suppose that Condition (B) is satisfied for the domain decomposition. Then there exists a constant C_3, independent of $u, \triangle t$, such that*

$$\|(I - R_p) \cdots (I - R_2)(I - R_1)u\|_A \leq C_3 \triangle t^{\frac{1}{2}} \|u\|_A, \forall u \in H_0^1(\Omega). \tag{18}$$

Theorem 3.2 *Suppose that Condition (C) is satisfied for the domain decomposition. Then there exists a constant C_4 independent of $\triangle t, u$ such that*

$$\|(I - R_p) \cdots (I - R_2)(I - R_1)u\|_A \leq C_4 \triangle t^{\frac{1}{2}} \|u\|_A, \forall u \in H_0^1(\Omega). \tag{19}$$

Let $\partial_t u^n = (u^n - u^{n-1})/\triangle t$, $e^n = U^n - u^n$, and $e_i^n = U_i^n - u^n$. Then $e^n = e_{mp}^n$ and

$$\begin{cases} (\frac{e_{jp+i}^n - \bar{e}^{n-1}}{\triangle t}, v) + a(e_{jp+i}^n, v) = -(\rho^n, v), & \forall v \in H_0^1(\Omega_i), \\ (e_{jp+i}^n - e_{jp+i-1}^n) = 0, & x \in (\Omega - \Omega_i). \end{cases} \tag{20}$$

Theorem 3.3 *Suppose that Condition (B) or Condition (C) hold. Then there exists a constant C_5, independent of $\triangle t$, such that*

$$\|e^n\|_A \leq (1 + C_5 \triangle t^m)\|e^{n-1}\|_A + C \triangle t(\triangle t^{\frac{m}{2}} + \triangle t) \tag{21}$$

where the constant C is independent of $\triangle t$.

Theorem 3.4 *Suppose that the solution of (1) is sufficiently smooth. Suppose also that Condition (B) or Condition(C) holds for the domain decomposition. For $m \geq 1$ and the solution of Scheme I, we have*

$$\|u^n - U^n\| \leq C(\triangle t + \triangle t\|\frac{\partial^2 u}{\partial \tau^2}\|_{L^\infty(J;L^2(\Omega))} + \triangle t^{\frac{m}{2}}),$$

where C denotes a generic constant which is independent of $\triangle t$.

For error estimates of Algorithm II, define $R_{h,i} : M_h \to M_h^0(\Omega_i)$ such that

$$A(R_{h,i}u - u, v) = 0, \quad \forall v \in M_h^0(\Omega_i). \tag{22}$$

Theorem 3.5 *Suppose that Condition (B) or condition (C) are satisfied for the domain decomposition. Then, there exists a constant C_6, independent of $u, h, \triangle t$, such that*

$$\|(I - R_{h,p}) \cdots (I - R_{h,2})(I - R_{h,1})u\|_A \leq C_6(h + \triangle t)^{\frac{1}{2}} \|u\|_A, \forall u \in M_h \tag{23}$$

Define an auxiliary function $\widetilde{u}_h^n \in M_h$ such that

$$a(\widetilde{u}_h^n - u^n, v) = 0, \forall v \in M_h, \tag{24}$$

and let $\eta = u^n - \widetilde{u}_h^n$, $E^n = W^n - \widetilde{u}_h^n$, $E_i^n = W_i^n - \widetilde{u}_h^n$, as in (18), (19) and (21). If Condition (B) or Condition (C) is satisfied, we can prove that

$$\|(I - R_{h,p}) \cdots (I - R_{h,2})(I - R_{h,1})\| \leq C(\Delta t + h)^{\frac{1}{2}}, \tag{25}$$

$$\|E^n\|_A \leq (1 + C_7(\Delta t + h)^m)\|E^{n-1}\|_A + C\Delta t(\Delta t + h^{r+1} + (\Delta t + h)^{\frac{m}{2}}), \tag{26}$$

where C_7 denotes a constant which satisfies $1 + C^2(\Delta t + h)^m \leq 1 + 2C_7(\Delta t + h)^m$.

Theorem 3.6 *Under the assumptions of Theorem 3.4, and for $h^m = O(\Delta t)$, there exists a constant C, independent of $h, \Delta t$, such that for the solution of Scheme II*

$$\|u^n - W^n\| \leq C(\Delta t + \Delta t\|\frac{\partial^2 u}{\partial \tau^2}\|_{L^\infty(J;L^2(\Omega))} + \Delta t^{-\frac{1}{2}}h^{m+1} + (\Delta t + h)^{\frac{m}{2}}). \tag{27}$$

Here C denote a generic constant independent of Δt and h.
Remark 2 If $b_1 = b_2 = 0$, the term $\Delta t^{-\frac{1}{2}}h^{m+1}$ can be removed and the error estimates are of optimal order.
Remark 3 Theorem 3.1, Theorem 3.2 and (25) tell us that the convergence rate for Algorithm I, Algorithm II is $\rho = C\Delta t^{\frac{1}{2}}$, $\rho = C(\Delta t + h)^{\frac{1}{2}}$ respectively. Since we do not use a coarse mesh triangulation, the constant C depends on the overlapping parts of subdomains.

Acknowledgement: The authors give their thanks to Professor Yuan Yirang for his helpful suggestion and guidance.

REFERENCES

[1] Bramble J. H., Pasciak J. E., Wang J. and Xu J. (1991) Convergence estimates for product iterative methods with application to domain decomposition, *Math. Comp.* , 57(195): 1-21.

[2] Cai X. C. (1991) Additive Schwarz algorithms for parabolic convection diffusion equations. *Numer. Math.*, 60: 41-61.

[3] Xu J. (1992) Iterative methods by space decomposition and subspace correction. *SIAM Rev.*, 34: 581-613.

[4] Ciarlet P. G. (1978) *The Finite Element Method for Elliptic Problems,* North-Holland.

[5] Dawson C. N., Dupont T. F. (1992) Explicit/implicit conservative Galerkin domain decomposition procedures for parabolic problems. *Math. Comp.*, 58(197): 21-34.

[6] Jim Douglas Jr. and Russell T. F. (1982) *SIAM J.Numer. Anal.*, 19: 871-885.

[7] Dryja M. and Widlund O.B. (1990) Towards a unified theory of domain decomposition algorithm for elliptic problems. in*3rd International Symposium on Domain Decomposition Methods for Partial Differential Equations* T. Chan, R. Glowinski, J. Periaux and O. B. Widlund, eds., SIAM, Philadelphia.

[8] Lions P.L. (1988) On the Schwarz alternating method I, in *First International Symposium on Domain Decomposition Methods for Partial Differential Equations,* R.Glowinski, G.H.Golub, G.A.Meurant and J.Periaux, eds., SIAM, Philadelphia.

[9] Lions P. L. (1989) On the Schwarz alternating method II, in *Second International Symposium on Domain Decomposition Methods for Partial Differential Equations.* T. Chan, R. Glowinski, J. Periaux and O. B. Widlund, eds., SIAM, Philadelphia.

A characteristic domain splitting method

Xue-Cheng Tai, Torbjørn O. Widnes Johansen, Helge K. Dahle, Magne S. Espedal [1]

Abstract. This work treats a linear convection-diffusion problem. Diffusion and convection may be equally important or convection may dominate the problem. The method of characteristics is combined with an overlapping domain decomposition technique so that domain decomposition is naturally combined with the time stepping. In each time step, the algorithm first determines the characteristic solution. Then a diffusion problem is solved in parallel on each subdomain with the characteristic solution as boundary conditions. No iteration is needed between the subdomain problems. Adaptive time steps are used for the characteristic tracing. The time steps used for the diffusion problems can be large.

Introduction

In this work, we consider the following linear convection-diffusion problem

$$\begin{cases} u_t - \epsilon \nabla \cdot (a(x)\nabla u) + \vec{b}(x) \cdot \nabla u = f, & \text{in } \Omega \subset \mathrm{R}^n, \ n = 1, 2, 3. \\ u(x,t) = 0 & \text{on } \partial\Omega \times [0, \mathrm{T}] , \\ u(x,0) = u_0(x) & \text{in } \Omega \text{ at } t = 0. \end{cases}$$

Assume that the problem has been suitably scaled such that a and $\vec{b}$ are of the same order. The parameter ϵ can be very small, but it can also be large. We are going to use the method of characteristics, see [DR] [Piro], to treat the convection part, and then use an overlapping domain decomposition technique to treat a symmetric diffusion problem. An important feature of this method is that no iterations may be required between the subdomain problems. This is due to the way domain decomposition is

[1] Department of Mathematics, University of Bergen, Alleg. 55, 5007, Bergen, Norway.
Email: Tai@mi.uib.no, widnes@mi.uib.no, reshd@mi.uib.no and resme@mi.uib.no.
The work is supported by the University of Bergen and by VISTA, a research cooperation between the Norwegian Academy of Science and Letters and Den norske stats oljeselskap a.s. (Statoil)

Domain Decomposition Methods in Sciences and Engineering, edited by R. Glowinski *et al.*

combined with the method of characteristics.

Domain decomposition methods have been used for nonsymmetrical convection-diffusion problems, see [BLP]–[CW], [W], [X1]-[XC], However, the methods proposed in these papers are efficient only for diffusion dominated problems. For problems where diffusion and convection are equally important or convection is dominating, special care must be taken. In a recent work by Rannacher and Zhou [RZ], the streamline diffusion method is used with an overlapping domain decomposition for a linear convection dominated problem. This work has been motivated by [RZ]. Here we use the method of characteristics to treat the convection term. By doing this, we get symmetrical problems when working with the diffusion part. Even for diffusion dominated problems or problems where diffusion and convection are equally important, the proposed algorithm will give better results than the conventional finite element method. This is due to the characteristic treatment of the convection term.

The Algorithm

At a given time t, and for a given x, let $X = X(x, t; t_0)$ be the solution of:

$$\begin{cases} \frac{dX}{dt} = \vec{b}(X), \\ X(x, t_0; t_0) = x. \end{cases} \tag{1}$$

If $\vec{b}$ is smooth, then there always exists a $\tau_0 > 0$ such that (1) has a unique solution for $|t - t_0| < \tau_0$. Let us choose an integer $N > 0$ such that $\Delta t = \frac{T}{N} \leq \tau_0$ and divide $[0, T] = \cup_{n=1}^{N}[t^{n-1}, t^n]$, $t^n = n\Delta t$. Backward tracing will be used to approximate the solution of (1). Moreover, adaptivity of the step used in the backward tracing will be needed. Thus let Δt_x be the step used in the backward tracing from the point x, i.e.

$$\Delta t = t^{n+1} - t^n = m_x \Delta t_x.$$

Then define points at the approximate characteristic (streamline) backwards from x by:

$$\tilde{x}^{n+1} = x,$$
$$\tilde{x}^{n+\frac{m_x-k}{m_x}} = \tilde{x}^{n+\frac{m_x-k+1}{m_x}} - \vec{b}\left(\tilde{x}^{n+\frac{m_x-k+1}{m_x}}\right) \cdot \Delta t_x , \qquad \text{for } k = 1, 2, \cdots, m_x.$$

When $k = m_x$, one finds that

$$\tilde{x}^n = \tilde{x}^{n-\frac{m_x-k}{m_x}} .$$

The characteristic solution is now given by

$$\tilde{u}^n = u^n(\tilde{x}^n) ,$$

where u^n is the computed solution at time level t^n. When $\tilde{x}^n$ falls outside of Ω, one takes $\tilde{u}^n = 0$. If nonhomogenous boundary conditions are used, we need to determine the point where the characteristic curve hits the boundary, and take $\tilde{u}^n$ to be the boundary value at that point.

The above iterative procedure is simply using the explicit one step backward tracing $\tilde{x} = x - \vec{b}(x) \cdot \Delta t$ over many local steps. For the domain decomposition, we assume that Ω has been divided into finite elements $\Omega = \cup_{e \in \mathcal{T}_h} e$. Let $\Omega_i, i = 1, 2, \ldots, m$ be a nonoverlapping decomposition of $\bar{\Omega}$, such that each Ω_i is the union of some elements. To each Ω_i, we associate an enlarged subdomain

$$\Omega_i^\delta = \{ e \in \mathcal{T}_h | \quad dist(e, \Omega_i) \leq \delta \} \ .$$

Hence, Ω_i^δ forms an overlapping domain decomposition with overlapping size δ. With each Ω_i^δ, we use $S_h^0(\Omega_i^\delta)$ to denote the linear finite element space with zero traces on Ω_i^δ. Note that the decomposition of Ω can be different from time level to time level in order to follow possible shock fronts.

Algorithm 1 *(The characteristic domain splitting algorithm).*

1. Choose $u_h^0 \in S_h^0(\Omega)$ to be an approximation for u_0.
2. If u^n is known, do characteristic tracing to find

$$\tilde{u}^n = u^n(\tilde{x}^n) \ .$$

This can be done in parallel for each of the nodal points at time level t^{n+1}.
3. On each subdomain Ω_i^δ, find u_i^{n+1} in parallel for $i = 1, 2, \ldots, m$ such that

$$\begin{cases} \left(\frac{u_i^{n+1} - \tilde{u}^n}{\Delta t}, v_i \right) + \epsilon (a \nabla u_i^{n+1}, \nabla v_i) = (f, v_i), \quad \forall v_i \in S_h^0(\Omega_i^\delta), \\ u_i^{n+1} = \tilde{u}^n \quad \text{on} \quad \partial \Omega_i^\delta \ . \end{cases}$$

4. From the patchwise solution u_i^{n+1}, a global single valued solution

$$u^{n+1} = \mathcal{C} \left(\{ u_i^{n+1} \}_{i=1}^m \right) \in S_h^0(\Omega)$$

is constructed such that

$$\| u^{n+1} \|_{L^2(\Omega)} \leq \sum \| u_i^{n+1} \|_{L^2(\Omega_i)}. \tag{2}$$

5. If $t^{n+1} < T$, go to the next time level.

For the above algorithm, detailed convergence analysis is given in [TDE]. We assume that the local time step satisfies

$$\Delta t_x \leq \frac{C(\vec{b})}{\| \nabla u^n \|_{\infty, V(x)}} \Delta t \ ,$$

where $C(\vec{b})$ is a constant depending on $\vec{b}$, u^n is the solution computed at the previous time level and $V(x)$ is a neighbourhood of x such that the characteristic curve starting from x and going backward is contained in $V(x) \times [t^n, t^{n+1}]$. Furthermore, if the overlapping size satisfies

$$\delta \geq c_0 \max(\sqrt{\Delta t \epsilon}, h) |\ln \Delta t| \ , \tag{3}$$

where c_0 is a constant depending on the maximum angles of the elements in the overlapping area, then the computed solution u^n satisfies

$$\| u(t^n) - u^n \|_{L^2(\Omega)} \leq C(h^2 + \Delta t) \ .$$

Above, C does not depend on ϵ, which means that we get the same accuracy in a region where the gradient of u is sharp.

Numerical Experiments.

As is shown by (3), for a given h and a given time step size Δt, we need to have sufficient overlap to guarantee the computational results to be of accuracy $O(h^2 + \Delta t)$.

As a test example, we consider a shock moving in the characteristic direction. It is known that $\phi(x, y, t) = \frac{1}{4\pi\epsilon t} e^{-\frac{x^2+y^2}{4\epsilon t}}$ satisfies the heat equation $\phi_t - \epsilon\Delta\phi = 0$. When ϵ is small, ϕ is singular near the point $x = y = 0$ for $t > 0$. Defining $u(x, y, t) = \phi(x - t, y - t, t + 0.1)$, one may easily verify that u is a solution of

$$\begin{cases} u_t - \epsilon\Delta u + u_x + u_y = 0, \\ u(x, y, 0) = \phi(x, y, 0.1). \end{cases}$$

This solution represents a shock moving in the direction of $\vec{b} = (1, 1)$. See Figure 1 for the location of the shock at different times.

In the computations, the domain Ω is taken as $\Omega = [0, 1] \times [0, 1]$. It is first divided into coarse rectangular subdomains with size $H = H_x = H_y$, and then each subdomain is divided into fine mesh rectangular elements with size $h = h_x = h_y$. Both the fine and the coarse meshes are uniform. A bilinear finite element space is used in the computations. Each coarse subdomain is extended by L elements into its neighbours. This defines the overlap. In the following tables, $m \times m$ is the number of subdomains and $n \times n$ is the number of elements in each subdomain. Hence $H = 1/m$, $h = 1/nm$. We let $\| \cdot \|_2$ denote the discrete L^2-norm.

The effect of varying the overlapping size L, and the number of subdomains m, is investigated in Table 1. Thus, the values of ϵ, Δt and h are fixed. Choosing $m = 1$ means that we are solving the global problem without domain decomposition. The computed solution u^n based on domain decomposition ($m > 1$) is compared with the exact solution $u(t^n)$ and the global solution u_G^n ($m = 1$), at $t^n = 0.5065$. From the table, one may observe that for different m, just one or two elements of overlap is needed to give the same accuracy as the global solution. However, by increasing the overlapping size L, u^n is getting closer and closer to the global solution u_G^n.

In Table 2, different values of ϵ is tested. When ϵ is getting smaller, the shock becomes sharper as one should expect. We observe that for large ϵ, more overlap is needed to retain the accuracy of the global solution. By decreasing Δt, a relatively small overlap may also retain the accuracy. When ϵ is small ($\epsilon = 0.01$), just one element of overlap is sufficient. Figure 1 shows the computed solution for $\epsilon = 0.01$ at different times. The figure for the analytical solution looks exactly the same.

Conclusion

The method of characteristics is combined with an overlapping domain decomposition technique to solve a convection-diffusion problem. When ϵ is small or Δt is small, one or two elements, are enough overlapping to get as accurate solution as the global solution. No iteration is needed between the subdomain problems. However, by increasing the overlapping size, the domain decomposition solution gets closer to the global finite element solution. When ϵ is large, more overlap is needed or one should decrease the time step size in order to decrease the needed overlapping size. By combining the domain decomposition method with the method of characteristics, the algorithm is able to capture sharp travelling shocks.

m	n	L	$\|u(t^n) - u^n\|_\infty$	$\frac{1}{h}\|u(t^n) - u^n\|_2$	$\|u_G^n - u^n\|_\infty$	$\frac{1}{h}\|u_G^n - u^n\|_2$
1	40	-	$6.96 \cdot 10^{-2}$	1.1817	-	-
4	10	1	$3.96 \cdot 10^{-2}$	0.3629	0.1062	1.5284
4	10	3	$5.86 \cdot 10^{-2}$	1.0120	$1.15 \cdot 10^{-2}$	0.1670
4	10	5	$6.83 \cdot 10^{-2}$	1.1621	$1.30 \cdot 10^{-3}$	0.0197
5	8	1	$5.33 \cdot 10^{-2}$	0.8033	0.1175	1.9780
5	8	2	$3.30 \cdot 10^{-2}$	0.5448	$3.72 \cdot 10^{-2}$	0.6389
5	8	4	$6.54 \cdot 10^{-2}$	1.1070	$4.32 \cdot 10^{-3}$	$7.48 \cdot 10^{-2}$
8	5	1	$1.33 \cdot 10^{-1}$	1.9804	0.2026	3.1568
8	5	2	$8.96 \cdot 10^{-3}$	0.1605	$6.39 \cdot 10^{-2}$	1.030
8	5	3	$4.84 \cdot 10^{-2}$	0.8264	$2.17 \cdot 10^{-2}$	0.3558

Table 1 $\epsilon=0.1$, $\Delta t=1/160$, $t^n=81/160$,
$u(t^n)$: known solution, u^n: computed solution, u_G^n: global finite element solution.

ϵ	Δt	nt	m	n	L	$\|u(t^n) - u^n\|_\infty$	$\frac{1}{h}\|u(t^n) - u^n\|_2$	$\|u_G^n - u^n\|_\infty$	$\frac{1}{h}\|u_G^n - u^n\|_2$
0.01	1/160	81	1	80	-	1.7365	21.8401	-	-
0.01	1/160	81	4	20	1	1.4983	18.9283	0.3439	3.3324
0.01	1/160	81	4	20	3	1.7226	21.6590	$2.05 \cdot 10^{-2}$	0.2012
0.01	1/160	81	4	20	5	1.7358	21.8293	$1.10 \cdot 10^{-3}$	$1.20 \cdot 10^{-2}$
0.01	1/160	81	8	10	1	1.1523	14.7338	0.7372	7.8819
0.01	1/160	81	8	10	3	1.7017	21.3977	$4.23 \cdot 10^{-2}$	0.4704
0.01	1/160	81	8	10	5	1.7347	21.8144	$2.20 \cdot 10^{-3}$	$2.74 \cdot 10^{-2}$
0.5	1/160	81	1	40	-	$7.89 \cdot 10^{-4}$	$1.86 \cdot 10^{-2}$	-	-
0.5	1/160	81	4	10	3	$1.63 \cdot 10^{-2}$	0.3306	$1.71 \cdot 10^{-2}$	0.3487
0.5	1/160	81	4	10	5	$5.46 \cdot 10^{-3}$	0.1087	$6.25 \cdot 10^{-3}$	0.1268
0.5	1/160	81	4	10	8	$7.19 \cdot 10^{-4}$	$1.28 \cdot 10^{-2}$	$1.50 \cdot 10^{-3}$	$3.04 \cdot 10^{-2}$
1.0	1/500	250	1	40	-	$1.74 \cdot 10^{-4}$	$4.07 \cdot 10^{-3}$	-	-
1.0	1/500	250	4	10	3	$2.26 \cdot 10^{-3}$	$4.54 \cdot 10^{-2}$	$2.43 \cdot 10^{-3}$	$4.94 \cdot 10^{-2}$
1.0	1/500	250	4	10	5	$5.64 \cdot 10^{-4}$	$1.09 \cdot 10^{-2}$	$7.38 \cdot 10^{-4}$	$1.49 \cdot 10^{-2}$
1.0	1/500	250	4	10	6	$2.40 \cdot 10^{-4}$	$4.37 \cdot 10^{-3}$	$4.14 \cdot 10^{-4}$	$8.35 \cdot 10^{-3}$
1.0	1/500	250	4	10	7	$5.87 \cdot 10^{-5}$	$9.00 \cdot 10^{-4}$	$2.33 \cdot 10^{-4}$	$4.69 \cdot 10^{-3}$

Table 2 nt: number of time steps, $u(t^n)$: known solution,
u^n: computed solution, u_G^n: global finite element solution.

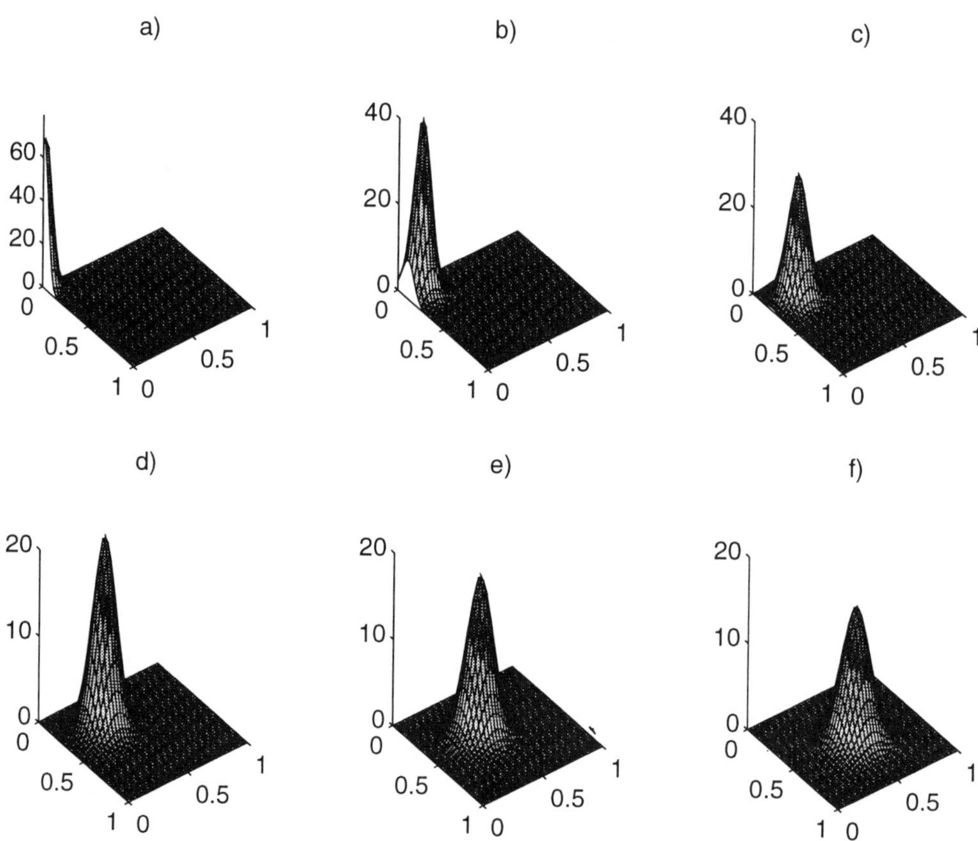

Figure 1 The computed solution with $H = 1/10, h = 1/200, L = 1, \epsilon = 0.01,$ $\Delta t = 1/100$, at different times: a) t=0.01, b) t=0.1, c) t=0.2 d) t=0.3 e) t=0.4 f) t=0.5.

REFERENCES

[BLP] J. H. Bramble, Z. Leyk and J. Pasiciak, Iterative schemes for nonsymmetric and indefinite elliptic boundary problems, Math. Comput., vol 60, pp. 1-22, 1993.

[Cai] X.-C. Cai, An additive Schwarz algorithm for nonselfadjoint elliptic equations, "Domain decomposition methods for partial differential equations, III", SIAM, Philadelphia, 1989, (T. F. Chan, R. Glowinski, J. Periaux and O. B. Widlund eds), pp. 232-244.

[CGK] X. C. Cai, W. D. Gropp and D. E. Keyes, A comparison of some domain decomposition and ILU preconditioned iterative methods for nonsymmetric elliptic problems, Preprint Report. 92-03, 1992.

[CW] X. C. Cai and O. B. Widlund, Domain decomposition algorithms for indefinite elliptic problems, SIAM J. Sci. Stat. Comput., vol. 13, 1992, pp. 243-258.

[DR] J. Douglas and T. F. Russell, Numerical methods for convection-dominated diffusion problems based on combining the method of characteristic with finite element or finite difference procedure, SIAM J. Numer. Anal., vol. 19, 1982, pp. 871-885.

[Piro] O. Pironeau, On the transport-diffusion algorithm and its application to the Navier-Stokes equations, Numer. Math., vol. 38, 1982, pp. 309-332.

[RZ] R. Rannacher and G. H. Zhou, Analysis of a domain–splitting method for nonstationary convection-diffusion problems, East-West J. Numer. Math., vol. 2, 1994, pp. 151-174.

[TDE] X.–C. Tai, H. K. Dahle and M. Espedal, A characteristic domain splitting method for time dependent convection-diffusion problems, in preparation, 1995.

[W] J. P. Wang, Convergence analysis of the Schwartz algorithm and multilevel decomposition iterative methods II: nonselfadjoint and indefinite elliptic problems, SIAM J. Numer. Anal., vol. 30, 1993, pp. 953–970.

[X1] J. C. Xu, A new class of iterative methods for nonselfadjoint or indefinite problems, SIAM J. Numer. Anal., vol. 29, 1992, pp. 303–319.

[X2] J. C. Xu, Iterative methods by SPD and small subspace solvers for nonsymmetric or indefinite problems, "the Fifth of this proceeding", (D. Keyes, T. Chan, G. Meurant, J. Scroggs and R. Voigt eds), SIAM, Philadelphia, 1992, pp. 106–118.

[XC] J. Xu and X.-C. Cai, A preconditioned GMRES method for nonsymmetric or indefinite problems, Math. Comput., vol. 59, 1992, pp. 311-320.

A Characteristic-Based Domain Decomposition and Space-Time Local Refinement Method for Advection-Reaction Equations with Interfaces

HONG WANG [1], JAN E. VÅG [2], and MAGNE S. ESPEDAL [2]

Abstract

In this paper we present a characteristic-based, noniterative, nonoverlapping, domain decomposition and space-time local refinement method to solve the initial-boundary value problems for advection-reaction equations with various interfaces.

1 Introduction

Advection-reaction partial differential equations arise in a variety of applications and often cause numerical difficulties. Conventional space-centered finite difference/element methods usually result in severe non-physical oscillatory solutions. While upstream weighting techniques can eliminate the oscillations, they generate numeri-

---

[1] Department of Mathematics, University of South Carolina, Columbia, SC 29208, USA
[2] Department of Mathematics, University of Bergen, Alleg. 55, 5007, Bergen, Norway

Domain Decomposition Methods in Sciences and Engineering, edited by R. Glowinski *et al.*

cal solutions with serious numerical dispersion [EW83]. Moreover, practical problems often have various physical and numerical interfaces that introduce further complexities. Physical interfaces arise when the media properties change abruptly, leading to advection-reaction equations with discontinuous coefficients. Numerical interfaces arise when domain decomposition or local refinement techniques are used. The solutions of advection-reaction equations are generally smooth outside some small regions and may have sharp fronts/discontinuity inside, which need to be resolved accurately free of oscillation or numerical dispersion in practice. In this case local refinement should be used within the sharp front regions. Domain decomposition should be used when the governing equations (especially in the case of strongly coupled systems) are imposed over large domains. One can see that numerical interfaces are introduced in either case.

It is more difficult to develop domain decomposition and local refinement techniques for advection-reaction equations than it is for elliptic and parabolic equations, because in the context of advection-reaction equations locally generated numerical errors at the boundaries/interfaces can be propagated into the domain and destroy the overall accuracy/stability of the method. In this paper we present a characteristic-based, noniterative, nonoverlapping, domain decomposition and space-time local refinement method for advection-reaction equations with various physical/numerical interfaces. To demonstrate the ideas, we consider the model problem

$$
\begin{aligned}
u_t + (V(x,t)\,u)_x + K(x,t)u &= f(x,t), & x &\in (a,b), \;\; t \in (0,T], \\
u(a,t) &= g(t), & t &\in (0,T], \\
u(x,0) &= u_0(x), & x &\in [a,b].
\end{aligned}
\tag{1}
$$

Here $V(x,t) > 0$ is a velocity field, $K(x,t)$ is a first-order reaction coefficient, $u_x = \frac{\partial u}{\partial x}$, $u_t = \frac{\partial u}{\partial t}$. $V(x,t)$ is continuously differentiable except at the interfaces d_l ($l = 1, 2, \ldots, L-1$ with $a = d_0 < d_1 < \ldots < d_{L-1} < d_L = b$) where $V(x,t)$ has a jump discontinuity in x. Problem (1) is closed by the following interface conditions

$$
V(d_l-,t)u(d_l-,t) = V(d_l+,t)u(d_l+,t), \quad t \in [0,T], \; l = 1,2,\ldots,L-1.
\tag{2}
$$

2   An ELLAM Scheme

In this section we present an ELLAM (Eulerian-Lagrangian localized adjoint method) scheme for problem (1) with smooth coefficients. Based on this scheme we develop a domain decomposition and local refinement method. ELLAM was originally developed for the solution of advection-diffusion equations with general boundary conditions [CRHE90]. Let I and N be two positive integers, define the spatial and temporal partitions $x_i = a + i\Delta x$ for $i = 0, 1, \ldots, I$ and $t_n = n\Delta t$ for $n = 0, 1, \ldots, N$ with $\Delta x = (b-a)/I$ and $\Delta t = T/N$. In addition, we introduce a local time refinement $t_{n,i}$ at the outflow boundary $\{b\} \times [t_n, t_{n+1}]$ by $t_{n+1} = t_{n,I} > t_{n,I+1} > \ldots > t_{n,I+IC} > t_{n,I+IC+1} = t_n$, whose exact definition will be given in Section 4. At time t_{n+1} (or the outflow boundary $x = b$), we define an approximate characteristic $X(\theta; x, t_{n+1})$, $\theta \in [t_n, t_{n+1}]$, (or $X(\theta; b, t)$, $\theta \in [t_n, t]$,) to be the tangent line emanating backward from (x, t_{n+1}) (or (b, t)). We also let $x^* = X(t_n; x, t_{n+1})$, $b^*(t) = X(t_n; b, t)$, and $\tilde{a}$ to satisfy $a = X(t_n; \tilde{a}, t_{n+1})$.

With any space-time test functions w that vanish outside of $[a, b] \times (t_n, t_{n+1}]$ and are discontinuous in time at time t_n, one can write a space-time variational formulation for the governing equation in (1) as follows

$$(u_{n+1}, w_{n+1})_{L^2(a,b)} + <[Vu]_c, w_c> \big|_{c=a}^{c=b} - <(u, [w_t + Vw_x - Kw])> \tag{3}$$

$$= (u_n, w_n^+)_{L^2(a,b)} + <(f, w)>,$$

where $(u_n, w_n)_{L^2(a,b)} = \int_a^b u(x, t_n) w(x, t_n) dx$, $<u_c, w_c> = \int_{t_n}^{t_{n+1}} u(c, t) w(c, t) dt$ for $c = a$ or b, $<(u, w)> = <(u, w)_{L^2(a,b)}>$, and $w_n^+ = \lim_{t \to t_n, t > t_n} w(x, t)$.

It is difficult to find the test functions w to satisfy $w_t + Vw_x - Kw = 0$ since one cannot track the characteristics exactly, in general. Nevertheless, the test functions w_i, which are defined by $w_i(X(\theta; x, t_{n+1}), \theta) = w_i(x, t_{n+1}) e^{-K(x, t_{n+1})(t_{n+1} - \theta)}$ for $\theta \in [t_n, t_{n+1}]$ and $i = 0, 1, \ldots, I$, and by $w_i(X(\theta; b, t), \theta) = w_i(b, t) e^{-K(b, t)(t - \theta)}$ for $\theta \in [t_n, t]$ and $i = I, I+1, \ldots, I+IC+1$, satisfy $w_t + Vw_x - Kw = 0$ approximately. Substituting w_i for w in (3), one can rewrite (3) as follows

$$(u_{n+1}, w_{n+1})_{L^2(a,b)} + <V_b u_b, w_b> - <(u, [w_t + Vw_x - Kw])>$$

$$= (\Psi_{n+1}^{(1)} u_n^*, w_{n+1})_{L^2(\tilde{a},b)} + <\Psi_b^{(2)} u_n^*, w_b> + (\Psi_{n+1}^{(3)} f_{n+1}, w_{n+1})_{L^2(a,b)} \tag{4}$$

$$+ <\Psi_b^{(4)} f_b, w_b> + <V_a g, w_a> + R(f, w).$$

Here $u_n^* = u(x^*, t_n)$ in $(\cdot, \cdot)$ with $x^* = X(t_n; x, t_{n+1})$ and $u_n^* = u(b^*(t), t_n)$ in $<\cdot, \cdot>$ with $b^*(t) = X(t_n; b, t)$. $R(f, w)$ is a truncation-error term resulting from the application of the backward Euler quadrature to the last term on the right-hand side of Equation (4). $\Psi_{n+1}^{(1)} = 1 + O(\Delta t)$, $\Psi_b^{(2)} = V(b, t)(1 + O(\Delta t))$, $\Psi_{n+1}^{(3)} = \Delta t(1 + O(\Delta t))$, $\Psi_b^{(4)} = \Delta t(1 + O(\Delta t))$ are Jacobian-related factors whose exact forms are omitted here.

In the numerical scheme the trial functions U are chosen to be piecewise-linear functions at the time t_{n+1} and at the outflow boundary. Note that $w_{it} + Vw_{ix} - Kw = 0$ approximately, the term $<(u, [w_t + Vw_x - Kw])>$ should be small and dropping it introduces negligible errors. Replacing u by U in (4) and dropping the last terms on both the left-hand and right-hand sides of Equation (4), one obtains an ELLAM scheme at $i = 1, 2, \ldots, I + IC$ as follows

$$(U_{n+1}, \hat{w}_{i,n+1})_{L^2(a,b)} + \; < V_b U_b, \hat{w}_{ib} >$$

$$= (\Psi_{n+1}^{(1)} U_n^*, \hat{w}_{i,n+1})_{L^2(\check{a},b)} + \; < \Psi_b^{(2)} U_n^*, \hat{w}_{ib} > + (\Psi_{n+1}^{(3)} f_{n+1}, \hat{w}_{i,n+1})_{L^2(a,b)} \tag{5}$$

$$+ \; < \Psi_b^{(4)} f_b, \hat{w}_{ib} > + \; < V_a g, \hat{w}_{ia} >,$$

where $\hat{w}_i = w_i$ for $i = 1, 2, \ldots, I + IC - 1$, $\hat{w}_1 = w_0 + w_1$, $\hat{w}_{I+IC} = w_{I+IC} + w_{I+IC+1}$. Since $U(a, t_{n+1}) = g(t_{n+1})$ is known, no equation is needed at $i = 0$. Thus, for $i = 1$ in Equation (5) we use $\hat{w}_1 = w_1 + w_0$ instead of w_1 on $[a, x_1]$. Similarly, because $U(b, t_n)$ is known from the computations at the previous time t_n, we choose $\hat{w}_{I+IC}$ instead of w_{I+IC} on $[t_n, t_{n+1,I+IC}]$ in Equation (5) for $i = I + IC$.

With the given boundary condition at the inflow boundary $x = a$ and the known solution at the time t_n, one can solve Equation (5) for the ELLAM approximation U at the time t_{n+1} and at the outflow boundary $x = b$. The scheme has a well-conditioned, symmetric and positive definite (tridiagonal in one dimension) coefficient matrix without any artificial boundary conditions added.

3 A Space-Time Local Refinement and Domain Decomposition Algorithm

Based on the scheme (5) we present a noniterative, nonoverlapping, domain decomposition and space-time local refinement method for problem (1) with interfaces at d_l $(l = 1, 2, \ldots, L - 1)$: Partition the time interval $[0, T]$ into K intervals $[T_{k-1}, T_k]$ with $0 = T_0 < T_1 < \ldots < T_{k-1} < T_k = T$.

(1) With the given initial and boundary conditions in (1), apply Equation (5) to solve problem (1) over the space-time domain $[a, d_1] \times [0, T_1]$. The solution $U(x, t)$ over this domain defines the left-limit $U(d_1-, t)$ for $t \in [0, T_1]$ and $U(x, T_1)$ for $x \in [a, d_1]$.

(2) When $V(x, t)$ is discontinuous at $x = d_1$, U is discontinuous at the same location too. The continuity condition (2) yields the right-limit $U(d_1+, t)$ for $t \in [0, T_1]$. With $U(d_1+, t)$ as the inflow boundary condition and the initial condition in (1), one applies Equation (5) to solve problem (1) over the domain $[d_1, d_2] \times [0, T_1]$, except that the $g(t)$ in the last term on the right-hand side of (5) should be replaced by $U(d_1+, t)$.

(3) With $U(x, T_1)$ $x \in [a, d_1]$ as the initial condition and the inflow boundary condition in (1), apply Equation (5) to solve problem (1) over $[a, d_1] \times [T_1, T_2]$.

(4) Next one applies Equation (5) to solve problem (1) over $[a, d_1] \times [T_2, T_3]$, $[d_1, d_2] \times [T_1, T_2]$, and $[d_2, d_3] \times [T_0, T_1]$. Repeating this process one can obtain the solution $U(x, t)$ over the global domain $[a, b] \times [0, T]$.

It is easy to see that steps 2 and 3 can be performed in parallel. In general, one can solve problem (1) over $[a, d_1] \times [T_{k-1}, T_k]$, $[d_1, d_2] \times [T_{k-2}, T_{k-1}]$, ..., $[d_{k-1}, d_k] \times [T_0, T_1]$, in parallel. Thus, this algorithm actually defines a characteristic-based, noniterative, nonoverlapping, parallelized, domain decomposition algorithm. Secondly, note that this algorithm is well-defined independent of the space-time grids defined on each subdomain $[d_{l-1}, d_l] \times [T_{k-1}, T_k]$. When the solution has a sharp front within $[d_{l-1}, d_l] \times [T_{k-1}, T_k]$ and is smooth outside, one can use refined space-time grids only within this subdomain and coarse grids outside. Thus, this algorithm also gives a space-time local refinement method, which can resolve the sharp front accurately with reasonable computational cost.

4   Conforming/Nonconforming Matching of Interfacial Nodes

In this section we briefly discuss the matching of interfacial nodes. At the current time slab $[d_{l-1}, d_l] \times [t_n, t_{n+1}]$, the following three different partitions of interfaces can be used:

PARTITION 1: One can partition the interface $\{d_l\} \times [t_n, t_{n+1}]$ based on the magnitude of the Courant number $Cu_b = V_b \Delta t / \Delta x$ with $V_b = \max_{t \in [t_n, t_{n+1}]} V(b, t)$. In this case one defines the nodes $t_{n,i} \equiv t_{n+1} - (i - I)\Delta t / Cu_b = t_{n+1} - (i - I)\Delta x / V_b$ for $i = I, I+1, \ldots, I+IC$ and $t_{n,I+IC+1} = t_n$, where IC is the integer part of Cu_b if Cu_b is not an integer and $IC = Cu_b - 1$ otherwise [CRHE90].

PARTITION 2: One can define a uniform partition at the interface $\{d_l\} \times [t_n, t_{n+1}]$ based on the Courant number Cu_b, which is essentially the same as Partition 1.

Notice that partition 1 or 2 defines $U(d_l-, t)$ ($t \in [t_n, t_{n+1}]$) to be a piecewise-linear function on the nodes $t_n = t_{n,I+IC+1} < t_{n,I+IC} < \cdots < t_{n,I+1} < t_{n,I} = t_{n+1}$. On the other hand, the last term on the right-hand side of (5) actually defines the integral of $U(d_l+, t)\hat{w}_i$ on the interval $[t^*_{n,i+1}, t^*_{n,i-1}]$ where $t^*_{n,i}$ is given by $a = X(t^*_{n,i}; x_i, t_{n+1})$ for $i = 0, 1, \ldots, IC_a$ and $t^*_{n,IC_a+1} = t_n$ with IC_a being the integer part of $(\tilde{a} - a)/\Delta x$. In other words, $t^*_{n,i}$ $(i = 0, 1, \ldots, IC_a)$ is the time such that the approximate characteristic extending backward from x_i at time t^{n+1} meets the inflow boundary $x = a$ at $t^*_{n,i}$. Since the $t^*_{n,i}$ $(i = 0, 1, \ldots, IC_a)$ at the interface $\{d_l\} \times [t_n, t_{n+1}]$ are different from $t_{n,i}$ $(i = I, I+1, \ldots, I+IC)$, in general, one has to interpolate $U(d_l-, t)$ on a shifted grid $t_n = t^*_{n,IC+1} < t^*_{n,IC} < \cdots < t^*_{n,1} < t^*_{n,0} = t_{n+1}$ when one used Equation (5) to solve problem (1) on $[d_l, d_{l+1}] \times [t_n, t_{n+1}]$. Thus, Partition 1 or 2 defines a nonconforming matching.

PARTITION 3: We define the nodes $t_{n,I+i}$ from the subdomain $[d_l, d_{l+1}] \times [t_n, t_{n+1}]$ to be equal to $t^*_{n,i}$ from the subdomain $[d_{l-1}, d_l] \times [t_n, t_{n+1}]$. When we use Equation (5) to solve problem (1) over $[d_{l-1}, d_l] \times [t_n, t_{n+1}]$, we obtain $U(d_l-, t)$ defined on the grid $t_n = t^*_{n,IC+1} < t^*_{n,IC} < \cdots < t^*_{n,1} < t^*_{n,0} = t_{n+1}$, which is the same grids over which we need to compute the last integral on the right-hand side of (5). Thus, this partition gives a conforming matching.

5 Numerical Example

In this section we apply Equation (5) to solve problem (1) with discontinuous coefficients. The data are given as follows: the domain $D_1 = (-1, 0)$ and $D_2 = (0, 1)$, the velocity $V_1(x) = 2$ on D_1 and $V_2(x) = 1$ on D_2, $g(t) = 0.0$, the initial condition $u_0(x) = 1000(x + 0.15)^4(x + 0.85)^4$ for $x \in (-0.85, -0.15)$ and 0 otherwise.

Because of the discontinuity of $V(x, t)$ at $x = 0$, the interface condition (2) now reduces to $2u(0-, t) = u(0+, t)$ for $t \in [0, T]$. It is easy to see that the analytical solution of this problem is given by $u(x, t) = u_0(x - 2t)$ for $x \in (-1, 0)$ and $u(x, t) = 2u_0(2(x - t))$ for $x \in (0, 1)$. In Figures 1–2 the numerical solutions are plotted against the analytical ones for time $t = 0.1$, 0.25 and 0.8, respectively. In Figure 1 we used coarser grids on D_1 and finer grids on D_2. The Courant number is 8. One can see that the numerical solutions have been quite accurate. Similar conclusions can be drawn in Figure 2. Our other experiments, which are omitted here, show that when the time step Δt is relatively large, Partition 3 produces a slightly better solution than Partitions 1 and 2 (about 20 % less errors). On the other hand, Partitions 1 and 2 are more feasible and convenient to implement especially for nonlinear problems or multi-dimensional problems.

Acknowledgments

This research was funded in part by DOE DE-ACO5-840R21400, Martin Marietta, Subcontract SK965C and SK966V, by ONR N00014-94-1-1163, by NSF DMS-8922865, by VISTA, a research cooperation between Statoil and the Norwegian Academy of Science and Letters, and by the Norwegian Research Council.

References

[CRHE90] Celia M., Russell T., Herrera I., and Ewing R. (1990) An Eulerian-Lagrangian localized adjoint method for the advection-diffusion equation. *Advances in Water Resources* 13: 187–206.
[EW83] Ewing R.E. (ed.): The Mathematics of Reservoir Simulation. Frontiers in Applied Mathematics, Vol. 1. SIAM, Philadelphia, 1983.

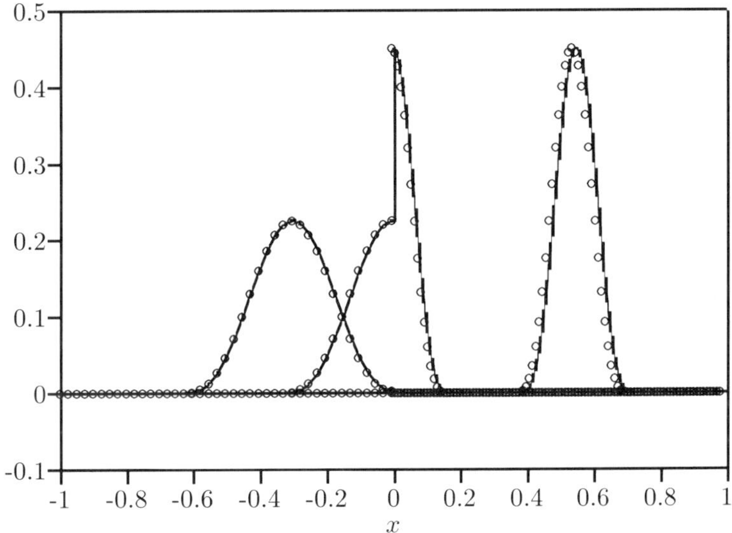

Figure 1 $\Delta t = 0.1$, $\Delta x_1 = 1/40$, $\Delta x_2 = 1/100$, Partition 3 used.

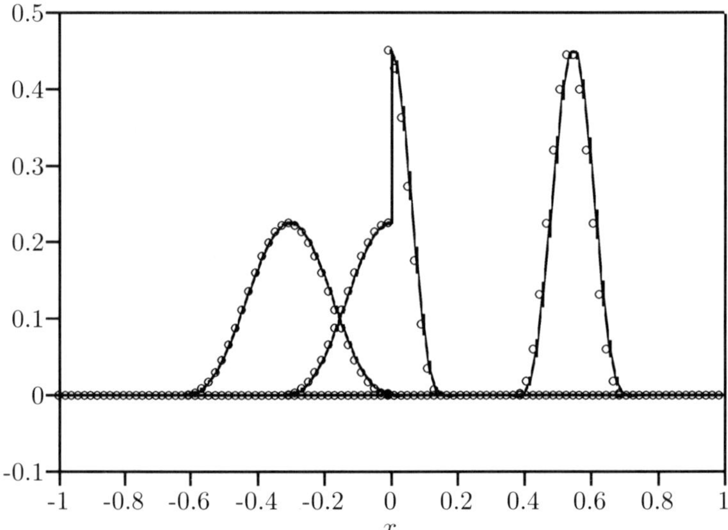

Figure 2 $\Delta t = 0.05$, $\Delta x_1 = 1/50$, $\Delta x_2 = 1/50$, Partition 1 used.

A Parallel Domain Decomposition Procedure for Convection Diffusion Problems

Junping Wang[1] and Ningning Yan [2]

1 Introduction

For simplicity we consider the following model convection-diffusion equation

$$
\begin{aligned}
-\nabla \cdot (\varepsilon \nabla u - \vec{\beta} u) + \alpha u &= f && \text{in } \Omega, \\
u &= g && \text{on } \Gamma = \partial\Omega,
\end{aligned}
\tag{1.1}
$$

where Ω is an open bounded domain on the plane, and α, f, g are given functions on Ω and Γ, $\vec{\beta}$ is a vector-valued function, and ε is a 2×2 matrix which is symmetric and positive definite.

In this paper, we assume that the problem (1.1) is convection-dominated so that the equation is of hyperbolic type and the solution possesses boundary and interior layers. Such problems are known to be difficult to discretize and to implement in practical computation.

Our objective here is to outline a stable finite element scheme and its parallel implementation by using domain decomposition techniques for (1.1). The finite element method is based on the standard discontinuous Galerkin procedure [10] combined with mixed finite element technique. The discretization scheme and some error estimates were discussed in [9]. We will report some new error estimates that have been derived in [12]. Our parallel domain decomposition algorithm [13] and its convergence analysis follow the work of Després, Joly, Robert [6] and Douglas, Paes Leme, Roberts, Wang [7] for the standard mixed finite element method.

The reader is referred to [12] and [13] for details of this research.

---

[0] This research was supported in part by NSF Grant No. INT-9309286.
[1] Department of Mathematics, University of Wyoming, Laramie, WY 82071
[2] Institute of Systems Science, Academia Sinica, Beijing 100080, China

Domain Decomposition Methods in Sciences and Engineering, edited by R. Glowinski *et al.*

2   A Mixed Discontinuous Galerkin Scheme

Without loss of generality, assume that the matrix-valued function ε is a (small) constant. Then a mixed formulation for (1.1) can be obtained by introducing a variable:

$$\vec{q} = -\varepsilon^{\frac{1}{2}} \nabla u.$$

The problem (1.1) is equivalent to seeking $(\vec{q}, u)$ with $u = g$ on Γ such that

$$\begin{aligned}
\vec{q} + \varepsilon^{\frac{1}{2}} \nabla u &= 0 && \text{in } \Omega, \\
\nabla \cdot (\varepsilon^{\frac{1}{2}} \vec{q}) + \nabla \cdot (\vec{\beta} u) + \alpha u &= f && \text{in } \Omega.
\end{aligned} \tag{2.1}$$

For each real parameter $h > 0$, let $\mathcal{T}_h$ be a finite element partition of $\bar{\Omega}$ consisting of triangles or quadrilaterals e with diameters bounded by h. To derive a variational form for (2.1), we introduce the following functional spaces:

$$\begin{aligned}
\mathbf{V} &= \{\vec{v} \in \left[L^2(\Omega)\right]^2, \nabla \cdot \vec{v} \in L^2(\Omega)\}, \\
W &= \{w \in L^2(\Omega), w|_e \in H^1(e), \forall e \in \mathcal{T}_h\}.
\end{aligned}$$

Also introduce the following linear and bilinear forms:

$$\begin{aligned}
A(\vec{q}, \vec{v}) &= (\vec{q}, \vec{v}), \vec{q}, \vec{v} \in \mathbf{V}, \\
B(\vec{q}, w) &= \varepsilon^{\frac{1}{2}} (\nabla \cdot \vec{q}, w), \vec{q} \in \mathbf{V}, w \in W, \\
D(u, w) &= \sum_{e \in \mathcal{T}_h} \left(- \int_e u \vec{\beta} \cdot \nabla w + \int_{\partial e_-} u_+ [w] \vec{n} \cdot \vec{\beta} \, ds\right) + (\alpha u, w) u, w \in W, \\
g(\vec{v}) &= -\varepsilon^{\frac{1}{2}} \int_\Gamma g \, \vec{v} \cdot \vec{n} \, ds \vec{v} \in \mathbf{V}, g \in H^{\frac{1}{2}}(\Gamma),
\end{aligned}$$

where $(\cdot, \cdot)$ is the standard inner product in $L^2(\Omega)$ or $\left[L^2(\Omega)\right]^2$ as appropriate, $\vec{n}$ is the outward normal direction on ∂e,

$$\partial e_- = \{l \in \partial e, \quad \vec{n} \cdot \vec{\beta}|_l < 0\},$$

and

$$[w] = w_+ - w_-, \quad w_+(\vec{x}) = \lim_{t \to 0^+} w(\vec{x} + t\vec{\beta}), \quad w_-(\vec{x}) = \lim_{t \to 0^-} w(\vec{x} + t\vec{\beta}).$$

With the above notation, a weak form for (2.1) which seeks $(\vec{q}, u) \in \mathbf{V} \times W$ such that

$$\begin{aligned}
A(\vec{q}, \vec{v}) - B(\vec{v}, u) &= g(\vec{v}), && \forall \vec{v} \in \mathbf{V}, \\
B(\vec{q}, w) + D(u, w) &= (f, w), && \forall w \in W.
\end{aligned} \tag{2.2}$$

It is not hard to verify that if the solution of (1.1) is smooth enough, then the problem (2.1) is equivalent to (2.2).

The Ritz-Galerkin procedure can be applied to yield a mixed discontinuous Galerkin method for (1.1). To this end, let $\mathbf{V}_h \times W_h \subset \mathbf{V} \times W$ be appropriately defined finite element spaces associated with $\mathcal{T}_h$. Then the Ritz-Galerkin approximation is the solution of the following linear system:

$$\begin{aligned}
A(\vec{q}_h, \vec{v}) - B(\vec{v}, u_h) &= g(\vec{v}), && \forall \vec{v} \in \mathbf{V}_h, \\
B(\vec{q}_h, w) + D(u_h, w) &= (f, w), && \forall w \in W_h.
\end{aligned} \tag{2.3}$$

Now we comment briefly on the construction of $\mathbf{V}_h \times W_h$. Since there is no continuity requirement for functions in W_h, it is natural to include in W_h piecewise polynomials only. Consequently, the standard mixed finite element spaces are good candidates for $\mathbf{V}_h$ in order to provide a stable and accurate approximation by using (2.3). If the diffusion effect is negligible, one could replace $\mathbf{V}_h$ by continuous piecewise polynomials. For more information on the construction of $\mathbf{V}_h \times W_h$, we refer to [5], [11], [4], [8]. Other possibilities can be found in [12].

If the mixed finite element spaces are employed in (2.3), then the following global error estimates can be derived [12]:

Theorem 2.1 *Let $(\vec{q}, u)$ be the unique solution of (2.2), and $(\vec{q}_h, u_h)$ be the finite element approximation by using the Raviart-Thomas element of order $i \geq 0$. Assume that $\vec{\beta} \in \mathbf{V}$, $\alpha \in L^\infty(\Omega)$, and $\alpha + \frac{1}{2}\nabla \cdot \vec{\beta} \geq \alpha_0 > 0$. Then,*

$$\|\vec{q} - \vec{q}_h\| + \|u - u_h\| \leq Ch^{i+1}(\|\vec{q}\|_{i+1} + h^{-\frac{1}{2}}\|u\|_{i+1}), \tag{2.4}$$

where $\|\cdot\|$ denotes the L^2-norm and $\|\cdot\|_{i+1}$ stands for the norm in $H^{i+1}(\Omega)$.

Interior error estimates are also available for the mixed discontinuous Galerkin method. Intuitively speaking, if the solution u does not change rapidly in Ω except in a small region of boundary layers, then one would have the following error estimate:

$$\|u - u_h\|_{L^2(D)} \leq C(\varepsilon^{1/2}h^{-1/2} + h^{1/2}),$$

where $D \subset \Omega$ is any subregion excluding the boundary layer of u and C is a constant independent of ε.

3 A Parallel Iterative Procedure

Our iterative algorithm can be considered as a modification of the parallel procedure studied in [7]. In this section, we outline the algorithm as well as some convergent results. Details of the analysis can be found in [13].

Let $\{\Omega_j, j = 1, \ldots, M\}$ be a partition of Ω:

$$\bar{\Omega} = \bigcup_{j=1}^{M} \bar{\Omega}_j : \qquad \Omega_j \cap \Omega_k = \emptyset, \ j \neq k.$$

Assume that $\partial\Omega_j, j = 1, \ldots, M$, is Lipschitz and that Ω_j is star-shaped. Set

$$\Gamma_j = \Gamma \cap \partial\Omega_j, \qquad \Gamma_{jk} = \Gamma_{kj} = \partial\Omega_j \cap \partial\Omega_k.$$

Let us consider decomposition of (2.1) or (2.2) over $\{\Omega_j\}$. In addition to requiring $\{\mathbf{q}_j, u_j\}, j = 1, \ldots, M$, to satisfy

$$\begin{aligned} \mathbf{q}_j + \epsilon^{\frac{1}{2}}\nabla u_j &= 0, & &\text{in } \Omega_j, \\ \epsilon^{\frac{1}{2}}\nabla \cdot \mathbf{q}_j + \nabla \cdot (\vec{\beta} u_j) + \alpha u_j &= f, & &\text{in } \Omega_j, \\ u_j &= g, & &\text{on } \Gamma_j, \end{aligned} \tag{3.1}$$

it is necessary to impose the consistency conditions

$$
\begin{aligned}
u_j &= u_k, & x \in \Gamma_{jk}, \\
\mathbf{q}_j \cdot \mathbf{n}_j + \mathbf{q}_k \cdot \mathbf{n}_k &= 0, & x \in \Gamma_{jk},
\end{aligned}
\tag{3.2}
$$

where $\mathbf{n}_j$ is the unit outward normal to Ω_j. It is convenient to replace (3.2) by the Robin boundary condition

$$
\begin{aligned}
-\eta \mathbf{q}_j \cdot \mathbf{n}_j + u_j &= \eta \mathbf{q}_k \cdot \mathbf{n}_k + u_k, & x \in \Gamma_{jk} \subset \partial\Omega_j, \\
-\eta \mathbf{q}_k \cdot \mathbf{n}_k + u_k &= \eta \mathbf{q}_j \cdot \mathbf{n}_j + u_j, & x \in \Gamma_{jk} \subset \partial\Omega_k,
\end{aligned}
\tag{3.3}
$$

where η is a positive (normally chosen to be a constant) function on $\bigcup \Gamma_{jk}$. Following an idea in [6] and [7], we can define a parallel iterative procedure for (3.1) by introducing Lagrange multipliers on the edges $\{\Gamma_{jk}\}$.

Let $\mathbf{V}_h \times W_h$ be a mixed finite element space over $\{\Omega_j\}$; any of the usual choices is acceptable, see [5], [11], [4], [3], [8]. Each of these spaces defined through local spaces $\mathbf{V}_j \times W_j = \mathbf{V}(\Omega_j) \times W(\Omega_j)$, and setting

$$
\begin{aligned}
\mathbf{V}_h &= \{\mathbf{v} \in H(div, \Omega) : \mathbf{v}|_{\Omega_j} \in \mathbf{V}_j\}, \\
W_h &= \{w : w|_{\Omega_j} \in W_j\}, \\
\Lambda_h &= \{\lambda : \lambda|_{\Gamma_{jk}} \in P_m(\Gamma_{jk}) = \Lambda_{jk}, \Gamma_{jk} \neq \emptyset\},
\end{aligned}
$$

where m is the order of polynomial on Γ_{jk} such that $\mathbf{q}_j \cdot \mathbf{n}_j|_{\Gamma_{jk}} \in \Lambda_{jk}$. Then the hybridized mixed discontinuous Galerkin method is given by seeking

$$
\{\mathbf{q}_j \in \mathbf{V}_j, u_j \in W_j, \lambda_{jk} \in \Lambda_{jk} : j = 1, \ldots, M; k = 1, \ldots, M\}
$$

such that for all $(\mathbf{v}, w, \mu) \in \mathbf{V}_j \times W_j \times \Lambda_{jk}$,

$$
(\mathbf{q}_j, \mathbf{v})_{\Omega_j} - \epsilon^{\frac{1}{2}}(\nabla \cdot \mathbf{v}, u_j)_{\Omega_j} + \epsilon^{\frac{1}{2}} \sum_k \langle \lambda_{jk}, \mathbf{v} \cdot \mathbf{n}_j \rangle_{\Gamma_{jk}} = -\epsilon^{\frac{1}{2}} \langle g, \mathbf{v} \cdot \mathbf{n}_j \rangle_{\Gamma_j},
$$

$$
\begin{aligned}
\epsilon^{\frac{1}{2}}(\nabla \cdot \mathbf{q}_j, w)_{\Omega_j} &+ d(u_j, w)_{\Omega_j} \\
&= (f, w)_{\Omega_j} - \langle g, \mathbf{n}_j \cdot \vec{\beta} w_+ \rangle_{\Gamma_{j-}} - \sum_k \langle u_{k-}, \mathbf{n}_j \cdot \vec{\beta} w_+ \rangle_{\Gamma_{jk-}}
\end{aligned}
\tag{3.4}
$$

$$
\langle \mu, \mathbf{q}_j \cdot \mathbf{n}_j + \mathbf{q}_k \cdot \mathbf{n}_k \rangle_{\Gamma_{jk}} = 0.
$$

Substitute u_j and u_k in (3.3) by λ_{jk} and λ_{kj}, so that

$$
\langle \lambda_{jk}, \mathbf{v} \cdot \mathbf{n}_j \rangle_{\Gamma_{jk}} = \langle \eta(\mathbf{q}_j \cdot \mathbf{n}_j + \mathbf{q}_k \cdot \mathbf{n}_k) + \lambda_{kj}, \mathbf{v} \cdot \mathbf{n}_j \rangle_{\Gamma_{jk}}.
\tag{3.5}
$$

Then, the iterative process can be defined as follows: let, for all j and k,

$$
\mathbf{q}_j^0 \in \mathbf{V}_j, \quad u_j^0 \in W_j, \quad \lambda_{jk}^0 \in \Lambda_{jk}
$$

arbitrarily. ($\lambda_{jk}^0 = \lambda_{kj}^0$ seems natural) and then compute $\{\mathbf{q}_j^n, u_j^n, \lambda_{jk}^n\} \in \mathbf{V}_j \times W_j \times \Lambda_{jk}$

recursively as the solution of the equations

$$(\mathbf{q}_j^n, \mathbf{v})_{\Omega_j} - \epsilon^{\frac{1}{2}}(\nabla \cdot \mathbf{v}, u_j^n)_{\Omega_j} + \epsilon^{\frac{1}{2}} \sum_k \langle \eta \mathbf{q}_j^n \cdot \mathbf{n}_j, \mathbf{v} \cdot \mathbf{n}_j \rangle_{\Gamma_{jk}}$$

$$= -\epsilon^{\frac{1}{2}} \sum_k \langle \eta \mathbf{q}_k^{n-1} \cdot \mathbf{n}_k + \lambda_{kj}^{n-1}, \mathbf{v} \cdot \mathbf{n}_j \rangle_{\Gamma_{jk}} - \epsilon^{\frac{1}{2}} \langle g, \mathbf{v} \cdot \mathbf{n}_j \rangle_{\Gamma_j}, \forall \mathbf{v} \in \mathbf{V}_j,$$

$$\epsilon^{\frac{1}{2}}(\nabla \cdot \mathbf{q}_j^n, w)_{\Omega_j} + d(u_j^n, w)_{\Omega_j}$$

$$= (f, w)_{\Omega_j} - \langle g, \mathbf{n}_j \cdot \vec{\beta} w_+ \rangle_{\Gamma_{j-}} - \sum_k \langle u_{k-}^{n-1}, \mathbf{n}_j \cdot \vec{\beta} w_+ \rangle_{\Gamma_{jk-}}, \forall w \in W_j,$$

$$\lambda_{jk}^n = \eta(\mathbf{q}_j^n \cdot \mathbf{n}_j + \mathbf{q}_k^{n-1} \cdot \mathbf{n}_k) + \lambda_{kj}^{n-1}.$$

$$(3.6)$$

Note that in (3.6), the first and the second equations are independent of λ_{jk}^n. Thus, λ_{jk}^n can be evaluated by the third equation after the determination of $\mathbf{q}_j^n$ and u_j^n.

The following convergence result has been obtained for the above algorithm.

Theorem 3.1. *If $\alpha + \frac{1}{2}\nabla \cdot \vec{\beta} \geq 0$ holds, then the iterate solution $\{\mathbf{q}_j^n, u_j^n, \lambda_{jk}^n\} \in \mathbf{V}_j \times W_j \times \Lambda_j$ of (3.6) converges to the solution $\{\mathbf{q}_j, u_j, \lambda_{jk}\}$ of the global hybridized mixed discontinuous Galerkin procedure (3.4) in the following senses:*

$$\begin{aligned}
\mathbf{q}_j^n \to \mathbf{q}_j &= \mathbf{q}^*|_{\Omega_j} &\quad in \quad & L^2(\Omega_j), \\
u_j^n \to u_j &= u^*|_{\Omega_j} &\quad in \quad & L^2(\Omega_j), \\
\lambda_{jk}^n \ and \ \lambda_{kj}^n &\to \lambda_{jk} &\quad in \quad & L^2(\Gamma_{jk}),
\end{aligned}$$

where $\{\mathbf{q}^*, u^*\} \in \mathbf{V}_h \times W_h$ is the solution of the global mixed discontinuous Galerkin method (2.3).

To estimate the rate of convergence, let $T_{f,g}$ be the affine mapping from $\mathbf{V}_h \times W_h \times \Lambda_h$ to itself such that, for any $(\mathbf{s}, p, \theta) \in \mathbf{V}_h \times W_h \times \Lambda_h$, $(\mathbf{r}, e, \mu) \equiv T_{f,g}(\mathbf{s}, p, \theta)$ is the solution of the following linear system:

$$(\mathbf{r}_j, \mathbf{v})_{\Omega_j} - \epsilon^{\frac{1}{2}}(\nabla \cdot \mathbf{v}, e_j)_{\Omega_j} + \epsilon^{\frac{1}{2}} \sum_k \langle \eta \mathbf{r}_j \cdot \mathbf{n}_j, \mathbf{v} \cdot \mathbf{n}_j \rangle_{\Gamma_{jk}}$$

$$= -\epsilon^{\frac{1}{2}} \sum_k \langle \eta \mathbf{s}_k \cdot \mathbf{n}_k + \theta_{kj}, \mathbf{v} \cdot \mathbf{n}_j \rangle_{\Gamma_{jk}} - \epsilon^{\frac{1}{2}} \langle g, \mathbf{v} \cdot \mathbf{n}_j \rangle_{\Gamma_j} \qquad \forall \mathbf{v} \in \mathbf{V}_j,$$

$$\epsilon^{\frac{1}{2}}(\nabla \cdot \mathbf{r}_j, w)_{\Omega_j} + d(e_j, w)_{\Omega_j}$$

$$= (f, w)_{\Omega_j} - \langle g, \mathbf{n}_j \cdot \vec{\beta} w_+ \rangle_{\Gamma_{j-}} - \sum_k \langle p_{k-}, \mathbf{n}_j \cdot \vec{\beta} w_+ \rangle_{\Gamma_{jk-}} \qquad \forall w \in W_j,$$

$$\mu_{jk} = \eta(\mathbf{r}_j \cdot \mathbf{n}_j + \mathbf{s}_k \cdot \mathbf{n}_k) + \theta_{kj}.$$

$$(3.7)$$

The convergence rate of the iterative scheme can be characterized by the spectral radius of the linear operator $T_{0,0}$ [7].

Theorem 3.2. *Let $\rho(T_0)$ be the spectral radius of $T_0 = T_{0,0}$. Assume that*

$$\alpha + \frac{1}{2}\nabla \cdot \vec{\beta} \geq 0,$$

then

$$\rho(T_0) < 1.$$

Thus, the iterative procedure (3.6) is convergent.

Theorem 3.3 *Assume that there exists an $\alpha_0 > 0$ such that*

$$\alpha + \frac{1}{2}\nabla \cdot \vec{\beta} \geq \alpha_0.$$

Assume that the partition $\{\Omega_j\}$ is quasiregular, and the parameter η in the iterative procedure (3.6) satisfies $\eta = \sqrt{h(C_1\epsilon^{-1} + \rho\alpha_0^{-1})}$. Then,

$$\rho(T_0) \leq 1 - \frac{Ch}{\sqrt{C_1 h + \epsilon h\rho\alpha_0^{-1} + \alpha_0^{-1}|\gamma|^{-2}}} = \gamma_0,$$

where $\rho = \max_j |\partial\Omega_j|/|\Omega_j|$, γ is the eigenvalue of T_0. The iteration (3.6) converges with an error in the nth iteration bounded asymptotically by $O(\gamma_0^n)$.

The following are some particular cases of the domain decomposition.

Case One: Assume the triangulation Ω_j of Ω into elements to be quasiregular and that the subdomains in the domain decomposition to coincide with $\{\Omega_j\}$. In addition, assume $\alpha_0 = O(1)$. Then, by choosing the parameter $\eta = O(\epsilon^{-\frac{1}{2}})$, it follows that (3.6) converges with rate bounded by $\gamma_0 = 1 - Ch$.

Case Two: Let us consider the convection-diffusion problem with "good" convective direction in the sense that any streamline passes through only a finite number of subdomains. Also assume that $\alpha_0 = O(1)$ and $|\Omega_j| = O(1)$. Then, choosing $\eta = O(\epsilon^{-\frac{1}{2}}h^{\frac{1}{2}})$, leads to the estimate $\gamma_0 = 1 - C\sqrt{h}$.

Case Three: Suppose $\epsilon \leq h^{1+\omega}$ with $\omega \geq 0$. If the convective direction $\vec{\beta}$ is "good", and $\alpha_0 = O(1)$, $|\Omega_j| = O(1)$. By choosing $\eta = O(\epsilon^{-\frac{1}{2}}h)$, one arrives at the following estimate

$$\gamma_0 = 1 - \frac{Ch}{\sqrt{h^2 + \epsilon}} = \begin{cases} 1 - Ch^{\frac{1-\omega}{2}}, & 0 \leq \omega < 1; \\ 1 - c\,(c < 1), & \omega > 1, \end{cases}$$

which shows a uniform convergence for the domain decomposition algorithm when $\epsilon \leq h^2$.

REFERENCES

[1] Babuška I. (1973) The finite element method with Lagrangian multipliers. *Numer. Math.*, 20:179–192.

[2] Brezzi F. (1974) On the existence, uniqueness, and approximation of saddle point problems arising from Lagrangian multipliers. *R.A.I.R.O.*, 8: 129–151.

[3] Brezzi F., Douglas J., Jr., Duran R., and Fortin M. (1987) Mixed finite elements for second order elliptic problems in three variables. *Numer. Math.*, 51:237–250.

[4] Brezzi F. , Douglas J. Jr., Fortin M., and Marini L. D. (1987) Efficient rectangular mixed finite elements in two and three space variables. *M^2AN*, 21:581–604.

[5] Brezzi F. and Fortin M. (1991) *Mixed and Hybrid Finite Element Methods,* Springer-Verlag, New York.

[6] Després B., Joly P. and Roberts J. E. (1990) Domain decomposition method for harmonic Maxwell's equations. *Proceedings of IMACS international symposium on iterative methods in linear algebra,* Elsevier, North Holland.

[7] Douglas J. Jr, Paes Leme P. J., Roberts J. E. and Wang J. (1993) A parallel iterative procedure applicable to the approximate solution of second order partial differential equations by mixed finite element methods. *Numer. Math.,* 65:95–108.

[8] Douglas J. Jr. and Wang J. (1993) A new family of mixed finite element spaces over rectangles. *Comp. Appl. Math.,* 12:183–197.

[9] Jaffre J. (1984) Decentrage et elements finis mixtes four les equations de diffusion-convection. *Calcolo,* XXI, 171–197.

[10] Johnson C. and Pitkränta J. (1986) An analysis of the discontinuous Galerkin method for scalar hyperbolic equation. *Math. Comp.,* 46:1–26.

[11] P.-A. Raviart and J.-M. Thomas (1977) A mixed finite element method for second order elliptic problems, Mathematical Aspects of the Finite Element Method. *Lecture Notes in Mathematics 606,* Springer-Verlag, Berlin and New York, 292–315.

[12] Wang J., Yan N., and Liu M. New finite element discretizations and error estimates for convection-dominated diffusion problems, in preparation.

[13] Wang J. and Yan N. A parallel domain decomposition iterative method for approximate solutions of convection-diffusion problems by finite element methods, in preparation.

A New Domain Decomposition Method for Convection–Dominated Problems

Guohui Zhou

Institut für Angewandte Mathematik, Universität Heidelberg, D–69120 Heidelberg, Germany. E–mail: zhou@gaia.iwr.uni-heidelberg.de

Summary. In this article, we present a new domain decomposition method for solving convection–diffusion problems with dominant convection. We combine the sequential algorithm with a parallel strategy to adapt to the special properties of the equations. The sequential algorithm is used in the downwind direction while the parallel algorithm is applied in the crosswind direction. In both algorithms, an overlapping domain decomposition process is introduced. In each patch, we solve a local convection–dominated problem with artificial boundary conditions by the streamline diffusion finite element method. A globally continuous discrete solution is constructed from all the patch-wise solutions. It is proven that this global solution converges to the exact solution with an order of $O(h^{3/2})$ in the L^2–norm as long as the overlapping width is kept to $O(h^{3/4}|\log h|)$ in the crosswind direction and to $O(h|\log h|)$ in the downwind direction. It is emphasized that we do not require the patch size to be sufficiently small and do not iterate the whole process as usually done in other domain decomposition approaches.

1 Introduction

Since the domain decomposition method was proposed years ago, many papers have been published for solving symmetric positive definite elliptic problems. This method has been demonstrated to have a good condition number as an iteration method and can also be implemented in a parallel way. For nonstationary convection–dominated problems, a domain splitting method was proposed and analyed in [6]. The proper convergence order was there achieved without conditions on the macroelements. For stationary nonsymmetric problems, there are also some papers, e.g., [5] and [1], in which the convection term is treated as a perturbation of the elliptic part. In those

Domain Decomposition Methods in Sciences and Engineering, edited by R. Glowinski *et al.*

papers, the coefficient of the diffusion is not allowed to be very small in comparison with that of the convection, i.e., the Peclet number should not be too large. They must enforce an inconvenient mesh condition that the coarse mesh must be fine enough. Looking at the analysis carefully, one can see that the coarse mesh size must be proportional to the inverse of the Peclet number. This is almost impossible for convection–dominated problems.

In Kapurkin and Lube [4], a modified Schwarz iteration method for convection–dominated problems was discussed. An overlapping domain decomposition technique was applied and a convergence proof was given for the continuous problems. Their numerical experiments already showed that the iteration number depends on the number of subdomains in the flow direction. From the theory of partial differential equations, it is obvious for convection–dominated problems that any change in the inflow boundary condition influences the solution at any point in the downwind direction decisively. The discrete solution in a subdomain must wait for its inflow boundary condition from the upstream–neighboring subdomain. Before the approximate inflow boundary condition comes from the neighboring subdomain, any artificial boundary condition in this subdomain gives only a false approximation and hence wastes time and effort.

For convection–dominated problems, we propose a new domain decomposition technique which avoids any unnecessary iterations. Simply speaking, we take sequential computations for subdomains in the downwind direction while we take parallel algorithms for subdomains in the crosswind direction. Since we cannot compute the discrete solution in the convection direction in a parallel way, the sequential algorithm in this direction is proposed only to reduce the size of the subproblems. In fact, we could use one whole strip from the inflow boundary to the outflow boundary if one would like to solve a comparatively bigger problem instead of several smaller ones.

In order to reduce the overlapping widths, we could also iterate the above process. One can investigate the dependence of the condition number on the iteration. This will be done in a forthcoming paper.

Test computations for model problems show that the proposed domain decomposition method works very well. The overlapping widths are usually limited to $3h$ in the downstream direction and to $h|\log h|$ in the crosswind direction. This logarithmic dependence occurs only in the crosswind direction. On the other hand, the test computations show that the overlapping sizes in the two directions are not influenced by the number of macro–elements, though we have linear dependence in our theoretic analysis. The discrete solution constructed by a global process converges usually with an order of $O(h^2)$. Since the sub–problems on each patch are small in comparison with the global one, the iterations for solving each algebraic system are significantly reduced.

2   A Domain Decomposition Technique

As an illustrative example, we consider the model problem

$$- \rho u_{xx} - \varepsilon u_{yy} + u_x + u = f, \qquad \text{in} \quad \Omega, \qquad (2.1.a)$$

$$u = g, \qquad \text{on} \quad \partial\Omega, \qquad (2.1.b)$$

where $\Omega = (0,1) \times (0,1)$ in R^2, ρ and ε are small non-negative parameters. If ρ or ε is zero, we should specify the boundary condition only on a part of the boundary $\partial\Omega$.

Since the standard finite element method gives an oscillating discrete solution if the boundary condition g or the source term f is not smooth, the streamline diffusion finite element method (SDFEM) was proposed to damp away the possible over– and undershootings. Let $\mathbf{V}_h \subset H^1(\Omega)$ be a finite element subspace, in which linear or bilinear functions are defined on each element. The functions in $\mathbf{V}_h^0$ have homogeneous boundary conditions. By modifying the test function as $W + \delta W_x$ with $\delta = O(h)$, the SDFEM reads: Find $V \in \mathbf{V}_h(\Omega)$, such that

$$\rho(V_x, W_x) + \varepsilon(V_y, W_y) + (V_x + V, W + \delta W_x) = (f, W + \delta W_x), \quad \forall W \in \mathbf{V}_h^0(\Omega),$$
$$V = G, \qquad\qquad \text{on } \partial\Omega,$$

where G is some projection of g on $\mathbf{V}_h$ and the term $\rho V_{xx} + \varepsilon V_{yy}$ is neglected for a linear or bilinear ansatz. Since the diffusion coefficient ε is very small, or even zero, we introduce an artificial diffusion $\varepsilon_m = O(h^\alpha)$ with $\frac{3}{2} \le \alpha \le 2$, following an idea of Johnson et al. [3]. The order of ε_m depends on the localization property of the scheme, which will be determined later. Defining the bilinear form

$$B(V, W) = ((\rho + \delta)V_x, W_x) + \varepsilon_m(V_y, W_y) + (V_x, W) + (V, W + \delta W_x),$$

and a linear functional
$$L(W) = (f, W + \delta W_x),$$

we work with the discrete problem of finding $V \in \mathbf{V}_h$ such that

$$B(V, W) = L(W), \qquad \forall W \in \mathbf{V}_h^0(\Omega), \tag{2.2.a}$$
$$V = G, \qquad \text{on } \partial\Omega. \tag{2.2.b}$$

We associate the bilinear form $B(\cdot, \cdot)$ with the following energy norm

$$\|W\|^2 = \delta\|W_x\|^2 + \varepsilon_m\|W_y\|^2 + \|W\|^2. \tag{2.3}$$

It is easy to verify that

$$B(W, W) \ge \frac{1}{2}\|W\|^2, \quad \forall W \in \mathbf{V}_h^0(\Omega), \tag{2.4}$$

which gives the global stability of the SDFEM:

$$\|V\| \le 2\|f\| + C\|g\|_{\partial\Omega}. \tag{2.5}$$

To distinguish the type of boundaries, we introduce the following definitions for the domain Ω. Let $\mathbf{n} = (n_x, n_y)$ be the outer normal along the boundary $\partial\Omega$. Its inflow boundary $\partial\Omega_-$, outflow boundary $\partial\Omega_+$ and stationary boundary $\partial\Omega_0$ are defined by

$$\partial\Omega_- = \{(x, y) \in \partial\Omega, \; n_x(x, y) < 0\},$$
$$\partial\Omega_+ = \{(x, y) \in \partial\Omega, \; n_x(x, y) > 0\},$$
$$\partial\Omega_0 = \{(x, y) \in \partial\Omega, \; n_x(x, y) = 0\}.$$

The solution of problems with dominant convection has significantly different properties along these types of boundaries.

Problem (2.1) depicts a motion from the left to the right with small diffusion. A discretization method should cover this property of the equation. For parallel computation, we divide the domain Ω in patches with two families of lines which are parallel and orthogonal to the characteristics, respectively. In the present case, these two families of lines are $\{x = \text{const.}\}$ and $\{y = \text{const.}\}$. We let

$$\Omega_{i,j} = \Big\{(x,y): \ x_{i-1} \leq x \leq x_i, \ y_{j-1} \leq y \leq y_j\Big\},$$

for $i = 1, \ldots, M$ and $j = 1, \ldots, N$. Here, we have $x_0 = 0$, $x_M = 1$, $y_0 = 0$ and $y_N = 1$. Define $\Omega_j = \cup_{i=1}^{M} \Omega_{i,j}$.

The main idea for solving convection–dominated problem is as follows. In the crosswind direction we solve subproblems in a parallel manner, while in the downwind direction we solve subproblems in a sequential manner. To understand this scheme, we assume that there are N processors. The j–th processor solves subproblems in the "big" patch Ω_j. In the strip Ω_j, we solve subproblems for $i = 1, \ldots, M$ sequentially. We take overlaps in the crosswind and downwind direction, but with different overlapping sizes. Let d_x^i and d_y^i be the overlapping sizes of $\Omega_{i,j}$ in the downwind and crosswind direction, respectively. Thus, we define the overlapping patches (see Figure 1) as

$$\Omega_{i,j}^d = \Big\{(x,y): \ x_i \leq x \leq x_{i+1} + d_x^i, \ y_j - d_y^i \leq y \leq y_{j+1} + d_y^i\Big\} \cap \Omega.$$

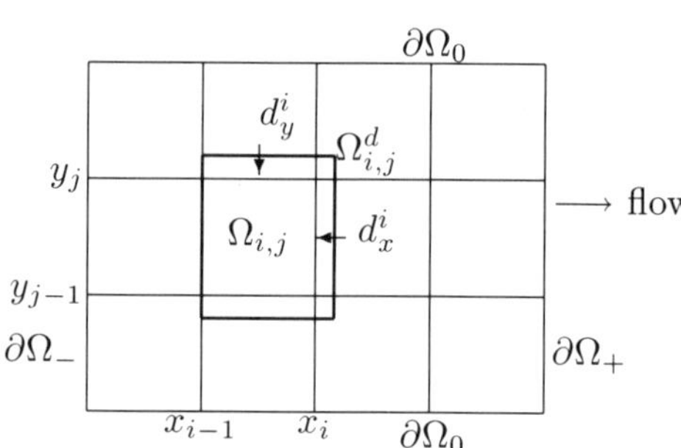

Figure 1 Subdomain $\Omega_{i,j}^d$

Now we can describe our domain decomposition scheme as follows.

Step 1: Assume that we have the discrete solution $U_{i-1,j}$ on the boundary $x = x_i$ with $U_{0,j} = G$. For $i = 1, \ldots, M$ and $j = 1, \ldots, N$, we find $U_{i,j} \in \mathbf{V}_h(\Omega_{i,j}^d)$, such that

$$B(U_{i,j}, W) = L(W), \qquad \forall W \in \mathbf{V}_h^0(\Omega_{i,j}^d), \tag{2.6.a}$$

$$U_{i,j} = U_{i-1,j}, \qquad \text{on } \partial(\Omega_{i,j}^d)_-, \tag{2.6.b}$$

$$U_{i,j} = 0, \qquad \text{on } \partial(\Omega_{i,j}^d)_+ \cup \partial(\Omega_{i,j}^d)_0. \tag{2.6.c}$$

Remark 1 *For $j = N$, $i = 1$ or $i = M$, we need to modify the homogeneous boundary conditions to the global boundary condition. We specify $U_{i,j}(x,y) = G(x,y)$ if (x,y) is located on $\partial\Omega$.*

Remark 2 *For a fixed i, we can compute the patchwise solutions in parallel with $j = 1, \dots, N$, but for a fixed j we must compute them sequentially with $i = 1, \dots, M$.*

Step 2: We assemble the patchwise solutions to a globally continuous solution $U \in \mathbf{V}_h(\Omega)$ by an averaging process in the nodal points z of T_h:

$$U(z) = U_{i,j}(z), \qquad \text{for } z \in \Omega_{i,j} \backslash \partial\Omega_{i,j}, \tag{2.7.a}$$

$$U(z) = \text{averaging}, \qquad \text{for } z \text{ on the common boundaries of } \Omega_{i,j} \tag{2.7.b}$$

3   Error Analysis for the DD Scheme

By Splitting $u - U = (u - V) + (V - U)$ with V defined in (2.2), we see that the estimate of the error $u - U$ between exact solution and the discrete solution of the domain decomposition scheme leads back to the error $U - V$, since the estimate of the first part can be found in a standard book (see, e.g., Johnson [2]). For the L^2–norm, we have

$$\|U - V\|^2 = \sum_{i,j} \|U - V\|_{\Omega_{i,j}}^2 \leq C \sum_{i,j} \|U_{i,j} - V\|_{\Omega_{i,j}}^2. \tag{3.1}$$

Obviously, we have $\mathbf{V}_h^0(\Omega_{i,j}^d) \subset \mathbf{V}_h^0(\Omega)$. Setting $E_{i,j} = V - U_{i,j}$, we subtract equation (2.6) from equation (2.2) and get

$$B(E_{i,j}, W) = 0, \qquad \forall W \in \mathbf{V}_h^0(\Omega_{i,j}^d), \tag{3.2.a}$$

$$E_{i,j} = E_{i-1,j}, \qquad \text{on } \partial(\Omega_{i,j}^d)_-, \tag{3.2.b}$$

$$E_{i,j} = V, \qquad \text{on } \partial(\Omega_{i,j}^d)_+ \cup \partial(\Omega_{i,j}^d)_0. \tag{3.2.c}$$

To estimate $E_{i,j}$ in (3.2), we state a lemma, the proof of which will be omitted.

Lemma 1 *Suppose that E satisfies*

$$B(E, W) = 0, \qquad \forall W \in \mathbf{V}_h^0(\omega^d), \tag{3.3.a}$$

$$E = G, \qquad \text{on } \partial\omega^d. \tag{3.3.b}$$

If the triangulation in $\omega^d \backslash \omega$ is quasi–uniform, then the choices of the streamline diffusion and the artificial diffusion must satisfy

$$\delta \geq Ch, \quad \varepsilon_m \geq Ch^{3/2}. \tag{3.4}$$

Then there exist constants $C > 0$ and $\theta > 0$, independent of h, ρ and ε, such that the function E in the subdomain ω admits the estimate

$$\|E\|_\omega^2 \leq C \left(\|G\|_{\partial\omega^d}^2 + e^{-\frac{\theta d_x}{\delta}} \|G\|_{\partial\omega_+^d}^2 + e^{-\frac{\theta d_y}{\sqrt{\varepsilon}m}} \|G\|_{\partial\omega_0^d}^2 \right). \tag{3.5}$$

Applying this lemma to (3.2) on the subdomain $\Omega_{i,j}^d$ and noting that $\partial(\Omega_{i,j})_-^d \subset \Omega_{i-1,j}$, we have

$$\|E_{i,j}\|_{\Omega_{i,j}}^2 \leq C \left(\|E_{i-1,j}\|_{\partial(\Omega_{i,j})_-^d}^2 + e^{-\frac{\theta d_x^i}{\delta}} \|V\|_{\partial(\Omega_{i,j})_+^d}^2 + e^{-\frac{\theta d_y^i}{\sqrt{\varepsilon}m}} \|V\|_{\partial(\Omega_{i,j})_0^d}^2 \right)$$

$$\leq C_0 h^{-1} \|E_{i-1,j}\|_{\Omega_{i-1,j}}^2 + C_1 \left(e^{-\frac{\theta d_x^i}{\delta}} \|V\|_{\partial(\Omega_{i,j})_+^d}^2 + e^{-\frac{\theta d_y^i}{\sqrt{\varepsilon_m}}} \|V\|_{\partial(\Omega_{i,j})_0^d}^2 \right).$$

Taking into account that $E_{0,j} \equiv 0$ for $j = 1, \ldots, N$, and this into (3.1), we get

$$\|U - V\|^2 \leq C \sum_{i=1}^{M} h^{-(M-i)} \left(e^{-\frac{\theta d_x^i}{\delta}} + e^{-\frac{\theta d_y^i}{\sqrt{\varepsilon_m}}} \right) \sum_{j=1}^{N} \|V\|_{\Omega_{i,j}}^2.$$

For any $\nu \geq 3/2$, we can clearly choose d_x^i and d_y^i as

$$d_x^i = (M - i + 2\nu)\theta^{-1}\delta|\log h|, \quad d_y^i = (M - i + 2\nu)\theta^{-1}\sqrt{\varepsilon_m}|\log h|, \tag{3.6}$$

and we obtain by the global stability estimate (2.5)

$$\|U - V\|^2 \leq Ch^{2\nu}\|V\|^2 \leq Ch^{2\nu}\left(\|f\|^2 + \|g\|_{\partial\Omega}^2\right). \tag{3.7}$$

Summarizing the proof above and combining the standard estimate for $u - V$, we can state our convergence result.

Theorem 1 *Suppose that the triangulations $T_h(\Omega_{i,j})$ are quasi–uniform for any i, j, and that the diffusion coefficients satisfy*

$$\rho \leq Ch, \quad \varepsilon \leq Ch^{3/2}.$$

We specify the parameters δ and ε_m in the SDFEM by

$$\delta = Ch, \quad \varepsilon_m = Ch^{3/2}.$$

For any $\nu \geq 3/2$, we take the overlapping widths in the streamline and crosswind directions as

$$d_x^i = c(M - i + 2\nu)h|\log h|, \quad d_y^i = c(M - i + 2\nu)h^{3/4}|\log h|.$$

Then the discrete solution achieved by the domain decomposition scheme (2.6) and (2.7) admits the error estimate

$$\|u - U\| \leq Ch^{3/2}\|u\|_{H^2(\Omega)} + Ch^{\nu}\left(\|f\| + \|g\|_{\partial\Omega}\right). \tag{3.8}$$

Remark 3 *If we orient the quadrilateral mesh in the streamline direction and take it almost equidistant in the direction, we can improve the convergence order in the L^2–norm to $O(h^2)$ and reduce the overlapping width in the crosswind direction to $O(h|\log h|)$ (see [7] for the techniques of the proof).*

4   Numerical Results

The theoretical results of the preceding sections have been verified through various test computations. It turns out that the overlapping widths d_x and d_y actually do not depend on the number of the macro elements, though theoretical convergence results show such a dependence. All tests have been performed on the unit square and the equidistant $M \times M$ macro–elements are used. To verify the order of convergence, we choose a smooth solution.

Example 1: We consider the model problem

$$-\varepsilon \Delta u + u_x + 2u = f, \qquad \text{in } \Omega, \qquad (4.1.\text{a})$$
$$u = 0, \qquad \text{on } \partial\Omega, \qquad (4.1.\text{b})$$

with the exact solution $u = 4x(1 - x)y(1 - y)$. We take $\varepsilon = 10^{-10}$ for all the computations.

Test computations show that the number of overlapping elements in the downstream direction (here, x) is independent of the size of the fine meshes, though we theoretically have a logarithmic dependence on the fine mesh size. At each refinement level, for fixed overlapping size in the crosswind direction, the error does not decrease when d_x goes over $3h_x$. This means that the choice $d_x = 3h$ suffices to control the error of overlapping in the streamline direction. However, this logarithmic dependence on the fine mesh size is visible in the number of overlapping elements in the crosswind direction. While refining the fine mesh, we must increase the crosswind–overlapping size d_y logarithmically in order to control the error of the overlappings. In the theoretical analysis we have also that d_x and d_y are proportional to the number of macro–elements in the streamline direction. But in the test computations, we have not seen any relations between the overlapping sizes and the number of macro–elements.

In order to illustrate the error of the domain decomposition method, we first give the L^2–error and the corresponding order of convergence with the SDFEM without domain decompositions. We set $h = 2^{-N}$ with the refinement level N.

N	5	6	7	8	9	10	11
Error	4.68(-4)	1.18(-4)	2.95(-5)	7.39(-6)	1.85(-6)	4.63(-7)	1.16(-7)
Order		1.99	2.00	2.00	2.00	2.00	2.00

Table 1 The error and the convergence order without DD

In the following tables, we specify the overlapping size $d_y = Lh$ in the crosswind direction with the constant L. Table 2 represents the error for different refinement levels. The constant L shown in the tables is the smallest necessary number of overlapping elements to suppress the error of the domain decomposition.

N	5	6	7	8	9	10
L	4	4	5	6	7	8
Error	4.68(-4)	1.18(-4)	2.95(-5)	7.40(-6)	1.85(-6)	4.63(-7)

Table 2 Results for a 2×2 decomposition

Table 3 shows that the error does not depend on the number of macro-elements and that the number of overlapping elements in the crosswind direction increases

N	7	8	9	10	11
L	5	7	7	8	8
Error	2.97(-5)	7.40(-6)	1.85(-6)	4.63(-7)	1.17(-7)

Table 3   Results for a 32×32 decomposition

logarithmically with the refinement level. Comparing the errors in Tables 2 and 3 with those given in Table 1 for a fixed N, we conclude that the extra error caused by the domain decomposition is fully suppressed by appropriate overlapping sizes d_x and d_y and that the convergence is of second order. We can surely construct a special triangular mesh which gives an $O(h^{3/2})$–convergence. This was discussed in a previous paper [8].

The domain decomposition method has another great advantage. One can refine a macro–element to cope with the local property of the solution. For some solutions with boundary layers or interior layers, we must use locally refined mesh to reduce the global error. This is complicated to implement if one uses only one mesh over the domain. Using the domain decomposition method, it is easily perform by specifying a much finer mesh in some subregions.

Acknowledgement The author thanks Professor R. Rannacher and Dr. Xue-cheng Tai for many useful discussions.

References

[1] Xiao-chun Cai, O.B.Widlund, Domain decomposition algorithms for indefinite elliptic problems, SIAM, J. Sci. Stat. Comput., 13, 243–258(1992)

[2] C.Johnson, Numerical Solution of Partial Differential Equations by the Finite Element Method, Cambridge University Press, Cambridge, 1987

[3] C.Johnson, A.H.Schatz, L.B.Wahlbin, Crosswind smear and pointwise errors in streamline diffusion finite element methods, Math. Comp., 49, 25-38(1987)

[4] A.Kapurkin, G.Lube, A domain decomposition for singularly perturbed elliptic problems, NAM-Bericht Nr. 70, Göttingen University, Germany, 1994

[5] Junping Wang, Convergence analysis of Schwarz algorithm and multilevel decomposition iterative methods II: nonselfadjoint and indefinite elliptic problems, SIAM, J. Numer. Anal., 30, 953–970(1993)

[6] R.Rannacher, Guohui Zhou, Analysis of a Domain–Splitting Method for Nonstationary Convection–Diffusion Problems, East–west J. Numer. Math., 2, 151–172 (1994)

[7] Guohui Zhou, R.Rannacher, Pointwise superconvergence of the streamline diffusion finite element method, Preprint No. 94–72, SFB 359, Heidelberg University, 1994, to appear in Numer. Methods of Par. Diff. Eqns.

[8] Guohui Zhou, How accurate is the streamline diffusion finite element method, Preprint 95–22, SFB 359, Heidelberg University, 1995, submitted.

Part V   Domain Decomposition Methods for
Flow Problems

A mortar element method for fluids

Yves ACHDOU[1] and Jean-Claude HONTAND[2] and Olivier PIRONNEAU[3]

ABSTRACT

Our interest is turned towards the Navier-Stokes equations. The general trend in computer architecture being towards message passing coarse grain MIMD parallelism, we wish to develop robust, fast and user-friendly algorithms for such machines.

Robustness is achieved by selecting well tested methods with known stability and error properties. Speed is achieved by block decomposition, fast solvers within blocks and minimal communication between blocks. Versatility is obtained by using non matching meshes between subdomains and user-friendliness will be obtained by using a high level language to drive the software as done in FreeFEM (see Pironneau,1994).

1 PROBLEM STATEMENT

The Navier-Stokes equations for incompressible flows are

$$D_t u + \nabla p - \nabla . \nu \nabla u = 0 \quad \nabla . u = 0$$

or, in stream function-vorticity formulation (for constant siscosity ν)

$$D_t \omega - \nu \Delta \omega = 0 \quad - \Delta \psi = \omega$$

where

$$D_t = \partial_t + u \cdot \nabla.$$

In most application ν is found through a turbulence model such as the $k - epsilon$ model:

$$D_t k - \nabla . \nu \nabla k - \frac{k^2}{\varepsilon} |\nabla U + \nabla U^T|^2 + \varepsilon = 0,$$

---

[1] CMAP,Ecole Polytechnique 91128 Palaiseau cedex: achdou@cmapx.polytechnique.fr
[2] University of Paris 6:hontand@ann.jussieu.fr
[3] University of Paris 6:pironneau@ann.jussieu.fr

Domain Decomposition Methods in Sciences and Engineering, edited by R. Glowinski *et al.*

$$D_t\varepsilon - \nabla.\nu\nabla\varepsilon - c_1 k|\nabla U + \nabla U^T|^2 + c_2\frac{\varepsilon^2}{k} = 0.$$

2   GENERAL DESCRIPTION OF THE ALGORITHM

2.1   Time discretization

Total derivatives are approximated by a finite difference formula in space and time, leading to a Eulerian-Lagrangian method:

$$D_t w = \partial_t w + u \cdot \nabla w \approx \frac{1}{\delta t}[w^{m+1} - w^m o X^m] \quad \text{where} \quad w^m o X(x) \approx w^m(x - \delta t u^m(x)).$$

For the velocity pressure formulation of the Navier-Stokes equations, a projection method for the pressure leads to

Step 1: solve

$$\Delta p^{m+1} = \frac{1}{\delta t}\nabla.u^m o X^m + \nabla.(\nabla.\nu\nabla u^m + f) \quad \text{in } \Omega, \quad \partial_n p|_{\partial\Omega} = 0.$$

Step 2: Solve

$$\frac{u^{m+1}}{\delta t} - \nabla\nu\nabla u^{m+1} = \frac{u^m o X^m}{\delta t} + \nabla p + f.$$

2.2   Spatial discretization

A finite element method with a feasible set of finite element space for velocity and pressure is used (LBB condition) or a stabilized term is added in the pressure equation if the $P^1 - P^1$ couple is used (Franca et al (1992)).

2.3   The Language Gfem

Such algorithms are fairly standard by now and the difficulty is not so much in the problem rather than in the infinite possibilities for coupling with other equations such as turbulence, temperature, electromagnetism.... We are experimenting with a macro-description of the problem with a dedicated language called Gfem. It used two basic blocks: an elliptic solver called by the key word "solve" and a convection operator instanced by the key word "convect". To demonstrate the power of the language we send the reader to http://www.ann.jussieu.fr/freefem/gfem.html

3   TWO BASIC BLOCKS

The first basic block is a general solver for linear second order PDE

$$au + b\partial_x u + c\partial_y u - \nabla.M\nabla u = f \quad \text{in} \quad \Omega \subset R^d,$$

$$u = u_0, \quad \text{or } \alpha u + \partial_\nu u = g \quad \text{on} \quad \partial\Omega.$$

So far only the symmetric case is implemented in parallel with mortars:

$$au - \nabla.M\nabla u = f \text{ in } \Omega, \quad u = u_0 \text{ or } \alpha u + \partial_\nu u = g \text{ on } \partial\Omega$$

where $M = M^T$ is an $R^d \times R^d$ matrix, not necessarily positive definite.

The second block is the convection operator

$$w(x) \rightarrow w(X(x))$$

where X(x) is the solution at $t - \delta t$ of

$$\frac{dX}{dt} = u(X,t), \quad X(t) = x.$$

4   NON MATCHING MESHES

Quadrilateral subdomains are used with cartesian meshes inside each subdomain. Q^1 or P^1 discretization inside each macro-element is used with possible discontinuities at interfaces. We do not use the mortar element in its generality, but restrict the macro-mesh to be geometrically conforming.

The subdomains will be large and few of them must be allowed to have at most one curved boundary if necessary. In this case we require the knowledge of a parametric description of the curved boundary from which we can then build a map G which transforms the unit square (the reference element R) into the element $Q = G(R)$.

Meshes in each macro elements are chosen independently. They do not match at the interfaces beetween subdomains. Therefore, in order to build a finite element space approaching $H^1(\Omega)$ one has to write a weak continuity constraint at the subdomain interfaces.

Within each macro element some adaption is done via a small number of parameters such as the mesh aspect ratio. Of course such mesh do not have the flexibility of unstructured meshes as used in adaptive FEM. But they allow fast linear solvers in each block.

5   CHARACTERISTICS IN PARALLEL

The problem is mapped to the reference macro element:

Given $x = G(y)$ u and $w(.)$ find $w(X(x)$ with $X(x) = G(Y(t - \delta t))$ and $M = G'^{-1}$

$$\frac{dY}{d\tau} = Mu(Y,\tau), \quad Y(t) = y.$$

A time stepping procedure such as Runge-Kutta or backward Euler is used:

$$Y^{k+1} = Y^k - \frac{\delta t}{K}Mu^k(Y^k),$$

where the time step δt is subdivided into K steps $\frac{\delta t}{K}$. To evaluate $Mu(Y^k, t^k)$ is not difficult because the mesh being cartesian the values of u can be found once the element which contains Y^k is known and that can be found by integer divisions.

However during these sub time steps, Y^k on the curve $Y(t)$ may leave the reference macro-element. Instead of continuing immediately the integration process on the neighbor element which contains the rest of $Y(t)$ it is much better for parallelism to store the boundary point $\{t^k, Y(t^k)\}$ for later use.

When all the characteristics have been partially computed in parallel, subdomain by subdomain, a new loop is made with for initial points all the $\{t^k, Y(t^k)\}$ corresponding to interrupted characteristics.

We summarize the process for a mesh made of L macro elements, each having $M_l \times N_l$ elements $Q_{m,n}$, with quadrature points $y_{m,n,l}$:

$$
\boxed{
\begin{aligned}
&\text{Set } t_{m,n,l} = \delta t \text{ for all } m, n, l \\
&\text{while some } t_{m,n,l} > 0 \\
&\quad \text{do } // \text{ l=1 to L} \\
&\qquad \text{for m} = 1 \text{ to } M_l \text{ and n} = 1 \text{ to } N_l \text{ do} \\
&\qquad \text{Loop on doing} \\
&\qquad\qquad y_{m,n,l} = y_{m,n,l} - \tfrac{\delta t}{K}\, M u^k(y_{m,n,l}) \\
&\qquad\qquad t_{m,n,l} = t_{m,n,l} - \tfrac{\delta t}{K} \\
&\qquad \text{till } t_{m,n,l} < 0 \text{ or } y_{m,n,l} \quad \text{not in} \quad]0,1[^2
\end{aligned}
}
$$

6   A PARALLEL ELLIPTIC SOLVER

6.1   The mortar element method

Consider the model elliptic problem

$$
u - \mu \Delta u = f|_\Omega, \quad au + \partial_n u = g|_{\partial\Omega}.
$$

Two finite element spaces are chosen, one for u and the other one for the Lagrange multipliers of the weak continuity constraint. In our case the functions restricted to a macro element Ω_k are continuous functions piecewise bilinear on the quadrilateral mesh of Ω_k : we call $Q_h(\Omega_k)$ this space of functions and we introduce the product

$$
Q_h = \prod_{k=1}^{K} Q_h(\Omega_k),
$$

which has an obvious canonical basis of shape functions.

Since the meshes do not match at the interfaces, we need to introduce a Lagrange multiplier space for the weak continuity constraint: for each interface $\Gamma_{kl} = \Omega_k \cup \Omega_l$ the Lagrange multiplier space is called $X_h(\Gamma_{kl})$ and the weak continuity constraint is

$$
\int_{\Gamma_{kl}} (u_{h|\Omega_k} - u_{h|\ \Omega_l})\lambda_h = 0 \quad \forall \lambda_h \in X_h(\Gamma_{kl}).
$$

The space $X_h(\Gamma_{kl})$ may be constructed either from $Q_h(\Omega_k)$ or $Q_h(\Omega_l)$. Assuming it is built from $Q_h(\Omega_k)$, let $(\lambda^i)_{i \in \{0,\ldots,s_{kl}\}}$ be the traces of the shape functions of $Q_h(\Omega_k)$ on Γ_{kl}, ordered according the abcissa on Γ_{kl} of the mesh nodes with which they are associated. The convergence proof of the method tells us that a good choice for $X_h(\Gamma_{kl})$ is

$$X_h(\Gamma_{kl}) = Span\{\lambda^0 + \lambda^1, \lambda^2, ..., \lambda^{s_{kl}-2}, \lambda^{s_{kl}-1} + \lambda^{s_{kl}}\},$$

(the space $Span(\lambda^i)_{i \in \{0,...,s_{kl}\}}$ would be too large for preserving the LBB condition).

Calling U the vector of coordinates of an element u_h of Q_h, the weak continuity constraint yields a set of linear equations for U:

$$BU = 0.$$

Similarly the weak form of the PDE is discretized by

$$\sum_k \int_{\Omega_k} [u_h w_h + \mu \nabla u_h \nabla w_h] + \int_{\partial\Omega} a u_h w_h = \sum_l \int_{\Omega_l} f w_h + \int_{\partial\Omega} g w_h$$

for all w_h with $BW = 0$.

This method has the same order of accuracy as its conforming analogon (Bernardi-Maday-Patera[1991], Le Tallec[1992]).

6.2 Solution of the linear system

At the discrete level the problem is of the type : find $U \in R^N$, such that $BU = 0$ and

$$W^T AU = F^T W, \quad \forall W \in R^N \text{ with } BW = 0,$$

or equivalently : find $U \in R^N$ and $\Lambda \in R^S$, $(S = \sum_{k,l}(s_{kl} - 2))$ such that

$$\begin{pmatrix} A & B \\ B^T & 0 \end{pmatrix} \begin{pmatrix} U \\ \Lambda \end{pmatrix} = \begin{pmatrix} F \\ 0 \end{pmatrix}.$$

Iterative solution

The system above is solved by an iterative method in a subspace of constraints. Preconditionning is necessary for efficiency (see Achdou-Kuznetsov in these proceedings).

7 APPLICATION TO THE NAVIER-STOKES EQUATIONS

The Navier-Stokes equations in their stream function-vorticity formulation are

$$D_t\omega - \nu\Delta\omega = 0, \quad -\Delta\psi = \omega, \quad \psi|_{\partial\Omega} = \psi_\Gamma, \quad \frac{\partial\psi}{\partial n} = g.$$

We discretize them in time by

$$\frac{1}{\delta t}[\omega^{m+1} - \omega^m o X^m] - \nu\Delta\omega^{m+1} = 0, \quad -\Delta\psi^{m+1} = \omega^{m+1}.$$

So at each time step we must solve

$$\omega - \mu\Delta\omega = f, \quad -\Delta\Psi = \omega, \quad \frac{\partial\Psi}{\partial n}| = g_1, \quad \Psi|_\Gamma = g_2.$$

As this is not in the standard form of our PDE block we use the following splitting: decompose ω and Ψ as $\omega = \omega^1 + \omega^0$, $\Psi = \Psi^1 + \Psi^0$ with

$$\omega^0 - \mu\Delta\omega^0 = f, \quad \omega^0|_\Gamma = 0, \text{ and } -\Delta\Psi^0 = \omega^0, \quad \Psi|_\Gamma = \psi_\Gamma,$$

and

$$\omega^1 - \mu\Delta\omega^1 = 0, \quad -\Delta\Psi^1 = \omega^1 \text{ and } \frac{\partial\Psi^1}{\partial n} = g - \frac{\partial\Psi^0}{\partial n}, \quad \Psi^1|_\Gamma = 0.$$

The first set of equations are 2 second order elliptic problems which can be solved successively and the last equations are a fourth order problem homogeneous in space. The traces of Ψ^1 and ω^1 can be obtained by a boundary element method (see Achdou, Pironneau). Alternatively, for laminar flows at high Reynolds number the following approximation can be used

$$\omega^1 - \mu\Delta\omega^1 = 0, \quad \omega^1|_\Gamma = -\frac{1}{\sqrt{\mu}}[g - \frac{\partial\Psi^0}{\partial n}] \text{ and } -\Delta\Psi^1 = \omega^1, \quad \Psi^1|_\Gamma = 0.$$

8 NUMERICAL TEST

The program is written in C and parallelized with PVM; all test have been made on a network of HP 735 connected by Ethernet.

8.1 Test 1

Test Parallelism efficiency on a single problem of Helmholtz type, (conjugate gradient without preconditioning for the Schur complement system on Λ).

$$\omega - \Delta\omega = 1 + \sin(x)\sin(y)$$

$$\omega|_{\partial\Omega} = \sin(x)\cos(y).$$

The grid has 36 macro elements and a total of 21 000 vertices. On a network of 6 machines the following has been measured:

processors	cpu time	% com / cpu
2	58 s	3 %
3	39 s	8 %
4	28 s	14 %
6	19 s	26 %

8.2 Test 2: The Navier-Stokes equations

A stream function-vorticity formulation is used with a decomposition at the boundary. The flow past a cylinder at $Re = 300$ is computed with 40 macro-elements, 20558 vertices on 3 processors (see figure 1 and 2).

The conjugate gradient method for the Schur complement system for Λ is much too slow; so a preconditionner developped by Achdou-Kuznetsov(1995) has been used.

The following performances are obtained with the grid of 36 macro elements and a total of 90 000 vertices. The number of iteration for the iterative algorithm (PCG) is 36. In the table below, the column *matrix* indicates the time needed for assembling the matrix and for factorizing the subdomains blocks, the column *pcg* indicates the time needed for solving one linear system.

proc	cpu	com	% com/cpu	matrix	precond.	pcg
2	486 s	4 s	1 %	416 s	7 s	67 s
3	330 s	6 s	2 %	289 s	20 s	38 s
4	262 s	8 s	3 %	227 s	28 s	32 s
6	197 s	28 s	14%	151 s	32 s	31 s

9   CONCLUSION

The Mortar element method allows the use of nonmatching meshes which is a key feature for time dependent problems with sliding subdomains and for parallel solvers with domain decomposition. It remains to see if this will make our implicit algorithms faster than their explicit competitors with unstructured mesh adaption (Mohammadi (1995) for instance), but it is well worth the programming effort to try it. A 3D implementation of the same ideas for the Navier Stokes equations is underway. With modifications, the same method can applied to the equations of electromagnetism and numerical tests will be presented in a forthcoming paper.

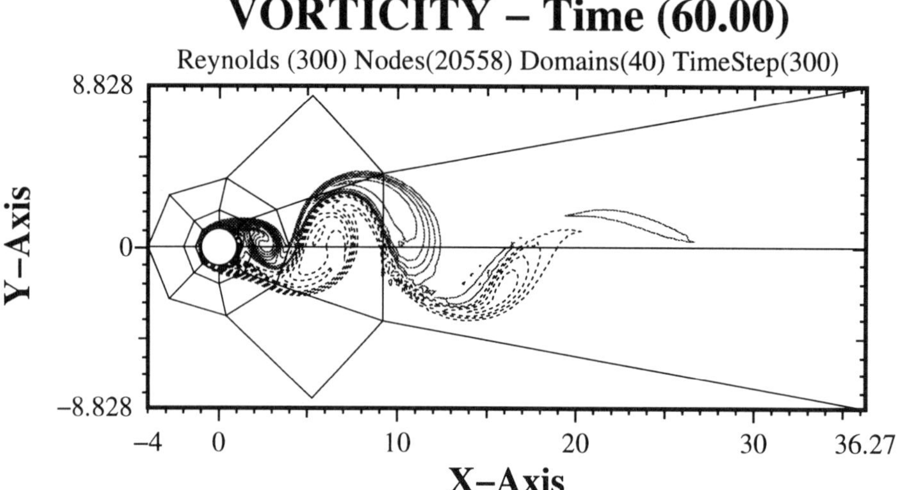

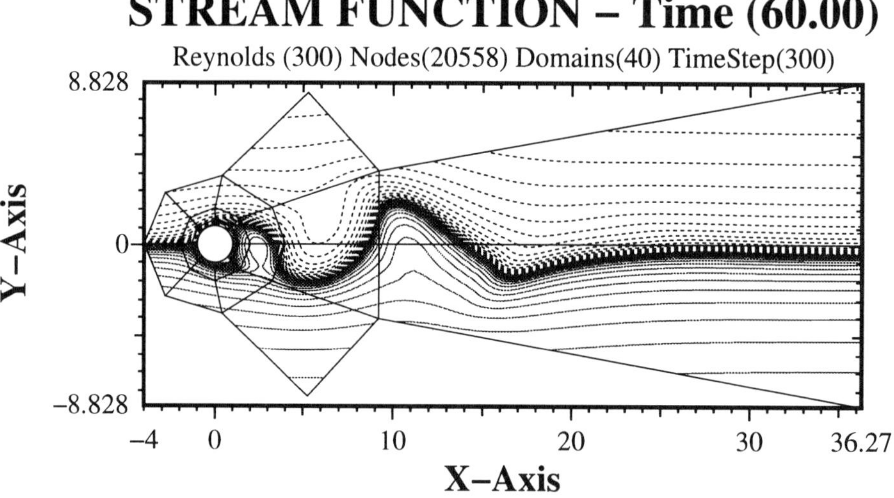

Figure 1 unsteady flow around a cylinder at Reynolds 300

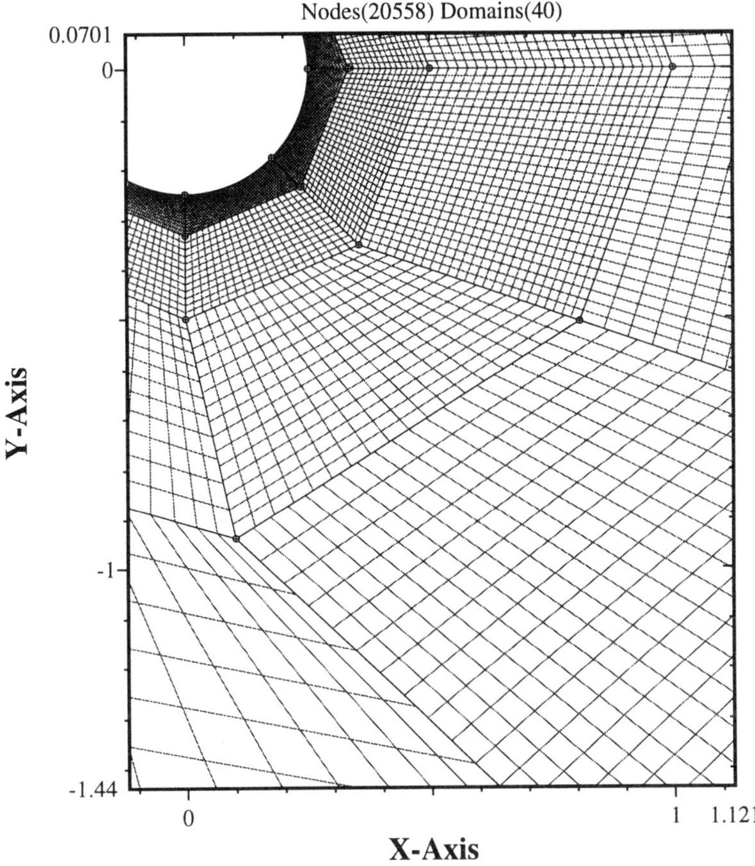

Figure 2   mesh used for computation of stream function and vorticity solutions

References

Y.Achdou, O.Pironneau, *Integral Equation for the Generalized Stokes Operator with Applications to Boundary layer Matching*, Comptes Rendus Acad. Sci. Paris 315 Série I (1992),pp 91-96.

Y.Achdou, O.Pironneau, *A fast solver for Navier-Stokes equations in the laminar regime using mortar finite element and boundary element method.*, SIAM journal of Numerical Analysis. Vol 32, No4, pp985-1016 (1995).

Y.Achdou, Yu.A.Kuznetsov,*Algorithms for the Mortar Element Methods* these proceedings.

O.Pironneau,*FreeFEM: A finite element software with a language and an integrated environment.* Colloque d'Analyse Numérique. Aussoix, 1994. http://www.ann.jussieu.fr/freefem/gfem.html.

B.Mohamadi,*Unstructured explicit solver for Navier-Stokes equations*, INRIA report (1993).

On a Fictitious Domain Method for Flow and Wave Problems

M. O. Bristeau[1], V. Girault[2], R. Glowinski[3], T.-W Pan[4],
J. Périaux[5], and Y. Xiang[1]

Summary

The main goal of this article is to discuss the application of a Lagrange multiplier based fictitious domain method to the solution of elliptic problems, then of the Navier-Stokes equations for incompressible viscous flow and finally of linear wave equations. This method is well suited to finite element approximations and takes advantage of the variational principles associated to large classes of partial differential problems. The results of numerical experiments validate this fictitious domain approach to the solution of partial differential equations with Dirichlet boundary conditions.

1 Introduction: Synopsis

Fictitious domain methods is a general term which covers in fact a large variety of solution methods for Partial Differential Equations; the literature on these methods is so large that we shall limit ourselves to [1]-[8] and the references therein. These methods are based however on the same following principle:

Suppose that one wants to solve the following boundary value problem

[1] INRIA, 78153 Le Chesnay, Paris, France
[2] Université P. et M. Curie, Paris, Toulouse, France
[3] Department of Mathematics, University of Houston, Houston, Texas 77204 USA, and Université P. et M. Curie, Paris, France
[4] Department of Mathematics, University of Houston, Houston, Texas 77204 USA
[5] Dassault Aviation, 92214 Saint–Cloud, France

Domain Decomposition Methods in Sciences and Engineering, edited by R. Glowinski *et al.*

$$A(u) = f \ in \ \omega, \tag{1.1}$$

$$B(u) = g \ on \ \gamma, \tag{1.2}$$

where in (1.1), (1.2), ω is (for simplicity) a bounded domain of $\boldsymbol{R}^d$, $\gamma = \partial\Omega$ is the boundary of ω, A and B are differential operators, and f, g are given functions. The idea is to *embed* ω in a larger domain Ω of very simple shape and to solve on Ω a problem whose solution $\tilde{u}$ satisfies

$$\tilde{u}|_\omega = u. \tag{1.3}$$

There is no miracle here since one has to take into account the boundary condition (1.2) in a way or another. Schematically we have boundary fitted fictitious domain methods (in which the finite element or finite difference mesh used to discretize the "extended problem" is locally distorted in the neighborhood of γ) or genuine fictitious domain methods in which the mesh used to discretize Ω is essentially independent of ω and γ.

In this article we shall concentrate on the second approach: it does not have the flexibility of general finite element methods, it is however well suited to the treatment of *Dirichlet boundary conditions* such as

$$u = g \ on \ \gamma, \tag{1.4}$$

via *Lagrange multipliers*. The possibilities of this fictitious domain methodology will be illustrated by the numerical treatment of the following problems: *elliptic equations* in Section 2, the *Navier-Stokes equations* for incompressible viscous flow in Section 3 and the wave equation in Section 4.

2   Fictitious Domain Methods for Linear Dirichlet Problems

2.1   Formulation of a Model Problem

Let ω be a *bounded* domain of $\boldsymbol{R}^d$, we denote by $\gamma = \partial\Omega$ the boundary of ω. The problem under consideration is the *Dirichlet* problem defined by:

$$\alpha u - \nu\Delta u = f \ in \ \omega, \tag{2.1}$$

$$u = g \ on \ \gamma, \tag{2.2}$$

where, in (2.1), we have $\alpha \geq 0$ and $\nu > 0$.

2.2   A Fictitious Domain Formulation of the Dirichlet Problem (2.1), (2.2)

We follow the approach in [5]. We embed therefore ω in Ω as shown in Figure 1, below, and we denote by Γ the boundary of Ω. We assume that f and g are reasonably smooth (typically $f \in L^2(\omega)$ and $g \in H^{1/2}(\gamma)$).

It follows from [5] that problem (2.1), (2.2) is equivalent to the following saddle-point problem:

$$Find \ \{\tilde{u}, \lambda\} \in V \times H^{-1/2}(\gamma) \ such \ that$$

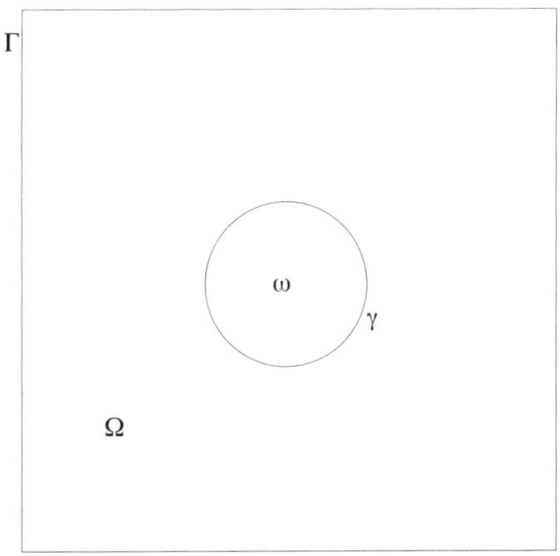

Figure 1. Embedding of ω

$$\int_\Omega (\alpha \tilde{u} v + \nu \nabla \tilde{u} \cdot \nabla v)\, dx = \int_\Omega \tilde{f} v\, dx + <\lambda, v>, \forall v \in V, \tag{2.3}$$

$$<\mu, \tilde{u} - g> = 0, \forall \mu \in H^{-1/2}(\gamma), \tag{2.4}$$

in the sense that $\tilde{u}|_\omega = u$; in (2.3), (2.4):

- $\tilde{f}$ denotes an $L^2(\Omega)$-extension of f such that $\tilde{f}|_\omega = f$;
- $<\cdot, \cdot>$ denotes the *duality pairing* between $H^{-1/2}(\gamma)$ and $H^{1/2}(\gamma)$ such that $<\mu, v> = \int_\gamma \mu v\, d\gamma$ if $\mu \in L^2(\gamma)$;
- V is either $H^1(\Omega)$, $H_0^1(\Omega)$, or $H_P^1(\Omega)$ where $H_P^1(\Omega)$ is the subspace of the functions of $H^1(\Omega)$ periodic at Γ.

Remark 2.1. The boundary "function" λ is a Lagrange multiplier associated with the boundary condition (2.2). Using Green's formula we can easily show that

$$\lambda = \nu \Big[\frac{\partial \tilde{u}}{\partial \mathbf{n}}\Big]\Big|_\gamma,$$

where $\big[\frac{\partial \tilde{u}}{\partial \mathbf{n}}\big]\big|_\gamma$ denotes the jump of $\frac{\partial \tilde{u}}{\partial \mathbf{n}}$ at γ. □

Remark 2.2. If one uses penalty, we can approximate problems (2.1), (2.2) and (2.3), (2.4) by

Find $\tilde{u}_\epsilon \in V$ such that

$$\int_\Omega (\alpha \tilde{u}_\epsilon v + \nu \nabla \tilde{u}_\epsilon \cdot \nabla v)\, dx + \frac{1}{\epsilon} \int_\gamma \tilde{u}_\epsilon v\, d\gamma = \int_\Omega \tilde{f} v\, dx + \frac{1}{\epsilon} \int_\gamma g v\, d\gamma, \ \forall v \in V, \tag{2.5}$$

with $\epsilon > 0$. It follows from, e.g., [9, Appendix 1] that

$$\lim_{\epsilon \to 0} \|\tilde{u}_\epsilon - u\|_{H^1(\omega)} = 0.$$

One of the main drawbacks of the above approach is that after approximation the condition number of the discrete analogue of problem (2.5) deteriorates as $\epsilon \to 0$, leading to an ill posed problem. $\square$

2.3 Iterative Solution of Problem (2.3), (2.4)

Problem (2.3) , (2.4) is a typical *linear saddle-point problem*; such problems can be solved by conjugate gradient variants of the Uzawa algorithm, like those discussed in, e.g., [5]. For the sake of completeness we shall describe such an algorithm. For simplicity, we assume that $\alpha > 0$. If $\tilde{u}$ is such that $\lambda \in L^2(\gamma)$, we can substitute this latter space to $H^{-1/2}(\gamma)$ in (2.3), (2.4) and solve the saddle-point problem by the following algorithm:

$$\lambda^0 \in L^2(\gamma) \ \textit{is given}; \tag{2.6}$$

solve

$$\textit{Find } u^0 \in V \textit{ such that}$$
$$\int_\Omega (\alpha u^0 v + \nu \nabla u^0 \cdot \nabla v)\, dx = \int_\Omega \tilde{f} v\, dx + \int_\gamma \lambda^0 v\, d\gamma, \ \forall v \in V, \tag{2.7}$$

and then

$$\textit{Find } g^0 \in L^2(\gamma) \textit{ such that}$$
$$\int_\gamma g^0 \mu\, d\gamma = \int_\gamma (u^0 - g)\mu\, d\gamma, \ \forall \mu \in L^2(\gamma), \tag{2.8}$$

and set
$$w^0 = g^0. \tag{2.9}$$

   For $n \geq 0$, assuming that λ^n, g^n, w^n are known, compute $\lambda^{n+1}, g^{n+1}, w^{n+1}$, as follows:

solve

$$\textit{Find } \bar{u}^n \in V \textit{ such that}$$
$$\int_\Omega (\alpha \bar{u}^n v + \nu \nabla \bar{u}^n \cdot \nabla v)\, dx = \int_\gamma w^n v\, d\gamma, \ \forall v \in V, \tag{2.10}$$

$$\textit{Find } \bar{g}^n \in L^2(\gamma) \textit{ such that}$$
$$\int_\gamma \bar{g}^n \mu\, d\gamma = \int_\gamma \bar{u}^n \mu\, d\gamma, \ \forall \mu \in L^2(\gamma), \tag{2.11}$$

and compute

$$\rho_n = \int_\gamma |g^n|^2\, d\gamma \bigg/ \int_\gamma w^n \bar{g}^n\, d\gamma, \tag{2.12}$$

and then

$$\lambda^{n+1} = \lambda^n - \rho_n w^n, \tag{2.13}$$
$$u^{n+1} = u^n - \rho_n \bar{u}^n, \tag{2.14}$$
$$g^{n+1} = g^n - \rho_n \bar{g}^n, \tag{2.15}$$

If $\|g^{n+1}\|_{L^2(\gamma)}/\|g^0\|_{L^2(\gamma)} \le \epsilon$, take $\lambda = \lambda^{n+1}$, $\tilde{u} = u^{n+1}$; if not compute

$$\gamma_n = \|g^{n+1}\|^2_{L^2(\gamma)}/\|g^n\|^2_{L^2(\gamma)} \tag{2.16}$$

and update w^n by

$$w^{n+1} = g^{n+1} + \gamma_n w^n. \tag{2.17}$$

Do $n = n + 1$ and go to (2.10).

Remark 2.3. Relations (2.8) and (2.11) clearly imply that $g^0 = u^0|_\gamma - g$ and $\bar{g}^n = \bar{u}^n|_\gamma$, respectively. We have chosen to write these solutions in integral form since this is better suited for the discrete problems where the discrete analogues of g^n and $u^n|_\gamma$ may live in different spaces. □

Remark 2.4. Algorithm (2.6)-(2.17) is easy to implement; however we can not expect it to have optimal convergence properties since the natural space for λ is not $L^2(\gamma)$ but $H^{-1/2}(\gamma)$. Iterating in this last space is more complicated, at least if $\omega \subset \mathbf{R}^d$ with $d \ge 3$; in the particular case where $\omega \subset \mathbf{R}^2$ we have shown in [5] that efficient iterative solvers can be obtained by preconditioning algorithm (2.6)-(2.17) (in fact its discrete analogue) by discrete variants of the boundary operator $(\frac{\alpha}{\nu}I - \frac{d^2}{d\gamma^2})^{1/2}$ (see [5] for implementation details). □

2.4 Finite Element Approximation of Problem (2.3), (2.4)

We consider *two dimensional problems* only (for three dimensional application see ref. [6]). In order to discretize the saddle-point problem (2.3), (2.4) we need to approximate both spaces V and $H^{-1/2}(\gamma)$. Let us denote by h the pair $\{h_\Omega, h_\gamma\}$. To approximate V we introduce a *regular triangulation* $\mathcal{T}_h$ of Ω (as shown in Figure 2) with h_Ω the largest length of the edges of $\mathcal{T}_h$; then, we (classically) approximate V by

$$V_h = \{v_h | v_h \in V \cap C^0(\bar{\Omega}), v_h|_T \in P_1, \forall T \in \mathcal{T}_h\}, \tag{2.18}$$

with P_1 the space of the polynomials in 2 variables of degree ≤ 1. Concerning $H^{-1/2}(\gamma)$ we divide γ in subarcs of maximal length h_γ and denote by $\mathcal{A}_h$ the collection of these subarcs; we approximate then $H^{-1/2}(\gamma)$ (and also $L^2(\gamma)$) by

$$\Lambda_h = \{\mu_h | \mu_h|_a = constant, \forall a \in \mathcal{A}_h\}. \tag{2.19}$$

Remark 2.5. Let us insist on the fact that the extremities of the arcs of $\mathcal{A}_h$ do not have to be at the intersection of γ with the edges of $\mathcal{T}_h$ (this can happen but it is not a necessity); actually we think that the partitioning of γ has to be mostly related to its intrinsic geometrical properties since this approach will facilitate the numerical treatment of those time dependent problems where ω (and therefore γ) is moving. □

The saddle-point problem (2.3), (2.4) is approximated by the following finite dimensional saddle-point problem

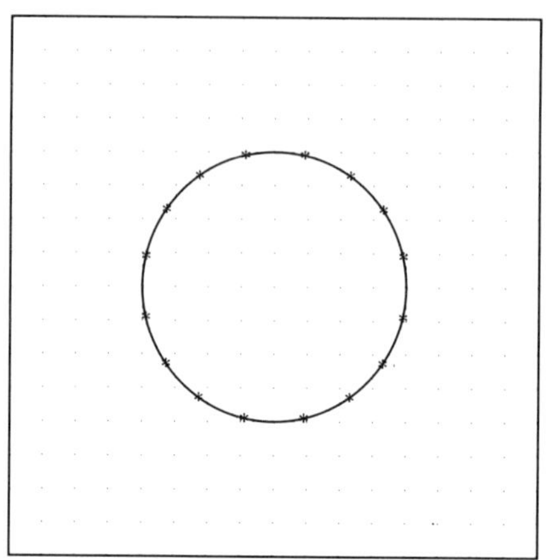

Figure 2. Visualization of $\mathcal{T}_h$ and $\mathcal{A}_h$

Find $\{u_h, \lambda_h\} \in V_h \times \Lambda_h$ *such that*

$$\int_\Omega (\alpha u_h v_h + \nu \nabla u_h \cdot \nabla v_h)\, dx = \int_\Omega \tilde{f} v_h\, dx + \int_\gamma \lambda_h v_h\, d\gamma,\ \forall v_h \in V_h, \tag{2.20}$$

$$\int_\gamma (u_h - g)\mu_h\, d\gamma = 0,\ \forall \mu_h \in \Lambda_h. \tag{2.21}$$

In the particular case where $g = 0$ it has been shown in [5] that

$$\lim_{h \to 0} u_h|_\omega = u\ \ in\ H^1(\omega), \tag{2.22}$$

where u is the solution of problem (2.1), (2.2); the convergence result (2.22) only requires that $\lim h_\Omega = \lim h_\gamma = 0$. In the cases where $g \neq 0$ it has been shown in [10] that

$$\lim_{h \to 0} \{u_h, \lambda_h\} = \{\tilde{u}, \lambda\}\ \ in\ H^1(\Omega) \times H^{-1/2}(\gamma) \tag{2.23}$$

if we assume that

$$c_1 h_\gamma \leq h_\Omega \leq c_2 h_\gamma, \tag{2.24}$$

with c_1 and c_2 well chosen positive numbers; actually in ref. [10] we shall also find a priori error estimates and values of c_1 and c_2 implying convergence.

To conclude this paragraph let us mention that numerical experiments for two and three dimensional test problems suggest (see [5] and [6]) that

$$\|u_h - u\|_{H^1(\omega)} \leq ch,$$
$$\|u_h - u\|_{L^\infty(\omega)} \leq ch^2,$$

which is better than the error estimates proved in [10].

Remark 2.5. It follows from (2.20), (2.21) that to obtain the linear system (associated to a symmetric indefinite matrix) equivalent to the above discrete saddle-point problem

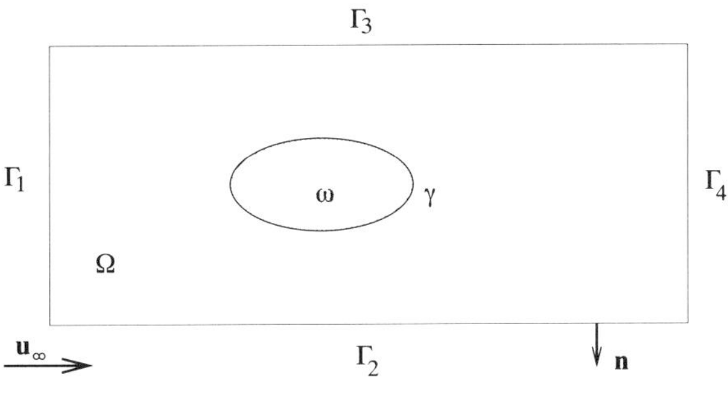

Figure 3.

we need the calculation of integrals like $\int_\gamma w_i \theta_k \, d\gamma$ where $\{w_i\}$ (resp., $\{\theta_k\}$) is a vector basis of V_h (resp., Λ_h). $\square$

3 Application to the Solution of the Navier-Stokes Equations for Incompressible Viscous Fluids

3.1 Generalities. Problem Formulations

The fictitious domain solution of the Navier-Stokes equations modelling incompressible viscous fluids by methods generalizing those described in Section 2 has been discussed at length in [6]-[8], [11] (see also [12] where one applies this methodology to the simulation of three-dimensional visco-elastic flow). In the present article (of survey nature) we shall briefly discuss the fictitious domain solution of external flow like the one associated with Figure 3; indeed, such a flow is modelled by

$$\frac{\partial \mathbf{u}}{\partial t} - \nu \Delta \mathbf{u} + (\mathbf{u} \cdot \nabla)\mathbf{u} + \nabla p = \mathbf{f} \; in \; \Omega \setminus \bar{\omega}, \tag{3.1}$$

$$\nabla \cdot \mathbf{u} = 0 \; in \; \Omega \setminus \bar{\omega}, \tag{3.2}$$

$$\mathbf{u} = \mathbf{u}_\infty \; on \; \Gamma \setminus \Gamma_4, \; \mathbf{u} = \mathbf{0} \; on \; \gamma, \tag{3.3}$$

$$\nu \frac{\partial \mathbf{u}}{\partial \mathbf{n}} - \mathbf{n}p = \mathbf{0} \; on \; \Gamma_4, \tag{3.4}$$

$$\mathbf{u}(\mathbf{x}, 0) = \mathbf{u}_0(\mathbf{x}), \quad \mathbf{x} \in \Omega \setminus \bar{\omega}, (with \; \nabla \cdot \mathbf{u}_0 = 0). \tag{3.5}$$

In (3.1)-(3.5),

- $\Omega \setminus \bar{\omega}$ is the flow region,
- $\mathbf{u} = \{u_i\}_{i=1}^d$ denotes the velocity field,
- p is the pressure,
- $\nu(> 0)$ is a viscosity coefficient,
- $\mathbf{n}$ is the outer normal unit vector at $\partial\Omega$,
- $\mathbf{a} \cdot \mathbf{b} = \sum_{i=1}^d a_i b_i, \; \forall \mathbf{a} = \{a_i\}_{i=1}^d, \; \forall \mathbf{b} = \{b_i\}_{i=1}^d,$
- $(\mathbf{v} \cdot \nabla)\mathbf{w} = \{\sum_{j=1}^{j=d} v_j \frac{\partial w_i}{\partial x_j}\}_{i=1}^{i=d}, \; \forall \mathbf{v} = \{v_i\}_{i=1}^d, \; \forall \mathbf{w} = \{w_i\}_{i=1}^d,$

- $\mathbf{u}(0)$ denotes the function $\mathbf{x} \to \mathbf{u}(\mathbf{x}, 0)$; more generally, we shall denote by $\mathbf{u}(t)$ the function

$$\mathbf{x} \to \mathbf{u}(\mathbf{x}, t), \ \forall t \geq 0.$$

3.2 An Equivalent Fictitious Domain Formulation

The above Navier-Stokes problem is equivalent to the following *variational* system:

Find $\{\mathbf{U}(t), P(t), \lambda(t)\}$ *such that for almost every* $t > 0$, *we have*

$$\int_\Omega \mathbf{U}_t \cdot \mathbf{v} \, d\mathbf{x} + \nu \int_\Omega \nabla \mathbf{U} \cdot \nabla \mathbf{v} \, d\mathbf{x} + \int_\Omega (\mathbf{U} \cdot \nabla) \mathbf{U} \cdot \mathbf{v} \, d\mathbf{x} - \int_\Omega P \nabla \cdot \mathbf{v} \, d\mathbf{x}$$

$$= \int_\Omega \tilde{\mathbf{f}} \cdot \mathbf{v} \, d\mathbf{x} + \int_{\Gamma_1} \mathbf{g}_1 \cdot \mathbf{v} \, d\Gamma + \int_\gamma \lambda \cdot \mathbf{v} \, d\gamma, \ \forall \mathbf{v} \in \mathbf{V}_0, \tag{3.6}$$

$$\int_\gamma \mu \cdot \mathbf{U} \, d\gamma = 0, \ \forall \mu \in (L^2(\gamma))^d, \tag{3.7}$$

$$\int_\Omega q \nabla \cdot \mathbf{U} \, d\mathbf{x} = 0, \ \forall q \in L^2(\Omega), \tag{3.8}$$

$$\mathbf{U} = \mathbf{u}_\infty \ on \ \Gamma \setminus \Gamma_4, \tag{3.9}$$

$$\mathbf{U}(0) = \mathbf{U}_0, \ such \ that \ \nabla \cdot \mathbf{U}_0 = 0 \ in \ \Omega, \ \mathbf{U}_0|_{\Omega \setminus \bar{\omega}} = \mathbf{u}_0. \tag{3.10}$$

In (3.6) the space $\mathbf{V}_0$ is defined by

$$\mathbf{V}_0 = \{\mathbf{v} | \mathbf{v} \in (\mathbf{H}^1(\Omega))^d, \mathbf{v} = \mathbf{0} \ on \ \Gamma \setminus \Gamma_4\}. \tag{3.11}$$

In this context the equivalence between (3.1)-(3.5) and (3.6)-(3.10) means that $\mathbf{U}|_{\Omega \setminus \bar{\omega}} = \mathbf{u}$, $P|_{\Omega \setminus \bar{\omega}} = p$. We remind that we are here in a situation where the condition on Γ_4, namely (3.4), implies the *uniqueness* of the pressure associated to $\mathbf{u}$.

3.3 Time Discretization by Operator Splitting

Operator splitting algorithms like those discussed in, e.g., [13]-[16] (see also [17]) provide a systematic and elegant way to decouple the various difficulties associated with the solution of problem (3.6)-(3.10); the resulting schemes are described in [6]-[8], [11] and the scheme below is from these references. It is a particular application of the θ-scheme that we have introduced in [18], [19] and which combines good stability and accuracy properties (particularly if $\theta = 1 - 1/\sqrt{2}$). With $\triangle t > 0$ a time discretization step, this scheme takes the following form:

$$\mathbf{U}^0 = \mathbf{U}_0; \tag{3.12}$$

for $n \geq 0$, *knowing* $\mathbf{U}^n$, *find* $\{\mathbf{U}^{n+\theta}, P^{n+\theta}\}$, $\mathbf{U}^{n+1-\theta}$, *and* $\{\mathbf{U}^{n+1}, P^{n+1}\}$ *via*

$$\int_\Omega \frac{\mathbf{U}^{n+\theta} - \mathbf{U}^n}{\theta \triangle t} \cdot \mathbf{v} \, d\mathbf{x} + \alpha \nu \int_\Omega \nabla \mathbf{U}^{n+\theta} \cdot \nabla \mathbf{v} \, d\mathbf{x}$$

$$- \int_\Omega P^{n+\theta} \nabla \cdot \mathbf{v} \, d\mathbf{x} - \int_\gamma \lambda^{n+\theta} \cdot \mathbf{v} \, d\gamma \tag{3.13}_1$$

$$= \int_{\Omega} \tilde{\mathbf{f}}^{n+\theta} \cdot \mathbf{v} \, d\mathbf{x} - \int_{\Omega} (\mathbf{U}^n \cdot \nabla)\mathbf{U}^n \cdot \mathbf{v} \, d\mathbf{x} - \beta\nu \int_{\Omega} \nabla\mathbf{U}^n \cdot \nabla\mathbf{v} \, d\mathbf{x}, \ \forall \mathbf{v} \in \mathbf{V}_0,$$

$$\int_{\Omega} \nabla \cdot \mathbf{U}^{n+\theta} q \, d\mathbf{x} = 0, \ \forall q \in L^2(\Omega), \tag{3.13}_2$$

$$\int_{\gamma} \mathbf{U}^{n+\theta} \cdot \mu = 0, \ \forall \mu \in (L^2(\gamma))^d, \tag{3.13}_3$$

$$\mathbf{U}^{n+\theta} = \mathbf{u}_\infty \text{ on } \Gamma \setminus \Gamma_4; \tag{3.13}_4$$

$$\int_{\Omega} \frac{\mathbf{U}^{n+1-\theta} - \mathbf{U}^{n+\theta}}{(1-2\theta)\triangle t} \cdot \mathbf{v} \, d\mathbf{x} + \beta\nu \int_{\Omega} \nabla\mathbf{U}^{n+1-\theta} \cdot \nabla\mathbf{v} \, d\mathbf{x}$$
$$+ \int_{\Omega} (\mathbf{U}^{n+\delta} \cdot \nabla)\mathbf{U}^{n+1-\theta} \cdot \mathbf{v} \, d\mathbf{x} \tag{3.14}_1$$
$$= \int_{\Omega} \tilde{\mathbf{f}}^{n+1-\theta} \cdot \mathbf{v} \, d\mathbf{x} + \int_{\gamma} \lambda^{n+\theta} \cdot \mathbf{v} \, d\gamma + \int_{\Omega} P^{n+\theta} \nabla \cdot \mathbf{v} \, d\mathbf{x}$$
$$- \alpha\nu \int_{\Omega} \nabla\mathbf{U}^{n+\theta} \cdot \nabla\mathbf{v} \, d\mathbf{x}, \ \forall \mathbf{v} \in \mathbf{V}_0,$$

$$\mathbf{U}^{n+1-\theta} = \mathbf{u}_\infty \text{ on } \Gamma \setminus \Gamma_4; \tag{3.14}_2$$

and finally,

$$\int_{\Omega} \frac{\mathbf{U}^{n+1} - \mathbf{U}^{n+1-\theta}}{\theta\triangle t} \cdot \mathbf{v} \, d\mathbf{x} + \alpha\nu \int_{\Omega} \nabla\mathbf{U}^{n+1} \cdot \nabla\mathbf{v} \, d\mathbf{x}$$
$$- \int_{\Omega} P^{n+1} \nabla \cdot \mathbf{v} \, d\mathbf{x} - \int_{\gamma} \lambda^{n+1} \cdot \mathbf{v} \, d\gamma \tag{3.15}_1$$
$$= \int_{\Omega} \tilde{\mathbf{f}}^{n+1} \cdot \mathbf{v} \, d\mathbf{x} - \int_{\Omega} (\mathbf{U}^{n+1-\theta} \cdot \nabla)\mathbf{U}^{n+1-\theta} \cdot \mathbf{v} \, d\mathbf{x}$$
$$- \beta\nu \int_{\Omega} \nabla\mathbf{U}^{n+1-\theta} \cdot \nabla\mathbf{v} \, d\mathbf{x}, \ \forall \mathbf{v} \in \mathbf{V}_0,$$

$$\int_{\Omega} \nabla \cdot \mathbf{U}^{n+1} q \, d\mathbf{x} = 0, \ \forall q \in L^2(\Omega), \tag{3.15}_2$$

$$\int_{\gamma} \mathbf{U}^{n+1} \cdot \mu = 0, \ \forall \mu \in (L^2(\gamma))^d, \tag{3.15}_3$$

$$\mathbf{U}^{n+1} = \mathbf{u}_\infty \text{ on } \Gamma \setminus \Gamma_4. \tag{3.15}_3$$

In (3.12)-(3.15) we have taken $\alpha = (1-2\theta)/(1-\theta)$ and $\beta = \theta/(1-\theta)$. In relation $(3.14)_1$, δ is equal to either θ or $1-\theta$, the second choice making problem (3.14)

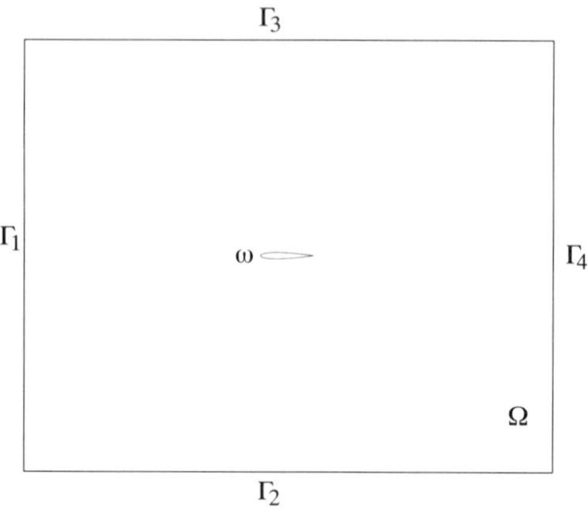

Figure 4.

nonlinear but increasing the stability of the scheme. We observe that the intermediate solution $\mathbf{U}^{n+1-\theta}$ is not required to satisfy the incompressibility condition and the no-slip condition on γ. The solution of the subproblems encountered in each time step is briefly discussed in the following paragraph.

3.4   Solution of the Subproblems

The solution of *advection-diffusion* problems such as (3.14) has been discussed at length in refs. [13]-[16]; we shall not return therefore on those topics. After discretization, problems (3.13) and (3.15) provide linear systems of the following form

$$\begin{cases} A\mathbf{U} + B^t P + C^t \lambda & = \mathbf{b}, \\ B\mathbf{U} & = \mathbf{c}, \\ C\mathbf{U} & = \mathbf{d}. \end{cases} \tag{3.16}$$

Problem (3.16) is a generaiized saddle-point problem. The solution of (3.16) has been discussed in [6]-[8], [11]; in ref. [11], in particular, we discuss a *one shot method*, of *conjugate gradient type*, which *simultaneously* adjusts the *incompressibility condition* $\nabla \cdot \mathbf{U} = 0$ in Ω and the *no-slip condition* $\mathbf{U} = \mathbf{0}$ on γ. The one shot method has good parallelization properties, but we think that there is still room for many improvements concerning the speed of convergence; in particular, we think, to apply to (3.16) the methods developed recently by Y. Kuznetsov for the fast solution of saddle-point problems (see [20]).

3.5   Numerical Experiments

We consider test problems where ω is a NACA0012 airfoil with zero degree and then 5 degrees angle of attack. The airfoil is centered at $(0,0)$ and its chord length is 0.35; finally ω is contained in Ω, where $\Omega = (-0.625, 0.625) \times (-0.5, 0.5)$ (see Figure 4). The boundary conditions are defined as follows:

$$\mathbf{u} = \mathbf{0} \; on \; \gamma, \; \mathbf{u} = (1 - e^{-ct})\{1, 0\} \; on \; \Gamma_1 \cup \Gamma_2 \cup \Gamma_3 (= \Gamma \setminus \Gamma_4), \tag{3.17}$$

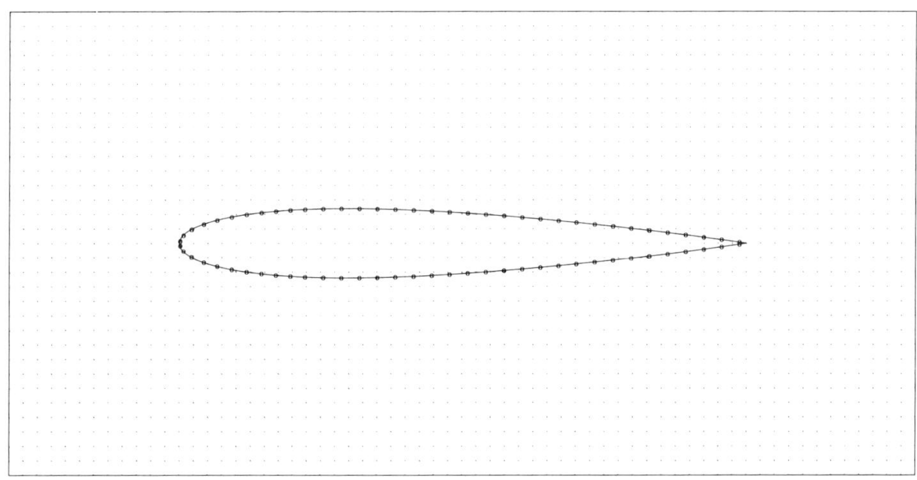

Figure 5. Mesh of γ where "o" are the mesh points on γ and part of the triangulation of Ω with meshsize $h = 1/128$.

$$\nu \frac{\partial \mathbf{u}}{\partial \mathbf{n}} - \mathbf{n}p = \mathbf{0} \ on \ \Gamma_4, \tag{3.18}$$

where c is a positive constant in (3.17).

As a finite dimensional subspace of $\mathbf{V}$, we choose

$$\mathbf{V}_h = \{\mathbf{v}_h | \mathbf{v}_h \in H_{0h}^1 \times H_{0h}^1\}, \tag{3.19}$$

where

$$H_{0h}^1 = \{\phi_h | \phi_h \in C^0(\bar{\Omega}), \phi_h|_T \in P_1, \ \forall T \in \mathcal{T}_h, \ \phi_h = 0 \ on \ \Gamma \backslash \Gamma_4\}, \tag{3.20}$$

with $\mathcal{T}_h$ a triangulation of Ω (see, e.g, Figure 5), P_1 the space of the polynomials in x_1, x_2 of degree ≤ 1. A traditional way of approximating the pressure is to take it in the space

$$H_{2h}^1 = \{\phi_h | \phi_h \in C^0(\bar{\Omega}), \phi_h|_T \in P_1, \ \forall T \in \mathcal{T}_{2h}\}, \tag{3.21}$$

where $\mathcal{T}_{2h}$ is a triangulation twice coarser than $\mathcal{T}_h$. Concerning the space Λ_h approximating Λ, we define it by

$$\Lambda_h = \{\mu_h | \mu_h \in (L^\infty(\partial\omega))^2, \mu_h \ is \ constant \ on \ the \ arc \ joining \tag{3.22}$$
$$2 \ consecutive \ mesh \ points \ on \ \gamma\}.$$

A particular choice for the mesh points on γ is shown on Figure 5. As a rule we have to put more mesh points at the leading edge; also mesh points have to be chosen carefully at the trailing edge. With a bad choice of mesh points on γ (e.g., a uniform mesh for a NACA0012), the Dirichlet boundary condition can not be matched very well for the case where the product $\nu\triangle t$ is of the order of 10^{-8}. The numerical

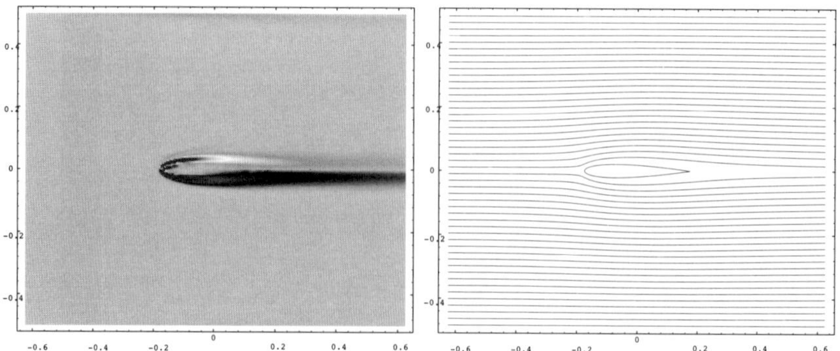

Figure 6. Vorticity density (left) and streamlines (right) for the flow past NACA0012 with zero degree angle of attack. Flow direction is from the left to the right, the Reynolds number is 1000, dimensionless time is 1.8.

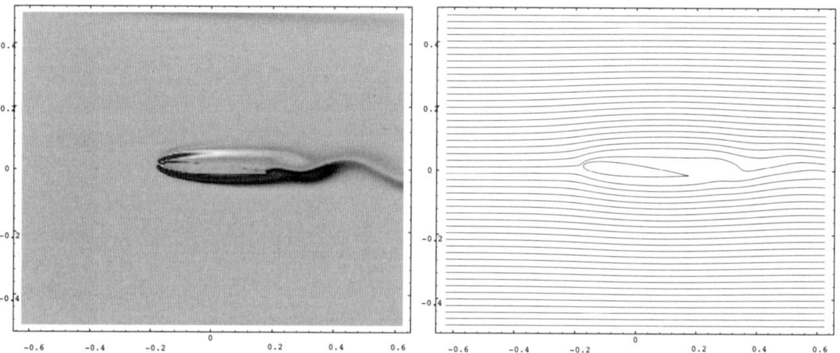

Figure 7. Vorticity density (left) and streamlines (right) for the flow past NACA0012 with 5 degrees angle of attack. Flow direction is from the left to the right, the Reynolds number is 1000, dimensionless time is 1.34.

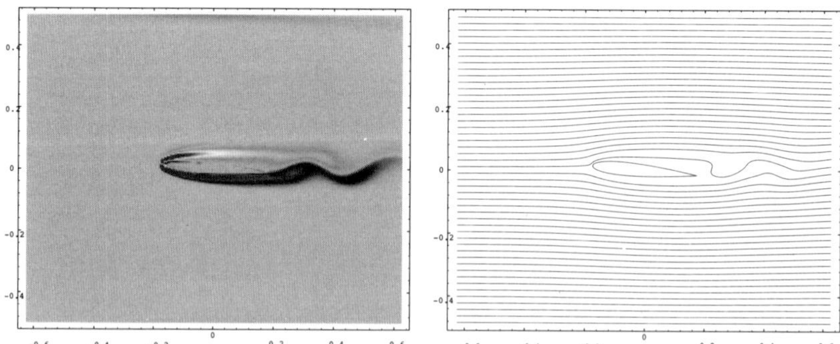

Figure 8. Vorticity density (left) and streamlines (right) for the flow past NACA0012 with 5 degrees angle of attack. Flow direction is from the left to the right, the Reynolds number is 1000, dimensionless time is 1.525.

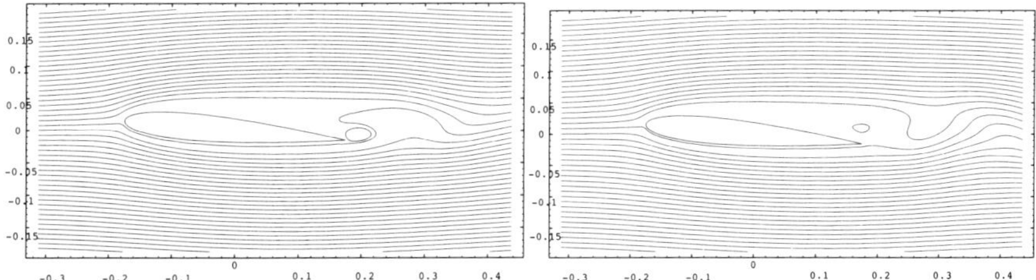

Figure 9. Local enlargement from Figures 7 (left) and 8 (right) of the streamlines distribution around NACA0012 with 5 degrees angle of attack.

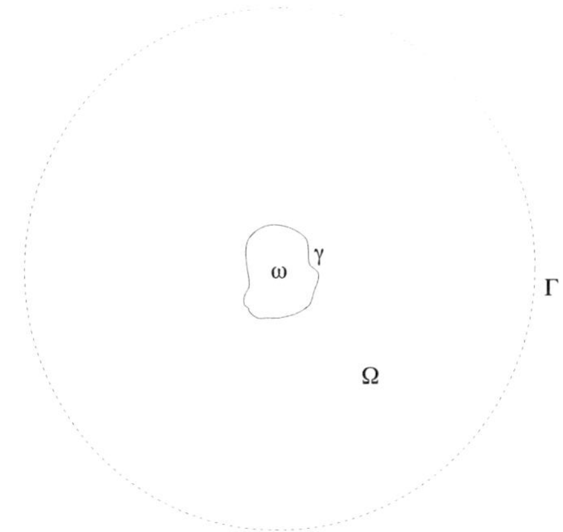

Figure 10

results presented here have been obtained for Re= 10^3 (taking the chord of the airfoil as characteristic length) with meshsizes $h_v = 1/256$ to approximate velocity and $h_p = 1/128$ to approximate pressure, time discretization step $\triangle t = 2.5 \times 10^{-3}$ and $c = 20$ in (3.17). In Figure 6 where the angle of attack is zero degree, the vorticity distribution and stream lines are almost symmetric with respect to the x_1 direction. For the case where the angle of attack is 5 degrees, *Von Kàrmàn vortex shedding* occurs (see Figures 7 and 8). The local enlargement of streamlines distribution around the airfoil is shown in Figure 9 for the 5 degrees incidence case.

4 Fictitious Domain Methods for Wave Problems

4.1 Formulation of a Model Wave Problem

We consider for simplicity a *wave equation* with constant coefficients but the method below can be generalized to variable coefficient wave equations and to systems such as the *Maxwell equations* and the *equations of Elasto-Dynamics*. The problem that we address is the solution of the *scattering* type problem associated with

$$\frac{\partial^2 u}{\partial t^2} - \Delta u = 0 \; in \; \Omega \setminus \bar{\omega}, \tag{4.1}$$

$$u = g \; on \; \gamma, \tag{4.2}$$

$$\frac{\partial u}{\partial t} + \frac{\partial u}{\partial \mathbf{n}} = 0 \; on \; \Gamma, \tag{4.3}$$

$$u(0) = u_0, \; \frac{\partial u}{\partial t}(0) = u_1. \tag{4.4}$$

The geometrical configuration is the one shown in Figure 10 where Ω is the larger domain in which ω has been embedded; actually $\Gamma(= \partial\Omega)$ will be used as an artificial boundary on which the (simple) *radiation condition* (4.3) has been specified.

4.2   *A Fictitious Domain Equivalent Formulation of the Wave Problem (4.1)-(4.4)*

Using, once again, a Lagrange multiplier approach it is fairly easy to show that problem (4.1)-(4.4) is equivalent to finding a pair $\{U, \lambda\}$ so that

$$\int_\Omega \frac{\partial^2 U}{\partial t^2} v \, d\mathbf{x} + \int_\Omega \nabla U \cdot \nabla v \, d\mathbf{x} + \int_\Gamma \frac{\partial U}{\partial t} v \, d\Gamma = \int_\gamma \lambda v \, d\gamma, \; \forall v \in H^1(\Omega), \tag{4.5}$$

$$\int_\gamma \mu(U - g) \, d\gamma = 0, \; \forall \mu \in L^2(\gamma), \tag{4.6}$$

$$U(0) = U_0, \; \frac{\partial U}{\partial t}(0) = U_1, \tag{4.7}$$

in the sense that $U|_{\Omega\setminus\bar{\omega}} = u$. The notation in (4.5), (4.6) is self-explanatory; in (4.7) U_0 and U_1 are well chosen extensions of u_0 and u_1, respectively (U_0 shall be a H^1-extension of u_0, while for u_1 a L^2-extension U_1 will be sufficient).

4.3   *Space and Time Discretization of Problem (4.5)-(4.7)*

Let us denote by V_h a finite dimensional finite element subspace of $H^1(\Omega)$ defined as follows:

$$V_h = \{v_h | v_h \in C^0(\bar{\Omega}), v_h|_T \in P_1, \; \forall T \in \mathcal{T}_h\}$$

with $\mathcal{T}_h$ a triangulation of Ω. Similarly we approximate $L^2(\gamma)$ by Λ_h, defined as in (3.22). We approximate then problem (4.5)-(4.7) by

$$\int_\Omega \frac{\partial^2 U_h}{\partial t^2} v_h \, d\mathbf{x} + \int_\Omega \nabla U_h \cdot \nabla v_h \, d\mathbf{x} + \int_\Gamma \frac{\partial U_h}{\partial t} v_h \, d\Gamma = \int_\gamma \lambda_h v_h \, d\gamma, \; \forall v_h \in V_h, \tag{4.8}$$

$$\int_\gamma \mu_h(U_h - g) \, d\gamma = 0, \; \forall \mu_h \in \Lambda_h, \tag{4.9}$$

$$U_h(0) = U_{0h}, \; \frac{\partial U_h}{\partial t}(0) = U_{1h}, \tag{4.10}$$

$$\{U_h(t), \lambda_h(t)\} \in V_h \times \Lambda_h \; for \; almost \; every \, t > 0. \tag{4.11}$$

Time discretizing problem (4.8)-(4.11) is quite easy, particularly if one uses the scheme below where we suppose that U_h^{n-1}, U_h^n are already known:

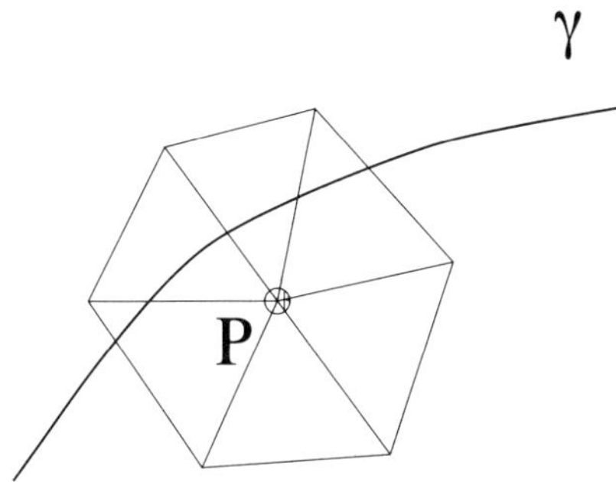

Figure 11

$$\int_\Omega \frac{U_h^{n+1} + U_h^{n-1} - 2U_h^n}{|\triangle t|^2} v_h \, d\mathbf{x} + \int_\Omega \nabla U_h^n \cdot \nabla v_h \, d\mathbf{x} \tag{4.12}$$

$$+ \int_\Gamma \frac{U_h^{n+1} - U_h^{n-1}}{2\triangle t} v_h \, d\Gamma = \int_\gamma \lambda_h^n v_h \, d\gamma, \ \forall v_h \in V_h,$$

$$\int_\gamma \mu_h (U_h^{n+1} - g^{n+1}) \, d\gamma = 0, \ \forall \mu_h \in \Lambda_h, \tag{4.13}$$

$$\{U_h^{n+1}, \lambda_h^n\} \in V_h \times \Lambda_h \ \forall n \geq 0, \tag{4.14}$$

$$U_h^0 = U_{0h}, \ U_h^1 - U_h^{-1} = 2\triangle t U_{1h}. \tag{4.15}$$

The above scheme is "almost" *explicit*; implicitness concerns only those values of U_h^{n+1} at the nodes of $\mathcal{T}_h$ whose associated shape function support intersects γ (as shown in Figure 11)

Assuming that we use the *trapezoidal rule* to compute the first and third integrals in the left hand side of (4.12), U_h^{n+1} and λ_h^n are obtained via the solution of a linear system of the following form

$$\begin{cases} A_h U_h + B_h^t \lambda_h = b_h, \\ B_h U_h \qquad\ = c_h, \end{cases} \tag{4.16}$$

where A_h is a $N_h \times N_h$ symmetric and positive definite matrix (with $N_h = \dim V_h =$ the number of vertices of $\mathcal{T}_h$) and where B_h is a $M_h \times N_h$ matrix (with $M_h = \dim \Lambda_h$). Both matrices A_h and B_h have a very simple structure since:

1. matrix A_h is diagonal,
2. the entry b_{ij} of B_h is zero unless

$$\int_\gamma \theta_i w_j \, d\gamma \neq 0$$

where θ_i (resp., w_j) is the ith (resp., j^{th}) basis function of Λ_h (resp., V_h); indeed most entries of B_h are zero.

From the above properties of A_h and B_h, the $M_h \times M_h$ matrix $B_h A_h^{-1} B_h^t$ is sparse, easy to compute and/or to invert; it is "located along γ". We can either compute it once for all and then compute its Cholesky factors or solve (4.16) by a Uzawa/conjugate gradient algorithm operating in $\mathbf{R}^{M_h}$.

Remark 4.1. Using energy techniques it is easy to show that scheme (4.12)-(4.15) to be stable has to satisfy a relation such as

$$\triangle t < ch \tag{4.17}$$

where, in (4.17), c is independent of γ. $\square$

4.4 Numerical Experiments

We shall report in this paragraph the solution of three test problems by the fictitious domain method (briefly) discussed in previous three subsections. These problems are *two-dimensional* and concerned by the scattering of harmonic planar waves, such as $e^{i(\alpha t + \mathbf{k} \cdot \mathbf{x})}$ (with $|\mathbf{k}| = 2\pi/\lambda$, λ being the wave length), by *perfectly reflecting* obstacles, namely a *disk*, a *NACA0012 airfoil* and a *semi-open cavity*. For all these test problems the artificial boundary Γ has been located at a distance of 3 λ, at least, from the scattering obstacle. We time integrate system (4.1)-(4.4) until we reach periodic solution.

(i) *First Test Problem: Scattering by a disk*

If we uses the notation of Figure 10, the scattering obstacle ω is a disk of radius .25 meter. The frequency is .6 GHz, implying (since the speed of light is $c = 3 \times 10^8$ meter/sec.) that the wave length is .5 meter. We suppose that ω is "illuminated" by a wave coming from the left and propagating in the horizontal direction. Concerning discretization we have used a triangulation $\mathcal{T}_h$ consisting of 25,088 triangles and 12,769 vertices with $h = \lambda/16$. We have taken $\triangle t = T/25$ (with $T = 2\pi/\alpha$).

For the present test problem the exact solution is known when Γ is located at infinity, i.e., when $\Omega = \mathbf{R}^2$. On Figures 12 to 17 we have compared the exact and computed real and imaginary parts of the scattered wave; we have compared the values taken by the exact and computed solutions on the half-lines containing the center of ω and parallel to the incidence direction (Figures 12 and 13), opposite to the incidence direction (Figures 14 and 15) and orthogonal to the incidence direction (Figures 16 and 17). Further calculations (to be reported elsewhere) show that, for h and $\triangle t$ sufficiently small, the main source of error is the replacement of the original unbounded domain $\mathbf{R}^2 \setminus \bar{\omega}$ by a bounded computational one with (4.3) specified on the artificial boundary Γ. Indeed the error decreases if for h and $\triangle t$ sufficiently small the distance from Γ to ω increases.

(ii) *Second Test Problem: Scattering by a NACA0012 airfoil*

The obstacle is a NACA0012 airfoil. The frequency is 3 GHz, implying that $\lambda = .1$ meter. The length of the airfoil chord is 4λ. The airfoil is illuminated by a wave coming from the left with a 45° angle of incidence. The computational domain Ω is the rectangle $(-.5, .5) \times (-.4, .4)$ shown on Figure 18 together with the triangulation $\mathcal{T}_h$ used for computation; $\mathcal{T}_h$ consists of 40,960 triangles and 20,769 vertices, with $h = \lambda/16$. Figure 19 shows a more detailed view of the mesh, close to the airfoil. For the time step we have used $\triangle t = T/25$. On Figures 20 and 21 we have represented the real and imaginary components of the *total* field. A shadow region clearly appears above the airfoil in the illumination direction.

(iii) *Third Test Problem: Scattering by a semi-open cavity*

For this test problem the obstacle Ω is a semi-open cavity like the one shown in Figure 22. We suppose that the frequency of the illuminating wave is 3 GHz, implying that $\lambda = .1$ meter. The illuminating wave is coming from the lower left of ω with a $30°$ angle of incidence. The internal length of ω is 4λ, its thickness is $\lambda/5$ and its external height is 1.4λ. On Figure 22 we have also shown part of the triangulation $\mathcal{T}_h$ used for the calculation: it consists of 60,384 triangles and 30,545 vertices. For our calculations we have used $h = \lambda/20$ and $\triangle t = T/50$. On Figures 23 and 24 we have represented the real and imaginary components of the scattered field, respectively.

5 Conclusion

Fictitious domain methods seem well suited to the numerical solution of linear and nonlinear partial differential equations of various types, associated to domains with curved boundaries. With respect to the methods discussed in this article we think that several further issues are worth investigating such as the speed up of the iterative methods described in this article and the use of higher order approximations in order to decrease the number of grid points.

Acknowledgements

We would like to acknowledge the helpful comments and suggestions of the following individuals: L. C. Cowsar, E. J. Dean, G. H. Golub, J. W. He, Y. Kuznetsov, W. Lawton, P. Le Tallec, J. Pasciak, M. Ravachol, P. Joly, J. Weiss, R. O. Wells, M. F. Wheeler, O. B. Widlund, and X. Zhou.

The support of the following corporations and institutions is acknowledged: AWARE, Dassault Aviation, INRIA, Texas Center for Advanced Molecular Computation, University of Houston, Université P. et M. Curie. We also benefited from the support of the NSF (Grants DMS 8822522, DMS 9112847, INT 8612680 and DMS 9217374), the Texas Board of Higher Education (Grants 003652156ARP and 003652146ATP), DRET (Grant 89424), DARPA (Contracts AFOSR F49620-89-C-0125 and AFOSR-90-0334) and again the NSF under the HPCC Grand Challenge Grant ECS-952.

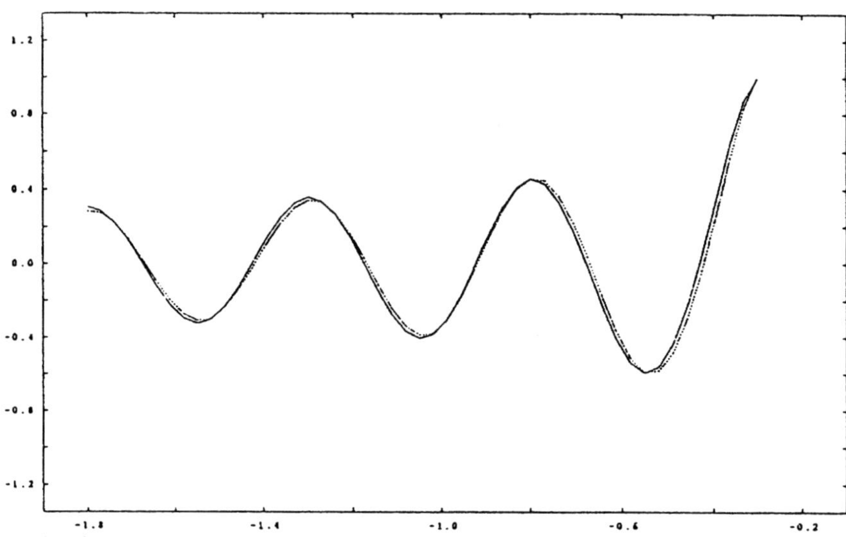

Figure 12

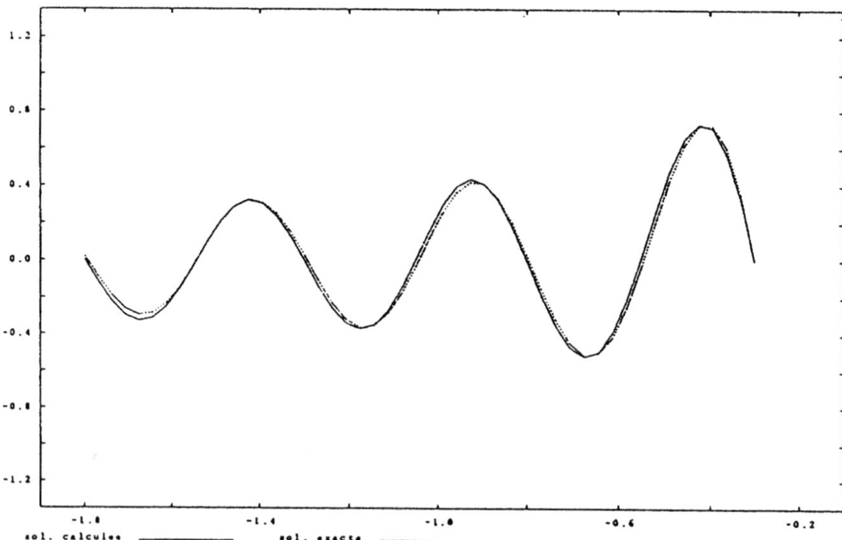

Figure 13

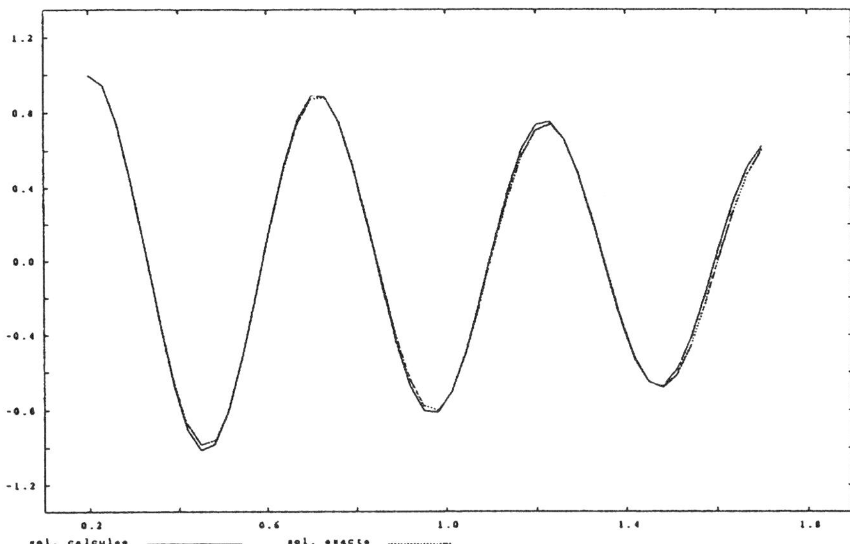

Figure 14

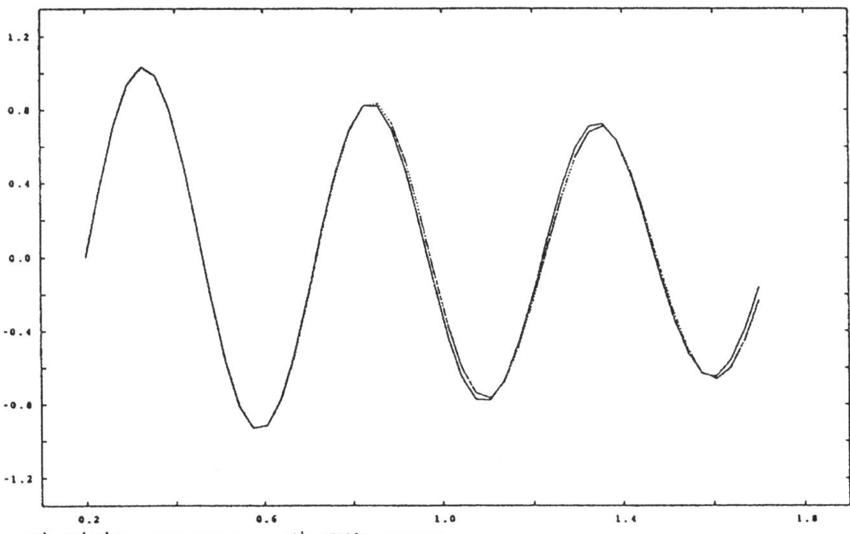

Figure 15

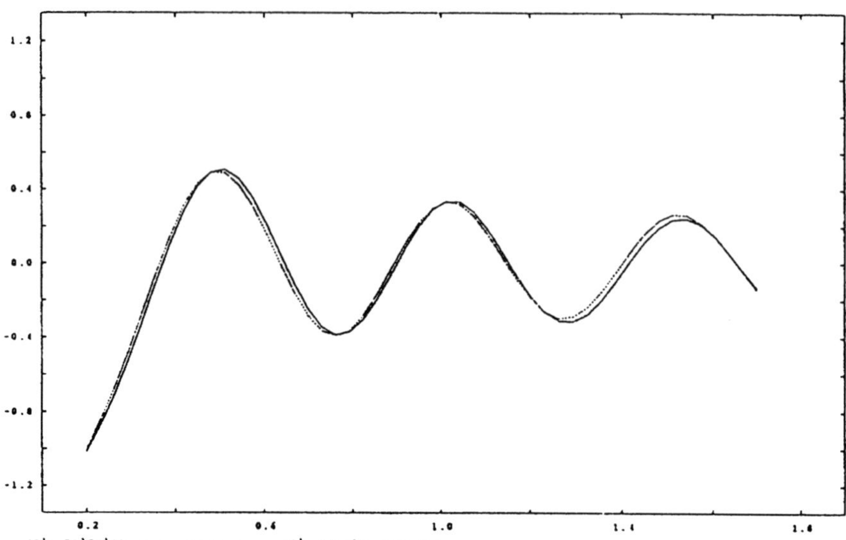

Figure 16

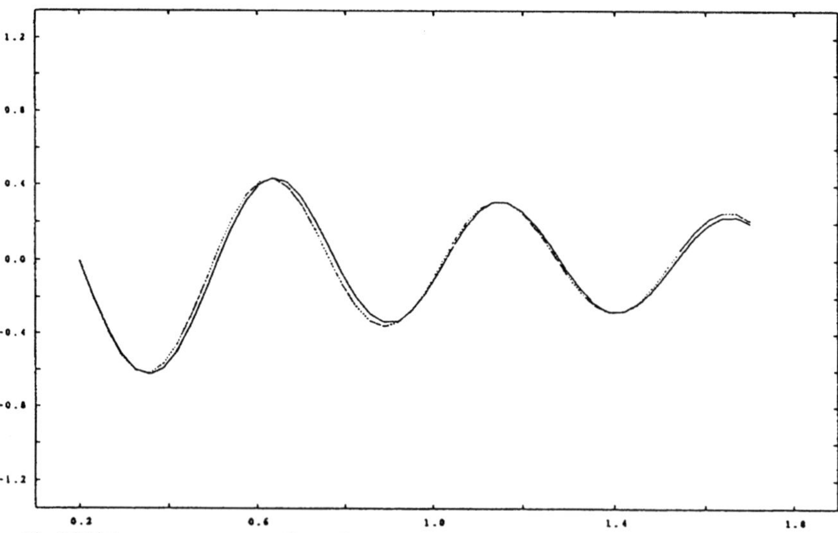

Figure 17

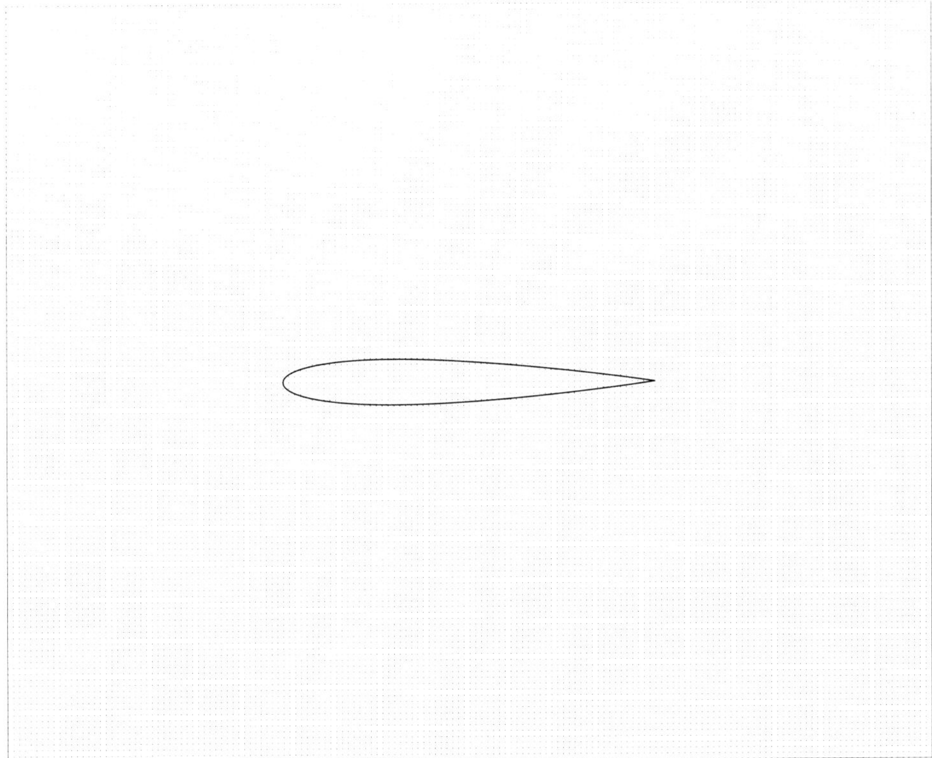

Figure 18

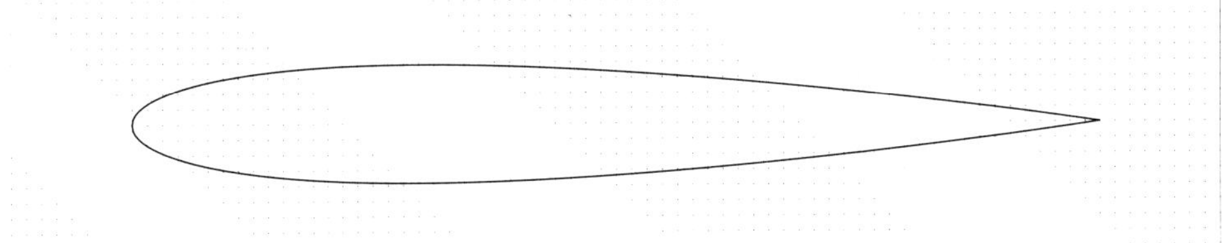

Figure 19

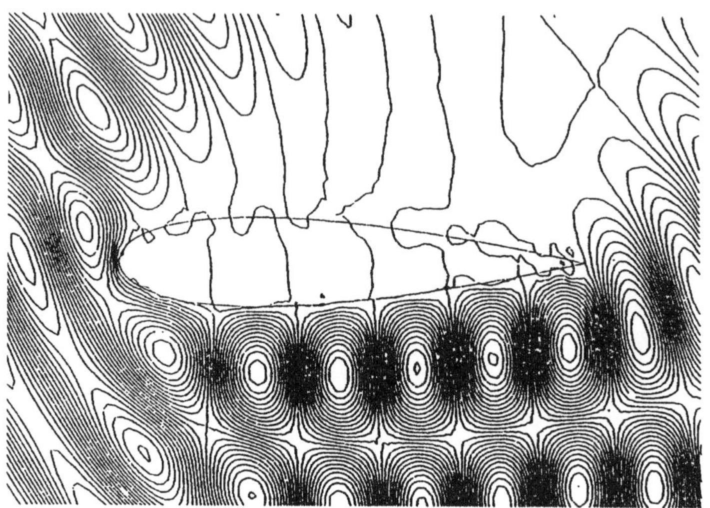

Figure 20

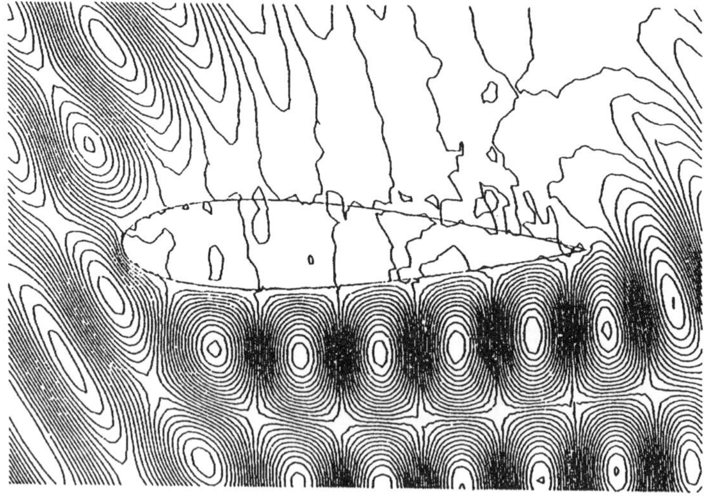

Figure 21

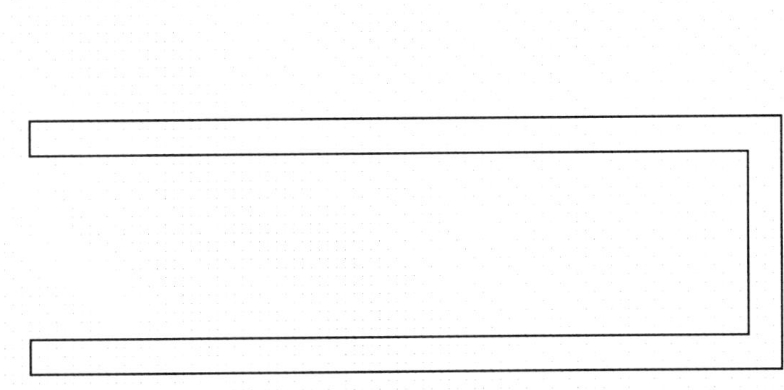

Figure 22

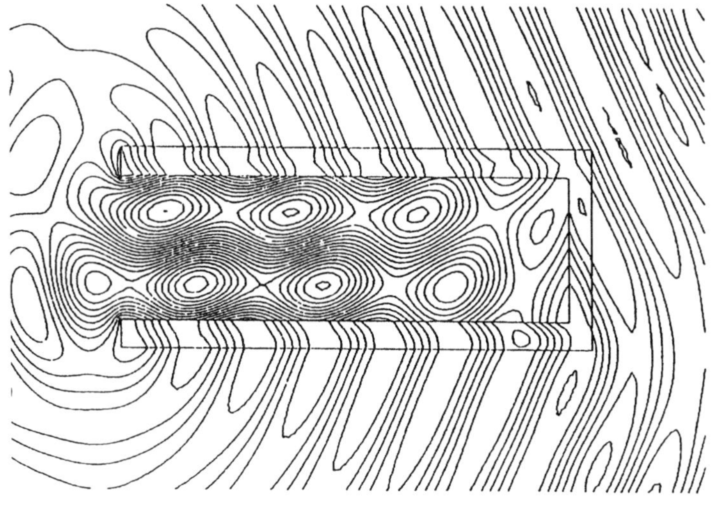

Figure 23

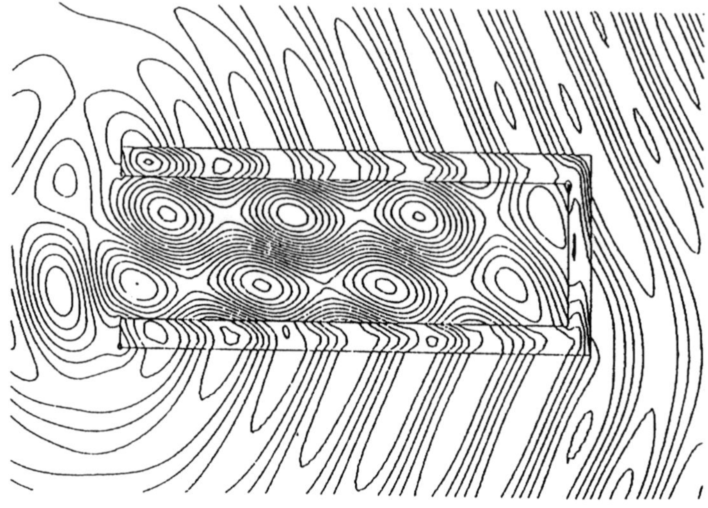

Figure 24

References

1. Proskurowski W., Widlund O. (1976) On the numerical solution of Helmholtz equation by the capacitance matrix method. *Math. Comp.*, 30: 433-468.
2. O'Leary D.P., Widlund O. (1979) Capacitance matrix methods for the Helmholtz equation on general three-dimensional regions. *Math. Comp.*, 33: 849-879.
3. Marchuk G.I., Kuznetsov Y.A., Matsokin A.M. (1986) Fictitious domain and domain decomposition methods. *Sov. J. Num. Anal. Math. Modeling*, 1: 3-35.
4. Finogenov S.A., Kuznetsov Y.A.(1988) Two-stage fictitious component methods for solving the Dirichlet boundary value problem. *Sov. J. Num. Anal. Math. Modeling*, 3: 301-323.
5. Glowinski R., Pan T.W., Periaux J. (1994) A fictitious domain method for Dirichlet problem and applications. *Comp. Meth. Appl. Mech. Eng.*, 111: 283-303.
6. Glowinski R., Pan T.W., Periaux J. (1994) A fictitious domain method for external incompressible viscous flow modeled by Navier-Stokes equations. *Comp. Meth. Appl. Mech. Eng.*, 112: 133-148.
7. Glowinski R., Pan T.W., Periaux J. (1995) A Lagrange multiplier/fictitious domain method for the Dirichlet problem. Generalization to some flow problems. *Japan J. of Industrial and Applied Mathematics*, 12: 87-108.
8. Glowinski R., Kearsley A.J., Pan T.W., Periaux J. (1995) Numerical simulation and optimal shape for viscous flow by fictitious domain method. *International J. for Numerical Methods in Fluids*, 20: 695-711.
9. Glowinski R. (1984) *Numerical methods for nonlinear variational problems*, Springer-Verlag, New York.
10. Girault V., Glowinski R. (1995) Error analysis of a fictitious domain method applied to a Dirichlet problem. *Japan J. of Industrial and Applied Mathematics*, 12: 487-514.
11. Glowinski R., Pan T.W., Periaux J. (1993) A one shot domain decomposition/fictitious domain method for the Navier-Stokes equations. In *Domain decomposition methods in scientific and engineering computing*, D.E. Keyes and J. Xu eds., AMS, Providence, RI, pp. 211-222.
12. Feng J., Joseph D.D., Glowinski R., Pan T.W. (1995) A three-dimensional computation on the force and moment on an ellipsoid settling slowly through a viscoelastic fluid. *J. of Fluid Mechanics*, 283: 1-16.
13. Bristeau M.O., Glowinski R, Periaux J. (1987) Numerical methods for the Navier-Stokes equations. Applications to the simulation of compressible and incompressible viscous flow. *Computer Phys. Rep.*, 6: 73-187.
14. Dean E.J., Glowinski R, Li C.H. (1989) Supercomputer solutions of partial differential equation problems in Computational Fluid Dynamics and in Control. *Computer Phys.*

Comm., 53: 401-439.

15. Glowinski R., Le Tallec P. (1989) *Augmented Lagrangian and operator splitting methods in nonlinear mechanics*, SIAM, Philadelphia.

16. Glowinski R. (1991) Finite element methods for the numerical simulation of incompressible viscous flow. Introduction to the control of the Navier-Stokes equations. In *Vortex Dynamics and Vortex Methods*, C.R. Andersonn and C. Greengard eds., Lectures in Applied Mathematics, **28**, AMS, Providence, R.I., pp. 219-301.

17. Marchuk G.I. (1990) Splitting and alternate direction methods. In *Handbook of Numerical Analysis, Vol. I*, P.G. Ciarlet and J.L. Lions eds., North-Holland, Amsterdam, pp. 197-462.

18. Glowinski R. (1985) Viscous flow simulation by finite element methods and related numerical techniques. In *Progress and Supercomputing in Computational Fluid Dynamics*, E. M. Murman and S. S. Abarbanel eds., Birkhauser, Boston, pp. 173-210.

19. Glowinski R. (1986) Splitting methods for the numerical solution of the incompressible Navier-Stokes equations. In *Vistas in Applied Mathematics*, A.V. Balakrishnan, A.A. Dorodnitsyn and J.L. Lions, eds., Optimzation Software, New York, pp. 57-95.

20. Kuznetsov Y.A., Wheeler M.F. (1995) Optimal order substructuring preconditioners for mixed finite element on nonmatching grids. *East-West J. of Num. Math.*, 3: 127-143.

Newton-Krylov-Schwarz:
An Implicit Solver for CFD

XIAO-CHUAN CAI[1], DAVID E. KEYES[2], and

V. VENKATAKRISHNAN[3]

1   ABSTRACT

Newton-Krylov methods and Krylov-Schwarz (domain decomposition) methods have begun to become established in computational fluid dynamics (CFD) over the past decade. The former employ a Krylov method inside of Newton's method in a Jacobian-free manner, through directional differencing. The latter employ an overlapping Schwarz domain decomposition to derive a preconditioner for the Krylov accelerator that relies primarily on local information, for data-parallel concurrency. They may be composed as Newton-Krylov-Schwarz (NKS) methods, which seem particularly well suited for solving nonlinear elliptic systems in high-latency, distributed-memory environments. We give a brief description of this family of algorithms, with an emphasis on domain decomposition iterative aspects. We then describe numerical simulations with Newton-Krylov-Schwarz methods on aerodynamics applications emphasizing comparisons with a standard defect-correction approach, subdomain preconditioner consistency, subdomain preconditioner quality, and the effect of a coarse grid.

2   INTRODUCTION

Several trends contribute to the importance of parallel implicit algorithms in CFD. Multidisciplinary analysis and optimization put a premium on the ability of algorithms to achieve low residual solutions rapidly, since analysis codes for individual components

---

[1] Department of Computer Science, University of Colorado-Boulder, Boulder, CO 80309-0430, USA. `cai@cs.colorado.edu`

[2] Department of Computer Science, Old Dominion University, Norfolk, VA 23529-0162 and ICASE, MS 132C, NASA LaRC, Hampton, VA 23681-0001, USA. `keyes@icase.edu`

[3] ICASE, MS 132C, NASA LaRC, Hampton, VA 23681-0001, USA. `venkat@icase.edu`

Domain Decomposition Methods in Sciences and Engineering, edited by R. Glowinski *et al.*

are typically solved iteratively and their results are often differenced for sensitivities. Problems possessing multiple scales provide the classical motivation for implicit algorithms and arise frequently in locally adaptive contexts or in dynamical contexts with multiple time scales, such as aero-elasticity. Meanwhile, the never slackening demand for resolution and prompt turnaround forces consideration of parallelism, and, for cost effectiveness, particularly parallelism of the high-latency, low-bandwidth variety represented by workstation clusters.

A Newton-Krylov-Schwarz (NKS) method combines a Newton-Krylov (NK) method such as nonlinear GMRES [BS90], with a Krylov-Schwarz (KS) method, such as additive Schwarz [DW87]. The key linkage is provided by the Krylov method, of which the restarted form of GMRES [SS86] is perhaps the best-known example for nonselfadjoint problems. From a computational point of view, the most important characteristic of a Krylov method for the linear system $Au = f$ is that information about the matrix A needs to be accessed only in the form of matrix-vector products in a small number (relative to the dimension of the matrix) of carefully chosen directions. NK methods are suited for nonlinear problems in which it is unreasonable to compute or store a true Jacobian. However, if the Jacobian A is ill-conditioned, the Krylov method will require an unacceptably large number of iterations. The system can be transformed into the equivalent form $B_1^{-1} A B_2^{-1} v = B_1^{-1} f$, where $v = B_2 u$, through the action of left and right preconditioners, B_1 and B_2. It is in the choice of preconditioning where the battle for low computational cost and scalable parallelism is usually won or lost.

In KS methods, the preconditioning is introduced on a subdomain-by-subdomain basis, which provides good data locality for parallel implementations over a range of granularities, and allows significant architectural adaptivity. The emphasis today is on operation count complexity and parallel efficiency, which means that Schwarz is usually employed with very modest subdomain overlap and in a two-level form, in which a small global problem is solved together with the local subdomain problems at each iteration. Mathematically, if $Au = f$ arises as the linearized correction step of a discretized PDE computation, Schwarz operates by:

1. Decomposing the space of the solution u: $\mathcal{U} = \sum_k \mathcal{U}_k$;
2. Finding the restriction of A to each $\mathcal{U}_k$: $A_k = R_k A R_k^T$, for some restriction operators $R_k : \mathcal{U} \to \mathcal{U}_k$ and extension operators $R_k^T : \mathcal{U}_k \to \mathcal{U}$;
3. Forming B^{-1} from the A_k^{-1}, where the inverse of A_k is well defined within the kth subspace.

In Schwarz-style domain decomposition, the subspace $\mathcal{U}_k$ corresponding to subdomain k is the span of nodal basis or other expansion functions with support over the subdomain. A practical Schwarz preconditioner is

$$B^{-1} \equiv \sum_k R_k^T (\tilde{A}_k)^{-1} R_k, \tag{1}$$

where $\tilde{A}_k$ is a convenient approximation to $A_k \equiv R_k A R_k^T$. In this paper, $\tilde{A}_k$ is usually an incomplete LU (ILU) factorization of A_k, with modest fill permitted. For $k = 1, 2, \ldots$, the R_k and R_k^T are simply gather and scatter operators, respectively, one for each subdomain with small overlap between the subdomains. For an optional

$k = 0$ term corresponding to the coarse space, R_0 represents a full-weighting restriction operator in the sense of multigrid, and R_0^T is the corresponding prolongation. We never actually assemble either A or B^{-1} globally. Rather, when their action on a vector is needed, a processor governing each subdomain executes local operations, after receiving a thin buffer of data required from its neighbors to complete stencil operations on the boundary of the subdomain. For the assembly and solution of the coarse-grid component of the preconditioner, data exchanges further than nearest neighbor must generally occur.

The two-level form of additive Schwarz can be proved to possess mesh-independent and granularity-independent condition number in elliptically dominated problems, including nonsymmetric and indefinite problems, when the coarse global and fine local operators are solved with sufficient precision. Ref. [CGK94] contains several examples demonstrating this optimality when exact subdomain solvers are used, and thus shows their superiority to global incomplete LU factorizations. Architecturally adaptive strategies for dealing with the coarse-grid component of the preconditioner are outlined in [GK89]. The collection [KX95] is representative of the state of the art of algorithms, applications, and parallel implementations.

The NKS technique is compared in this paper against a defect correction algorithm common to many implicit codes. The objective of either algorithm is to solve the steady-state conservation equations $f(u) = 0$ through the pseudo-transient form $\frac{\partial u}{\partial t} + f(u) = 0$, where the time derivative is approximated by backwards differencing, with a time step that ultimately approaches infinity. A standard defect correction approach employs an accurate right-hand side residual discretization, $f_{high}(u)$, and a convenient left-hand side Jacobian approximation, $J_{low}(u)$, based on a low-accuracy residual $f_{low}(u)$, to compute a sequence of corrections, $\delta u \equiv u^{n+1} - u^n$. Computational short-cuts are employed in the creation of the left-hand side matrix, which may, for instance, be stabilized by a degree of first-order upwinding that would not be acceptable in the discretization of the residual itself.

The so-called "defect" is $f_{high}(u) - f_{low}(u)$, and the nonlinear defect correction scheme to drive $f_{high}(u)$ to zero is to solve approximately for u^{n+1} in

$$f_{low}(u^{n+1}) = f_{low}(u^n) - f_{high}(u^n), \tag{2}$$

which may be linearized as

$$J_{low}(u^n)\delta u = -f_{high}(u^n). \tag{3}$$

In the case of pseudo-transient computations, the approximate Jacobian J_{low} is based on a low-accuracy residual:

$$J_{low} = \frac{D}{\delta t} + \frac{\partial f_{low}}{\partial u}, \tag{4}$$

where D is a scaling matrix. It is required either to solve with J_{low}, itself, or with some further algebraic or parallel approximation, $\tilde{J}_{low}$. Inconsistency between the left- and right-hand sides prevents the use of large time steps, δt, and prevents (3) from being a true Newton method.

A Newton-Krylov approach employs a (nearly) consistent left-hand side obtained by directionally differencing the actual residual, f_{high}:

$$J_{high}(u^n)\,\delta u = -f_{high}(u^n), \tag{5}$$

in which the action of J_{high} on a vector is obtained through directional differencing, for instance,

$$J_{high}(u^n)v \approx \frac{1}{h}\left[f_{high}(u^n + hv) - f_{high}(u^n)\right],\tag{6}$$

where h is a small parameter. The operators on both sides of (5) are based on consistent high-order discretizations; hence time steps can be advanced to values as large as linear conditioning permits, recovering a true Newton method in the limit.

In practice, the convergence of the method is sensitive to the choice of h in (6), which is not entirely trivial. When u and v are comparably scaled, it should ideally sit near the square-root of the machine unit roundoff, $\sqrt{\varepsilon_{mach}}$, or around 10^{-7}–10^{-8} in 64-bit precision. Smaller values improve the Taylor approximation upon which (6) is based. Larger values preserve more significant digits when the perturbed residuals on the right-hand side of (6) are differenced in finite precision. In a nondimensionalized formulation, the elements of u^n in (6) will have an RMS of approximately unity, but the elements of v will have an RMS smaller than unity by a factor of $\sqrt{n}$, where n is the dimension of the discrete unknown vector, since GMRES calls the matrix-vector evaluation routine with $||v||_2 = 1$. We therefore set h to be $\sqrt{n \cdot \varepsilon_{mach}}$. When less is known about the scaling of u^n and v, a reasonable choice is $\sqrt{\varepsilon_{mach}} \cdot (u^n, v)/||v||^2$, with guard code to set $h = \sqrt{\varepsilon_{mach}}$ if $||v||$ is too small. For a fuller discussion, see [BS90]. For numerical experiments demonstrating the importance of the relative scaling of h in the CFD context, see [KM93, NWAK95].

Preconditioning (5) by $\tilde{J}_{low}$, for instance on the left, as in

$$(\tilde{J}_{low})^{-1}J_{high}(u^n)\, \delta u = -(\tilde{J}_{low})^{-1}f_{high}(u^n),\tag{7}$$

shifts the inconsistency from the nonlinear to the linear aspects of the problem. This should be contrasted with the customary preconditioned form of (3),

$$(\tilde{J}_{low})^{-1}J_{low}(u^n)\, \delta u = -(\tilde{J}_{low})^{-1}f_{high}(u^n).\tag{8}$$

At this level of abstraction, it is not clear which is better — many nonlinear steps with cheap subiterations (8), or a few nonlinear steps with expensive subiterations (7). Execution time comparisons are more practical arbiters than are rates of convergence for the steady-state residual norm, but running times are sensitive to parametric tuning as well as to architectural parameters. We present a comparison of (7) and (8) in Section 4.

A more comprehensive set of comparisons of this type, comparing (7), (8), and

$$(\tilde{J}_{high})^{-1}J_{high}(u^n)\, \delta u = -(\tilde{J}_{high})^{-1}f_{high}(u^n)\tag{9}$$

may be found in [JF95]. Of course, (9) relies on possessing the full high-order Jacobian, and is not a matrix-free method.

We might mislead if we closed this section while failing to emphasize the importance of a globalization strategy when using any of the methods (7–9). Pseudo-transient continuation is usually recommended when a Newton-like method is used on flow problems in primitive variables. In such cases, the steady-state nonlinear residual norm should not be expected to decrease monotonically, and step-selection strategies should not be geared to monotonic decrease. Potential- or streamfunction-based formulations can more confidently be posed directly as steady-state problems, but in this case damping strategies for λ^n in $u^{n+1} = u^n + \lambda^n \delta u$ may be critical to convergence. Strategies for λ^n that provide as a minimum that $\|f(u^{n+1})\| \leq \|f(u^n)\|$ may need to be supplemented by feasibility checks on the components of u^{n+1} and/or a strategy that prevents spuriously large (though feasible) fluctuations in the components. In addition, artificial continuation parameters, such as upwinding strength, may enhance the convergence of globally divergent or slowly convergent iterations with near-discontinuities in the solution. For transonic potential computations, we have found invaluable the practical advice on "viscosity damping" in Section 8 of [YMB+91].

3 PARALLEL SCALABILITY OF KRYLOV-SCHWARZ

Practical scalable parallelism is one of the major motivations for research on and implementation of domain decomposition methods in CFD, the operative buzzwords being "faster, bigger, and cheaper." Though it would be premature to attempt to draw conclusions about the optimal algorithm/architecture combination for a given CFD analysis, we offer some experimental evidence for the excellent scalability of Krylov-Schwarz methods on a simple problem for which it is relatively easy to isolate the factors that trade off against each other in any such study. Without digressing into the refined nomenclature of scability analysis [Hwa93], we loosely call an algorithm/architecture combination "scalable" if its parallel efficiency is constant asymptotically, in any of several coordinated limits of discrete problem size n and parallel granularity p.

For an iterative numerical method, in which the total execution time $T(n,p)$ is the product of an iteration count $I(n,p)$ with an average cost-per-iteration $C(n,p)$, it is useful to separate the parallel efficiency into two factors: numerical efficiency and implementation efficiency. Numerical efficiency η_n measures the degradation of the convergence rate as the problem is scaled, and implementation efficiency η_i measures the degradation in the cost per iteration as the problem is scaled. For instance, we may take $\eta_n \equiv I(n,1)/I(n,p)$ and $\eta_i \equiv C(n,1)/[p \cdot C(n,p)]$. The numerical efficiency is usually very difficult to predict for a nonlinear problem, particularly when refining the grid (increasing n) resolves new physics. However, for certain domain decomposition methods applied to model linear problems with smooth solutions, the relative numerical efficiency $I(n_1,p_1)/I(n_2,p_2)$, with $p_2 > p_1$ and $n_1/p_1 = n_2/p_2$, can be proved to be 100% asymptotically.

The proof relies on the link between the rate of convergence of Krylov methods and the condition number of the (preconditioned) operator $B^{-1}A$, and on the link between the condition number and the extremal eigenvalues in the symmetric case, which can be estimated by Rayleigh quotients. Upper and lower bounds on the condition number of $B^{-1}A$ may be constructed that are independent of the mesh cell diameter h and the subdomain diameter H, or that depend only upon their ratio. In turn, h and H can be inversely related to n and p in simple problems. The theory, which has evolved over a decade to cover nonsmooth, nonsymmetric and indefinite problems, as well as nonnested spaces, is digested among other places in [CM94] and [SBG95]. Two such numerically optimal methods for the scalar self-adjoint elliptic problem are the two-level additive Schwarz method [DW87] and BPS-I [BPS86], which is a wire-

Table 1   Fixed-size scalability results – Poisson problem

# grid cells	# proc.	# iter.	seconds	sec./iter.
262,144	1	1*	183.5	183.5
262,144	4	8	325.7	40.7
262,144	16	12	96.7	8.1
262,144	64	12	17.6	1.5

Table 2   Fixed-memory-per-node scalability results – Poisson problem

# grid cells	# proc.	# iter.	seconds	sec./iter.
16,384	1	1*	8.75	8.75
65,536	4	7	58.4	8.34
262,144	16	12	96.7	8.06
1,048,576	64	11	91.2	8.29

basket (Schur complement-based) method. Before presenting parallel CFD results, we illustrate the performance achievable by such methods on contemporary parallel systems. Implementation efficiencies have improved substantially since we conducted our first such study in 1985.

A message-passing code for the Poisson problem on a unit square described in [KG87] was converted to MPI [MPI94] and ported to several machines. With convergence defined as five orders of magnitude reduction in (unpreconditioned) residual, as in [KG87], we tested both fixed-size scalability and fixed-memory-per-node scalability on an Intel Paragon with 1, 4, 16, or 64 subdomains, with one subdomain per processor. The results for a fixed-size 512×512 grid are shown in Table 1. Table 2 is based on a problem size that grows from 16K to 1M unknowns, with a 128×128 subdomain problem on every processor. (The third row is common to both tables.)

On each subdomain, a direct FFT-based method is employed, so that only one iteration is required in the uni-processor case. Asymptotically, approximately 12 iterations are required, independent of the granularity. As seen in the column "sec./iter." in Table 2, the implementation efficiency is near perfect in this granularity range. Problems can be solved in constant time as resolution and processing power are increased in proportion. (We expect implementation efficiency to degrade eventually at higher granularity, due to a sequential bottleneck in the coarse-grid part of the preconditioner.) Consulting the "sec./iter." column of Table 1, we note a super-unitary implementation efficiency – as p increases by a factor of 4, runtime decreases by more than a factor of 4. This is attributable to the cacheing or paging advantages of domain-based array blocking, which are clearly more important than communication effects in this range of n and p.

In general CFD applications, finding a cost-effective coarse-grid operator is not straightforward, and one is often resigned to a Schwarz-preconditioned operator that

Table 3   Inter-architecture comparison – Poisson problem

Machine	# proc.	seconds
IBM SP2	16	65.2
Intel Paragon	64	91.2
SPARC Cluster	16	124.5

deteriorates in numerical efficiency as the granularity of the decomposition increases. In such cases, optimizing execution time as a function of granularity is difficult, apart from numerical experimentation.

The largest problem solved on the Paragon (a grid of 1024×1024) was also run on 16 nodes of an IBM SP2 and on 16 SPARCstations connected by Ethernet (during a relatively quiescent period of the CPUs and the network). The overall runtime results are shown in Table 3. It is apparent that large memory-per-node workstations on an Ethernet are competitive with more expensive machines built around proprietary dedicated-link interconnects (a mesh for the Paragon, a multi-stage bi-directional switch for the SP2). This is due to the relatively small communication-to-computation ratio for Krylov-Schwarz methods, and is encouraging for cost-effective large-scale CFD computations, at least for dedicated clusters. Parallel workstation cluster implementations [CGKT94] of structured-grid Euler problems reveal the vulnerability of highly synchronous algorithms, including Krylov methods with their frequent inner product calls, to a non-dedicated environment. The other three main sources of communication inefficiency in parallel algorithms, namely load imbalance, latency, and finite bandwidth, are believed to impose much less serious limits on the number of workstations that can be clustered together to solve PDEs than frequent synchronization of non-dedicated resources.

4   AERODYNAMICS APPLICATIONS

In this section, we present parallel numerical results for two different formulations of inviscid, subsonic compressible external flow over two-dimensional airfoils using Newton-Krylov-Schwarz. The formulation with greater fidelity to the flow physics is the set of Euler equations:

$$\nabla \cdot (\rho \mathbf{v}) = 0 \tag{10}$$

$$\nabla \cdot (\rho \mathbf{v}\mathbf{v} + pI) = 0 \tag{11}$$

$$\nabla \cdot ((\rho e + p)\mathbf{v}) = 0 \tag{12}$$

where ρ is the fluid density, $\mathbf{v}$ the velocity, p the pressure, and e the specific total energy, together with the ideal gas law, $p = \rho(\gamma - 1)(e - |\mathbf{v}|^2/2)$, where γ is the ratio of specific heats.

Under additional restrictive assumptions of irrotationality ($\mathbf{v} = \nabla \Phi$) and isentropy

$(\nabla(p/\rho^\gamma) = 0)$, one can derive [Hir90] a scalar equation for the velocity potential Φ:

$$\nabla \cdot (\rho \nabla \Phi) = 0. \tag{13}$$

We report on a model computation based on (13), which has the advantage of being simple in structure, and hence relatively easy to test algorithmic varieties upon. We then report on a computation based on a state-of-the-art unstructured grid solver for (10–12).

4.1 Full Potential Flow

A nonlinear full potential code has been built as a "laboratory" for parallel algorithms for nonlinear elliptic problems. A two-level Schwarz preconditioner was implemented on top of an early version of the PETSc [GMS95] library, offering variable-fill incomplete factorization, a variety of subdomain preconditionings, variable subdomain overlap, and a coarse grid of variable density. The coarse grid is not necessarily nested in the underlying decomposition into subdomains, which would be an impractical restriction in real world problems with adaptively refined meshes. Full details may be found in [CGK95]; we summarize some representative trends.

Consider the scalar nonlinear BVP

$$\nabla \cdot (\rho(||\nabla\Phi||)\nabla\Phi) = 0,$$

where

$$\rho = \rho_\infty \left(1 + \frac{\gamma - 1}{2} M_\infty^2 (1 - \frac{q^2}{q_\infty^2})\right)^{\frac{1}{(\gamma-1)}},$$

and where $q \equiv ||\nabla\Phi||$. $M_\infty \equiv q_\infty/a_\infty$ is the free-stream Mach number. Boundary conditions of Neumann or Dirichlet type are derived from inviscid boundary conditions for the velocity. The simplest possible problem of aerodynamic interest is a thin nonlifting symmetric airfoil at zero angle of attack. The airfoil lies along the x-axis, where its shape is parameterized by $f = y(x)$. So-called transpiration boundary conditions permit a uniform grid to be employed on a rectilinear domain. A complete set of boundary conditions, adequate for testing the nonlinear algebraic solver, if not for extracting accurate results about the underlying continuous problem, is:

- Upstream and far field: $\Phi = q_\infty \cdot x$,
- Downstream: $\Phi_{,n} = q_\infty$,
- Symmetry: $\Phi_{,n} = 0$,
- On parameterized airfoil $(y = f(x))$: $\Phi_{,n} = -q_\infty f'(x)$.

See [YMB$^+$91], which motivates our present study, for a more refined treatment of boundary conditions.

We study convergence rate and parallel efficiency as functions of the accuracy of the subdomain solvers, the overlap of the subdomains, the density of the coarse grid component of the preconditioner, and the granularity of the decomposition. All tests presented below are on a uniform grid of 250K unknowns (512×512) for flow over a NACA0012 airfoil at a free-stream Mach number of 0.5. In this problem, it is never necessary to use pseudo-time-stepping to reach the domain of convergence of Newton's method. The initial iterate is the simple uniform flow $\Phi(x, y) = \int_{x_0}^{x} q_\infty dx$.

Table 4 Effect of preconditioner level of fill – Full Potential problem

k	0	1	2	3	4	5
Newton	11	7	6	*5*	5	5
GMRES	494	277	222	*169*	167	166
Time	891.44	512.16	417.37	*325.01*	327.30	332.86

Table 5 Effect of subdomain overlap – Full Potential problem

$ovlp$	1	2	3	4	5
Newton	*5*	5	*5*	5	5
GMRES	117	92	*77*	66	64
Time	278.23	246.10	*222.75*	211.65	213.34

Each of Tables 4 through 6 examines the sensitivity of the convergence to a single preconditioner parameter with all others controlled. The decomposition of the domain into eight rectangular subdomains is also controlled. The total number of outer Newton steps, the accumulated number of inner GMRES steps, and the overall running time of the parallel computation are tabulated.

Table 4 shows the effect of varying the level of fill k in ILU(k). As k varies from 0 to the full discrete dimension of the local Jacobian matrix, the subdomain solves gain increasing exactness. However, there is a law of diminishing returns in convergence rate, and overall execution time actually increases after a minimum, as the cost per iteration begins to rise more rapidly than the number of iterations falls. It has been observed in other contexts for the same full potential equation [YAM$^+$93] that Schwarz methods are forgiving of inexactness in the individual subdomain solves. Furthermore, a factorization with a fixed level of fill is increasingly accurate as the subdomain over which it is defined gets smaller. Thus, for fine-grained computations, it is not cost-effective to work too hard on the individual subdomains. In this table, the overlap is fixed at $3h$ and the coarse grid is 4×5 — fewer than one coarse-grid vertex for every 10,000 fine-grid vertices.

Table 5 shows the effect of varying the overlap between subdomains. The $ovlp$ listed is the distance that each subdomain is extended into its neighbors; hence the overall zone of overlap is twice as wide. As overlap varies from one mesh cell diameter, h, to roughly half the subdomain diameter, convergence rate improves. (For additive methods, too much overlap can lead to deteriorating condition number, however.) Even in the regime of monotonically improving convergence with increase in overlap there is a law of diminishing returns, and overall execution time increases after a minimum, as the cost per iteration rises. In this table, an exact solver is used on all subdomains, and the coarse grid is 8×9.

Table 6 shows the effect of varying the coarse grid density, with exact subdomain solves and subdomain overlap of $3h$. Exceedingly modest coarse grids provide a major

Table 6 Effect of coarse-grid density – Full Potential problem

Coarse Grid	0×0	2×3	4×5	6×7	8×9
Newton	5	4	5	*5*	5
GMRES	164	95	98	*89*	77
Time	373.32	240.82	259.60	*245.02*	243.24

Table 7 Fixed-size scalability results – Full Potential code on an IBM SP2
(three different preconditioners)

# proc.	$k = 3$ ILU fill level		$3h$ subdomain overlap		6×7 coarse grid density	
	sec./iter.	rel. eff.	sec./iter.	rel. eff.	sec./iter.	rel. eff.
8	1.92	(1.00)	2.89	(1.00)	2.75	(1.00)
16	1.01	0.95	1.39	1.04	1.59	0.86
32	0.55	0.87	0.73	0.99	0.89	0.77

improvement over no coarse grid at all; however, this is a relatively smooth problem. As with the other parameters, the marginal benefit of effort spent on the coarse grid decreases rapidly and ultimately reverses as the cost per iteration overtakes improved convergence rate.

Table 7 explores the implementation scalability of the NKS method, as seen over the range of two successive doublings, from 8 to 16 and from 16 to 32 processors, for three fixed-size preconditioner combinations (corresponding to the parameter choices in the italicized columns of the first three tables). Dividing elapsed computation times by the number of GMRES inner iterations, to obtain an average cost per iteration on this fixed-size problem, yields the scalability results shown. There is a superunitary relative efficiency for one of the preconditioners, attributable to more favorable cache blocking. The fixed-size parallel efficiencies remain above 75% throughout the preconditioner parameter and granularity range considered.

4.2 Euler Flow

The problem of inviscid incompressible flow around a two-dimensional four-element airfoil in landing configuration was studied in terms of convergence rate and parallel performance in [Ven94], and the same code was converted to NKS form for the present study. The details of the discretization are left to the original reference. From [Ven94] we consider the vertex-based discretization with a first-order Roe scheme on the left (out of which we form $\tilde{J}_{low}^{-1}$), and a second-order Roe scheme on the right (which defines f_{high}). The flow is subsonic (Ma = 0.2), with an angle of attack of $5°$. Adaptively placed unstructured grids of approximately 6,000 and 16,000 vertices were decomposed into from 1 to 128 load-balanced subdomains, including all power-of-two granularities in between. We report below on the problem of 6,019 vertices, with

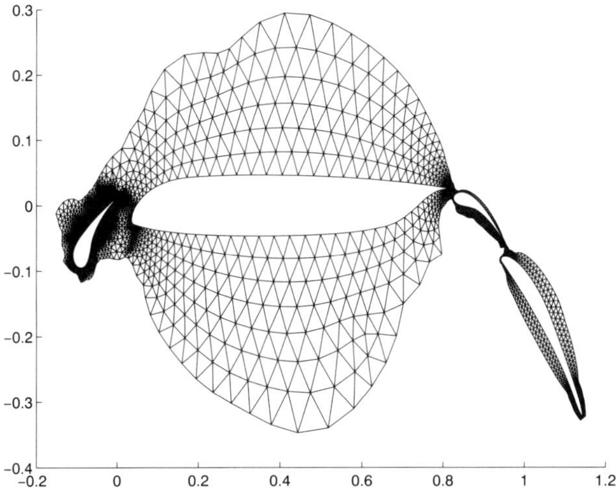

Figure 1   Zoom of the unstructured grid cells in the near field.

four degrees of freedom per vertex (giving 24,076 as the algebraic dimension of the discrete problem). This is certainly small by parallel computational standards, though it is probably reasonably adequate in two dimensions from a physical modeling point of view, since the unstructured grid is not restricted to quasi-uniformity, and mesh cells are concentrated into small regions between the airfoils requiring the greatest refinement. The clustering can be seen in Fig. 1, which shows just a near field subset of the grid. (The grid recedes into the far field with smoothly increasing cell sizes. If the entire grid is is scaled to the page size, the flaps are too small to be visible.)

Figure 2 compares the convergence histories of the defect correction and NKS solvers, over a range of time sufficient to permit the reduction of the residual of the NKS method to drop to within an order of magnitude of ε_{mach}. Both solvers utilize a residual-adaptive setting of the CFL number (related to the size of the time step δt in the pseudo-transient code), known as "switched evolution/relaxation" (SER) [ML85]. Starting from some small initial CFL number, CFL is adaptively advanced according to:

$$\mathrm{CFL}^{l+1} = \mathrm{CFL}^{l} \cdot \frac{||f(u)^{l-1}||}{||f(u)^{l}||}.$$

As $||f(u^{l})|| \to 0$, $\delta t \to \infty$. In practice, it is wise to bound the relative growth of CFL in any one step by some factor and/or to bound it asymptotically in the range of $10^{3} - 10^{6}$ to preserve a modest diagonal dominance for the linear subiterations. Since convergence is not generally monotonic in $||f(u)||$, CFL may also adaptively decrease, and it should be ratcheted away from too large a relative decrease, as well.

Both solvers use the same Schwarz preconditioner, namely one-cell overlap and point-block ILU(0) in each subdomain. NKS is clearly superior to defect correction in convergence rate, though the cost per iteration is sufficiently high that defect correction is faster in execution time up to a modest residual reduction. (The cross-over point in the right plot is at about a reduction of 10^{4} of the initial residual. A polyalgorithm,

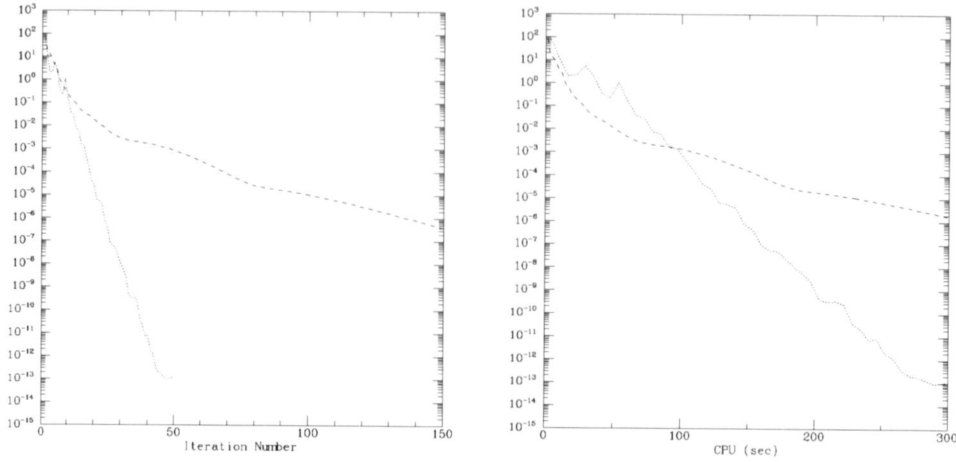

Figure 2   Norm of steady-state residual vs. iterations (left) and vs. execution
time on 32 nodes of the Intel Paragon (right) for the defect correction scheme
(dashed), and the NKS method (dotted).

initially defect correction then switched to NKS when defect correction prohibits fast
growth in CFL, may ultimately be much faster than either pure algorithm exclusively,
as demonstrated for a related problem in [NWAK95], and as found in preliminary
experiments for the present problem.) The asymptotic convergence rate is shown to
be linear, since we truncated the Newton iterations well above the tolerances necessary
to guarantee superlinear or quadratic convergence. Improving the constant in linear
convergence is reason enough to use the matrix-free split discretization method (5).
Table 8 compares the performance of the NKS version of the solver across three
doublings of the processor force of the Intel Paragon for this fixed-size problem. The
second-order evaluation of fluxes in $f_{high}(u)$ requires that first conserved variables, and
later their fluxes, be communicated across subdomain boundaries each time the routine
to evaluate the nonlinear residual is called. This imposes an extra communication
burden per iteration on the matrix-free NKS solver, relative to a method that explicitly
stores the elements of the Jacobian. Nevertheless, for residual norm reductions of
more than a few orders of magnitude, the parallelized NKS solver is faster than the
parallelized defect correction solver. The number of subdomains matches the number
of processors, so convergence rate of the preconditioned system degrades slowly with
increasing granularity, as coupling is lost in the preconditioner. However, the number
of Krylov vectors per Newton iteration is bounded (at 2 restart cycles of 25 each),
so the data translates directly to parallelization efficiency of the truncated Newton
method.

In this example, no coarse grid is used, but [Ven94] compares the defect correction
form of the algorithm with and without a coarse grid. The coarse grid appears
multiplicatively rather than additively, as in (1). The restriction operator consists
of summing subdomain boundary fluxes and the prolongation operator is essentially

Table 8   Wall-clock performance and relative parallel efficiency for
unstructured Euler code on an Intel Paragon.

# proc.	sec./iter.	rel. eff.
4	36.09	(1.00)
8	19.21	0.94
16	10.65	0.85
32	6.25	0.72

piecewise constant subdomain extension followed by a boundary relaxation process. On the original platform of the Intel iPSC/2, the convergence rate advantage of the coarse grid is nearly completely cancelled by the sequential bottleneck. The coarse grid aspect of the preconditioner demands further attention.

5   CONCLUSIONS AND RELATED EXTENSIONS

We have shown that steady aerodynamics problems in two different formulations (full potential and Euler) can be effectively solved, and cost-effectively solved in parallel, by NKS methods.

The NK technique has been compared with V-cycle multigrid on Euler and Navier-Stokes problems without parallelizing the preconditioning in [Key95, NWAK95]. For a subsonic unstructured grid example, NK trails multigrid in execution time by a factor of only about 1.5. This penalty can be accepted when it is realized that the NK method has the advantage of doing all of its computation without generation of a family of coarse unstructured grids (which is difficult for three-dimensional unstructured grids). This work has been extended to three-dimensional problems in [NWAK95].

Large-scale time-dependent problems suffering from multiple scales often require parallel implicit algorithms. The KS technique has been shown effective in the unsteady Navier-Stokes context in [CFS96]. In [CFS96], two of the same parameters explored herein (level of fill in the local ILU factorizations and subdomain overlap) are varied to produce a Schwarz preconditioner whose strength can be adjusted to adapt to the varying time-evolving ill-conditioning of the linear system arising at each implicit time step.

A variety of CFD applications are (or have inner) nonlinear elliptically-dominated problems amenable to solution by NKS algorithms, which are characterized by low storage requirements (for an implicit method) and locally concentrated data dependencies with small overlaps between the preconditioner blocks. The addition of a global coarse grid in the Schwarz preconditioner is often effective, where architecturally convenient. A deterrent to the widespread adoption of NKS algorithms is the large number of parameters that require tuning. Each component (Newton, Krylov, and Schwarz) has its own set of parameters, the most important of which, in our experience, is the convergence criterion for the inner Krylov subiterations. In large-scale, poorly preconditioned problems, including the test problems of this paper, tunings that guarantee quadratic convergence lead to unacceptable inner iteration counts and/or memory consumption. However, the plethora of parameters can be exploited, in principle, to produce optimal tradeoffs in space and time for a given problem class. Though parametric tuning is important to performance, conservative robust choices are not difficult.

References

[BPS86] Bramble J. H., Pasciak J. E., and Schatz A. H. (1986) The construction of preconditioners for elliptic problems by substructuring, I. *Mathematics of Computation* 47: 103–134.

[BS90] Brown P. and Saad Y. (1990) Hybrid Krylov methods for nonlinear systems of equations. *SIAM Journal of Scientific and Statistical Computing* 11: 450–481.

[CFS96] Cai X.-C., Farhat C., and Sarkis M. (1996) Schwarz methods for the unsteady compressible Navier-Stokes equations on unstructured meshes. In Glowinski R., Periaux J., Shi Z., and Widlund O. (eds) *Domain Decomposition Methods in Sciences and Engineering*. Wiley, Chichester.

[CGK94] Cai X.-C., Gropp W. D., and Keyes D. E. (1994) A comparison of some domain decomposition and ILU preconditioned iterative methods for nonsymmetric elliptic problems. *Journal of Numerical Linear Algebra and Applications* 1: 477–504.

[CGK95] Cai X.-C., Gropp W. D., and Keyes D. E. (1995) Parallel implementation of Newton-Krylov-Schwarz algorithms for the transonic full potential equation. ICASE Technical Report, in preparation.

[CGKT94] Cai X.-C., Gropp W. D., Keyes D. E., and Tidriri M. D. (1994) Newton-Krylov-Schwarz methods in CFD. In Hebeker F. and Rannacher R. (eds) *Proceedings of an International Workshop on Numerical Methods for the Navier-Stokes Equations*, pages 17–30. Vieweg Verlag, Braunschweig.

[CM94] Chan T. F. and Mathew T. (1994) Domain decomposition algorithms. *Acta Numerica*, pages 61–143.

[DW87] Dryja M. and Widlund O. B. (1987) An additive variant of the Schwarz alternating method for the case of many subregions. Technical Report 339, Courant Institute, NYU.

[GK89] Gropp W. D. and Keyes D. E. (1989) Domain decomposition on parallel computers. *Impact of Computing in Science and Engineering* 1: 421–439.

[GMS95] Gropp W. D., McInnes L. C., and Smith B. F. (1995) PETSc 2.0 users guide. Technical Report ANL 95/11, Argonne National Laboratory.

[Hir90] Hirsch C. (ed) (1990) *Numerical Computation of Internal and External Flows, Vol. 2.* Wiley, Chichester.

[Hwa93] Hwang K. (1993) *Advanced Computer Architectures: Parallelism, Scalability, and Programmability (Chap. 3).* McGraw-Hill, New York.

[JF95] Jiang H. and Forsyth P. A. (1995) Robust linear and nonlinear strategies for solution of the transonic Euler equations. *Computers & Fluids* 24: 753–770.

[Key95] Keyes D. E. (1995) Aerodynamic applications of Newton-Krylov-Schwarz solvers. In Deshpande S., Desai S., and Narasimha R. (eds) *Proceedings of the 14th International Conference on Numerical Methods in Fluid Dynamics*, pages 1–20. Springer Verlag, New York.

[KG87] Keyes D. E. and Gropp W. D. (1987) A comparison of domain decomposition techniques for elliptic partial differential equations and their parallel implementation. *SIAM Journal of Scientific and Statistical Computing*

8: s166–s202.

[KM93] Knoll D. A. and McHugh P. R. (1993) Inexact Newton's method solutions to the incompressible Navier-Stokes and energy equations using standard and matrix-free implementations. Technical Report 93-3332, AIAA.

[KX95] Keyes D. E. and Xu J. (eds) (1995) *Proceedings of the Seventh International Conference on Domain Decomposition Methods.* Number 180 in Contemporary Mathematics. AMS, Providence.

[ML85] Mulder W. and Leer B. V. (1985) Experiments with implicit upwind methods for the Euler equations. *Journal of Computational Physics* 59: 232–246.

[MPI94] MPI Forum (1994) MPI: A message-passing interface standard. *International Journal of Supercomputer Applications* 8(3/4).

[NWAK95] Nielsen E. J., Walters R. W., Anderson W. K., and Keyes D. E. (1995) Application of Newton-Krylov methodology to a three-dimensional unstructured Euler code. Technical Report 95-1733, AIAA.

[SBG95] Smith B. F., Bjorstad P. E., and Gropp W. D. (1995) *Domain Decomposition: Parallel Multilevel Algorithms for Elliptic Partial Differential Equations.* Cambridge Univ. Press, Cambridge.

[SS86] Saad Y. and Schultz M. H. (1986) GMRES: A generalized minimal residual algorithm for solving nonsymmetric linear systems. *SIAM Journal of Scientific and Statistical Computing* 7: 856–869.

[Ven94] Venkatakrishnan V. (1994) Parallel implicit unstructured grid Euler solvers. *AIAA Journal* 32: 1985–1991.

[YAM+93] Young D. P., Ashcraft C. C., Melvin R. G., Bieterman M. B., Huffman W. P., Johnson F. T., Hilmes C. L., and Bussoletti J. E. (1993) Ordering and incomplete factorization issues for matrices arising from the TRANAIR CFD code. Technical Report BCSTECH-93-025, Boeing Computer Services.

[YMB+91] Young D. P., Melvin R. G., Bieterman M. B., Johnson F. T., Samant S. S., and Bussoletti J. E. (1991) A locally refined rectangular grid finite element method: Application to computational fluid dynamics and computational physics. *Journal of Computational Physics* 92: 1–66.

Variational Inequalities for Navier-Stokes Flows
Coupled with Potential Flow through Porous Media

H.Kawarada*, H.Fujita** and H.Kawahara***

     *   University of Chiba

    **   Meiji University

   ***   Kawasaki Steel System Development

1. Introduction

The domain is made of two parts; In one part, the flow obeys the Navier-Stokes equations. In the second part, the flow is potential and driven by Darcy's law. On some parts of the interface between them, the fluid flows into both sides under the control of the threshold to the potential of friction type. This threshold describes the effect of surface tension of the fluid in capilaries which are regularly arrayed in the neighbourhood of the boundary. We show how this problem is formulated by a variational inequality: existence of solution, approximation by means of fictitious domain method via singular perturbations and numerical simulations are studied and presented.

2. Formulation

2.1. The geometry

The problem is discussed in R^2. The geometry is the rectangle $(0, L) \times (0, a)$ for the Navier-Stokes part and the rectangle $(0, L) \times (a, 2a)$ for the Darcy part, which we denote by Ω_0 and Ω_1 respectively. $(0, L) \times \{a\}$ is an interface between Ω_0 and Ω_1, which we denote by γ . The Darcy boundary is the upper horizontal one plus the halves of the vertical one, the union of which we denote by Γ_1.

The Navier-Stokes boundary is composed of the inflow boundary Γ_{in} and the outflow boundary Γ_{out}, which are the remaining two halves of the vertical boundaries, and the lower horizontal boundary Γ_w.

2.2. Notations

p : pressure ($p = p_k$ in $\Omega_k, k = 0, 1$)
u: velocity ($u = u_k$ in $\Omega_k, k = 0, 1$)
ϕ : potential for the flow in Ω_1 ($\phi = p_1 + \rho g x_2$, $\rho g = 1$)
n : unit outward normal vector to the boundary of Ω_0
n_1 : unit outward normal vector to the boundary of Ω_1
ν : effective viscosity (inverse of Reynolds number)
k : permeability coefficient of the flow in Ω_1
c : resistance coefficient of the flow in Ω_1 (inverse of k)

Domain Decomposition Methods in Sciences and Engineering, edited by R. Glowinski *et al.*
© 1997 John Wiley & Sons, Ltd.

g_n : threshold parameter (> 0) controlling the occurence of penetration on γ
f : external force acted in Ω_0 ($f = f_0 - \nabla x_2$)
β : velocity profile of Poiseuille type defined on Γ_{in}
ϵ : singular perturbation parameter (> 0)
u_n : normal component of u
u_t : tangential component of u
$S_n = \nu \frac{\partial u_{0n}}{\partial n} - p_0 + p_1$: difference between normal stress of the flow in Ω_0 and
$\qquad\qquad\qquad\qquad$ pressure of the flow in Ω_1
$S_t = \nu \frac{\partial u_{0t}}{\partial n}$: tangential stress of the flow in Ω_0
χ : characteristic function of Ω_1 in Ω.

2.3. The model problem

The model problem is described as follows ;

Find

$$u = \begin{cases} u_0 & \text{in } \Omega_0, \\ u_1 & \text{in } \Omega_1 \end{cases}$$

and

$$p = \begin{cases} p_0 & \text{in } \Omega_0, \\ p_1 & \text{in } \Omega_1 \end{cases}$$

such that

$$(2.1) \qquad -\nu \Delta u_0 + (u_0 \cdot \nabla)u_0 + \nabla p_0 = f \quad \text{in } \Omega_0$$

$$(2.2) \qquad \nabla \cdot u_0 = 0 \quad \text{in } \Omega_0$$

$$(2.3) \qquad c\, u_1 + \nabla p_1 = -\nabla x_2 \quad \text{in } \Omega_1$$

$$(2.4) \qquad \nabla \cdot u_1 = 0 \quad \text{in } \Omega_1$$

$$(2.5) \qquad |S_n| \le g_n \quad \text{on } \gamma$$

$$(2.6) \qquad g_n\, |u_n| + S_n \cdot u_n = 0 \quad \text{on } \gamma$$

$$(2.7) \qquad S_t = 0 \quad \text{on } \gamma$$

$$(2.8) \qquad u_{0n} = u_{1n} \quad \text{on } \gamma$$

$$(2.9) \qquad u_0 = \beta \quad \text{on } \Gamma_{in}$$

$$(2.10) \qquad u_0 = 0 \quad \text{on } \Gamma_w$$

$$(2.11) \qquad \nu \frac{\partial u_{on}}{\partial n} - p_0 = 0 \quad \text{on } \Gamma_{out}$$

$$(2.12) \qquad u_{ot} = 0 \quad \text{on } \Gamma_{out}$$

$$(2.13) \qquad u_{1n_1} = 0 \quad \text{on } \Gamma_1.$$

Hereafter we denote (2.1) - (2.13) by (PDEF).

Remark 2.1. The physical meaning of (2.5) and (2.6) is following [1], [3] , [4];
$\quad$ If $|S_n| < g_n$, then $u_n = 0$ on γ ,
$\quad$ If $|S_n| = g_n$, then $u_n = 0$ or $u_n \ne 0, S_n \cdot u_n < 0$ on γ.

One easily check that
(i) $u_n \to 0$ as $g_n \to +\infty$;
(ii) $S_n \to 0$ as $g_n \to 0$.

Remark 2.2. Let us note that (2.2), (2.11) and (2.12) mean $\frac{\partial u_{0n}}{\partial n} = 0$. Then (2.11) becomes $p_0 = 0$.

Remark 2.3. Obviously we see

$$(2.14) \qquad u_1 = -k\nabla\phi \quad \text{in} \quad \Omega_1 .$$

Substituting (2.14) into (2.4), we have

$$(2.15) \qquad \Delta\phi = 0 \quad \text{in} \quad \Omega_1 .$$

(2.8) and (2.13) mean

$$(2.16) \qquad -k\frac{\partial\phi}{\partial n} = u_{on} \quad \text{on } \gamma$$

$$(2.17) \qquad \frac{\partial\phi}{\partial n_1} = 0 \quad \text{on } \Gamma_1 .$$

Furthermore, ϕ satisfies the related one with respect to (2.5) and (2.6) on γ . Here we should note that the solvability of the problem (2.15), (2.16) and (2.17) leads to $\int_\gamma u_{on}\, ds = 0$.

If one solves this problem, the solution ϕ_0 obtained has inevitably an uncertainty of the constant .

Let normalize ϕ_0 such that $\int_{\Omega_1} \phi_0\, dx = 0$ and let represent ϕ as the solution of the coupled system (PDEF) by $\phi = \phi_0 + d$, where d is an unknown constant to be determined according to (2.11). One method to fix d is shown in the algorithm solving $(NM)_1$.

3. Reformulation by variational inequalities

3.1. Coupled formulation

Define

$$a_k(u, v) = \int_{\Omega_k} \nabla u \cdot \nabla v\, dx,$$

$$(u, v)_k = \int_{\Omega_k} u \cdot v\, dx \quad \text{for } k = 0, 1$$

and

$$J(v) = \int_\gamma g_n\, |v_n|\, d\Gamma .$$

Let $K_{0\sigma} = \{\, v \mid v = \beta \text{ on } \Gamma_{in},\ v = 0 \text{ on } \Gamma_w,\ v_t = 0 \text{ on } \Gamma_{out}\, \} \cap H^1_\sigma(\Omega_0)$

where $H^1_\sigma(\Omega_0)$ is the solenoidal subspace of $H^1(\Omega_0)$ for vector functions and $P = H^1(\Omega_1) = P_0 + \{1\}$ where $P_0 = \{\, \eta \in H^1(\Omega_1) \mid \int_{\Omega_1} \eta \, dx = 0 \,\}$.

If we couple (2.1), (2.2) and (2.15) under the boundary conditions prescribed on γ, we have

Find $u \in K_{0\sigma}$ and $\phi = \phi_0 + d \in P$ ($\phi_0 \in P_0$, $d \in R$) such that

$$(3.1) \qquad \nu \, a_0(u, v - u) + J(v) - J(u) + \Big((u \cdot \nabla)u, v - u \Big)_0$$

$$\geq - \int_\gamma \phi(v_n - u_n) \, d\Gamma + \int_\gamma a(v_n - u_n) \, d\Gamma - (f, v - u)_0 \ , \quad \forall v \in K_{0\sigma}$$

$$(3.2) \qquad k \, (\nabla \phi, \, \nabla \eta)_1 = \int_\gamma u_n \, \eta \, d\Gamma \ , \quad \forall \eta \in P \ .$$

We denote (3.1) and (3.2) by $(VIF)_1$.

Remark 3.1. When the Reynolds number is sufficiently small, the existence and uniqueness theorem on $(VIF)_1$ is obtained .

Remark 3.2. If we use the fictitious domain method stated in [2], (3.1) is approximated by singularly perturbed problem defined in Ω. In this case, Ω_1 plays a role of the fictitious domain in place of the Darcy part.

3.2. Uncoupled formulation
Let

$$K_\sigma = \{\, v \in L^2_\sigma(\Omega) \cap H^1_\sigma(\Omega_0) \mid v = \beta \ \text{on} \ \Gamma_{in}, \ v = 0 \ \text{on} \ \Gamma_w,$$

$$v_t = 0 \ \text{on} \ \Gamma_{out}, \ v_{n1} = 0 \ \text{on} \ \Gamma_1 \,\} \ .$$

Where $L^2_\sigma(\Omega)$ is the solenoidal subspace of $L^2(\Omega)$ for vector functions.

Find $u \in K_\sigma$ such that

$$(3.3) \qquad \nu \, a_0(u, \, v - u) + J(v) - J(u) + \Big((u \cdot \nabla)u, v - u \Big)_0 + (c \, u, v - u)_1$$

$$\geq (f, v - u)_0 - (\nabla x_2, v - u)_1 \ , \quad \forall v \in K_\sigma \ .$$

We denote (3.3) by $(VIF)_2$.

Remark 3.3. One can check the equivalence relation among (PDEF), $(VIF)_1$ and $(VIF)_2$ and the unique existence of the solution of $(VIF)_2$ by use of the standard arguments in the theory of the variational inequalities [3] .

4.   Singularly perturbed approximation for $(VIF)_2$

Define

$$J_\epsilon(v) = \int_\gamma g_n \, \psi_\epsilon(u_n) \, ds \; ,$$

where $\psi_\epsilon(t) = \int_0^t \tanh(\frac{s}{\epsilon}) \, ds$.

Let

$$\widetilde{K_\sigma} = \{ \, v \in H_\sigma^1(\Omega) \mid v = \beta \text{ on } \Gamma_{in}, \; v = 0 \text{ on } \Gamma_w, \; v_t = 0 \text{ on } \Gamma_{out}, \; v = 0 \text{ on } \Gamma_1 \, \} \; .$$

Find $u \in K_\sigma$ such that

$$(4.1) \quad \nu \, a_0(u, v-u) + \epsilon \, a_1(u, v-u) + \Big((u \cdot \nabla)u, v-u\Big)_0 + J_\epsilon(v) - J_\epsilon(u) + (c \, u, v-u)_1$$

$$\geq \; (f, v-u)_0 - (\nabla x_2, v-u)_1, \quad \forall v \in \widetilde{K_\sigma}.$$

We denote (3.4) by $(VIF)_{2\epsilon}$ which reduces to the following formulation $(PDEF)_\epsilon$ [6] ;

$$(4.2) \qquad\qquad -\nu \, \Delta u_0 + (u_0 \cdot \nabla)u_0 + \nabla p_0 = f \quad \text{in } \Omega_0,$$

$$(4.3) \qquad\qquad \nabla \cdot u_0 = 0 \quad \text{in } \Omega_0,$$

$$(4.4) \qquad\qquad -\epsilon \, \Delta \, u_1 + cu_1 + \nabla p_1 = -\nabla x_2 \quad \text{in } \Omega_1,$$

$$(4.5) \qquad\qquad \nabla \cdot u_1 = 0 \quad \text{in } \Omega_1,$$

$$(4.6) \qquad \nu \cdot \frac{\partial u_{on}}{\partial n} - p_0 = \epsilon \cdot \frac{\partial u_{1n}}{\partial n} - p_1 + g_n \tanh(\frac{u_n}{\epsilon}) \quad \text{on } \gamma$$

$$(4.7) \qquad\qquad \nu \cdot \frac{\partial u_{ot}}{\partial n} = \epsilon \cdot \frac{\partial u_{1t}}{\partial n} \quad \text{on } \gamma$$

$$(4.8) \qquad\qquad u_0 = u_1 \quad \text{on } \gamma$$

$$(4.9) \qquad\qquad u_0 = \beta \quad \text{on } \Gamma_{in}$$

$$(4.10) \qquad\qquad u_0 = 0 \quad \text{on } \Gamma_w$$

$$(4.11) \qquad\qquad \nu \cdot \frac{\partial u_{0n}}{\partial n} - p_0 = 0 \quad \text{on } \Gamma_{out}$$

$$(4.12) \qquad\qquad u_{ot} = 0 \quad \text{on } \Gamma_{out}$$

$$(4.13) \qquad\qquad u_1 = 0 \quad \text{on } \Gamma_1.$$

Remark 4.1.   By virtue of a well-known arguments, we see that $(VIF)_{2\epsilon}$ has a unique solution u_ϵ in $\widetilde{K_\sigma}$, when the Reynolds number is sufficiently small. Let $\epsilon \to 0$, then there exists u_0 in K_σ such that

$$u_\epsilon \to u_0 \quad \text{in } H_2^1(\Omega_0) \text{ weakly },$$

$$u_\epsilon \to u_0 \quad \text{in } L^2(\Omega_1) \text{ weakly}$$

$$\text{and } u_0 \text{ satisties } (VIF)_2 \quad (see[2], [5], [7]).$$

5. Numerical models

Here we present two types of numerical model for (PDEF).

5.1.

$(VIF)_1$ leads to the one numerical model by itself, which we denote by $(NM)_1$.

5.2.

If we apply the distribution theoretic approach to fictitious domain method for Neumann problems [2] to $(PDEF)_\epsilon$, then we have the other numerical model $(NM)_2$ as follows ;

$$(5.1) \quad -\nabla \cdot \{\nu(1-\chi) + \epsilon\chi\}\nabla u + (1-\chi)(u \cdot \nabla)u + c\chi u + g_n \cdot \tanh(\frac{u_n}{\epsilon}) \cdot \nabla\chi + \nabla p$$

$$= (1-\chi)f - \chi \cdot \nabla x_2 \quad \text{in } \Omega ,$$

$$(5.2) \qquad\qquad\qquad\qquad \nabla \cdot u = 0 \quad \text{in } \Omega$$

and u satisfies (4.9)-(4.13).

6. Numerical results

We state an algorithm to solve $(NM)_1$.

Algorithm

Step 0.   Put the initial guess of ϕ to be $\phi^0 \equiv 0$ in Ω_1 .

Step 1.   Solve (3.1) of $(VIF)_1$ by putting $\phi = \phi_0$.

Let the solution obtained be u^0 and p^0 .

Step 2.   Compute $\delta^0 = \frac{1}{L} \int_\gamma u_n^0 \, ds$ (L is the length of γ).

Step 3.   Solve (3.2) putting $u_n = u_n^0 - \delta^0$ on γ .

Let the solution be ϕ^1 and normalize it as $\widetilde{\phi^1} \in P_0$.

Step 4.   Replace ϕ in (3.1) by $\widetilde{\phi^1} + k^1$, where $k^1 = \delta^0$.

Then return to Step 1.

Remark 6.1 In the n-th iteration, k^n is defined as follows ;

$$k^n = k^{n-1} + \delta^{n-1} = \sum_{j=0}^{n-1} \delta^j .$$

Where $\delta^j = \frac{1}{L} \int_\gamma u_n^j \, ds, \ (j = 0, 1, \cdots, n-1).$

Numerical experiments showed the following asymphotic property of δ^n and k^n ;

$$\delta^n \to 0 \quad (n \to \infty) ,$$

and

$$k^n \to d \quad (n \to \infty)$$

where d is a constant to be found out.

Remark 6.2. As the resolution of (3.1) in $(VIF)_1$, we adopted to solve the approximate equations stated in the Remark 3.2.

Finally several numerical results based on $(NM)_1$ and $(NM)_2$ are shown in the following. Figures 1 and 2 show the numerical results obtained by use of $(NM)_1$ and $(NM)_2$, respectively. Figure 3 shows the numerical result of the case that there is a tiny hole at $(L,\ 2a)$, which was obtained by use of $(NM)_1$.

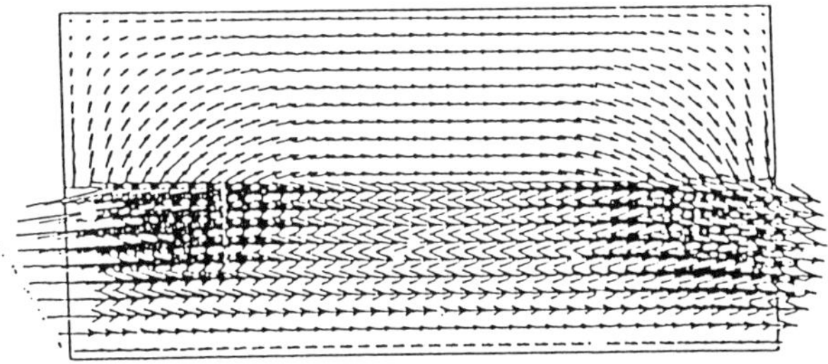

F i g u r e 1 .

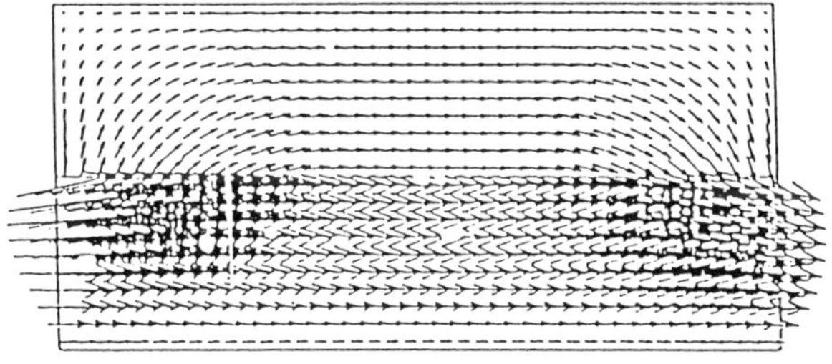

F i g u r e 2 .

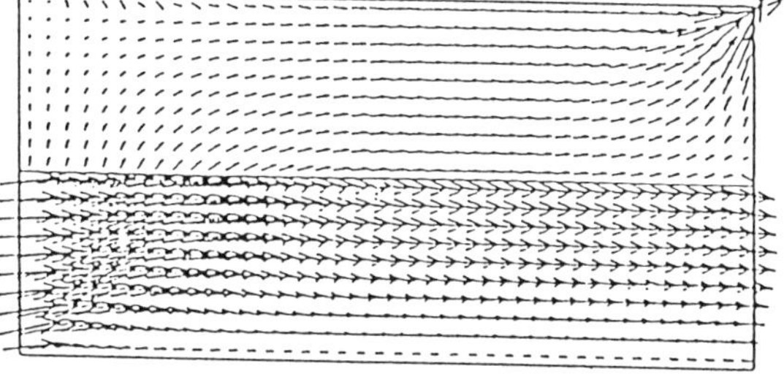

F i g u r e 3 .

References

1. G.Duvaut and J.L.Lions, Les Inequations en Mecanique et en Physique (Dunod, Paris 1972).

2. H.Fujita, H.Kawahara and H.Kawarada, Distribution theoretic approach to fictitious domain method for Neumann problems, East-West J.Numer.Math., Vol.3, No.2, pp.111-126 (1995).

3. R.Glowinski, J.L.Lions and R.Tremolieres, Numerical Analysis of Variational Inequalities (North-Holland, Amsterdam 1981).

4. R.Glowinski, Numerical Methods for Nonlinear Variational Problems (Springer, New York 1984).

5. J.L.Lions, Quelques methods de resolution des problems aux limites non lineares (Dunod Gauthier Villars, 1969).

6. J.L.Lions, Une Application de la Dualite aux Regularises et Penalises, Colloq. Rio-de-Janeiro,Aermann,Paris, pp.199-205 (1972).

7. R.Temam, Theory and Numerical Analysis of the Navier-stokes Equations (North-Holland, Amsterdam 1977).

Adaptive Multimodel Domain Decomposition in Fluid Mechanics

PATRICK LE TALLEC, FRANÇOIS MALLINGER *

1   INTRODUCTION

Many practical problems of Fluid Mechanics require the simultaneous use of different physical models inside a given computational domain : local kinetic models must be used in shock or boundary layers when simulating rarefied flows, refined meshes are needed in recirculation regions or next to solid boundaries, wall laws are required when calculating turbulent flows around obstacles. For such problems, one must identify the domain where the local enriched model should be used, write adequate interface conditions for matching the local and the global models, and propose an adequate iterative algorithm for solving the resulting coupled problem without rewriting a full new computer code.

This paper presents a general domain decomposition strategy for tackling such situations, based on the following steps :

- approximate preliminary solution of the Navier-Stokes equations on the full computational domain;

- calculation of generalized residuals for detecting the regions where the present Navier-Stokes solution is inaccurate. For standard problems, these residuals are estimated by the approximate values of the second derivatives of basic physical quantities. For polyatomic rarefied flows, the residuals are obtained by plugging the Navier-Stokes solution into a generalized fourteen moments Grad equation specially developed for the occasion;

- adaptive construction and meshing of the different subdomains, finer models and finer grids being used in the regions of large residuals;

- development of adequate interface conditions. These interface conditions match inflow and outflow fluxes when coupling a local Boltzmann model with a global Navier-Stokes equation. They use mortar elements when matching convection dominated Navier-Stokes equations discretized on two nonoverlapping and nonmatching grids;

- solution of the coupled problem by a Dirichlet Neumann algorithm.

*INRIA, Domaine de Voluceau, 78153 Le Chesnay Cedex, France. Email: patrick.letallec@inria.fr
This work has been supported by the Hermes Research program

Domain Decomposition Methods in Sciences and Engineering, edited by R. Glowinski *et al.*

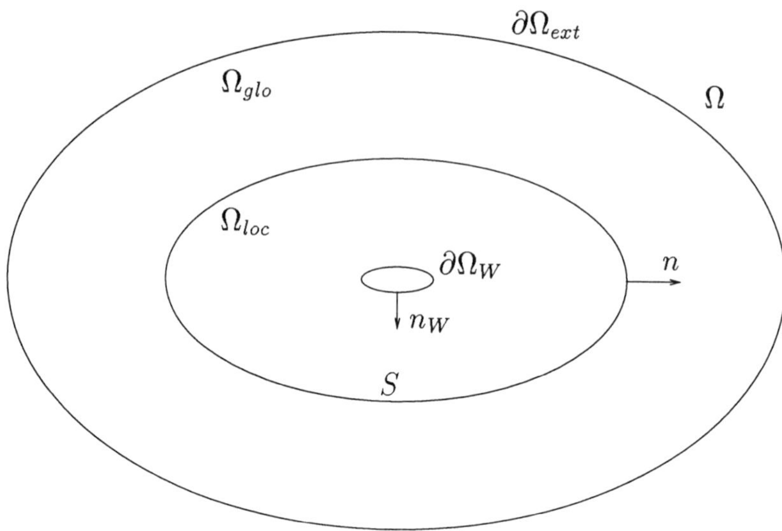

Figure 1: Decomposition of the domain in two subregions

This global strategy is presented below, with a particular emphasis on the simulation of rarefied gas flows.

2   THE PHYSICAL MODELS

2.1   Geometric description

We consider the flow of a gas in a global domain Ω around a solid body of boundary $\partial\Omega_W$ (Figure 1). We wish to split this domain into several nonoverlapping domains and to use the best avalaible numerical model inside each subdomain. In general, one has at least to separate two different regions Ω_{glo} and Ω_{loc}. The domain Ω_{loc} is a local domain which contains the obstacle, with internal boundary $\partial\Omega_W$ and external boundary S. The domain Ω_{glo} is a large domain, with an internal boundary S which surrounds the body and an external boundary $\partial\Omega_{ext}$ which is the external boundary of the computational domain Ω.

The subdomains Ω_{glo} and Ω_{loc} can be fixed arbitrarily at the beginning of the calculation or can be automatically adapted to the physical characteristics of the solution.

2.2   Basic Equations

On any subdomain Ω_i where the flow can be reasonably described by these equations, the Navier-Stokes equations will be used with appropriate boundary conditions imposed on the body or on the different interfaces as specified later. These equations will be numerically solved by a SUPG solver written in entropic variables operating on a given finite element triangulation of each local domain Ω_i [8], [11]. The different finite element grids do not necessarily match at the subdomain interfaces.

Unfortunately, these equations cease to be valid at high altitudes corresponding to semi-rarefied regimes. At this level, slip effects can be observed in the boundary layer, shock begin to be thicker and the gas gets rarefied in the wake. Such local phenomena must then be described by kinetic models, such as the Boltzmann equation.

Let f be the density of gas particles at position x with velocity v, and internal energy I. The Boltzmann equation of rarefied gas dynamics characterizes this density as the solution of the integrodifferential equation ([4])

$$\frac{\partial f}{\partial t} + v \cdot \frac{\partial f}{\partial x} = Q(f, f).$$

The collision operator counts the particles which are gained or lost through intermolecular collisions. For molecular gas having internal degrees of freedom, it is defined by

$$Q(f, f) = \int_\Delta \left(\frac{f' f'_*}{(I' I'_*)^{\delta-1}} - \frac{f f_*}{(II_*)^{\delta-1}} \right) B(II_*^{\delta-1}) d\Delta,$$

$$d\Delta = dv_* dI_* [r(1-r)]^{\frac{\delta}{2}-1} dr R^2 (1-R^2)^{\delta-1} dR d\omega.$$

As usual we have used the notation

$$f = f(v, I), f' = f(v', I'), f_* = f(v_*, I_*), f'_* = f(v'_*, I'_*),$$

with (v, I) and (v_*, I_*) the velocities and internal energies of the colliding particles, and (v', v'_*) and (I', I'_*) the post collision velocities and internal energies. As in the monoatomic case, the collision direction $\omega \in S^2$ is fixed and in a collision, we transform the vector (v, v_*, I, I_*) with $v, v_* \in \mathrm{IR}^3, I, I_* \geq 0$, by setting

$$e^2 = \frac{1}{4}|v - v_*|^2 + I^2 + I_*^2 = \text{ total energy of the collision},$$

$$g = v - v_* = \text{ relative velocity},$$

and by defining the post collision velocities (v', v'_*) and energies (I', I'_*) by

$$v' + v'_* = v + v_*,$$

$$g' = v' - v'_* = 2Re\{g - 2\omega g.\omega\}/|g|,$$

$$I'^2 = r(1 - R^2)e^2, \quad I'^2_* = (1 - r)(1 - R^2)e^2.$$

The factors $R, r \in [0, 1]$ introduced in the collision operator determine the quantity of energy which is exchanged between internal and kinetic energy and between the two internal energies ([1], [3]). The term $(II_*/I' I'_*)^{\delta-1}$ is introduced to give the right value of

$$\gamma = \frac{\delta + 5}{\delta + 3},$$

in the limiting hydrodynamic equation of state $p = (\gamma - 1)\rho e$.

The collision cross section B measures the probability of collision of particles (v, I) and (v_*, I_*) with parameters (ω, r, R). In the general case, it is a function of all collision invariants. In our simulations, we have used the classical Variable Hard Sphere model (VHS)

$$B = C|g|^{-2\alpha}|g.\omega|R^{1-2\alpha},$$

which is the simplest model compatible with a Sutherland type viscosity law

$$\mu = KT^{\frac{1}{2}+\alpha},$$

at the Navier-Stokes limit. Here C and K are some constants.

This equation must be complemented by boundary conditions imposing the distribution of incoming particles. In the case of perfect accomodation on the body's surface, we would have

$$f(x, v, I, t) = \rho_\infty M_{u_\infty, T_\infty}(v, I) \quad \text{if} \quad v \cdot n < 0 \text{ at infinity,}$$

$$f(x, v, I, t) = k M_{u_w, T_w}(v, I) \quad \text{if} \quad v \cdot n_W < 0 \text{ on the body's surface,}$$

$$\int f(x, v, I, t) v \cdot n_W \, dv dI = 0 \text{ on the body's surface,}$$

with $M_{u,T}$ denoting the Maxwellian distribution with mean velocity u and temperature T

$$M_{u,T}(v, I) = \lambda_\delta \frac{\rho I^{\delta-1}}{T^{(3+\delta)/2}} exp\left(-\frac{|v-u|^2 + 2I^2}{2T}\right).$$

More elaborate boundary conditions can also be introduced ([4]).

2.3　Transition between Boltzmann and Navier-Stokes

When the gas is dense, solving Boltzmann equation is very expensive or out of reach. Therefore, we wish to solve Boltzmann equations on very small subdomains, and to be able to realise smooth transitions or couplings between these Boltzmann solutions and cheaper Navier-Stokes solutions. For this purpose, the best way is to introduce an asymptotic kinetic model, approximating the Boltzmann equation, and degenerating to the Navier-Stokes model when the gas gets dense.

For monoatomic gases, this is classically achieved by introducing the so-called Chapman-Enskog or Grad expansions [6]. For more realistic polyatomic gas having internal energies, we first need to extend these approaches.

To obtain such a polyatomic generalisation of Grad's thirteen moments expansion, we first assume that f can be represented as a linear combination of Hermite polynomials, given here for a diatomic gas ($\delta = 2$), by

$$f = f^{(0)}\left[1 + \frac{p_{ij}}{2pR_gT}\left((v-u)_i(v-u)_j - I^2\delta_{ij}\right)\right.$$

$$\left. + \frac{1}{14}\frac{S_i(v-u)_i}{pR_gT}\left(\frac{|v-u|^2}{R_gT} + \frac{2I^2}{R_gT} - 7\right)\right],$$

$$f^{(0)} = \frac{\sqrt{2}\rho I^{\delta-1}}{(2\pi)^{\frac{3}{2}}(R_gT)^{(3+\delta)/2}} exp\left(-\left(\frac{|v-u|^2 + 2I^2}{2R_gT}\right)\right).$$

This ansatz can be either postulated or derived as the trace of the Grad's expansion of a monoatomic gas in dimension $(3 + \delta)$ ([7]). From the orthogonality of the Hermite polynomials, the thermodynamic coefficients ρ (density), p (pressure), p_{ij} (the opposite of the viscous stress tensor), $S_i = 2q_i$ (two times the heat conduction vector) correspond to the usual definition

$$\rho = \int f \, dv dI,$$

$$\rho u_i = \int v_i f \, dv dI,$$

$$p = \rho R_g T = \frac{1}{\delta + 3} \int (\frac{|v - u|^2}{2} + I^2) f dv dI,$$

$$\sigma_{ij} = - \int (v - u)_i (v - u)_j f \, dv dI,$$

$$q_i = \int (v - u)_i (\frac{|v - u|^2}{2} + I^2) f dv dI$$

$$p_{ij} = -\sigma_{ij} - p \delta_{ij},$$

$$S_i = 2q_i.$$

To identify these unknown thermodynamic moments, we now report the above expression of f into the Boltzmann equation and integrate the resulting equation against the 14 moments

$$K_{14} = \begin{pmatrix} 1 \\ v_i \\ v_i v_j \\ 2I^2 \\ v_i(|v|^2 + 2I^2) \end{pmatrix}.$$

This yields the final reduced varitational kinetic system

$$\frac{\partial}{\partial t} \int K_{14} f \, dI dv + \frac{\partial}{\partial x_k} \int K_{14} v_k f \, dI dv = \int K_{14} Q(f) \, dI dv. \tag{1}$$

If we calculate the different integrals in (1) and take into account the particular expression of f, after lengthy algebraic calculations, we can rewrite this variational equation as the following Grad system of partial differential equations [7]

- Navier-Stokes Conservation laws

$$\frac{\partial \rho}{\partial t} + u_k \cdot \frac{\partial \rho}{\partial x_k} + \rho \frac{\partial u_k}{\partial x_k} = 0,$$

$$\frac{\partial u_i}{\partial t} + u_k \cdot \frac{\partial u_i}{\partial x_k} - \frac{1}{\rho} \frac{\partial \sigma_{ik}}{\partial x_k} = 0,$$

$$\frac{\partial p}{\partial t} + \frac{2}{5} p_{ik} \frac{\partial u_i}{\partial x_k} + u_k \frac{\partial p}{\partial x_k} + \frac{7}{5} p \frac{\partial u_k}{\partial x_k} + \frac{1}{5} \frac{\partial S_k}{\partial x_k} = 0,$$

- differential generalisation of Navier-Stokes constitutive laws

$$\frac{\partial p_{ij}}{\partial t} + \frac{\partial u_k\, p_{ij}}{\partial x_k} + \frac{1}{7}\left(\frac{\partial S_i}{\partial x_j} + \frac{\partial S_j}{\partial x_i} - \frac{2}{5}\frac{\partial S_k}{\partial x_k}\,\delta_{ij}\right)$$

$$+\, p_{jk}\frac{\partial u_i}{\partial x_k} + p_{ik}\frac{\partial u_j}{\partial x_k} - \frac{2}{5}\, p_{kl}\frac{\partial u_l}{\partial x_k}\,\delta_{ij}$$

$$+\, p\left(\frac{\partial u_i}{\partial x_j} + \frac{\partial u_j}{\partial x_i} - \frac{2}{5}\frac{\partial u_k}{\partial x_k}\,\delta_{ij}\right) = J_{ij}^{(2)},$$

$$\frac{\partial S_i}{\partial t} + \frac{\partial u_k\, S_i}{\partial x_k} + \frac{9}{7}\, S_k\frac{\partial u_i}{\partial x_k} + \frac{2}{7}\, S_i\frac{\partial u_k}{\partial x_k} + \frac{2}{7}\, S_k\frac{\partial u_k}{\partial x_i}$$

$$+\, \frac{2}{\rho}\, p_{il}\frac{\partial \sigma_{lk}}{\partial x_i} + 2 R_g T\, p_{il}\frac{\partial p_{ik}}{\partial x_k} + 9\, p_{ik}\frac{\partial R_g T}{\partial x_k}$$

$$+\, 7\, p\,\frac{\partial R_g T}{\partial x_i} = J_i^{(3)} - 2\, u_l\, J_{il}^{(2)}.$$

Right Hand Sides $J_{ij}^{(n)}$ are explicit integrals of collision cross-sections and can be expressed as functions of the different thermodynamic variables ([9]).

To derive the Navier-Stokes equation, we then suppose in addition that the fluid is close to equilibrium so that moments p_{ij} or S_i and gradients of thermodynamic variables remain small. At first order, the differential constitutive laws then reduce to the standard laws

$$p_{ij} = -\,\mu\left(\frac{\partial u_i}{\partial x_j} + \frac{\partial u_j}{\partial x_i}\right) - \lambda\frac{\partial u_k}{\partial x_k}\,\delta_{ij},$$

$$S_i = -\,2\,\kappa\,\frac{\partial T}{\partial x_i},$$

with transport coefficients

$$\mu = \frac{30\,(R_g T)^{1/2}}{C\,\pi^{1/2}},$$

$$\kappa = \frac{1323}{570}\, R_g\,\mu,$$

$$\lambda = -\left(\frac{2}{5} + \frac{4}{375}\right)\mu.$$

With this simplification, we get the Navier-Stokes equations as the limiting case of a Grad's fourteen moments expansion. As a byproduct, which will be the key for closing our hierarchy of models and constructing our coupling strategy, we also obtain the polyatomic kinetic distribution associated to the Navier-Stokes solution as

$$f_{NS} = f^{(0)}\left[1 - \frac{2\mu}{pR_g T}\left((v_i - u_i)(v_j - u_j) - \left(\frac{2}{5} + \frac{4}{375}\right)|v - u|^2\delta_{ij}\right.\right.$$

$$\left. +\left(1 - \left(\frac{3}{5} + \frac{2}{375}\right)\right) I^2\delta_{ij}\right)\frac{\partial u_i}{\partial x_j}$$

$$\left. -\,\frac{\kappa}{7}\frac{(v - u)_i}{pR_g T}\left(\frac{|v - u|^2}{R_g T} + \frac{2I^2}{R_g T} - 7\right)\frac{\partial T}{\partial x_i}\right].$$

This distribution generalises the classical monoatomic distribution originally introduced by Chapman-Enskog and Grad.

Remark 1 *The Grad's equation introduced above as an intermediate step between Boltzmann and Navier-Stokes is not a good model by itself. Its physical relevance is not garanteed, and mathematically it loses the positivity and hyperbolicity properties of the underlying Boltzmann equation. What is expected nevertheless, and what is verified in our numerical test, is that the difference between Grad and Navier-Stokes is a good measure of the amount of desequilibrium in the fluid, and therefore of the local lack of validity of our Navier-Stokes solution.*

3 ADAPTIVE CONSTRUCTION OF THE SUB-DOMAINS

Our basic problem is now to find the legitimate Navier-Stokes domain, or equivalently to find the Boltzmann zone Ω_{loc} where the Navier-Stokes solution f_{NS} is not an adequate solution of our problem, and where we should therefore solve the kinetic Boltzmann equation. For this purpose, we use our above model hierarchy and simply observe the quality of the Navier-Stokes distribution as solution of the finer Boltzmann equation. This original distribution is obtained by solving the Navier Stokes equations on the whole domain with say wall laws or slip boundary conditions at the wall. Plugging this solution into Boltzmann equation leads to the following variational residual

$$\int \varphi(v, I)[\frac{\partial f_{NS}}{\partial t} + v \cdot \frac{\partial f_{NS}}{\partial x} - Q(f_{NS}, f_{NS})]dvdI$$

$$= \int \varphi(v, I)R(f_{NS})dvdI = R_\varphi(f_{NS})(x, t), \qquad \forall \varphi, \forall x \in \Omega.$$

As such, this residual is still too complex to compute. Therefore, we restrict ourselves to the test functions φ used in Grad's 14 moments derivation. With this choice, we can compute one local residual per finite element function ψ_N given by

$$R_N = \frac{1}{\text{VolCell}_N}\left\|\int_\Omega -u_k\, p_{ij}\frac{\partial \psi_N}{\partial x_k}\right.$$
$$-\frac{1}{7}\left(S_i\frac{\partial \psi_N}{\partial x_j} + S_j\frac{\partial \psi_N}{\partial x_i} - \frac{2}{5}S_k\frac{\partial \psi_N}{\partial x_k}\delta_{ij}\right)$$
$$+\left(p_{jk}\frac{\partial u_i}{\partial x_k} + p_{ik}\frac{\partial u_j}{\partial x_k} - \frac{2}{5}p_{kl}\frac{\partial u_l}{\partial x_k}\delta_{ij}\right)$$
$$\left.+p\left(\frac{\partial u_i}{\partial x_j} + \frac{\partial u_j}{\partial x_i} - \frac{2}{5}\frac{\partial u_k}{\partial x_k}\delta_{ij}\right) - J_{ij}^{(2)}\right)\psi_N\right\|.$$

This residual is simply the norm of the residual of Grad's equation in the p_{ij} moment, when used with the available Navier-Stokes solution. From our previous remark, this difference should be a good indicator of desequilibrium in the fluid, and therefore of the local relevance of the Navier-Stokes equations.

We now choose the Boltzmann region Ω_{loc} as the region of Ω where this residual is the largest.

In practice, recovering the original domain by an unstructured adapted Navier-Stokes grid and a cartesian Boltzmann grid, we identify this region by the following sequence of operations :

1. calculation of a global, numerically accurate Navier-Stokes solution;

2. calculation of the above residual on each Navier-Stokes node;

3. interpolation and smoothing of this residual on the cartesian grid;

4. definition of the Boltzmann domain Ω_{loc} as the set of cartesian cells where the residual is above a user defined threshold;

5. calculation of the internal Navier-Stokes interface by spline interpolation of the jagged Boltzmann interface;

6. automatic meshing of the external Navier-Stokes zone by a Voronoi algorithm, the size of the elements being equidistributed with respect to the Hessian of characteristic flow variables.

We illustrate below this strategy on the following example

$$\begin{aligned}
\gamma &= \tfrac{7}{5}, Pr = \tfrac{18}{25} \\
\text{Angle of attack} &= 30\,\text{degrees} \\
M_\infty &= 20 \\
Reynolds &= 5000/m\,,\,50000/m \\
T_\infty &= \frac{1}{\gamma\,R_g\,M_\infty^2} \\
u_\infty &= 1 \\
\rho_\infty &= 1 \\
T_W &= 6\,T_\infty \\
u_W &= 0
\end{aligned}$$

The Navier-Stokes solution is calculated by a SUPG code provided by Dassault, using a mesh of 6882 nodes. We present below the original Navier-Stokes mesh, the calculated residual isolines defining Boltzmann and Navier-Stokes regions, that we compare to the amount of rarefied desequilibrium predicted by a global Boltzmann calculation. We observe that apart from boundary problems induced by the noslip boundary condition in the Navier-Stokes solution, both the Grad's residual and the global kinetic solution predict the same geometry for the rarefied domain (figures 2, 3, 4, 5). Similar results can be found for denser flows in ([9]).

4   THE COUPLED PROBLEM

Once knowing the adaptive partition of our original domain between a Navier-Stokes and a Boltzmann domain, we need to formulate and solve the resulting coupled problem. For this purpose, we will first introduce the coupling strategy in the general hyperbolic case of a Boltzmann-Boltzmann coupling and then degenerate one Boltzmann domain to the limiting Navier-Stokes model.

Let us for the moment consider the domain as splitted into two nonoverlapping subdomains Ω_1 and Ω_2 with external unit normal vector n_1 and n_2, and interface Γ. By restricting Boltzmann equation to each subdomain and decomposing the interface continuity condition $(f_1 - f_2)v.n_i = 0$ along incoming subdomain characteristics, we can decompose the original Boltzmann equation into the coupled set of well-posed subproblems

$$\mathrm{div}_x(vf_1) = Q(f_1, f_1) \text{ in } \Omega_1 \times \mathrm{I\!R}^3,$$
$$f_1(x, v, I) = f_2(x, v, I) \text{ on } \Gamma \text{ if } v.n_1 < 0,$$

$$\mathrm{div}_x(vf_2) = Q(f_2, f_2) \text{ in } \Omega_2 \times \mathrm{I\!R}^3,$$
$$f_2(x, v, I) = f_1(x, v, I) \text{ on } \Gamma \text{ if } v.n_2 < 0.$$

This coupled problem can then be solved by the fixed point Dirichlet-Neumann type iterative procedure [10]

- guess $f_2(x, v, I)$ on incoming characteristics $v.n_1 < 0$,

- solve the resulting subproblem on Ω_1,

- with the resulting value $f_1(x, v, I)$ imposed on the incoming characteristics $v.n_2 < 0$, solve the subproblem on Ω_2,

- use the result to update $f_2(x, v, I)$ on the interface and reiterate.

This numerical strategy seems to be quite efficient, although a few theoretical results are yet unsolved : well-posedness of each subproblem (this can be proved for a BGK kinetic problem, not for Boltzmann), cost and noise control of each local numerical Boltzmann solution when using Monte-Carlo techniques, convergence of the fixed point iterations. For the time being, such convergence is only proved in the absence of collision terms [5].

Despite these theoretical weaknesses, we can easily extend the above strategy to the Navier-Stokes Boltzmann coupling, simply by replacing f_1 on Ω_1 by its Navier-Stokes approximation

$$f \approx f_{NS}.$$

The interface boudary condition is then transformed into

$$f_2(v, I) = f_{NS}(v, I) \text{ on } \Gamma \text{ if } v.n_2 < 0,$$
$$f_{NS}(v, I) = f_2(v, I) \text{ on } \Gamma \text{ if } v.n_1 < 0.$$

The first condition is imposed as boundary condition in the Boltzmann domain. The second condition is imposed as a boundary condition in the Navier-Stokes region by taking its velocity average

$$Flux(U, n) = Flux_+(U, n) + Flux_-(U, n), \tag{2}$$

$$Flux_+(U, n) = \int_{v.n>0} \begin{bmatrix} 1 \\ v \\ v^2/2 + I^2 \end{bmatrix} v.n f_{NS}(x, v, I) dv dI,$$

$$Flux_-(U, n) = \int_{v.n<0} \begin{bmatrix} 1 \\ v \\ v^2/2 + I^2 \end{bmatrix} v.n f_2(x, v, I) dv dI.$$

With this new boundary condition, our Dirichlet-Neumann loop reduces to the following sequence of operations already described in [2]

1. Solve few steps of the local Boltzmann problem

$$\frac{f^n - f^{n-1}}{\Delta t} + \text{div}_x(v f^n) = Q(f^n, f^n)(x, v, I) \text{ on } \Omega_2,$$
$$f^n(x, v, I) = f_{NS}(x, v, I) \text{ on } \Gamma \text{ if } v.n_2 < 0,$$
$$f^n(x, v, I) = \text{ Maxwellian on body}.$$

2. From f^n, compute the half fluxes $Flux_-$ entering the internal boundary of Navier-Stokes region

3. With these imposed fluxes $Flux_-$, using the mixed boundary condition (2) on the interface and usual boundary conditions at infinity, integrate the Navier-Stokes equations a few steps in time on the external domain and go back to the Boltzmann problem.

In all our numerical tests, we have used 5 such global iterations.

We have first tested this algorithm by computing the two-dimensional flow around an ellipse with the physical data

$$
\begin{aligned}
\gamma &= \tfrac{5}{3}, Pr = \tfrac{1}{3} \\
\text{Angle of attack} &= 30 \text{ degrees} \\
M_\infty &= 20 \\
Reynolds &= 5000/m \\
T_\infty &= 167.3K \\
u_\infty &= 5672m/s \\
\rho_\infty &= 1 \\
T_W &= 5.6 T_\infty \\
u_W &= 0
\end{aligned}
$$

The corresponding results are then validated by comparing them to the solution of a global Boltzmann simulation (figures 6, 7 and 8). We observe that the coupled solution is quite smooth at the interface and coincides with the global Boltzmann solution, including at the wall, and this is achieved at a smaller computational cost.

5 CONCLUSION

The proposed multimodel coupling numerical strategy appears to be operational for polyatomic gases while using adaptive domain splitting. For industrial flexibility, it was implemented by chaining independent Navier-Stokes and Boltzmann codes within an unique shell script.

The work to be done concerns the cost and noise control in the numerical solution of the local Boltzmann problem, the improvement of the rarefied model hierarchy by deriving more consistent asymptotic expansions (Levermore) preserving positivity and hyperbolicity of the asymptotic limit, and the generalisation to the numerical treatment of turbulent boundary layers via asymptotic wall laws.

References

[1] J.F. Bourgat, L. Desvillettes, P. Le Tallec, B. Perthame, "Microreversible collisions for polyatomic gases and Boltzmann's theorem" ,*J. Mech., B/Fluids*, 13, 1994.

[2] J.F. Bourgat, P. Le Tallec, F. Mallinger, B. Perthame, Y.Qiu, "Couplage Boltzmann Navier-Stokes", *Rapport de Recherche Inria N 2281*, Mai 1994.

[3] C. Borgnakke, P.S. Larsen, Statistical collision model for Monte-Carlo simulation of polyatomic gas mixtures. J. C.P. 18, 405-420 (1975).

[4] C. Cercignani, The Boltzmann equation and its applications. Springer-Verlag, 1975.

[5] F. Gastaldi, L. Gastaldi, "On a domain decomposition approach for the transport equation: theory and finite element approximations", *IMA Journal of Numerical Analysis*, Vol. 14, 1993.

[6] H. Grad, On the Kinetic Theory of Rarefied Gases. Comm. Pure and Appl. Math., 49, 331-407, 1949.

[7] P. Le Tallec, F. Mallinger, "Modélisation d'un gaz Polyatomique, Equations de Grad Généralisées et Validité de la Solution des Equations de Navier-Stokes" *Rapport de contrat final - Hermes* , April 1995.

[8] M. Mallet, "A finite element method for computational fluid dynamics", *Ph.D. Thesis*, Stanford University, 1985 .

[9] F. Mallinger, "Couplage adaptatif des équations de Boltzmann et de Navier-Stokes", *Thèse d'Université*, Paris Dauphine, à paraitre 1995.

[10] L.D. Marini, A. Quarteroni, " A relaxation procedure for Domain Decomposition Methods using Finite Elements", Num. Math., 55, 575-598, 1989.

[11] F. Shakib, "Finite element analysis of the compressible Euler and Navier-Stokes equations", *Ph.D. Thesis*, Stanford University, 1988 .

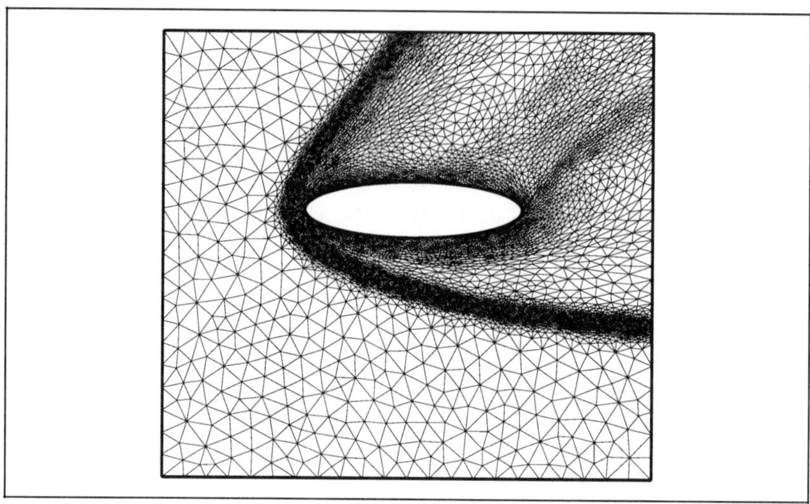

Figure 2: Navier-Stokes mesh: 6882 nodes. The Navier-Stokes solution is computed with no-slip boundary condition on the wall.

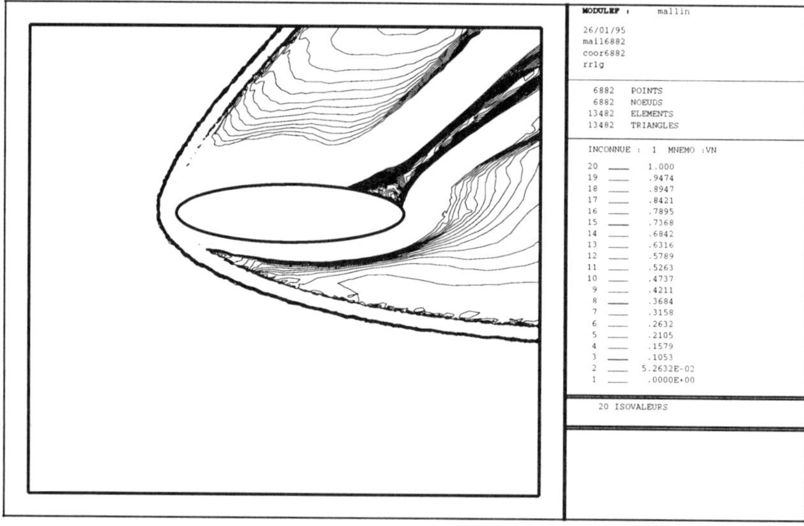

Figure 3: Grad criterion: Re=5000/m. We draw only the isolines between 0 and 1. The white zone defines the Boltzmann domain.

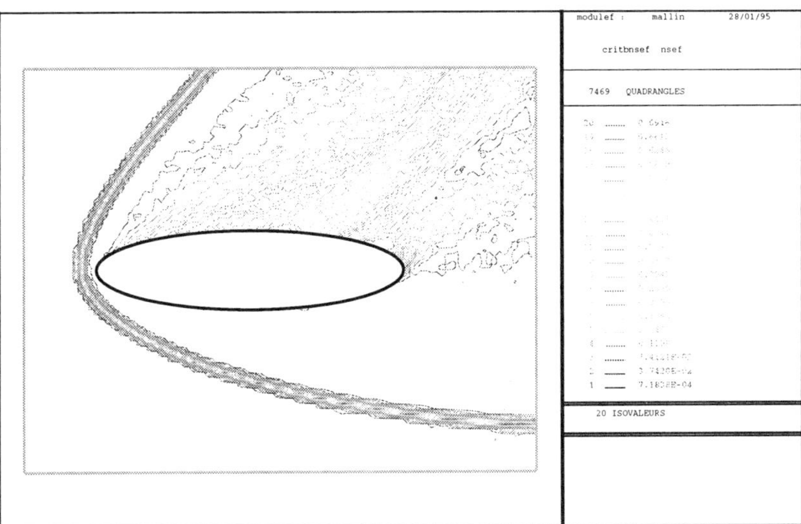

Figure 4: Boltzmann criterion estimated by a kinetic simulation. We verify that the Boltzmann region here is quite similar, except under the body. This difference is due to the bad boundary condition use in the Navier-Stokes computation.

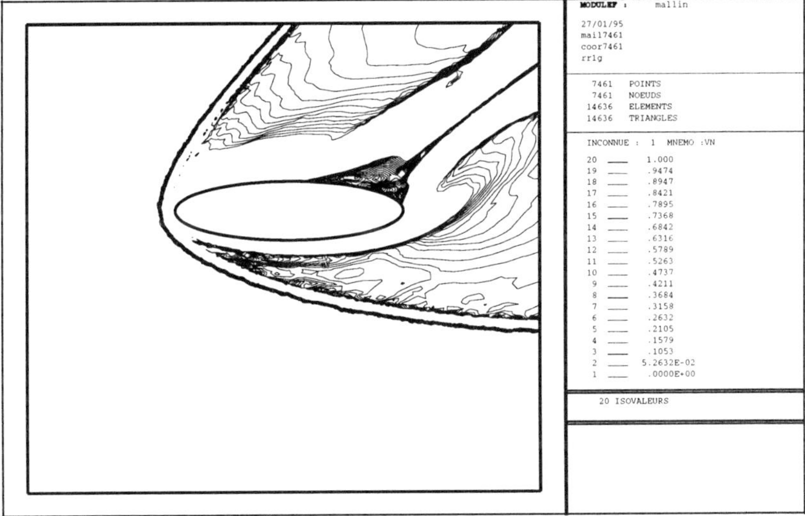

Figure 5: Grad criterion: Re=50000/m.

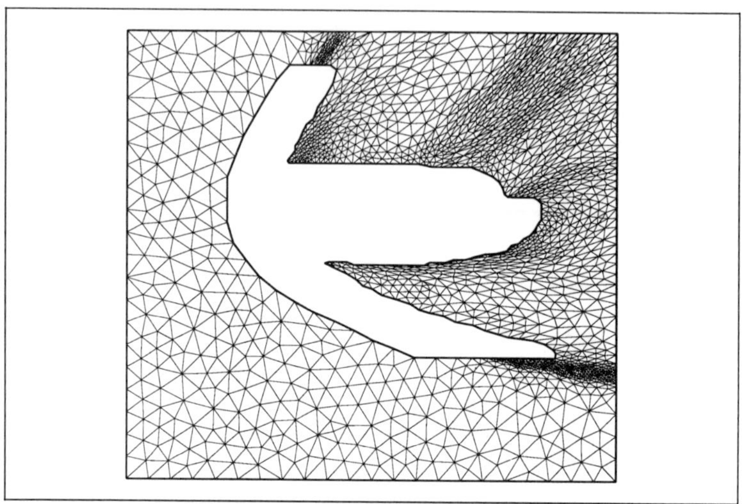

Figure 6: Navier-Stokes mesh to compute a coupled solution. The internal boundary is obtained thanks to the criterion. The mesh is adapted with the global initial Navier-Stokes solution.

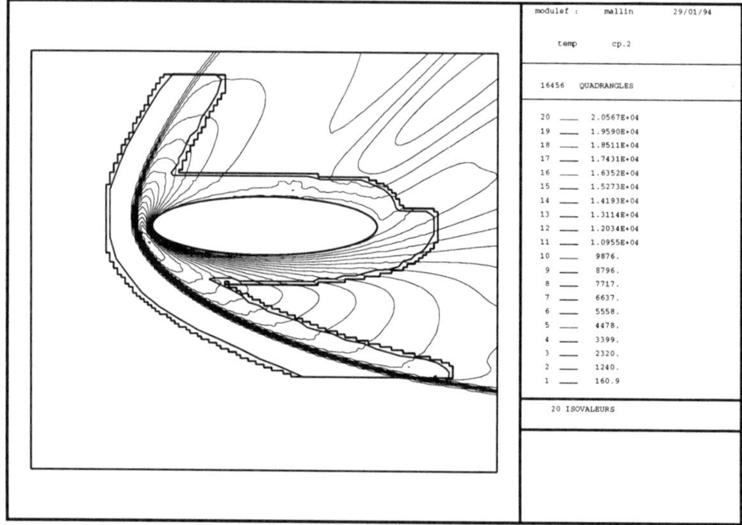

Figure 7: Temperature isolines of the coupled solution. The continuity of the isolines at the coupling interface is very good.

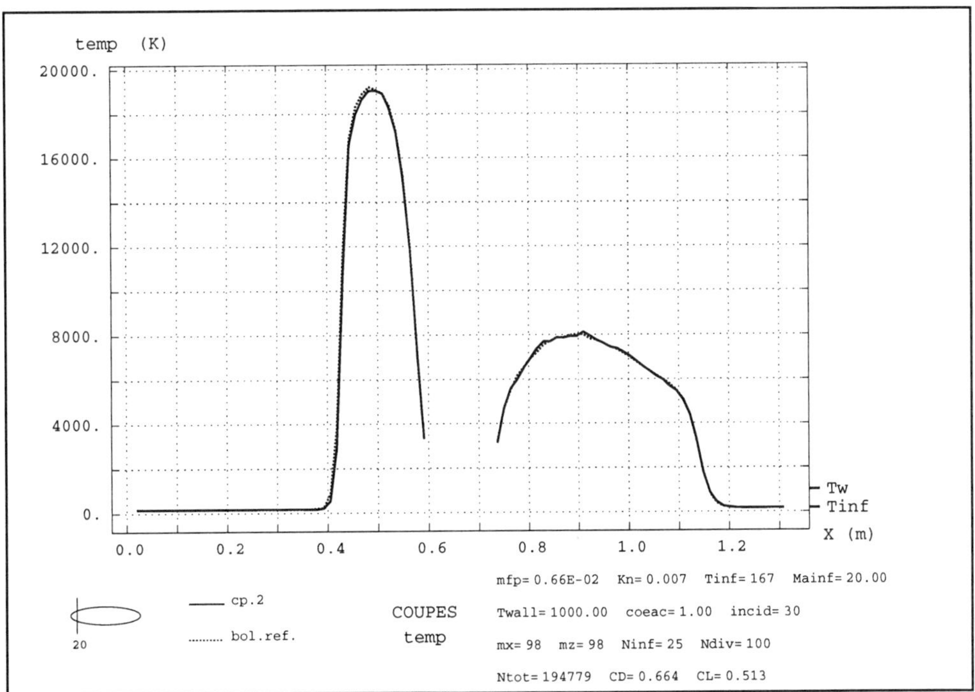

Figure 8: Cross section of the temperature. Superposition of the Boltzmann solution (dotted line) and the coupled solution (continuous line). The coupled solution is very good compared to the Boltzmann reference solution and recovers the right temperature jumps at the wall.

Domain Decomposition Method using Genetic Algorithms for Solving Transonic Aerodynamic Problems

J. PERIAUX [1] and H. Q. CHEN [2]

1 Introduction

Parallel computers and multiprocessor supercomputers are becoming more readily available; therefore highly parallel algorithms for solving practical complex problems are of great interest. Domain decomposition techniques [1], traced to Schwarz alternating procedure, are often used in parallel environments and genetic algorithms (GAs) are reported to be highly parallel [2]. The application of GAs to domain decomposition problems can be expected to have high parallelism. This paper explores GAs for solving 2-D aerodynamic problems through Schwarz's domain splitting with overlapping in order to develop a new kind of domain decomposition approach with high parallelism.

In the present method, GAs are the key techniques through the definition of a fitness function based on solutions on overlapping subdomains. The Schwarz alternating sequences are then guided by genetic algorithms. Binary codings for multiparameters and small population size are used in genetic iterations. The method presented is tested for 2-D transonic flows in a nozzle. The numerical results show that the method presented successfully finds the near optimum solutions within the finite genetic generations.

[1] Dassault Aviation, France
[2] Nanjing University of Aeronautics and Astronautics, P. R. of China

Domain Decomposition Methods in Sciences and Engineering, edited by R. Glowinski *et al.*

2   Description of the Problem

The aerodynamic test problem we treat here is the flows in a nozzle(see Fig. 2.1)

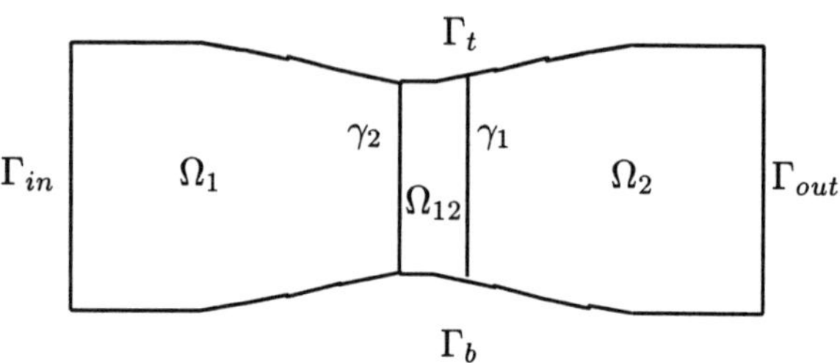

Fig. 2.1

which is modeled by

$$a^2 \nabla^2 \Phi - \frac{1}{2} \nabla \Phi \bullet \nabla(|\nabla \Phi|^2) = 0 \text{ in } \Omega \quad (\Omega_1 \bigcup \Omega_2)$$

$$\partial\Phi/\partial n = 0 \qquad \text{on} \qquad \Gamma_t \bigcup \Gamma_b$$

$$\Phi = g_{in} \qquad \text{on} \qquad \Gamma_{in}$$

$$\Phi = g_{out} \qquad \text{on} \qquad \Gamma_{out}. \tag{2.1}$$

Here Φ is a flow potential, n denotes the unit outward normal vector and a, the speed of sound given by

$$a^2 = \frac{1}{M_\infty^2} + \frac{\gamma - 1}{2}(1 - |\nabla \Phi|^2), \tag{2.2}$$

where γ is the ratio of specific heats and M_∞ represents the reference Mach number at infinity.

As shown in Fig. 2.1, we decompose the computational domain Ω into two subdomains Ω_1 and Ω_2 with overlap Ω_{12}; the interfaces are denoted by γ_1 and γ_2. We shall take values

$$g_1 \quad \text{on} \quad \gamma_1 \quad \text{and} \quad g_2 \quad \text{on} \quad \gamma_2$$

as extra boundary conditions in order to solve (2.1) in each subdomain. Using domain decomposition techniques, the problem can be reduced to minimizing the following function:

$$J(g_1, g_2) = \frac{1}{2} \| \Phi_1 - \Phi_2 \|^2,$$

where Φ_1 and Φ_2 are the solutions in the overlapping subdomain Ω_{12}.

3 Algorithm

In this section, we present an implementation of genetic algorithms for the domain decomposition problem described above.

Genetic algorithms (GAs) [2] are search algorithms which are different from the conventional search procedures encountered in engineering optimization. The core structure of the algorithm is based upon the principle of natural selection and natural genetics. For the function being optimized, the variables are rewritten in a code to form a structured string that the GAs can operate directly on. For the problem described above, binary codings for multiparameters are used, and we only code g_1, which can be given by

$$g_{1i}, i = 1, \cdots n \, (n \text{ parameters });$$

each g_{1i} is coded in l_i bits, thus the length of the strings

$$L = \sum_{i}^{n} l_i.$$

Let us consider a population size 25 (i.e. 25 strings). GAs decode each string to return the values of g_{1i}. With the g_{1i} known, we can compute the solutions in the domain Ω_1, namely Φ_1. Like Schwarz's alternative method, we take g_2 based on Φ_1, thus the solution, Φ_2, in the domain Ω_2 can be calculated. Now, we send both values in the overlapping region to the cost function:

$$J(g_1) = \frac{1}{2} \parallel \Phi_1 - \Phi_2 \parallel^2 .$$

Thus, each string has a cost value. A new generation of strings can be produced by performing GA operation [2], such as reproduction, crossover and mutation. By the principle of the survival of the fittest, strings with better cost values will be selected and structures containing the desired characteristics will survive while others die off as successive generations are produced. This process continues until convergence is achieved.

4 Numerical Results and Analysis

The approach presented has been tested for the domain decomposition problem described in the section 2. Fig. 4.1 shows the mesh used to solve the test problem and the overlap Ω_{12} whose interface γ_1 is located just by the throat.

The interface γ_1 has 6 grid points (i.e. 6 parameters). During the GA iteration, the length of the string, 30, is used with 5 bits for each parameter. The probability of crossover $Pc = 0.85$ and the probability of mutation $Pm = 0.015$ are not carefully selected but are fixed for the flow test cases.

As mentioned above, each calculation of $J(g_1)$ requires the solution of the full potential flow equation in each subdomain, which results in a computing cost which is directly proportional to the number of CFD evaluation. Thus a small population

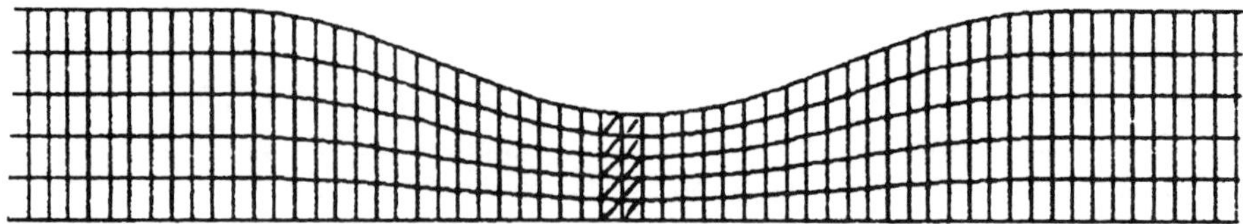

Fig. 4. 1 grid of the nozzle

size is used. The presented method begins by randomly generating a population of 25 strings. We keep in mind that each string represents one possible solution g_1 to the problem. But the values on the interface γ_1 are very sensitive for the transonic case due to the limitation of the sonic throat. If g_1 is given beyond this physical limitation, the CFD solver [3] will fail to converge. This is why traditional Schwarz alternative method may fail. Using GAs the individuals may have the same problem, but with the genetics population, the information before divergence can be still used.

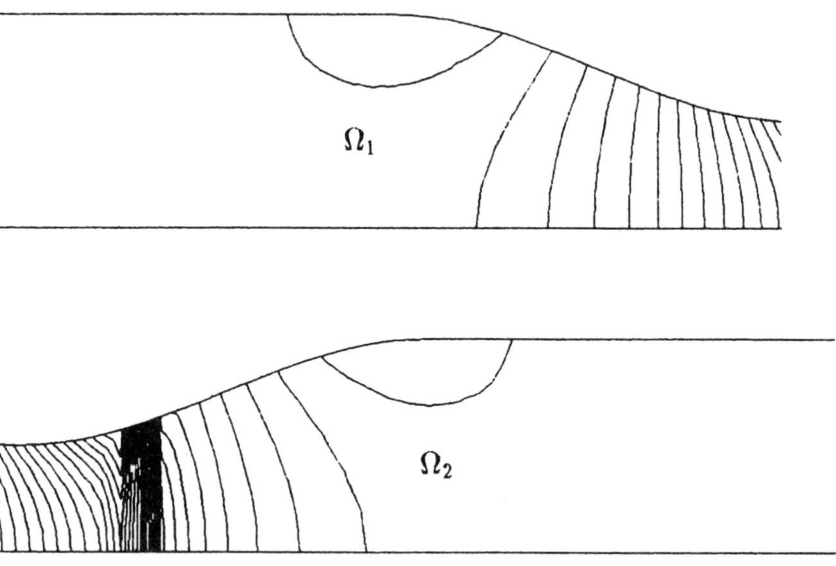

Fig. 4. 2

With the boundary conditions, $g_{in} = -1.8$ and $g_{out} = 1.8$, we have tested different flow cases. Here one typical example for the transonic case with $M_\infty = 0.70$ is presented. The corresponding isomach lines for each subdomain are also presented in Fig. 4.2. Fig. 4.3 shows the computed fitness function $J(g_1)$ for all the examined g_1.

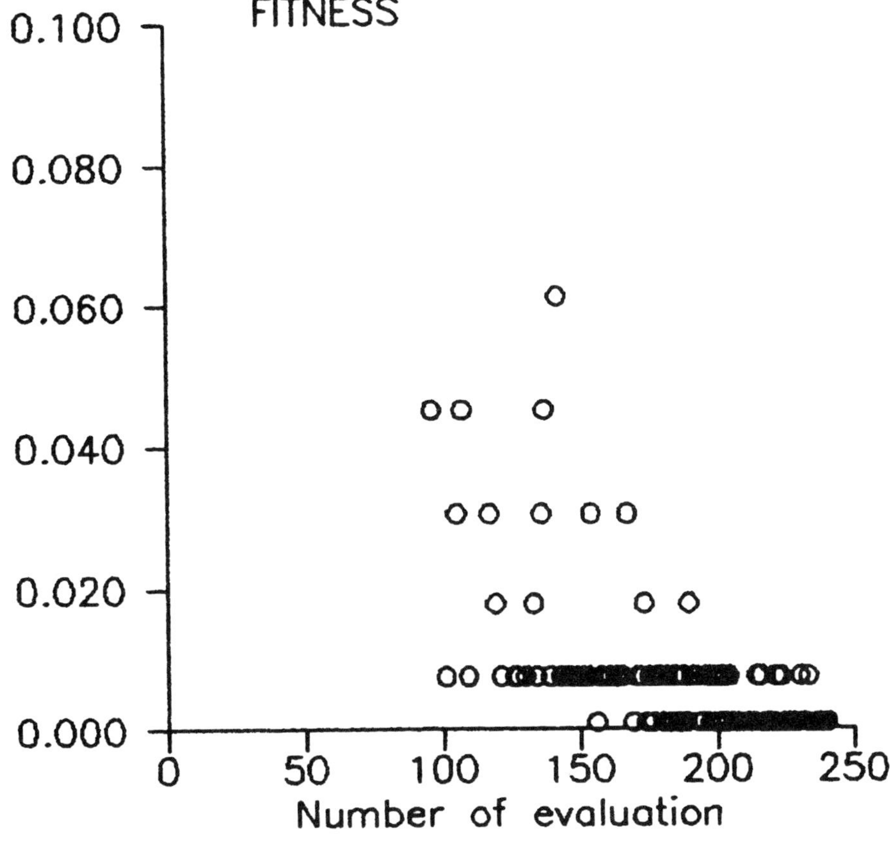

Fig. 4. 3 Isomach lines

It can be noted that the method presented can realize near optimal solutions within the finite genetic generations.

REFERENCES

[1] Goldberg D. E.(1989) *Genetic Algorithms in Search, Optimization, and Machine Learning.* Addison-Wesley, Reading, MA.

[2] Dinh Q. V., Glowinski R., Periaux J. and Terrasson G.(1988) On the Coupling of Viscous and Inviscid Models for Incompressible Fluid Flows via Domain Decomposition. *First International Symposium on Domain Decomposition Methods for PDE, SIAM.*

[3] Chen H. Q. and Huang M. K (1992) An AF3 Algorithm for the Calculation of Transonic Nonconservative Full Potential Flow over Wings or Wing-Body Combinations. *Chinese J. of Aero.,* 5(3).

Parallel Implementation of Multidomain Fourier Algorithm for 2-D and 3-D Navier-Stokes Equations[1]

0.1 Introduction

For long time integration of PDEs and resolving fine features of their solutions it is preferable to use high order methods, in particular spectral methods. But the *global* nature of spectral methods makes these methods difficult to parallelize with good performance. In particular, it can destroy the possibility of achieving scalability. Most numerical parallel algorithms based on spectral methods require global data transfers (such as transpose of a global matrix) and global communication. For massively parallel algorithm this may cause communication and synchronization bottlenecks.

A low communication parallel algorithms based on spectral methods were developed in [IVA1]-[IVA3] for the solution of non-linear time-dependent PDEs. The parallelization is achieved by domain decomposition. Elemental solutions in subdomains are constructed using the Local Fourier Basis (LFB) method. The idea of this method is to decompose a global problem into smooth local subproblems, using a collection of overlapping bell functions. Then, the Fourier method is applied in each local interval with spectral accuracy (without exhibiting the Gibbs phenomenon). It is shown in [IVA1]-[IVA3] that the multidomain local Fourier (MDLF) method is especially efficient for large problems using high resolutions in space. In this case, the relative number of operations required to perform the projection is insignificant in comparison with what is needed to execute the Fast Fourier Transform (FFT). Moreover, this extra work is reduced as the size of the problem grows in each domain[IVA4]. Therefore, for computation of large problems, which requires many degrees of freedom, the MDLF method is almost as efficient as the conventional Fourier method. An important advantage of this method, when compared to other multidomain spectral techniques, is that it enables a great simplification of the matching relations for the interface unknowns. For problems with constant coefficients and in simple domains, the use of the Fourier basis enables to fulfill the matching of each harmonic separately, and thus to eliminate the global coupling of the interface unknowns. Since then, the method was extended to handle complex geometries [IVA5]. The decomposition strategy of the the MDLF method can easily be changed to reflect different resolutions in each subdomain, which makes it valuable as an adaptive algorithm.

In [IVA4] the MDLF method was applied to the parallel solution of the model Navier-Stokes problem investigated previously in [T]. This model does not contain the pressure term, therefore, the time discretization procedure results in the Helmholtz equation to be solved in each time step. It was shown in [IVA4], that due to the locality properties of Helmholtz operator, application of the MLDF method to such problems is highly efficient. Matching of the solutions in subdomains requires only local communications between neighboring processors. Hence, the algorithm is fully

---

[1] A. Averbuch   K. Ruvinsky
School of Mathematical Sciences, Tel Aviv University, Tel Aviv 69978, Israel
M. Israeli   L. Vozovoi
Faculty of Computer Science, Technion, Haifa 32000, Israel

Domain Decomposition Methods in Sciences and Engineering, edited by R. Glowinski *et al.*

scalable. In [NS] a parallel algorithm, based on MDLF method for solution of the 2-D Navier-Stokes equations, was presented.

In the present paper the algorithm [NS] is generalized for 3-D Navier-Stokes equations and is applied to the parallel solution of the 2-D and 3-D Navier-Stokes equations. The treatment and the solution of the global Poisson equation for the pressure has the potential of degrading the performance and the speedup of the parallel solution. We show that for elliptic equation of Poisson type the size of the required global data transfer can essentially be reduced. The reason is that only the lowest harmonics of the pressure are treated and matched globally. Therefore, most of the communication that is required for parallelization of the Navier-Stokes equations is mainly local between adjacent subdomains (processors). The 2-D and 3-D Navier-Stokes equations are implemented on MIMD message-passing multiprocessor using PVM software package[PVM] and it achieves an almost linear speedup.

0.2 Governing Equations and Numerical Schemes

We are interested in the parallel solution of the Navier-Stokes equations, which governs the incompressible viscous flows with constant properties

$$\frac{\partial \mathbf{v}}{\partial t} = Re^{-1}\nabla^2 \mathbf{v} - \mathbf{N}(\mathbf{v}) - \nabla\Pi + \mathbf{F} \tag{0.1}$$

$$\nabla\Pi = \nabla\Pi' + q, \quad \Gamma(x + 2\pi, y + 2\pi, z + 2\pi) = \Gamma(x, y, z), \quad \Gamma = (\Pi', \mathbf{v}), \quad q = \text{const},$$

in the periodic domain $\Omega = [0, 2\pi]^3$. Here $\mathbf{v}(\mathbf{x}, t) = (u, v, w)$ is the velocity subject to the incompressibility constrain

$$\nabla \cdot \mathbf{v} = 0 \qquad \text{in} \quad \Omega, \tag{0.2}$$

$\Pi = p + \frac{1}{2}|\mathbf{v}|^2$ is the total pressure (p is the hydrostatic pressure), $\mathbf{F}$ is the external forcing and Re is Reynolds number. The nonlinear term is written in rotational form

$$\mathbf{N}(\mathbf{v}) = \mathbf{v} \times (\nabla \times \mathbf{v}). \tag{0.3}$$

The numerical solution of the problem (0.1)-(0.3) with periodic boundary conditions requires discretization in both time and space.

0.2.1 Discretization in Time

The discretization in time is performed via a splitting algorithm investigated in [KIO]:

$$\frac{\hat{\mathbf{v}} - \sum_{q=0}^{J_i - 1} \alpha_q \mathbf{v}^{n-q}}{\Delta t} = \sum_{q=0}^{J_q - 1} \beta_q (\mathbf{N}(\mathbf{v}^{n-q}) - \mathbf{F}^{n-q}), \tag{0.4}$$

$$\frac{\hat{\hat{\mathbf{v}}} - \hat{\mathbf{v}}}{\Delta t} = -\nabla\Pi^{n+1}, \tag{0.5}$$

$$\frac{\gamma_0 \mathbf{v}^{n+1} - \hat{\hat{\mathbf{v}}}}{\Delta t} = Re^{-1}\nabla^2 \mathbf{v}^{n+1}. \tag{0.6}$$

From (0.6) it follows that $\nabla \hat{\mathbf{v}} = 0$. Therefore, Eq. (0.5) can be rewritten in the form

$$\nabla^2 \Pi^{n+1} = \frac{1}{\Delta t} \nabla \cdot \hat{\mathbf{v}}, \tag{0.7}$$

This algorithm consists of an explicit advection step (0.4), a global pressure adjustment for incompressibility (0.5), (0.7) and implicit viscous step (0.6).

0.2.2 Discretization in Space

The splitting algorithm in time results in two types of elliptic equations: the Helmholtz equation

$$\nabla^2 u - \lambda^2 u = f(\mathbf{x}), \tag{0.8}$$

and the Poisson equation

$$\nabla^2 u = g(\mathbf{x}), \tag{0.9}$$

that have to be solved repeatedly (for each time step). The parameter λ in (0.8) is related to the time-stepping increment Δt, $\lambda^2 = \gamma_0 Re/\Delta t$. We consider computational domain $\Omega = [0, 2\pi]^d$, where $d = 2$ or 3 decomposed into parallel strips or slabs (further on called strips also) and the boundary conditions are periodic in all directions. We discretize the local problems in subdomains on a uniform grid. Then, we construct the local solutions using the multidomain local Fourier (MDLF) method of [IVA1]-[IVA3].

0.2.3 Solution of the Helmholtz Equation

After the application of the Fourier transform on Eq. (0.8) in the periodical directions y and z we get the 1-D equation in the nth strip:

$$\frac{d^2 \hat{u}^{(n)}}{dx^2} - \tilde{\lambda}^2 \hat{u} = \hat{f}^{(n)}(x). \tag{0.10}$$

In the 2-D case: $\hat{u}^{(n)} = \hat{u}_k^{(n)}$, $\hat{f}^{(n)} = \hat{f}_k^{(n)}$, $\tilde{\lambda}^2 = \lambda^2 + k^2$; in 3-D case: $\hat{u}^{(n)} = \hat{u}_{km}^{(n)}$, $\hat{f}^{(n)} = \hat{f}_{km}^{(n)}$, $\tilde{\lambda}^2 = \lambda^2 + k^2 + m^2$. Here $k = -\frac{N_y}{2}, ..., \frac{N_y}{2}$ and $m = -\frac{N_z}{2}, ..., \frac{N_z}{2}$ are harmonic numbers in y and z directions. According to the MDLF technique of [IVA1]-[IVA3] the solution of equation (0.10) in the whole domain consists of two main steps: Construction of the elemental particular solutions, matching step.

To estimate the gain of the local matching for the partial pressure harmonics in the 3-D case we compare the performance of the parallel solution of the Poisson equation (0.8) in two cases: complete global matching and local matching for partial pressure harmonics. We take the forcing in (0.8) to be $g(\mathbf{x}) = \cos kx \cdot e^{(\cos ky)(\cos kz)}$. The results are presented in Table 0.1. Local matching for partial pressure harmonics produces solution with order of accuracy of $\epsilon \sim 10^{-8}$ in comparison with complete solution that utilizes global matching. We need the following notations:

Notations:

N_g	size of square matrix of interface values of the lowest Fourier coefficients in y and z directions (included mean value) that are globally matched
T_g	executable time for the parallel algorithm where all the pressure harmonics are matched globally
T_g^c	communication time of the parallel algorithm where all the pressure harmonics are matched globally
$T_{g/l}$	executable time for the local matching of partial pressure harmonics
$T_{g/l}^c$	communication time of the local matching of partial pressure harmonics
$S_{g/l} = \frac{T_g}{T_{g/l}}$	speedup

Table 0.1 indicates that we achieve considerable savings in executable time by using local matching with partial pressure harmonics while preserving the accuracy of the solution (columns 6 and 11 in Table 0.1). This gain grows with the increase of the machine size and the amount of transferred data. This is due to the fact that the bottlenecks that are part of the complete global matching of pressure harmonics. We

P	$N_y = N_z = 128$					$N_y = N_z = 256$				
	$T_{g/l}^c$	$T_{g/l}$	T_g^c	T_g	$S_{g/l}$	$T_{g/l}^c$	$T_{g/l}$	T_g^c	T_g	$S_{g/l}$
4	17	54	41	80	1.5	73	182	168	289	1.6
6	19	50	65	91	1.8	75	185	253	405	2.2
8	20	51	89	130	2.6	74	183	362	553	3
10	22	53	121	164	3.1	75	195	507	747	3.8

Table **0.1**   Timing (in ticks) and speedup $S_{g/l} = T_g/T_{g/l}$ for parallel algorithms for the Poisson equation: $n = 48$, $N_x = n \cdot P$, $N_\epsilon = 8$, $N_g = 8$.

would like to emphasize the fact that even in the case when global communication is required, a careful choice of the method may results in a lesser communication. Thus, the overhead is not increased.

0.3   Computational Algorithm for 2-D and 3-D Navier-Stokes Equations

In this section we present the algorithm that was used for solving 2-D and 3-D Navier-Stokes equations using strip topology. Initially, we define the velocity $\mathbf{U}_d(\mathbf{x}, t)$ and the pressure $P_d(\mathbf{x})$ distributions in space (see Sections 0.5). Then, we compute the "forcing" $\mathbf{F}$ using Eq. (0.1). By using time marching algorithm (Eqs. (0.4) - (0.6)) we compute the velocity evolution $\mathbf{v}(\mathbf{x}, t)$ under the "forcing" $\mathbf{F}$ where the initial condition is $\mathbf{U}_d(\mathbf{x}, t)$. It is clear, that in this case deviation of the computed solution $\mathbf{v}(\mathbf{x}, t)$ from $\mathbf{U}_d(\mathbf{x}, t)$ in each time step have to be small and depends on the time stepping increment Δt and the number of collocation points (for example, in our 3-D computations the maximum relative error is $\epsilon \sim 10^{-10}$ for double precision at

$\Delta t = 10^{-3}$ where the number of collocation points $N_x = 128, N_y = 32, N_z = 32$, see Sec. 0.5).

Our algorithm consists of eight main steps (for detailed description of the algorithm see [AVIV]):

Step 1: Initialization

Step 2: Explicit advection step: beginning of time marching

Step 3: Calculation of the RHS of the Poisson equation

Step 4: Solution of the Poisson equation

Step 5: Matching of Π_p^{n+1} and $\partial\Pi_p^{n+1}/\partial x$ in the whole domain (partial global matching)

Step 6: Computation of the velocity $\hat{\mathbf{v}}$

Step 7: Implicit viscous step

Step 8: Matching of the particular solutions $\mathbf{v}_p^{n+1}$:Completion of the time step

After each step (from 2-6) we have to exchange data between processors. But only two data exchanges, before and after step 5, have some globality of data transfer: in step 5, we match particular solutions for the pressure Π and its derivative $\partial\Pi/\partial x$ in the whole domain. In this step only the lowest few harmonics have to be matched globally. The other four data exchanges are purely local. According to MDLF methodology the data transfer takes place between neighboring subdomains (processors).

0.4 Implementation

The proposed algorithms are appropriate for scalable distributed memory MIMD multiprocessors. Currently, the code is ported onto a distributed memory farm of DEC 3000 Model 400/400S AXP system called, from now on, as *Alpha Farm*. Each node in the farm is a high-performance desktop system that uses Digital DECchip 21064 RISC microprocessor running at 133.33 MHz. It is based on Digital Alpha AXP architecture and has ~75MB of memory space available to the user. This *Alpha Farm* consists of 10 computers connected through FDDI (Fiber Distributed Data Interface). Effective data and messages transfer through this interface is realized via *Giga Switch*, with averaged data transfer rate 3-4 MB/s.

To implement the above algorithms we used the Parallel Virtual Machine (PVM)[PVM] software package. Implementation results for Navier-Stokes equations on the *Alpha Farm* using PVM are presented in the next section.

0.5 Performance analysis

In this section we present the performance analysis of our algorithm for the solution of Navier-Stokes equations (Eq. (0.1)). The algorithm was tested according to the methodology described in Sec. 0.3. We identify the evolution of the solution under external force that has complicated behaviour in space and time. This algorithm can be easily adapted to the solution of other real problems of turbulence. For example, turbulence decay [BC], long time evolution of instable exact solution of Navier-Stokes equation [BMO], etc. Elimination of the "forcing" term is the only thing that has to be done for this adaption.

After the completion of the extension and folding of the MDLF method [IVA1, IVA2] we apply the Fourier transform to even functions in the subdomains. Therefore, we can use the discrete cosine transform (DCT) instead of the FFT (as was done in [IVA4]). It saves 40-50% of the overall time needed for the application of the Fourier transform.

Detailed performance analysis of MDLF algorithm for 2-D and 3-D Navier- Stokes equations are presented in [AVIV]. Here we dwell briefly on the results for 3-D Navier-Stokes equations.

Notations:

$N_x, \ N_y, \ N_z$	number of collocation points in x, y and z directions
P	number of processors
$n = N_x/P$	number of points in a processor in x
N_ϵ	number of points on an overlapped interval ϵ
$N = 2(n + 2N_\epsilon)$	the size of the local FFT
T_s	best serial time
T_{DCT}	parallel time in each processor using the Discrete Cosine Transform (DCT)
$S = T_s/T_{DCT}$	speedup
T_c	communication time
$E = S/P, \ 0 < E_p < 1$	efficiency

Parallel timing

The parallel measurements for the 3-D Navier-Stokes equations are presented in Tables 0.2.

Scalability: From Table 0.2 we see that the 3-D algorithm is highly scalable: when we increase the number of processors and the problem size in each processor remains the same (i.e. the problem size increases $\sim P$), the computation time of the whole problem stays the same (T_{DCT} in columns 2 and 10 in Table 0.2. This is due to the fact that $\sim 90\%$ of the communication between subdomains is local.

Communication time: Our results indicate that the total computation time, T_{DCT}, strongly depends on the data transfer rate between processors. Actually, from Table 0.2(columns 3 and 11) we see that the contribution of the communication time T_c to T_{DCT} varies from 30% to 50% depending upon the problem size. In our particular *Alpha Farm* configuration the average internal data transfer rate V_c is about 3 MB/s. It should be mentioned that, currently, the data transfer rate for internal network can be ten times higher than what is available by the Giga Switch of our *Alpha Farm*. So we extrapolate our results by taking $V_c = 9$ MB/s. The results of the extrapolation are presented in Table 0.2 (under $V_c = 9$ MB/s) and in the Figs. 0.1 From this tables we can see that by increasing V_c to 9 MB/s results in 10-3 0% increase in the speedup (S) and efficiency (E).

P	$n = 488$ $N = 1024$							
---	$V_c = 3$ MB/s				$V_c = 9$ MB/s			
	T_{DCT}	T_c	S	E	T_{DCT}	T_c	S	E
4	18	13	0.93	0.23	9	4	1.9	0.45
6	16	11	1.63	0.27	8	4	3.3	0.55
8	18	15	2	0.25	8	5	4.5	0.56
10	17	12	2.7	0.27	9	4	5.1	0.51
	$n = 2024$ $N = 4096$							
4	34	14	2.3	0.58	25	5	3.1	0.79
6	39	15	3.1	0.52	29	5	4.2	0.7
8	37	15	4.5	0.56	27	5	6.1	0.77
10	33	11	6.4	0.64	25	3	8.4	0.84

P	$n = 1000$ $N = 2048$							
---	$V_c = 3$ MB/s				$V_c = 9$ MB/s			
	T_{DCT}	T_c	S	E	T_{DCT}	T_c	S	E
4	22	15	1.7	0.42	12	5	3	0.77
6	22	15	2.6	0.43	12	5	4.8	0.8
8	20	13	3.9	0.49	12	4	6.5	0.81
10	20	13	4.9	0.5	12	4	8.2	0.82
	$n = 4072$ $N = 8192$							
4	67	14	2.5	0.63	58	5	2.9	0.72
6	67	20	3.8	0.64	54	7	4.8	0.79
8	67	20	5.2	0.65	54	7	6.5	0.81
10	65	15	6.8	0.68	55	5	8	0.8

Table 0.2   3-D timing (in ticks), speedup and efficiency for 10 iterations: $N_y = 8$, $N_z = 8$, $N_x = n \cdot P$, $N = 2(n + 2N_\epsilon)$, $N_\epsilon = 16$.

Speedup: From Tables 0.2 and Figs. 0.1 we can see that the achieved speedup grows almost linearly with the increase of the number of processors P and this is done while the problem size in each processor stays invariant. This linear growth is explained by the fact that algorithm is highly scalable. When the number of processors increases and, therefore, the size of the problem also increases, the parallel computation time T_{DCT} stays almost the same, but the serial computation time T_s increases.

Efficiency: The efficiency of the algorithm is high and increases with the increase of the problem size. Thus, it varies from 0.3 to 0.7-0.9 (Table 0.2).

Local matching of partial pressure harmonics: In [AVIV] we had a detailed explanation that for 2-D and 3-D Poisson equation only few (lowest) harmonics of the pressure in the directions along the strips have to be matched globally. The number of globally matched

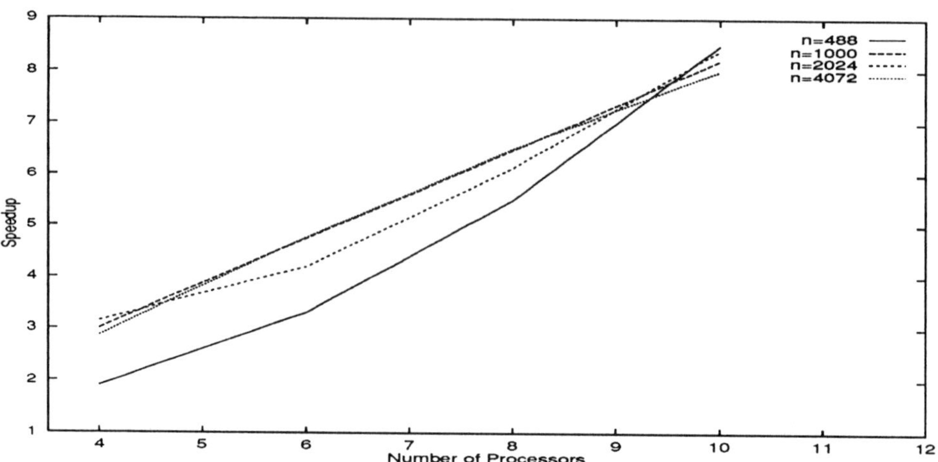

Figure 0.1 Achieved speedup when 3-D domain is decomposed into parallel strips for $N_y = 8$, $N_z = 8$.

harmonics depends on the required accuracy of the solution. Here we verify this observation on highly nonlinear solution with complex dynamics in space and time. To investigate the advantage of using local matching of partial pressure harmonics we take another extreme case, in comparison with the results in Tables 0.2. For this extreme case, we take the maximum number of harmonics (the only constraint is the memory size) that are being matched along strips. The results of these computations are presented in Table 0.3. The notations are the same as in Table 0.1. All the results reflect local matching of partial pressure harmonics at distinct N_g produce the same order of accuracy ($\epsilon \sim 10^{-4}$) as was achieved by using complete global matching of pressure harmonics. From this table we see that we need only the three lowest harmonics and mean value of the pressure to be matched globally. This results in $\sim 20\%$ saving in the executable time in comparison with the complete (full version) of the global algorithm. We get almost the same savings when the number of globally matched harmonics is increased (row 2 at $N_g = 32$ at Table 0.3). This is due to the fact that a packet size in the *Alpha Farm* is 16KB. Therefore, the transfer time of 4 and 32 harmonics is almost the same. From Table 0.3 we see that algorithm with local matching of partial pressure harmonics is scalable. Execution time of complete global (full version) algorithm increases with the number of processors. Global exchange of data in global algorithms can degrade the performance. Therefore, the speedup $S_{g/l}$ increases with machine size (see Table 0.3).

Thus, we show that local matching of partial pressure harmonics leads to a reduction in executable time while preserving the accuracy of the solution

P	N_g	$T_{g/l}^c$	$T_{g/l}$	T_g^c	T_g	$S_{g/l}$
10	4	600	681	694	792	1.16
4	32	568	660	621	726	1.1
6	32	546	637	632	726	1.14
8	32	559	652	680	762	1.17
10	32	562	651	694	792	1.21

Table **0.3** Timing (in ticks) and speedup $S_{g/l} = T_g/T_{g/l}$ for parallel algorithms for ten iterations; $n = 8$, $N_x = n \cdot P$, $N_y = N_z = 128$, $N_\epsilon = 4$.

Stability and Numerical Accuracy : Our implementation is stable. For example for parameters $P = 3$, $N_x = 128$, $N_y = 32$, $N_z = 32$, $N_\epsilon = 16$ and the time step $\Delta t = 10^{-3}$ the numerical error $\varepsilon = \max_{0 < x,y,z < 2\pi} |\mathbf{v} - \mathbf{v}_{ex}| \sim 10^{-9}$ during 7500 time units.

REFERENCES

[IVA1] Israeli M., Vozovoi L., Averbuch A., Spectral multi-domain technique with Local Fourier Basis, *J. Scientific Computing*, **8**, No. 2, (1993), pp. 181-195.

[IVA2] M. Israeli, L. Vozovoi, A. Averbuch, Parallelizing Implicit algorithms for time-dependent problems by parabolic domain decomposition, *J. Scientific Computing*, **8**, No. 2, (1993), pp. 197-212

[IVA3] Vozovoi L., Israeli M., Averbuch A., Spectral multidomain technique with Local Fourier Basis II: decomposition into cells, *J. Scientific Computing*), **9**, No. 3, (1994), pp. 311-326.

[IVA4] Averbuch A., Israeli M., Vozovoi L., Parallel implementation of non-linear evolution problems using parabolic domain decomposition, *Parallel Computing*, **21**, No. 7 (1995), pp. 1151-1183.

[IVA5] L. Vozovoi, M. Israeli, A. Averbuch, Multidomain local Fourier method for PDEs in complex geometries, to appear in *J. of Comput. Applied Math.*

[NS] L. Vozovoi, M. Israeli, A. Averbuch, Multidomain Fourier Algorithms for Parallel solution of the Navier-Stokes equation, *Contemporary Mathematics*, **180**, (1994), pp.539-546.

[KIO] G.E. Karniadakis, M. Israeli, S.A. Orszag, High order splitting methods for the incompressible Navier-Stokes Equations, *J. Computational Physics*, **97**, No. 2, (1991), pp. 414-443.

[PVM] A. Geist, A. Beguelin et. al., PVM: parallel virtual machine. *The MIT Press*, Cambridge, (1994).

[BC] G.L. Browning, H.O. Kreiss, Comparison of numerical methods for the calculation of two-dimensional turbulence, *Mathematic of Computation*, **52**, No. 186, (1989), pp. 369-388.

[BMO] M. Brachet, D.I. Meiron, S. A. Orszag et. al. Small-scale structure of the Taylor-Green vortex, *J. Fluid Mech.*, **130**, (1983), pp. 411-452.

[T] R.Temam, Stability Analysis of the Nonlinear Galerkin Method, *Math. of Computation*, **57**, No. 196, (1991), p. 477-505.

[AVIV] A. Averbuch, K. Ruvinsky, M. Israeli, L. Vozovoi, Highly scalable 2-D and 3-D Navier-Stokes Parallel Solver on MIMD Multiprocessors, *submitted.*

Domain decomposition for the incompressible Navier-Stokes equations: solving subdomain problems accurately and inaccurately[1]

E. Brakkee, C. Vuik, P. Wesseling

Abstract:

For the solution of practical flow problems in arbitrarily shaped domains, simple Schwarz domain decomposition methods with minimal overlap are quite efficient, provided Krylov subspace methods, such as the GMRES method, are used to accelerate convergence. With accurate subdomain solution, the amount of time spent in solving these problems may be quite large. To reduce computing time, inaccurate solution of subdomain problems is considered, which requires a different, GCR based, acceleration technique. Much emphasis is put on the multiplicative domain decomposition algorithm since we also want an algorithm which is fast on a single processor.

[1] Erik Brakkee, Applied Analysis Group, Delft University of Technology, Mekelweg 4, 2628 CD Delft, The Netherlands, e.brakkee@math.tudelft.nl

Domain Decomposition Methods in Sciences and Engineering, edited by R. Glowinski *et al.*

1   Introduction

For the solution of the incompressible Navier-Stokes equations in domains of arbitrary shape, we use a finite volume method on structured boundary fitted grids. References [MWSK91, CP92, WSvK+92, SWV+92, ZSW95a, ZSW95b] describe the discretization in detail and [OWSB93, ZSW95a] discuss the capability of the method to accurately solve a number of laminar and turbulent flows. A Schwarz type domain decomposition iteration [Sch69, BW94, BSK95] in combination with GMRES [SS86] acceleration is used. In [BW94], it was shown that significant reductions in computing time can be obtained using the GMRES acceleration procedure.

The method described in [BW94] requires accurate solution of subdomain problems, As a result of this, the computing time can be much larger than with single-block solution for the same number of unknowns. Also, it is not known beforehand how accurate the subdomain problems must be solved. The required subdomain solution accuracy may be quite high, especially when grid cells are very much stretched near block interfaces. A possible solution to both problems is to abandon the assumption of exact subdomain solution and to allow (very) inaccurate subdomain solution. Since the preconditioner may now vary in each iteration, GMRES acceleration may no longer be applied. Instead, a method based on GCR [EES83] is used.

Considerable reductions in computing time can be obtained in this way, see [BVW95b, BVW95a]. The present paper is an abstract from [BVW95a]. Theoretical results and numerical experiments are presented to illustrate the effect of inaccurate solution of subdomain problems for the incompressible Navier-Stokes equations.

Parallel computing is of increasing importance. Therefore it is important to compare the parallel (additive) domain decomposition algorithms with the best multiplicative algorithms, which are known to be faster than additive algorithms. Therefore, we pay much attention to multiplicative algorithms.

2   Discretization

For the spatial discretization, we use a finite volume method employing a staggered grid and central discretization, see [MWSK91, CP92, WSvK+92, SWV+92, ZSW95a, ZSW95b]. For the time discretization, the implicit Euler method is used. With V^n and P^n representing the algebraic vectors of velocity and pressure unknowns at time t^n, we get

$$\frac{V^{n+1} - V^n}{\Delta t} = M(V^n, P^n)V^{n+1} - GP^{n+1}, \tag{1}$$

$$DV^{n+1} = 0, \tag{2}$$

where (1) represents the momentum equations and (2) represents the incompressibility condition $\operatorname{div} u = 0$. The matrix M represents the linearized spatial discretization of the Navier-Stokes equations around time level n, G is the discretized gradient operator and D is the discretized divergence operator on a staggered grid.

The pressure correction method [HW65, Cho68, Van86] is used to solve (1) and (2). It consists of three steps. In the first step, an estimate V^* of V^{n+1} is computed by solving (1) with the pressure fixed at the old time level:

$$\frac{V^\star - V^n}{\Delta t} = M(V^n, P^n)V^\star - GP^n. \tag{3}$$

In the second step, the pressure correction ΔP is solved from

$$DG\Delta P = \frac{DV^\star}{\Delta t}. \tag{4}$$

Dirichlet boundary conditions are prescribed for ΔP on boundaries where the normal velocity is given, and in all other cases Neumann boundary conditions are used. The last step consists of correcting the pressure: $P^{n+1} = P^n + \Delta P$ and computing V^{n+1} satisfying the incompressibility condition (2)

$$V^{n+1} = V^\star - \Delta t G\Delta P. \tag{5}$$

3   Domain decomposition

Domain decomposition amounts to the solution of (3) and (4) using an alternating Schwarz method with minimal overlap, see [BW94, BSK95]. It is written as a block iteration to solve a system $Au = f$, with the blocks defined by the subdomains

$$u^{m+1} = (I - N^{-1}A)u^m + N^{-1}f, \tag{6}$$

with N^{-1} an approximation to the inverse of the block diagonal or block lower-triangular matrix of A. The subdomain problems are solved using GMRES [SS86, Vui93].

With accurate subdomain solution, u^{m+1} in (6) only depends on the components of u^m corresponding to unknowns on or near the block interfaces, see [BW94, BSK95]. Collecting these unknowns in a vector v and defining the trivial injection operator $u = Qv$, which extends v by zeroes to the full vector length, we get

$$u^{m+1} = (I - N^{-1}A)Qv^m + N^{-1}f, \tag{7}$$

By premultiplying (7) with Q^T and taking the stationary solution of the iteration process we get

$$Q^T N^{-1}AQv = Q^T N^{-1}f. \tag{8}$$

The system (8) is a system concerning only unknowns on or near the interfaces (similar to Schur's complement). The GMRES acceleration then solves (8); details are in [BW94, BSK95].

Inaccurate subdomain solution means that we replace (6) by

$$u^{m+1} = u^m + \tilde{N}^{-1}(f - Au^m), \tag{9}$$

where $\tilde{N}$ represents inaccurate subdomain solution. Because GMRES is used for subdomain solution, $\tilde{N}$ varies in each iteration step, and we may no longer use GMRES acceleration.

The GCR [EES83] method for solving $Ax = f$ can be easily adapted to cope with variable preconditioners. The GCR method seeks to minimize the residual $r_k = f - Ax_k$ over a search space $S_k = < s_1, s_2, \ldots, s_k >$. For this purpose, a subspace $V_k = < v_1, v_2, \ldots, v_k >$ with $As_i = v_i$ is stored. Gramm-Schmidt orthogonalization of V_k is used to project f onto V_k. The search directions are updated during orthogonalization such that the property $As_i = v_i$ is preserved. This enables a simple construction of the optimal solution x_k. By extending the search space with appropriate search directions $s_{k+1}, s_{k+2}, \ldots$, GCR reduces the residual further. With the choice $s_{k+1} = r_k$, the method is equivalent to GMRES [SS86].

GCR acceleration of (9) uses $s_{k+1} = \tilde{N}^{-1} r_k$, which corresponds to a single domain decomposition iteration. For the case of a single subdomain, the method simplifies to GMRESR [vdVV94].

Another Krylov method that enables variable preconditioners is FGMRES [Saa93]. In [Bör89], for instance, this algorithm was used together with inaccurate subdomain solution. The emphasis in [Bör89] was not on reduction of computing time but on restrictions on the subdomain solution accuracy to retain the h-independent convergence of Neumann-Dirichlet methods.

4   Theoretical motivation

Inaccurate solution of subproblems reduces the amount of work in each domain decomposition iteration at the cost of some additional iterations of the outer domain decomposition iteration. Therefore, a reduction of computing time is only possible if the number of additional iterations is not too large. A simple analysis of the condition number of the postconditioned matrix $A\tilde{N}^{-1}$ confirms this statement.

Each iteration involves solving $Nu = g$ with N the matrix from (6). With inaccurate solution of subdomains, we solve a problem $\tilde{N}\tilde{u} = g$ with $\tilde{N}$ as in (9). All subproblems are solved using a relative accuracy.

Condition 1 *Each subproblem $A_{ii}u_i = g_i$ is solved using initial guess 0 and with a relative accuracy of ϵ so that $\|g_i - A_{ii}\tilde{u}_i\| \leq \epsilon\|g_i\|$ in the Euclidean norm.*

Theorem 1 relates N and $\tilde{N}$.

Theorem 1 *If condition 1 holds for all subdomains and all possible right-hand sides g_i, then*

(a) $\|I - N_{gs}\tilde{N}_{gs}^{-1}\| \leq C\epsilon$, for some constant $C > 0$.

(b) $\|I - N_{jac}\tilde{N}_{jac}^{-1}\| \leq \epsilon$.

Proof:

Proof of (a): Combination of Condition 1 with $\tilde{A}_{ii}\tilde{u}_i = g_i$ (inaccurate subdomain solution) gives $\|g_i - A_{ii}\tilde{A}_{ii}^{-1}g_i\| = \|(I - A_{ii}\tilde{A}_{ii}^{-1})g_i\| \leq \epsilon\|g_i\|$ for all g_i. From the definition of a matrix norm it follows that $\|I - A_{ii}\tilde{A}_{ii}^{-1}\| \leq \epsilon$.

Without loss of generality we take two subdomains. We get

$$I - N\tilde{N}^{-1} = \begin{bmatrix} I - A_{11}\tilde{A}_{11}^{-1} & \emptyset \\ -(I - A_{22}\tilde{A}_{22}^{-1})A_{21}\tilde{A}_{11}^{-1} & I - A_{22}\tilde{A}_{22}^{-1} \end{bmatrix} \quad (10)$$

Partition $x = \begin{bmatrix} x_1 \\ x_2 \end{bmatrix}$ and note that for the Euclidean norm

$$\|x\| \leq \left\| \begin{bmatrix} x_1 \\ 0 \end{bmatrix} \right\| + \left\| \begin{bmatrix} 0 \\ x_2 \end{bmatrix} \right\| = \|x_1\| + \|x_2\|, \text{ then we have}$$

$\|I - N\tilde{N}^{-1}\| = \sup_{\|x\| \leq 1} \|(I - N\tilde{N}^{-1})x\| \leq \sup_{\|x\| \leq 1} \big(\|(I - A_{11}\tilde{A}_{11}^{-1})x_1\| + \|(I - A_{22}\tilde{A}_{22}^{-1})A_{21}\tilde{A}_{11}^{-1}x_1\| + \|(I - A_{22}\tilde{A}_{22}^{-1})x_2\| \big)$.
Furthermore, for the Euclidean norm $\|x\| \leq 1$ implies $\|x_1\| \leq 1$ and $\|x_2\| \leq 1$ so that finally (a) follows with $C = 2 + \|A_{21}\tilde{A}_{11}^{-1}\|$.
Proof of (b): For any block diagonal matrix $B = \text{diag}(D_1, D_2, \ldots, D_n)$, we have: $\|B\| = \sqrt{\rho(B^T B)} = \sqrt{\rho(\text{diag}(D_1^T D_1, \ldots, D_n^T D_n))} = \max\{\sqrt{\rho(D_1^T D_1)}, \ldots, \sqrt{\rho(D_n^T D_n)}\} = \max\{\|D_1\|, \ldots \|D_n\|\}$. If we use the additive post-conditioner, then $I - N\tilde{N}^{-1}$ is a block diagonal matrix with blocks $D_i = I - A_{ii}\tilde{A}_{ii}^{-1}$, so that $\|I - N\tilde{N}^{-1}\| = \max_i \|I - A_{ii}\tilde{A}_{ii}^{-1}\| \leq \epsilon$.
Therefore, (b) holds.

$\square$

Theorem 1 enables us to give a relation between the condition numbers of $A\tilde{N}^{-1}$ and AN^{-1}.

Theorem 2 *Under the conditions of Theorem 1 and $C\epsilon < 1$, the condition number of $A\tilde{N}^{-1}$ satisfies*

$$\kappa(A\tilde{N}^{-1}) \leq \frac{1 + C\epsilon}{1 - C\epsilon} \cdot \kappa(AN^{-1}). \quad (11)$$

Proof:

Application of Theorem 1, and noting that $\| \star \|$ is a least upper bound norm, gives $\|N\tilde{N}^{-1}\| = \|N\tilde{N}^{-1} - I + I\| \leq 1 + C\epsilon$ and $\|(N\tilde{N}^{-1})^{-1}\| = \|(N\tilde{N}^{-1})^{-1}(I - N\tilde{N}^{-1}) + I\| \leq 1 + \|(N\tilde{N}^{-1})^{-1}\|C\epsilon$.
Since $C\epsilon < 1$, $\kappa(N\tilde{N}^{-1}) = \|N\tilde{N}^{-1}\| \cdot \|(N\tilde{N}^{-1})^{-1}\| \leq \frac{1+C\epsilon}{1-C\epsilon}$.
Inequality (11) follows from $\kappa(A\tilde{N}^{-1}) = \kappa(AN^{-1}N\tilde{N}^{-1}) \leq \kappa(AN^{-1}) \cdot \kappa(N\tilde{N}^{-1})$.

$\square$

Theorem 2 shows that the subdomain solution accuracy has only a small effect on the condition number of the postconditioned matrix. This means that, at least for symmetric problems, the number of outer iterations will not increase (significantly) when the subdomain accuracy is lowered. The sensitivity of outer loop convergence to ϵ is given by the constant C in Theorem 1, which can be chosen 1 for the additive algorithm, independent of the number of subdomains. For multiplicative algorithms this sensitivity constant C will probably also be small and independent of the number of subdomains, however, sharper bounds may require a much more detailed analysis.

The theorems only hold for constant $\tilde{N}$, but the results in Section 6 show that the conclusions also hold in case $\tilde{N}$ varies in each iteration.

5   The model problem

We consider flow around a cylinder in a wall-bounded shear flow. Figure 1 shows the geometry and decomposition of the domain, coarse versions of the multi-block and single-block grids, and a description of the boundary conditions. The multi-block and single-block grids consist of 12240 and 10800 grid cells respectively.

The cylinder has diameter $a = 2$. The Reynolds number is defined as $\mathrm{Re} = \frac{1}{2}\left(\frac{au^\star}{\nu}\right)^2$ with $u^\star = \sqrt{\frac{\tau_0}{\rho}}$, where $\tau_0 = \mu\partial u/\partial y$ is the shear stress associated with the linear inlet velocity profile. Typical Reynolds numbers for this problem are $\mathrm{Re} = 1 - 5$. Our results are given for $\mathrm{Re} = 2$. In the computation we have used $L = H = 10$. For more details on this computation, see [BW94].

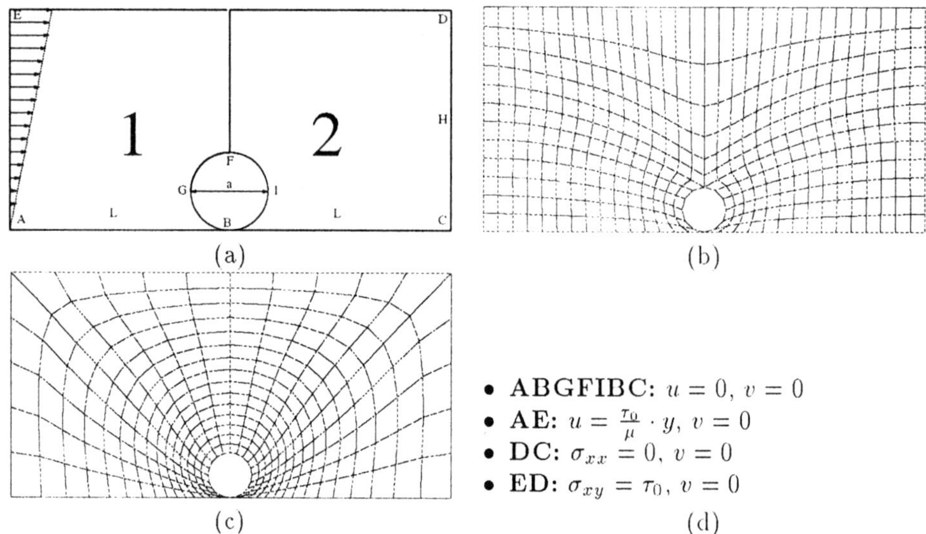

Figure 1   (a) Geometry and decomposition of the domain, (b) multi-block grid, (c) single-block grid, (d) boundary conditions

6   Results and conclusions

The time measurements in this section are given for only the first 10 time steps to avoid excessive computing times.

Table 1 lists the computing times (no. of iterations in brackets) for different subdomain solution accuracies; the top row is for the algorithm which assumes accurate subdomain solution. As the subdomain solution accuracy is lowered from 10^{-4} to 10^{-1}, the number of outer GCR iterations shows only a small increase, which, because of the reduced work in solving subproblems, results in a reduction of total computing time. This is in accordance with Theorem 2. In these experiments, the most significant reductions are obtained for the pressure equations.

ϵ	*Tot.*	*Mom.*	*Pres.*
$10^{-4\star}$	449.1	78.6(38)	315.2(154)
10^{-4}	465.2	58.4(37)	351.6(155)
10^{-2}	292.1	43.5(38)	193.4(168)
10^{-1}	230.3	47.5(48)	127.6(210)

Table 1 Results with varying accuracy of subdomain solution for the cylinder problem, multiplicative algorithm, 2 subdomains

Tot.	*Mom.*	*Pres.*
128.0	31.6	48.9

Table 2 Single-block solution times

Table 2 lists the computing times for single block solution. Comparing Tables 2 and 1, we see that solution time for the momentum equations comes close to single block solution time. For the pressure equations, solution time is still a factor $2 - 3$ larger. With accurate subdomain solution this was approximately a factor 7.

Because of the significant reduction in computing time for the pressure equations, the total computing time shows a reduction of a factor of 2 by inaccurate subdomain solution. Still, the computing time with multi-block solution is almost twice as large as with single-block solution, and is dominated by the pressure solution time. A coarse grid correction, e.g. [BPS89, BS92], could be implemented for the pressure equations to reduce computing time further.

Inaccurate solution of subdomain problems combined with GCR acceleration removes the restriction inherent in GMRES solution of interface equations (8) that subdomain problems should be solved accurately (enough). The GCR based algorithm is therefore in general more reliable than the GMRES algorithm for solving interface equations.

References

[Bör89] Börgers C. (1989) The Neumann-Dirichlet domain decomposition method with inexact solvers on the subdomains. *Numer. Math.* 55: 123–136.

[BPS89] Bramble J., Pasciak J., and Schatz A. (1989) The construction of preconditiones for elliptic problems by substructuring IV. *Math. Comp.* 53: 1–24.

[BS92] Bjørstad P. and Skogen M. (1992) Domain decomposition algorithms of Schwarz type, designed for massively parallel computers. In Keyes D. E., Chan T. F., Meurant G., Scroggs J. S., and Voigt R. G. (eds) *Proc. of the Fifth International Symposium on Domain Decomposition methods for Partial Differential Equations*, pages 362–375. SIAM, Philadelphia.

[BSK95] Brakkee E., Segal A., and Kassels C. (December 1995) A parallel domain decomposition algorithm for the incompressible Navier-Stokes equations. *Journal of Simulation Practice and Theory* To appear.

[BVW95a] Brakkee E., Vuik C., and Wesseling P. (1995) Domain decomposition for the incompressible Navier-Stokes equations: solving subdomain problems accurately and inaccurately. Report 95-37, Faculty of Technical Mathematics and Informatics, Delft University of Technology, Delft. Available from anonymous ftp://ftp.twi.tudelft.nl/TWI/publications/tech-reports/1995/DUT-TWI-95-37.ps.gz, submitted to Int. J. of Num. Methods in Fluids.

[BVW95b] Brakkee E., Vuik C., and Wesseling P. (1995) An investigation of Schwarz domain decomposition using accurate and inaccurate solution of subdomains. Report 95-18, Faculty of Technical Mathematics and Informatics, Delft University of Technology, Delft. Available from anonymous ftp://ftp.twi.tudelft.nl/TWI/publications/tech-reports/1995/DUT-TWI-95-18.ps.gz, Submitted to Numerical Linear Algebra with Applications.

[BW94] Brakkee E. and Wesseling P. (1994) Schwarz domain decomposition for the incompresssible Navier-Stokes equations in general coordinates. Report 94-84, Faculty of Technical Mathematics and Informatics, Delft University of Technology, Delft. Available from anonymous ftp://ftp.twi.tudelft.nl/TWI/publications/tech-reports/1994/DUT-TWI-94-84.ps.gz, submitted to Int. J. of Num. Methods in Fluids.

[Cho68] Chorin A. (1968) Numerical solution of the Navier-Stokes equations. *Math. Comp.* 22: 745–762.

[CP92] C.W.Oosterlee and P.Wesseling (1992) A multigrid method for an invariant formulation of the incompressible Navier-Stokes equations in general coordinates. *Comm. Applied Num. Methods* 8: 721–725.

[EES83] Eisenstat S., Elman H., and Schultz M. (1983) Variational iterative methods for nonsymmetric systems of linear equations. *SIAM Journal of Numerical Analysis* 20: 345–357.

[HW65] Harlow F. and Welch J. (1965) Numerical calculation of time-dependent viscous incompressible flow of fluid with a free surface. *The Physics of Fluids* 8: 2182–2189.

[MWSK91] Mynett A., Wesseling P., Segal A., and Kassels C. (1991) The ISNaS incompressible Navier-Stokes solver: invariant discretization. *Applied Scientific Research* 48: 175–191.

[OWSB93] Oosterlee C., Wesseling P., Segal A., and Brakkee E. (1993) Benchmark solutions for the incompressible Navier-Stokes equations in general coordinates on staggered grids. *International Journal for Numerical Methods in Fluids* 17: 301–321.

[Saa93] Saad Y. (1993) A flexible inner-outer preconditioned GMRES algorithm. *SIAM J. Sci. Stat. Comp.* 14: 461–469.

[Sch69] Schwarz H. (1869) Über einige Abbildungsaufgaben. *Journal für Reine und Angewandte Mathematik* 70: 105–120.

[SS86] Saad Y. and Schultz M. (1986) GMRES: a generalized minimal residual algorithm for solving non-symmetric linear systems. *SIAM J. Sci. Stat. Comp.* 7: 856–869.

[SWV$^+$92] Segal A., Wesseling P., Van Kan J., Oosterlee C., and Kassels C. (1992) Invariant discretization of the incompressible Navier-Stokes equations in boundary fitted coordinates. *International Journal for Numerical Methods in Fluids* 15: 411–426.

[Van86] Van Kan J. (1986) A second-order accurate pressure correction method for viscous incompressible flow. *SIAM J. Sci. Stat. Comp.* 7: 870–891.

[vdVV94] van der Vorst H. and Vuik C. (1994) GMRESR: a family of nested GMRES methods. *Numerical Linear Algebra with Applications* 1(4): 369–386.

[Vui93] Vuik C. (1993) Solution of the discretized incompressible Navier-Stokes equations with the GMRES method. *International Journal for Numerical Methods in Fluids* 16: 507–523.

[WSvK$^+$92] Wesseling P., Segal A., van Kan J., Oosterlee C., and Kassels C. (1992) Invariant discretization of the incompressible Navier-Stokes equations in general coordinates on staggered grids. *Comput. Fluids Dyn. J.* 1: 27–33.

[ZSW95a] Zijlema M., Segal A., and Wesseling P. (1995) Finite volume computation of incompressible turbulent flows in general coordinates on staggered grids. *Int. J. of Num. Meth. in Fluids* 20: 621–640.

[ZSW95b] Zijlema M., Segal A., and Wesseling P. (1995) Invariant discretization of the k-ε model in general coordinates for prediction of turbulent flows in complicated geometries. *Computers and Fluids* 24: 209–225.

Schwarz Methods for the Unsteady Compressible Navier-Stokes Equations on Unstructured Meshes

Xiao-Chuan Cai[1] Charbel Farhat [2] and Marcus Sarkis [3]

1 Introduction

Overlapping Schwarz is a family of preconditioners for solving large sparse linear systems arising from the discretization of partial differential equations, see e.g. [CS96, DW94, SBG95]. Here we report on our preliminary experiences on using it in the implicit solution of unsteady Navier-Stokes (N.-S.) equations discretized on two-dimensional unstructured meshes. One of the advantages of implicit methods is that they allow the time steps to be determined solely based on the physics of the fluid flow, not on the stability property of the time discretization scheme, [Ven95, VM95]. To advance in time, a large linear system of equations must be solved. Depending of the size of the time step, and other flow parameters, the conditioning of the matrix may change drastically from time step to time step. To solve these systems iteratively, it is necessary to have a family of preconditioners whose strength can be controlled. Overlapping Schwarz methods do have these properties, for examples, they have

[1] Dept. of Comp. Sci., Univ. of Colorado, Boulder, CO 80309. *cai@cs.colorado.edu*. The work was supported in part by NSF ASC-9457534, NSF ASC-9217394 and NASA NAG5-2218.

[2] Dept. of Aerospace Eng., Univ. of Colorado, Boulder, CO 80309. *charbel@colorado.edu*. The work was supported in part by NSF ASC-9217394, NSF ASC-9217394 and NASA NAG5-2218.

[3] Dept. of Comp. Sci., Univ. of Colorado, Boulder, CO 80309. *msarkis@cs.colorado.edu*. The work was supported in part by NSF ASC-9406582, NSF ASC-9217394 and NASA NAG5-2218.

Domain Decomposition Methods in Sciences and Engineering, edited by R. Glowinski *et al.*
© 1997 John Wiley & Sons, Ltd.

adjustable strength, controlled by (1) using the inexact solution techniques for solving local problems; (2) including or excluding the coarse preconditioner; (3) changing the size of the coarse mesh.

In this paper, we investigate the difference between the Schwarz family of preconditioners and the global ILU preconditioners. Within the Schwarz preconditioners, we try to understand the role of the overlapping size between subdomains, the effect of the number of subdomains and inexact subdomain solvers. At each time step, we solve the resulting global linear system by the preconditioned GMRES method, and in the preconditioning stage, we solve the local subdomain problems again by the preconditioned GERES method, with different preconditioners and stopping conditions. Since the construction of the preconditioner is very expensive, we explore the possibility of re-using the preconditioner for several time steps. For steady state problems, some studies can be found in [GKM94]. For other recent development in unsteady calculations, we refer the reader to [BL95, Ven95, VM95].

2 Governing Equations

Let $\Omega \subset \Re^2$ be the flow domain and Γ its boundary. The conservative form of the N.-S. equations is given by

$$\frac{\partial}{\partial t}W + \nabla_{\vec{x}} \cdot \mathcal{F}(W(\vec{x},t)) = \frac{1}{\mathrm{Re}}\nabla_{\vec{x}} \cdot \mathcal{R}(\mathcal{W}(\vec{x},t)), \tag{1}$$

where $W = (\rho, \rho u, \rho v, E)^T$, and $\vec{x}$ and t denote the spatial and temporal variables. The detailed definition of $\mathcal{F}$ and $\mathcal{R}$ can be found in [FFL93]. In the above expressions, ρ is the density, $\vec{U} = (u, v)^T$ is the velocity vector and E is the total energy per unit volume.

We are interested in unsteady, external flows around an airfoil as pictured in Fig. 1. The domain boundary is $\Gamma = \Gamma_\omega \cup \Gamma_\infty$ and the far field velocity is $\vec{U}_\infty$.

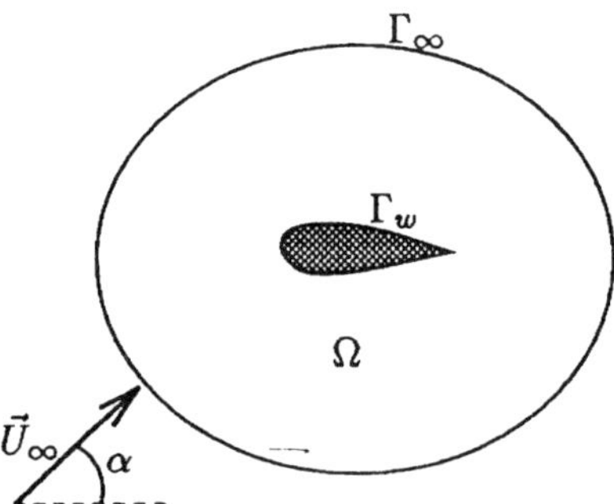

Figure 1 The computational domain

On the wall boundary Γ_ω, a no-slip condition on $\vec{U}$ and a Dirichlet condition on the temperature T are imposed, i.e., $\vec{U} = \vec{0}$ and $T = T_\omega$. No boundary conditions are specified for the density. In the far field, the viscous effect is assumed to be negligible, therefore a uniform free-stream velocity $\vec{U}_\infty$ is imposed on Γ_∞. More precisely, $\rho = \rho_\infty, \vec{U}_\infty = (cos\alpha, sin\alpha)^T$, and the pressure $p_\infty = 1/(\gamma M_\infty^2)$, where α is the angle of attack and M_∞ is the free-stream Mach number.

3 Discrete Formulation

Let the temporal variable t be discretized as $t^{n+1} = t^n + \delta t^n$, where δt^n is the discrete time increment. We also consider the increment $\delta W^{n+1} = W^{n+1} - W^n$, where W^n is an approximation of $W(\cdot, t^n)$. We note that when an algorithm is written in the "delta" form, the increment δW^{n+1} is the unknown variable rather than W^{n+1}. Here, we use a first-order accurate temporal difference approximation, namely, the backward Euler scheme given as

$$\frac{\partial W^{n+1}}{\partial t^{n+1}} + (\nabla_W(\mathcal{F}^n) \cdot \nabla_{\vec{x}})\delta W^{n+1} - \frac{1}{\text{Re}}(\nabla_W(\mathcal{R}^n) \cdot \nabla_{\vec{x}})\delta W^{n+1} = \nabla \cdot (-\mathcal{F}^n) + \frac{1}{\text{Re}}\nabla \cdot R^n,$$

$$(2)$$

where $\mathcal{F}^n$ and $\mathcal{R}^n$ are approximations of $\mathcal{F}(W(\cdot, t^n))$ and $\mathcal{R}(W(\cdot, t^n))$, respectively.

The computational domain is discretized by an unstructured, triangular grid. We locate the variables at the vertices of the grid. This gives rise to a cell-vertex scheme. The space of solutions is taken to be the space of piecewise linear functions. The discrete system is obtained via a mixed Galerkin finite element/finite volume formulation; see e.g. Farhat et al. [FFL93], and Fezoui and Stoufflet [FS89]. In short, the discrete system for (2) is obtained by using a "mass-lumping" technique for the time derivative, a first-order MUSCL scheme with a Roe approximate Riemann solver for the convective terms of the LHS, and a Galerkin finite element (first-order quadrature integration) for the diffusive terms of the LHS. For the RHS, we use a second-order MUSCL scheme with a Roe approximate Riemann solver and Van Albada's limiting procedure for the convective terms, and a Galerkin finite element method for the diffusive terms.

For the initial values, we assume that W_0 satisfies strongly the wall boundary conditions on Γ_ω and the far field boundary conditions at infinity. In the interior nodes of Ω, W_0 takes the free-stream boundary condition.

4 Algebraic Schwarz Algorithms

At each time step, we solve a linear system, $Au = f$, where A is a nonsymmetric sparse matrix with symmetric non-zero pattern. Each element of A can be considered as a 4×4 matrix, and each unknown of the vector u is a '4-size' vector. Thus, there is a bijection between unknowns and vertices. We denote the set of vertices (or nodes) by $\mathcal{N} = \{1, \ldots, n\}$, where n represents the total number of nodes (or unknowns). To define algebraic Schwarz algorithms [CS96], we first partition the set $\mathcal{N}$ into n_0 nonoverlapping subsets $\mathcal{N}_i$ whose union is $\mathcal{N}$. We use the TOP/DOMDEC mesh partitioning package of Farhat et al. [FLS95] to obtain sets $\mathcal{N}_i$. The recursive spectral bisection method with certain optimization is used in the partitioning. The number of nodes in each $\mathcal{N}_i$ is roughly the same. To generate an overlapping partition, we further expand each subgrid $\mathcal{N}_i$ by $ovlp$ number of neighboring nodes, denoted as $\tilde{\mathcal{N}}_i$.

We denote by L_i the vector space spanned by the set $\tilde{\mathcal{N}}_i$. For each subspace L_i we define an orthogonal projection operator I_i as follows: I_i is a $n \times n$ matrix whose diagonal elements are set to 4×4 identity matrices if the corresponding nodes belong to $\tilde{\mathcal{N}}_i$ and to 4×4 zero matrices otherwise. With this we define $A_i = I_i A I_i$, which is an extension to the whole subspace, of the restriction of A to L_i. Note that although A_i is not invertible, we can invert its restriction to the subspace spanned by $\tilde{\mathcal{N}}_i$, and define $A_i^{-1} \equiv I_i((A_i)_{|L_i})^{-1} I_i$. The additive and multiplicative Schwarz algorithms can now be simply described as follows: Solve the equation $MAu = Mf$ by a Krylov subspace method, where $M = A_1^{-1} + \ldots + A_{n_0}^{-1}$, for the additive Schwarz algorithm, and $MA = I - (I - A_1^{-1}A)\ldots(I - A_{n_0}^{-1})$ for the multiplicative Schwarz algorithm. We remark that in the algorithms discussed above all subproblems are assumed to be solved exactly; e.g., with sparse Gaussian elimination. In our numerical experiments we also consider inexact solvers.

5   Numerical Results

The main goal of this section is to compare the effectiveness of various preconditioners for unsteady subsonic and transonic flows. The experiments were performed using PETSc [GSM95] on a DEC Sable workstation.

We consider flows past a NACA0012 airfoil at an angle of attack of 30 degrees and Reynolds number 800. Problem 1 corresponds to a Mach number 0.1 and CFL 100, Problem 2 to a Mach number 0.8 and CFL 100, and Problem 3 to a Mach number 0.8 and CFL 25. The iteration numbers and CPU times that we report are for one single time step and for a non-dimensionalized time far from the initial transient regime, i.e. time equals to 3.0. In the CPU time, we do not include the time for constructing the preconditioner since the same preconditioner can be frozen for several time steps. We used left preconditioners and we stopped the iterations when the l_2 norm of the preconditioned residuals were reduced by a factor of 10^{-6}.

We found that we can take CFL up to 100 without losing much accuracy for the unstructured grid with 12280 nodes; see the left figure in Fig. 2. Therefore, using the implicit methods has a clear advantage over the explicit one in which the CFL must be less than 1.0.

Table 3 illustrates the results for additive and multiplicative Schwarz methods. The multiplicative OSM had better convergence properties although it is not as parallelizable as the additive OSM. We can also see that a small overlap gave generally less GMRES iterations than zero overlap. For Problem 2 and 3, we detected some pathological cases in which an increase in the overlap resulted in more GMRES iterations. We note, however, that when the subdomains were relatively large, a small overlap resulted in a significant decrease in the CPU time. We will expect promising results when we run our code on parallel machines with coarse granularity. We observed a slight increase in the number of GMRES iterations when we increased the number of subdomains, thus, it is not clear that for unsteady problems an additional coarse space would decrease the CPU time.

Table 4 represents the behavior of the GMRES/OSM iteration numbers with different inexact local solvers. We found that if we increased the overlap, the number

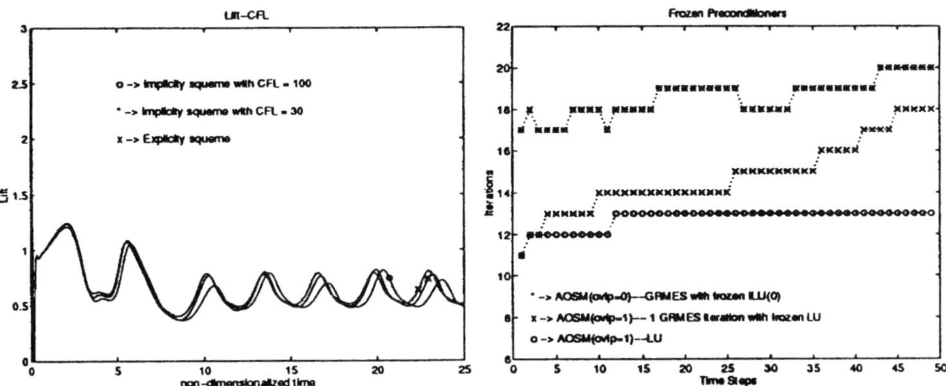

Figure 2   The left figure compares the solutions obtained by both explicit and implicit methods. The right figure shows the number of GMRES iterations when preconditioners are frozen for a number of time steps.

of GMRES iterations increased when the local problems are not solved to a certain tolerance. We tested several cases of additive OSM (8 subdomains) with GMRES/ILU (*fill*) as the inexact solvers. The best performance in terms of CPU time was zero overlap and zero *fill* with local stopping tolerance 1.0^{-2}; see Table 1. We noted also that as a local solvers, using LU gave smaller CPU time than using GMRES/ILU(0). We note, however, that for larger problems we might obtain different conclusions.

Table 1   GMRES iteration number and CPU time in seconds using Additive OSM with zero overlap. Number of nodes is 12280. We use GMRES/ILU(0) as local solvers reducing the local preconditioned residual by $= 2.0e^{-1}$.

	Problem 1	Problem 2	Problem 3
SUB $= 8$	17 (31.9)	14 (22.5)	8 (11.4)
SUB $= 16$	20 (29.3)	13 (17.6)	8 (9.9)
SUB $= 32$	23 (33.5)	15 (19.3)	10 (11.8)
SUB $= 64$	25 (33.3)	17 (21.1)	10 (11.6)
SUB $= 128$	29 (37.5)	20 (23.6)	12 (13.5)
SUB $= 256$	35 (43.9)	22 (25.5)	13 (14.0)
SUB $= 512$	41 (49.9)	27 (30.2)	16 (16.7)

We studied the behavior of other preconditioners; see Table 2. We observed that the 4×4 block Jacobi iterative method required a very large number of iterations. The family of global ILU (*fill*) gave the best results but it would not be trivial to code them on parallel machines.

Finally, we examined the effect of using the same preconditioner for several time steps, thus, we did not need to form the preconditioner or factorize it at every time step; see the right figure in Fig. 2. We considered a case in which we used LU local solver in one time step and then used this LU factorization as a preconditioner for local

Table 2   Iteration numbers and (CPU time in seconds) to reduce the preconditioned residual to $1.0e^{-6}$ using GMRES/ILU (*fill*) and Richardson/(4×4 block Jacobi).

	Problem 1	Problem 2	Problem 3
4×4 BJ	219 (44.4)	190 (40.3)	98 (18.0)
FILL = 0	34 (29.1)	21 (17.1)	10 (7.7)
FILL = 1	23 (21.4)	13 (13.6)	7 (6.8)
FILL = 2	19 (20.1)	12 (13.5)	6 (6.9)
FILL = 3	9 (10.9)	6 (8.18)	4 (5.7)
FILL = 4	7 (9.1)	5 (7.5)	3 (4.8)
FILL = 5	6 (8.4)	4 (6.5)	3 (5.1)
FILL = 6	6 (8.7)	4 (6.7)	2 (3.6)

Table 3   Iteration numbers and CPU time per time step GMRES/OSM (tolerance $= 1.0e^{-6}$) for Problems 1, 2 and 3 using 12280 nodes. As local solvers we use LU.

Problem 1	ovlp 0	ovlp 1	ovlp 2	ovlp 3
SUB = 8 (a)	15 (27.8)	11 (22.4)	10 (22.5)	9 (21.8)
(m)	8 (17.4)	5 (12.5)	4 (11.4)	4 (12.3)
SUB = 16(a)	17 (27.4)	13 (23.9)	13 (26.9)	12 (27.9)
(m)	10 (18.9)	6 (13.6)	5 (13.0)	5 (14.7)
SUB = 32(a)	19 (25.4)	16 (26.5)	15 (30.1)	15 (36.0)
(m)	11 (17.5)	7 (14.4)	6 (15.2)	5 (15.5)
SUB = 64(a)	21 (25.6)	20 (31.1)	20 (40.0)	19 (47.7)
(m)	12 (16.8)	9 (17.4)	7 (17.9)	6 (19.7)
Problem 2				
SUB = 8(a)	13 (24.0)	9 (18.3)	7 (16.1)	8 (19.5)
(m)	7 (15.6)	3 (8.2)	3 (9.1)	3 (9.7)
SUB = 16(a)	13 (20.8)	9 (16.7)	10 (20.8)	10 (23.3)
(m)	7 (13.5)	4 (9.5)	4 (10.8)	3 (9.6)
SUB = 32(a)	14 (18.4)	12 (19.8)	11 (22.0)	12 (28.6)
(m)	8 (12.9)	5 (10.6)	4 (10.7)	3 (10.1)
SUB = 64(a)	15 (17.7)	15 (23.6)	15 (30.6)	15 (38.6)
(m)	9 (12.6)	5 (10.2)	4 (11.0)	4 (13.9)
Problem 3				
SUB = 8(a)	8 (15.1)	7 (14.6)	6 (14.0)	6 (15.0)
(m)	5 (11.4)	2 (6.1)	2 (6.7)	2 (7.2)
SUB = 16(a)	8 (13.0)	7 (13.3)	8 (16.9)	8 (19.1)
(m)	5 (10.0)	3 (7.6)	3 (8.5)	2 (7.1)
SUB = 32(a)	10 (13.3)	9 (15.2)	8 (16.5)	9 (21.8)
(m)	6 (9.9)	3 (7.0)	3 (8.5)	2 (7.6)
SUB = 64(a)	10 (11.8)	11 (17.0)	11 (22.1)	12 (30.3)
(m)	6 (8.7)	3 (6.7)	3 (8.8)	3 (11.1)

ovlp = 0	Problem 1	Problem 2	Problem 3
SUB = 128	25 (26.7)	17 (17.5)	11 (11.3)
SUB = 256	31 (29.7)	20 (18.1)	13 (11.4)
SUB = 512	39 (35.7)	26 (21.5)	15 (11.7)

Table 4 GMRES iteration numbers to reduce the preconditioned residual of Problem 1 (12280 nodes) to $1.0e^{-6}$ using GMRES/(Additive OSM) with 8 subdomains. We use GMRES/ILU(0) inexact local with different local solvers stopping criteria.

tolerance	ovlp = 0	ovlp = 1	ovlp = 2	ovlp = 3
1 iteration	35	43	46	48
$5.0e^{-1}$	23	32	33	33
$3.0e^{-1}$	18	20	19	19
$1.0e^{-1}$	16	15	14	14
$1.0e^{-2}$	15	12	11	10
exact	15	11	10	9

problems (with one Richardson iteration) for the following time steps. We found that for additive OSM with 8 subdomains and 1 overlap the convergence did not deteriorate for 25 time steps.

6 Conclusions

We report the performance of GMRES/ILU (*fill*) and GMRES/OSM methods for solving systems that arise from the discretization of unsteady, compressible N.-S equations. The best results in terms of CPU time were obtained for the GMRES/ILU (*fill*) methods, where *fill* is around 5. However, the ILU (*fill*) is not easy to parallelize. We then tested several additive OSM. For problems in which the size of the subdomains were not too small, we observed that we reduced the CPU time if we used a small overlap rather than zero overlap. We note that the local problems were solved by Gaussian elimination. When using inexact local solvers, we noted that an overlap did not improve performance, and that the inexact local solver GMRES/ILU(0) gave the smallest CPU time; we must mention however that this observation may change if we can run problems with large subdomains. We also found that a slight increase in the number of GMRES iterations occurred when we increased the number of subdomains. We suspect that an improvement in performance might be obtained if we were to add a coarse space.

REFERENCES

[BL95] Barth T.J. and Linton S.W. (Jan. 1995) An unstructured mesh Newton solver for compressible fluid flow and its parallel implementation. *AIAA Paper 95-0221*.
[CS96] Cai X.-C. and Saad Y. (1996) Overlapping domain decomposition algorithms for general sparse matrices. *Numer. Lin. Alg. Applics* To appear.
[DW94] Dryja M. and Widlund O.B. (1994) Domain decomposition algorithms with small overlap. *SIAM J. Sci. Comp.* 15(3): 604–620.
[FFL93] Farhat C., Fezoui L., and Lanteri S. (1993) Two-dimensional viscous flow computation on the Connection Machine: Unstructured meshes, upwind schemes

and parallel computation. *Comput. Methods Appl. Mech. Engrg.* 102: 61–88.

[FLS95] Farhat C., Lanteri S., and Simon H. (1995) TOP/DOMDEC: A software tool for mesh partitioning and parallel processing and applications to CSM and CFD computations. *Comput. Sys. Engrg.* 6(1): 13–26.

[FS89] Fezoui L. and Stoufflet B. (1989) A class of implicit upwind schemes for Euler simulations with unstructured meshes. *J. Comp. Phys.* 84: 174–206.

[GKM94] Gropp W.D., Keyes D.E., and Mounts J.S. (1994) Implicit domain decomposition algorithms for steady, compressible aerodynamics. In Quarteroni A., Periaux J., Kuznetsov Y.A., and Widlund O.B. (eds) *Sixth Conference on Domain Decomposition Methods for Partial Differential Equations*. AMS, Providence, RI.

[GSM95] Groop W.D., Smith B.F., and McInnes L.C. (1995) PETSc 2.0 User's Manual. Technical Report ANL-95/11, Argonne National Laboratory.

[SBG95] Smith B.F., Bjørstad P.E., and Gropp W.D. (1995) *Domain Decomposition: Parallel Multilevel Methods for Elliptic Partial Differential Equations*. Cambridge University Press. To appear.

[Ven95] Venkatakrishnan V. (Feb. 1995) A perspective on unstructured grid flow solvers. Technical Report ICASE Report No. 95-3, ICASE, NASA Langley Research Center.

[VM95] Venkatakrishnan V. and Mavriplis D.J. (Aug. 1995) Implicit method for the computation of unsteady flows on unstructured grids. Technical Report ICASE Report No. 95-60, ICASE, NASA Langley Research Center.

Domain Decomposition Algorithms for a Generalized Stokes Problem

Deling Chu and Xiancheng Hu

1 Introduction

Domain decomposition methods are appropriate to address fluid dynamical problems, especially in complex physical regions and within parallel computational environments. The zonal approach generally consists of partitioning the whole region into subregions of simpler shape, and then reduces the given flow problem to a sequence of subproblems which, to some extent, can be solved simultaneously.

Recently, a mathematical analysis has been carried out in [1] for hyperbolic systems of conservation laws (e.g., Euler equations for compressible inviscid flows). A generalized Stokes problem and an inviscid generalized Stokes problem are addressed in [2] and [3] in the framework of a domain decomposition method, in which the physical computational region Ω is partitioned into two subdomains Ω_1 and Ω_2, and correct transmission conditions across the interface Γ between Ω_1 and Ω_2 are provided. An iterative procedure involving the successive solution of two subproblems is proposed, and the convergence for the iteration-by-subdomain algorithms, which are associated with various domain decomposition approaches, is proved.

Quarteroni et al [3] consider the following generalized Stokes equation

$$\begin{cases} \alpha\sigma + \mathrm{div}u = g \text{ in } \Omega \\ \alpha u - \gamma\Delta u + \beta\nabla\sigma = f \text{ in } \Omega \end{cases} \tag{1.1}$$

where Ω is an open two-dimensional domain, α, β, γ are positive constants, and g and f are given scalar and vector functions, respectively. The unknowns are σ, which is a scalar related to the flow density, and the velocity vector field u.

In this paper, we consider the problem (1.1), and solve it by domain decomposition methods partitioning the computational domain Ω into m subdomains ($m \geq 2$), and propose and analyze some parallel domain decomposition algorithms, and prove that these algorithms are convergent. Furthermore, their convergence rates are nearly optimal if problem (1.1) is discreted by finite elements, which shows that these algorithms are very suitable for solving problem (1.1).

In this paper, c will be a generic constant.

[1] Department of Applied Mathematics, Tsinghua University, Beijing 100084, P. R. China

Domain Decomposition Methods in Sciences and Engineering, edited by R. Glowinski *et al.*
© 1997 John Wiley & Sons, Ltd.

2   The differential formulation of the problem (1.1)

We first give some notation. Let $C^0(\bar{\Omega})$ be the space of continuous functions on $\bar{\Omega}$, let $H^s(\Omega)$ and $H^s(\Gamma)(s \in \Re, \Gamma$ is a curve) be the usual Sobolev spaces, endowed with the norm $\|\cdot\|_{s,\Omega}$ and $\|\cdot\|_{s,\Gamma}$, respectively. For the seminorms, we use the notations $|\cdot|_{s,\Omega}$ and $|\cdot|_{s,\Gamma}$, for an open curve Γ_0 contained in a closed curve Γ. Let $H_{00}^{1/2}(\Gamma_0)$ be the set of (generalized) functions such that their extension by zero onto Γ belongs to $H^{1/2}(\Gamma)$ and let $(u,v)_\Omega = \int_\Omega uvdxdy$.

We restate the generalized Stokes problem as follows

$$\begin{cases} \alpha\sigma + \text{div}u = g \text{ in } \Omega \\ \alpha u - \gamma\Delta u + \beta\nabla\sigma = f \text{ in } \Omega \end{cases} \tag{2.1}$$

This is an elliptic system. The boundary conditions for problem (2.1) are

$$\begin{cases} u \equiv 0 \text{ on } \Gamma_B \\ u \equiv u_\infty \text{ on } \Gamma_\infty^- \\ S(u,\sigma) \equiv \gamma\dfrac{\partial u}{\partial n} - \beta\sigma n = 0 \text{ on } \Gamma_\infty^+ \end{cases} \tag{2.2}$$

where $\partial\Omega = \Gamma_B \cup \Gamma_\infty^- \cup \Gamma_\infty^+$.

The following result was proved in [2].

Theorem 2.1. *Assume that* $f \in H^{-1/2}(\Omega)$, $g \in L^2(\Omega)$, $u_\infty \in H^{1/2}(\Gamma_\infty^-)$. *The problem (2.1). (2.2) has a unique solution* $(u,\sigma) \in H^1(\Omega) \times L^2(\Omega)$ *and*

$$\|u\|_{1,\Omega} + \|\sigma\|_{0,\Omega} \leq c(\|f\|_{-1/2,\Omega} + \|g\|_{0,\Omega} + \|u_\infty\|_{1/2,\Gamma_\infty^-}).$$

Let

$$V_0 \equiv \{v \in H^1(\Omega) : v = 0 \text{ on } \Gamma_B \cup \Gamma_\infty^-\}, \ \Sigma_0 \equiv L^2(\Omega).$$

Then (2.1), (2.2) can be formulated as

$$\begin{cases} (u - U_\infty) \in V_0, \ \sigma \in \Sigma_0 \\ a[(u,\sigma),(v,\varphi)] = (f,v)_\Omega + \beta(g,\varphi)_\Omega, \ \forall(v,\varphi) \in V_0 \times \Sigma_0 \end{cases} \tag{2.3}$$

where U_∞ *is a suitable extension of* u_∞ *to* Ω *satisfying* $U_\infty \in H^1(\Omega)$, $U_\infty = 0$ *on* Γ_B *and* $\|U_\infty\|_{1,\Omega} \leq c\|u_\infty\|_{1/2,\Gamma_\infty^-}$, $a[(u,\sigma),(v,\varphi)] = \int_\Omega(\alpha\beta\sigma\varphi + \beta\varphi\text{div}u + \alpha uv + \gamma\nabla u\nabla v - \beta\sigma\text{div}v)dxdy$

Remark 2.1. $a[(u,\sigma),(v,\varphi)]$ is continuous and coercive in $V_0 \times \Sigma_0$.

Now, we solve problem (2.3) by domain decomposition methods. Ω is decomposed into two nonoverlapping subdomains, and we assume for simplicity that $u_\infty = 0$, $\Gamma_\infty^- \cup \Gamma_B = \partial\Omega$. If this is not the case, we can argue in a similar way.

Define

$V_i \equiv \{v \in H^1(\Omega_i) : v = 0 \text{ on } \partial\Omega_i \cap (\Gamma_\infty^- \cup \Gamma_B)\}$

$V_{0,i} \equiv \{v \in V_i : v = 0 \text{ on } \Gamma\}, \ i = 1,2$

and the bilinear forms

$$a_i[(u_i,\sigma_i),(v_i,\varphi_i)] \equiv \int_{\Omega_i} (\alpha\beta\sigma_i\varphi_i + \beta\varphi_i\text{div}u_i + \alpha u_iv_i + \gamma\nabla u_i\nabla v_i - \beta\sigma_i\text{div}v_i)dxdy$$

which are continuous and coercive in $V_i \times L^2(\Omega_i)$.

The first domain decomposition algorithm studied in this paper can be described as
Algorithm 1.
Given $\lambda^0 \in H_{00}^{1/2}(\Gamma)$
Step 1. Compute $(u_i^n, \sigma_i^n) \in V_i \times L^2(\Omega_i)$ by solving the equations $(i = 1, 2)$

$$
\begin{cases}
a_i[(u_i^n, \sigma_i^n), (v, \varphi)] = (f, v)_{\Omega_i} + \beta(g, \varphi)_{\Omega_i} \, (v, \varphi) \in V_{0,i} \times L^2(\Omega_i) \\
u_i^n = \lambda^n \text{on } \Gamma
\end{cases}
$$

Step 2. Compute $(\omega_i^n, \psi_i^n) \in V_i \times L^2(\Omega_i)$ by solving the equation $(i=1, 2)$

$$
a_i[(\omega_i^n, \psi_i^n), (v, \varphi)] = \frac{1}{2} \sum_{j=1}^{2} \{ a_j[(u_j^n, \sigma_j^n), (v_j, \varphi_j)] - (f, v_j)_{\Omega_j} - \beta(g, \varphi_j)_{\Omega_j} \},
$$

$\forall (v, \varphi) \in V_i \times L^2(\Omega_i), (v_j, \varphi_j) \in V_j \times L^2(\Omega_j), v_j|_\Gamma = v|_\Gamma, j = 1, 2$.
Step 3. $\lambda^{n+1} = \lambda^n - \rho(\omega_1^n + \omega_2^n)$ on Γ, where ρ is a positive constant.
Remark 2.2. Obviously, it is true that

$$
\begin{aligned}
& a_j[(u_j^n, \sigma_j^n), (v_1, \varphi_1)] - (f, v_1)_{\Omega_j} - \beta(g, \varphi_1)_{\Omega_j} \\
& = a_j[(u_j^n, \sigma_j^n), (v_2, \varphi_2)] - (f, v_2)_{\Omega_j} - \beta(g, \varphi_2)_{\Omega_j},
\end{aligned}
$$

$\forall (v_1, \varphi_1), (v_2, \varphi_2) \in V_j \times L^2(\Omega_j), v_1|_\Gamma = v_2|_\Gamma, j = 1, 2$. This implies that (ω_i^n, ψ_i^n), given in Step 2, is unique.

It is easy to see that Algorithm 1 can be implemented in parallel. In the following, we analyze its convergence.

Theorem 2.2. *There exists a positive constant c such that if $0 < \rho < c$, then $\{(u_1^n, \sigma_1^n), (u_2^n, \sigma_2^n)\}$ generated by Algorithm 1 converge to the solution of equation (2.3).*

3 Domain decomposition algorithms for the finite element problem

Let us first briefly recall a finite element approximation for the problem (2.3); see [3]. Let T_h be a family of decompositions of Ω (which will be assumed hereafter to be a polygon) into triangles Δ. Assume that T_h is regular, i.e. there exists $\tau_0 > 0$ independent of h such that

$$
h_\Delta / \rho_\Delta \le \tau_0, \quad \forall \Delta \in T_h
$$

where $h_\Delta \equiv \text{diam} \Delta$, $\rho_\Delta \equiv \sup\{2\rho | \exists x_0 \in \Delta : B(x_0, \rho) \subset \Delta\}$, $h \equiv \max h_\Delta$.
Define the following finite element spaces for $r \ge 1$

$$
\begin{cases}
V_{0,h} = \{ v \in C^0(\bar{\Omega}) : v|_\Delta \in P_r, \forall \Delta \in T_h, v = 0 \text{ on } \Gamma_\infty^- \cup \Gamma_B \} \\
\Sigma_{0,h} = \{ \varphi \in L^2(\Omega) : \varphi|_\Delta \in P_{r-1}, \forall \Delta \in T_h \}
\end{cases}
\tag{3.1}
$$

where P_r denotes the space of polynomials of degree $\le r$ on Δ, and consider the following finite dimensional approximation to (2.3)

$$
\begin{cases}
(u_h, \sigma_h) \in V_{0,h} \times \Sigma_{0,h} \\
a[(u_h, \sigma_h), (v_h, \varphi_h)] = (f, v_h)_\Omega + \beta(g, \varphi_h)_\Omega, \quad \forall (v_h, \varphi_h) \in V_{0,h} \times \Sigma_{0,h}.
\end{cases}
\tag{3.2}
$$

Since $a[(u_h, \sigma_h), (v_h, \varphi_h)]$ is coercive on $V_{0,h} \times \Sigma_{0,h}$, problem (3.2) has a unique solution, and the following error estimate holds (see [16])

$$\|u - u_h\|_{1,\Omega} + \|\sigma - \sigma_h\|_{0,\Omega} \le ch^s(|u|_{s+1,\Omega} + |\sigma|_{s,\Omega}), \quad 0 < s \le r$$

where (u, σ) is the solution of problem (2.1). By assuming $f \in H^{-1/2}(\Omega)$, $g \in H^s(\Omega)$, we obtain $u \in H^{1+s}(\Omega)$, $\sigma \in H^s(\Omega)$ for each $0 \le s \le 1/2$. Higher regularity cannot be expected for the solution of problem (2.1) endowed with the mixed Dirichlet-Neumann boundary conditions (2.2). Hence it is safe to choose $r = 1$ in (3.1) and (3.2). So, in the following, we set $r = 1$.

Decompose Ω into nonoverlapping subdomains $\Omega_1, \ldots, \Omega_m$, and let

$$\Gamma \equiv \cup_{i=1}^{m} \partial\Omega_i \backslash \partial\Omega, \quad \Phi_h \equiv \{v|_\Gamma : v \in V_{0,h}\}$$

$$V_{i,h} \equiv \{v|_{\Omega_i} : v \in V_{0,h}\}, \quad \Sigma_{i,h} \equiv \{\varphi|_{\Omega_i} : \varphi \in \Sigma_{0,h}\}, \quad i = 1, \ldots, m$$

$$V_{i,h}^0 \equiv \{v|_{\Omega_i} : v = 0 \text{ on } \partial\Omega_i \cap \Gamma, v \in V_{0,h}\}, \quad i = 1, \ldots, m.$$

Assume that the nonoverlapping subdomains $\{\Omega_i\}_{i=1}^{m}$ satisfy
A1. $\bar\Omega = \cup_{i=1}^{m} \bar\Omega_i$, the sides of $\Omega_i(i = 1, \ldots, m)$ follow the finite element mesh lines of Ω;
A2. $\{\Omega_i\}_{i=1}^{m}$ are quasi-uniform quadrilaterals with size H.

For any i, trace average operator α_i is defined by

$$\begin{cases} \forall v_h \in V_{0,h}, \alpha_i(v_h) \in \Phi_h \text{ and} \\ \alpha_i(v_h)(x) = \dfrac{1}{2}v_h(x), \ \forall \text{ nodes } x \text{ on } \partial\Omega_i \backslash \Omega, \ x \text{ which are not a vertex of } \Omega_i \\ \alpha_i(v_h)(x) = \dfrac{1}{N}v_h(x), \ \forall \text{ common vertex } x \text{ of } N \text{ subdomains}, \ x \notin \partial\Omega \\ \alpha_i(v_h)(x) = 0 \ \forall \text{ node } x \text{ on } \Gamma \backslash \partial\Omega_i \end{cases}$$

obviously

$$\sum_{i=1}^{m} \alpha_i(v_h) = Tr(v_h), \quad \forall v_h \in V_{0,h} \tag{3.3}$$

where $Tr(v_h)$ is the trace of v_h on Γ.

Now, we describe our parallel algorithm.

Algorithm 2.

Given $\lambda_h^0 \in \Phi_h$

Step 1. Compute $(u_{i,h}^n, \sigma_{i,h}^n) \in V_{i,h} \times \Sigma_{i,h}$ by solving the equations

$$\begin{cases} a_i[(u_{i,h}^n, \sigma_{i,h}^n), (v_h, \varphi_h)] = (f, v_h)_{\Omega_i} + \beta(g, \varphi_h)_{\Omega_i} \quad \forall(v_h, \varphi_h) \in V_{i,h}^0 \times \Sigma_{i,h} \\ u_{i,h}^n = \lambda_h^n \text{ on } \Gamma \cap \partial\Omega_i \end{cases}$$

Step 2. Compute $(\omega_{i,h}^n, \psi_{i,h}^n) \in V_{i,h} \times \Sigma_{i,h}$ by solving the equations

$$a_i[(\omega_{i,h}^n, \psi_{i,h}^n), (v_h, \varphi_h)] = \sum_{j=1}^{m}\{a_j[(u_{j,h}^n, \sigma_{j,h}^n), (v_{j,h}, \varphi_{j,h})] - (f, v_{j,h})_{\Omega_j} - \beta(g, \varphi_{j,h})_{\Omega_j}\},$$

$\forall (v_h, \varphi_h) \in V_{i,h} \times \Sigma_{i,h}, (v_{j,h}, \varphi_{j,h}) \in V_{j,h} \times \Sigma_{j,h}, v_{j,h}|_\Gamma = \alpha_i(v_h), j = 1, \ldots, m.$

Step 3. $\lambda_h^{n+1} = \lambda_h^n - \rho \sum_{j=1}^{m} \alpha_j(\omega_{j,h}^n)$ on Γ, where ρ is a positive constant.

Remark 3.1. $(\omega_{i,h}^n, \psi_{i,h}^n)$ in Step 2 has a unique solution, and Algorithm 3 can be implemented in parallel.

Theorem 3.1. *For Algorithm 2, there exists a constant* τ

$$1 \leq \tau \leq O((1 + H^{-2})(1 + \ln^2(H/h))$$

such that if $\rho < 2/\tau$, *then*

$$\sum_{i=1}^{m} \|(\varepsilon_i^{n+1}, \delta_i^{n+1})\|_{a_i}^2 \leq (1 - 2\rho + \tau\rho^2) \sum_{i=1}^{m} \|(\varepsilon_i^n, \delta_i^n)\|_{a_i}^2.$$

4 The analysis of the preconditioner corresponding to Algorithm 2

In Section 3, we have analyzed Algorithm 2, which is a preconditioned Richardson iterative method. Our purpose in this section is to analyze the preconditioner related to Algorithm 2, and to estimate the condition number of the preconditioned system.

It is known that $Q = S(I - P)^{-1}$ is the preconditioner in the iterative scheme $x^{n+1} = Px^n + q$ for solving system $Sx = b$. From the point of view of parallel computation, the preconditioner Q should satisfy

1) Q^{-1} should be easy to obtain in parallel;

2) The condition number of $Q^{-1}S$ should not be large.

Define discrete Steklov-Poincaré operator S_h as

$$(S_h\lambda_h, v_h)_\Omega \equiv \sum_{i=1}^{m} a_i[E_{i,h}\lambda_h, E_{i,h}v_h], \quad \forall (\lambda_h, v_h) \in \Phi_h$$

and let $(u_{i,h}^*, \sigma_{i,h}^*) \in V_{i,h}^0 \times \Sigma_{i,h}$, be the solution of

$$a_i[(u_{i,h}^*, \sigma_{i,h}^*), (v_h, \varphi_h)] \equiv (f, v_h)_{\Omega_i} + \beta(g, \varphi_h)_{\Omega_i}, \quad \forall (v_h, \varphi_h) \in V_{i,h}^0 \times \Sigma_{i,h}$$

$$(b_h, v_h)_\Omega \equiv \int_\Omega ((f, E_h v_h)_\Omega - a[(u_h^*, \sigma_h^*), E_h v_h]) dx dy, \quad \forall v_h \in \Phi_h$$

where

$$(E_h v_h)|_{\Omega_i} \equiv E_{i,h} v_h, \quad (u_h^*, \sigma_h^*)|_{\Omega_i} = (u_{i,h}^*, \sigma_{i,h}^*)$$

and

$$S_h \lambda_{h,\Gamma} = b_h. \tag{4.1}$$

Here $b_h \in (\Phi_h)'$ is an element in the space dual to Φ_h. For S_h, we have

Lemma 4.1. *The discrete Steklov-Poincaré operator* S_h *is symmetric and positive definite, and the solution to (3.2) is given by* $(u_h, \sigma_h) = E_h \lambda_{h,\Gamma} + (u_h^*, \sigma_h^*)$.

According to Lemma 4.1 and considering $(u_{i,h}^*, \sigma_{i,h}^*)$, b_h can be obtained in parallel. So, if $\lambda_{h,\Gamma}$ can be solved, then we can easily obtain (u_h, σ_h), which is the solution of equation (3.2).

Algorithm 2 is an iterative algorithm to solve (4.1). Define the discrete Steklov-Poincaré operator $S_{i,h}$ by

$$(S_{i,h}\lambda_h, v_h)_\Omega \equiv a_i[E_{i,h}\lambda_h, E_{i,h}v_h], \quad \forall \lambda_h, v_h \in \Phi_h.$$

As S_h, $S_{i,h}$ is symmetric and positive definite. Assume that R_i is the matrix form of the trace average operator α_i(see Section 3). P is the iterative matrix of the iterative scheme from λ^n to λ^{n+1} in Algorithm 2. It can be verified that

$$P = I - \rho(\sum_{i=1}^{m} R_i S_{i,h}^{-1} R_i^T) S_h.$$

It can be shown that Q has the form

$$Q^{-1} = \rho \sum_{i=1}^{m} R_i S_{i,h}^{-1} R_i^T \tag{4.2}$$

which is the preconditioner contained in Algorithm 3.

(4.2) shows that Q is symmetric and positive definite, and that Q satisfies property 1).

In order to show that Q satisfies property 2), we prove the following theorem.

Theorem 4.1. *There exists a positive constant c, which is independent of H and h, such that*

$$\rho \leq \frac{(S_h Q^{-1} S_h \lambda_h, \lambda_h)_\Omega}{(S_h \lambda_h, \lambda_h)_\Omega} \leq c\rho(1 + H^{-2})(1 + \ln^2(H/h)), \quad \forall \lambda_h \in \Phi_h.$$

Equivalently, the condition number of matrix $Q^{-1}S_h$ satisfies

$$\kappa(Q^{-1}S_h) \leq O((1 + H^{-2})(1 + \ln^2(H/h))).$$

It is well-known that the conjugate gradient method(CG) is an efficient technique for symmetric positive definite system, but its convergence rate depends on the condition number of the coefficient matrix. So, Theorem 4.2 shows that the matrix Q satisfies property 2). The above analysis shows that Q is an efficient preconditioner of S_h. It will converge very fast when system (4.1) is solved by preconditioned CG.

5   Conclusions

In the above sections, we have studied the generalized Stokes problem using multi-subdomain decomposition methods, and proposed and analyzed Algorithm 1, Algorithm 2, and Algorithm 3, and estimated the condition number of the preconditioned system. The derived result shows that these algorithms are nearly optimal and are efficient for solving the generalized Stokes problem.

Our algorithms are also suitable for domains with $\Gamma_\infty^+ \neq \emptyset$ and the domain studied in [2] and [3], with boundary condition (2.2).

REFERENCES

[1] Gastaldi F., Quarteroni A. (1989) On the coupling of hyperbolic and parabolic systems: Analytical and numerical approach. *Appl. Numer. Math.,* Vol. 6:3-31.

[2] Quarteroni A., Valli A. (1990) Domain decomposition for a generalized Stokes problem, In: J. Manley et al., eds, *Third ECMI Proceedings,* Kluwer, Dordrecht and Teubner, Stuttgart, 59-74.

[3] Quarteroni A., Sacchi Landriani G. and Valli A. (1991) Coupling of viscous and inviscid Stokes equations via a domain decomposition method for finite elements. *Numer. Math.,* Vol. 59: 831-859.

[4] Chu D. L. (1990) *Some problems in domain decomposition methods for partial differential problems.* Ph. D thesis, Tsinghua University, Beijing.

Domain decomposition methods for a three dimensional extrusion model

Merete S. Eikemo [1] and Magne S. Espedal [2]

Abstract. An extrusion process in three dimensions is modelled, and gives rise to indefinite block systems. The multi-scaled complex geometry involved in such a process is one of the main motivations for making use of domain decomposition methods for this problem. An overlapping additive Schwarz method will be used, which allows the possibility of local refinement at locations where the system experiences large gradients. For a nonlinear parabolic equation an alternating Schwarz method will be presented.

INTRODUCTION

The thermo-mechanical properties of aluminium during an extrusion process is described by a coupled set of nonlinear partial differential equations. The model consists of a temperature equation, a continuity equation and Navier-Stokes equations with a nonlinear Zener-Holloman material law for velocities and pressure. In order to support practical applications and being able to take into account the behaviour of the aluminium as it flows through the complex geometry of a die, the model has to be three dimensional.

The first three sections will briefly introduce an aluminium process, present the

---

[1] Department. of Mathematics, University of Bergen, Johs. Brunsgt.12, 5008 Bergen, Norway, merete.eikemo@mi.uib.no

[2] Department. of Mathematics, University of Bergen, Johs. Brunsgt.12, 5008 Bergen, Norway, resme@mi.uib.no

Domain Decomposition Methods in Sciences and Engineering, edited by R. Glowinski *et al.*

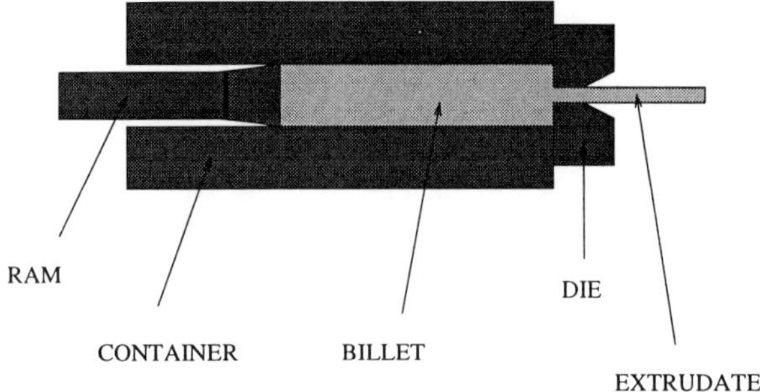

Figure 1   Aluminium extrusion process.

model equations and describe the solution procedure. The domain decomposition ideas
for the model and some numerical results will be given in the last two sections.

ALUMINIUM EXTRUSION PROCESS

A preheated aluminium billet is placed in a container. The typical initial temperature
is 450°C and the length of the billet is half a meter. At one end a die is placed, and as
a result of a slowly moving ram in the other end and the shape of the die, the billet
is deformed into a complex thin-walled profile of specified design, [HSH92]. Figure 1
gives a schematic picture of such an extrusion process with a simple die shape.

MODEL

The governing nonlinear equations are the pressure-velocity equations

$$
\begin{aligned}
\mathrm{Re}(\frac{\partial \mathbf{u}}{\partial t} + \mathbf{u} \cdot \nabla \mathbf{u}) &= -\nabla p + \nabla \cdot \tau & \text{in } \Omega, \\
\nabla \cdot \mathbf{u} &= 0 & \text{in } \Omega,
\end{aligned}
\tag{1}
$$

together with the heat balance equation

$$
\mathrm{Pe}(\frac{\partial \theta}{\partial t} + \mathbf{u} \cdot \nabla \theta) = \nabla^2 \theta + \beta \epsilon : \tau \quad \text{in } \Omega,
\tag{2}
$$

where Ω is the physical domain where the extrusion process takes place. The primary
variables are the velocity $\mathbf{u}$, the pressure p and the temperature θ. Further, $\tau = 2\mu\epsilon$
is the stress tensor, where $\mu = \bar{\tau}/(3\bar{\epsilon})$ is the nonlinear viscosity coeffisient. Here
$\bar{\tau}(\bar{\epsilon}, \theta) = \alpha^{-1}\mathrm{arcsinh}((Z/K)^{\frac{1}{m}})$ is the Zener-Holloman material model with the
parameter $Z = Z(\bar{\epsilon}, \theta) = \bar{\epsilon}\exp(Q/(R\theta))$. The viscosity coeffisient also consists of
the effective strain rate $\bar{\epsilon} = (\frac{2}{3}\epsilon : \epsilon)^{\frac{1}{2}}$, where $\epsilon = \frac{1}{2}(\nabla \mathbf{u} + \nabla \mathbf{u}^t)$ is the strain rate
tensor. The parameters α, m, K, Q, R and β are problem- and material dependent
with typical values [Sæv93]. The Reynolds number and the Péclét number are, in the
equations, denoted by Re and Pe, respectively.

From specific material- and problem-dependent parameters we get the Reynolds number to be very small, typical of magnitude 10^{-8}. We may therefore neglect the left side of the vector equation in (1), but note that the equation still is strongly nonlinear because of the term $\nabla \cdot \tau$.

$$
\begin{aligned}
-\nabla \cdot \tau + \nabla p &= 0 \qquad \text{in } \Omega, \\
\nabla \cdot \mathbf{u} &= 0 \qquad \text{in } \Omega.
\end{aligned}
\tag{3}
$$

SOLUTION PROCEDURE

The pressure and velocity are slowly varying over a timestep relative the variation of the temperature. To decouple the heat balance equation from the other equations, we therefore, on each timelevel, use a sequential iterative solution procedure. In the first step the pressure-velocity system (3) is solved, using the temperature solution from the previous timestep. In the second step the temperature equation (2) is solved, using the velocity from the first step. The further iterations between these two steps are performed using the most recently calculated temperature for the solving of the pressure-velocity system. Each equation is linearized by Picard iterations.

The discretization is done by using a mixed finite element approch with hexahedral elements and a triquadratic approximation for velocity and trilinear approximation for pressure and temperature. Carried out on (3), this results in a linear block system of the form

$$
\left\{
\begin{array}{cccc}
A_{11} & A_{12} & A_{13} & B_1^t \\
A_{21} & A_{22} & A_{23} & B_2^t \\
A_{31} & A_{32} & A_{33} & B_3^t \\
B_1 & B_2 & B_3 & 0
\end{array}
\right\}
\left\{
\begin{array}{c} X \\ Y \end{array}
\right\}
=
\left\{
\begin{array}{cc} A & B^t \\ B & 0 \end{array}
\right\}
\left\{
\begin{array}{c} X \\ Y \end{array}
\right\}
=
\left\{
\begin{array}{c} F \\ G \end{array}
\right\},
\tag{4}
$$

where B^t is the transpose of B. Bramble and Pasciak, [BP94], give a method for solving systems of this kind. The system (4) is reformulated in such a way that the new coeffisient matrix $\tilde{M}$ is positive definite,

$$
\tilde{M}
\left\{
\begin{array}{c} X \\ Y \end{array}
\right\}
=
\left\{
\begin{array}{cc}
A_0^{-1} A & A_0^{-1} B^t \\
B A_0^{-1}(A - A_0) & B A_0^{-1} B^t
\end{array}
\right\}
\left\{
\begin{array}{c} X \\ Y \end{array}
\right\}
=
\left\{
\begin{array}{c}
A_0^{-1} F \\
B A_0^{-1} F - G
\end{array}
\right\}.
\tag{5}
$$

Here A_0 is a preconditioner for A.

Let

$$
\tilde{M}_0 =
\left\{
\begin{array}{cc} I & 0 \\ 0 & \mathcal{K} \end{array}
\right\},
\tag{6}
$$

where I is the identity matrix and $\mathcal{K}$ is the preconditioner for the Schur complement $B A^{-1} B^t$,

$$
\mathcal{K} = N_h + h^2 I,
\tag{7}
$$

where N_h is the solution operator on the pressure grid for a finite element approximation to a Neumann problem and h is the spatial resolution. Each evaluation of the preconditioner $\mathcal{K}$ then requires solving a discrete Neumann problem on a grid of mesh size h. This preconditioner, [BP94], gives rise to convergence rates which can be bounded independently of h.

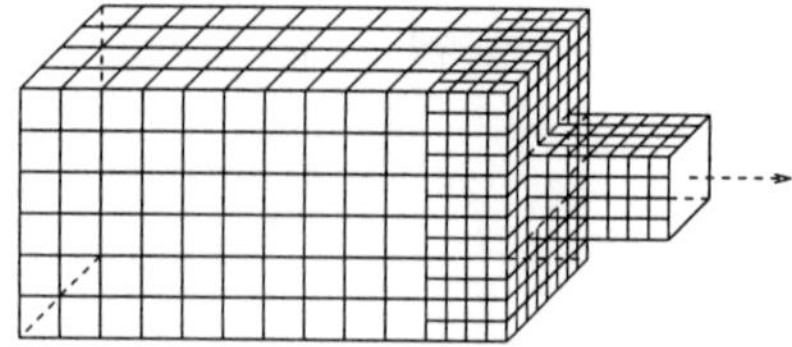

Figure 2   Full domain with refined outlet area.

A new inner product can be defined,

$$\left[\left\{ \begin{array}{c} U \\ V \end{array} \right\}, \left\{ \begin{array}{c} X \\ Y \end{array} \right\} \right] \equiv (AU, X) - (A_0 U, X) + (V, Y), \tag{8}$$

in which the matrix $\tilde{M}$ is symmetric. In addition to the matrix $\tilde{M}$ being positiv definite, it can also be shown, [BP94], that the condition number of $\tilde{M}_0 \tilde{M}$ is uniformly bounded. Therefore the preconditioned conjugate gradient method is used on the system (5) preconditioned by (6).

The temperature distribution is determined by first solving the hyperbolic part of the heat balance equation (2) by the method of characteristics, [DES92], and then solving the elliptic part by the finite element method. The resulting linear system is solved by a preconditioned conjugate gradient method.

DOMAIN DECOMPOSITION

As pointed out earlier, the extrusion problem is rich on localized phenomena, especially near the outlet through the die. The computational domain is divided into overlapping subdomains and refined at locations with large gradients. An example of a domain with a refined area is shown in Figure 2, where a sudden contraction is used to mimic the effect of a die. The pressure-velocity system is elliptic, and can be solved by conventional domain decomposition methods, [CW92], [DW89]. We will be using an additiv Schwarz method.

To illustrate the domain decomposition idea, the classical alternating Schwarz method is used for the heat balance equation. The total computational domain Ω is divided into two subdomains Ω_1 and Ω_2 of equal size and resolution and has a fullsized outlet, see Figure 3. The two subdomains overlap with a given number of elements. The individual boundaries are denoted $\partial\Omega_i$, in which the interior surfaces Γ_i are included, $i = 1, 2$. The initial-boundary value problem for the temperature is stated below, where the vector-valued function $\mathbf{f}$ is a given velocity field, the function g is a boundary function and θ_{t_0} is an initial profile.

$$\begin{aligned}
\mathrm{Pe}(\frac{\partial\theta}{\partial t} + \mathbf{u} \cdot \nabla\theta) &= \nabla^2\theta + \beta\epsilon : \tau && \text{in } [0, T] \times \Omega \subset R^3, \\
\mathbf{u}(x, y, z) &= \mathbf{f}(x, y, z) && \text{in } \Omega, \\
\theta(x, y, z, t) &= g(x, y, z) && \text{on } \partial\Omega \ \forall t, \\
\theta(x, y, z, 0) &= \theta_{t_0}(x, y, z) && \text{in } \Omega \text{ at } t = 0.
\end{aligned} \tag{9}$$

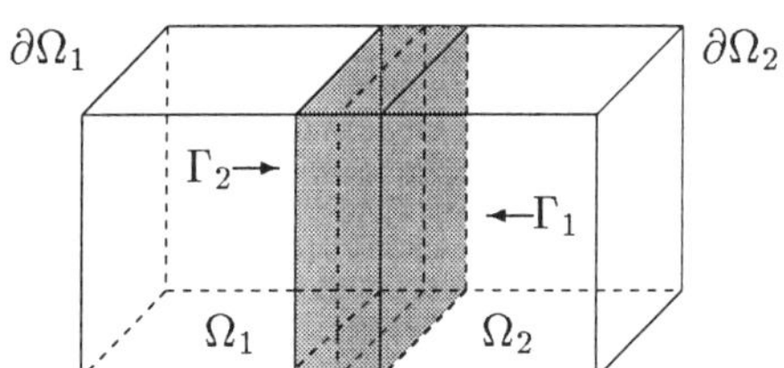

Figure 3 Two equally sized overlapping subdomains.

Before we present the algorithm we introduse some notation. Let $\theta_{i,k}^n$ denote the solution in $\overline{\Omega}_i$ after k Schwarz iterations at timelevel n, and $\theta_{2,k}^n|_{\Gamma_1}$ be the restriction of $\theta_{2,k}^n$ to Γ_1; similarly $\theta_{1,k}^n|_{\Gamma_2}$ is the restriction of $\theta_{1,k}^n$ to Γ_2. The part of the true boundary that surround each of the subdomains except for the interior surface is denoted $\partial\Omega_i\backslash\Gamma_i$, $i = 1, 2$. The characteristic solution at timelevel n is denoted by $\bar{\theta}^n$, and the subscript on the nonlinear part $(\beta\epsilon : \tau)_{i,k}^n$ indicates that the iterate $\theta_{i,k}^n$ is used to linearize. Note that $k = 0$ indicates the characteristic solution.

Algorithm:

1. Solve the hyperbolic part for the characteristic solution $\bar{\theta}^n$:

$$\frac{\partial\theta}{\partial t} + \mathbf{u}\cdot\nabla\theta = 0 \qquad \text{in } \Omega.$$

2. Solve for $\theta_{1,1}^n$:

$$\theta_{1,1}^n = \frac{\Delta t}{\text{Pe}}\Delta\theta_{1,1}^n + \frac{\Delta t}{\text{Pe}}(\beta\epsilon : \tau)_{1,0}^n + \bar{\theta}^n \qquad \text{in } \Omega_1,$$

$$\theta_{1,1}^n = \bar{\theta}^n \qquad \text{on } \Gamma_1,$$

$$\theta_{1,1}^n = g \qquad \text{on } \partial\Omega_1\backslash\Gamma_1.$$

3. Solve for $\theta_{2,1}^n$:

$$\theta_{2,1}^n = \frac{\Delta t}{\text{Pe}}\Delta\theta_{2,1}^n + \frac{\Delta t}{\text{Pe}}(\beta\epsilon : \tau)_{2,0}^n + \bar{\theta}^n \qquad \text{in } \Omega_2,$$

$$\theta_{2,1}^n = \bar{\theta}^n \qquad \text{on } \Gamma_2,$$

$$\theta_{2,1}^n = g \qquad \text{on } \partial\Omega_2\backslash\Gamma_2.$$

4. For $k = 2, 3, \ldots$ solve iteratively for $\theta_{1,k}^n$ and $\theta_{2,k}^n$:

$$\begin{cases} \theta_{1,k}^n = \dfrac{\Delta t}{\text{Pe}}\Delta\theta_{1,k}^n + \dfrac{\Delta t}{\text{Pe}}(\beta\epsilon : \tau)_{1,k-1}^n + \bar{\theta}^n & \text{in } \Omega_1, \\[2mm] \theta_{1,k}^n = \bar{\theta}_{2,k-1}^n|_{\Gamma_1}, \\[2mm] \theta_{1,k}^n = g & \text{on } \partial\Omega_1\backslash\Gamma_1, \end{cases}$$

$$\begin{cases} \theta_{2,k}^n = \dfrac{\Delta t}{\text{Pe}}\Delta\theta_{2,k}^n + \dfrac{\Delta t}{\text{Pe}}(\beta\epsilon : \tau)_{2,k-1}^n + \bar{\theta}^n & \text{in } \Omega_2, \\[2mm] \theta_{2,k}^n = \bar{\theta}_{1,k}^n|_{\Gamma_2}, \\[2mm] \theta_{2,k}^n = g & \text{on } \partial\Omega_2\backslash\Gamma_2. \end{cases}$$

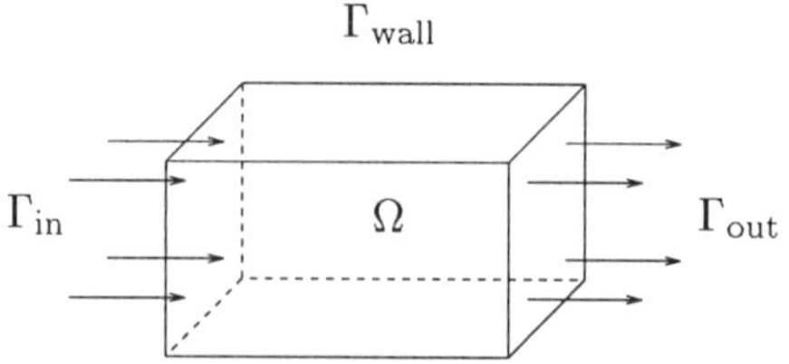

Figure 4   Computational domain, note that $\Gamma_{\text{wall}} = \partial\Omega \backslash (\Gamma_{\text{in}} \cup \Gamma_{\text{out}})$.

NUMERICAL EXAMPLES

In this section we show a solution of the system (2)-(3) for the simple geometry given above.

Consider Ω to be a bounded domain in R^3 with a boundary $\Gamma = \Gamma_{\text{in}} \cup \Gamma_{\text{out}} \cup \Gamma_{\text{wall}}$, i.e. Γ is a union of an in-boundary, an out-boundary and the rest of the total boundary, called the wall-boundary, see Figure (4). In the calculations presented, Γ_{out} is regarded as a free boundary in respect to velocity and temperature.

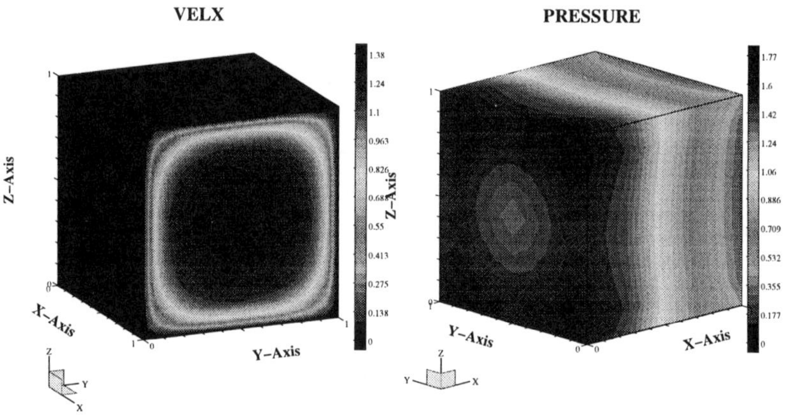

Figure 5   (a) x-component of velocity with a domal profil due to the no-slip condition, and (b) matching pressure with the largest values at Γ_{in} and especially near the corners.

Figure 5 shows the solution to the equations (3) with the boundary conditions

$$
\begin{aligned}
\mathbf{u} &= [1 \ 0 \ 0]^t && \text{on } \Gamma_{\text{in}}, \\
\mathbf{u} &= 0 && \text{on } \Gamma_{\text{wall}}, \\
p &= 0 && \text{on } \Gamma_{\text{out}}.
\end{aligned}
$$

The parabolic heat balance equation is solved by the alternating Schwarz method with overlap, following the algorithm discussed above. Results from two different cases are presented. For better to visualize the solutions in the interior of the domain, it is split in two along the flow-direction, and only one half of the domain is shown.

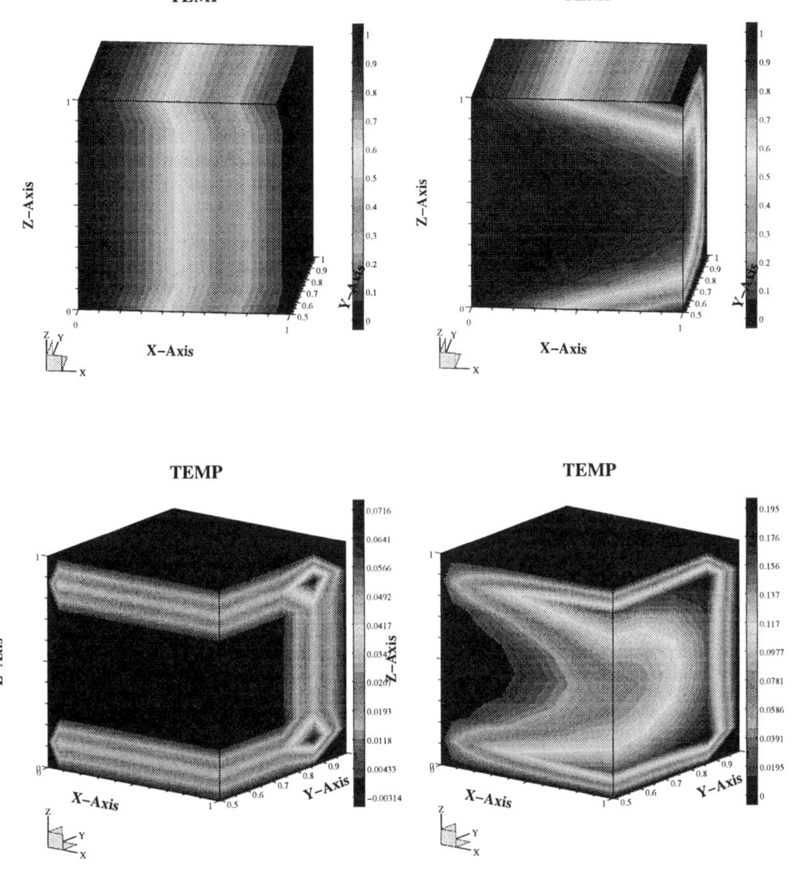

Figure 6 (a) Temperature after 5 timesteps, and (b) after 20 timesteps for the first case. (c) Temperature after 5 timesteps, and (d) after 20 timesteps for the second case.

In Figure 6, (a) and (b), the temperature distribution resulting from solving (9) with

$$\mathbf{f}(x, y, z) = [1\ 0\ 0]^t \qquad \text{in } \overline{\Omega},$$

and linear boundary and initial conditions,

$$g(x, y, z) = 1 - \tfrac{1}{2}x \qquad \text{on } \partial\Omega\backslash\Gamma_{\text{out}},$$
$$\theta_{t_0} = 1 - \tfrac{1}{2}x \qquad \text{in } \Omega,$$

is shown. In Figure 6, (c) and (d), we have a no-slip condition causing friction,

$$\mathbf{f}(x, y, z) = \begin{cases} [1\ 0\ 0]^t & \text{in } \overline{\Omega}\backslash\Gamma_{\text{wall}}, \\ 0 & \text{on } \Gamma_{\text{wall}}, \end{cases}$$

and zero conditions on the temperature,

$$g(x, y, z) \;=\; 0 \qquad \text{on } \partial\Omega\backslash\Gamma_{\text{out}},$$
$$\theta_{t_0} \;=\; 0 \qquad \text{in } \Omega.$$

The main difference between these to cases is due to the no-slip condition in the one case. In the first case the velocity is the same in the hole domain, and the heat is smoothly spreading into the middle of the domain. Looking at a cross section at some point in the flow-direction, the largest values will always be located in the middle. The next case also show smooth spreading into the middle of the domain, but in a cross section similar to the one mentioned above, the largest values will be in the corners and along the edges.

CONCLUSIONS

The tests carried out indicate that the domain decomposition algorithm presented above performs well for the nonlinear parabolic equation compared to the global solution on one domain. The method will, however, be further investigated.
For the nonlinear elliptic pressure-velocity system an additiv Schwarz method gives good results, but further experiments must be (and will be) carried out to support such a statement.

ACKNOWLEDGEMENT

The support provided by Norsk Hydro is gratefully acknowledged.

REFERENCES

[BP94] Bramble J. and Pasciak J. (1994) Iterative techniques for time dependent stokes problems. *Math. Comp.* .

[CW92] Cai X. and Widlund O. (1992) Domain decomposition algorithms for indefinite elliptic problems. *SIAM J. Stat. Comput.* 13: No.1, pp. 243–258.

[DES92] Dahle H., Espedal M., and Sævareid O. (1992) Characteristic local grid refinement for reservoir flow problems. *Int. j. numer. meth. eng.* 34: pp. 1051–1069.

[DW89] Dryja M. and Widlund O. (1989) Towards a unified theory of domain decomposition algorithms for elliptic problems. In *3rd International Symposium on Domain Decomposition Methods for Partial Differential Equations*.

[HSH92] Holthe K., Støren S., and Hansen L. (1992) Numerical simulation of the aluminium extrusion process in a series of press cycles. In *4th International Conference on Numerical Methods in Industrial Forming Processes*.

[Sæv93] Sævareid O. (1993) Simulation of extrusion combining local grid refinement and lagrangian time-marching. In *8th International Conference on Finite Elements in Fluids*.

A Domain Decomposition Method for Heterogeneous Reservoir Flow

Brit Gunn Ersland [1] and Magne S. Espedal [2]

ABSTRACT

The main objective for this work is to study two phase flow (oil, water) in a heterogeneous porous media.

To solve the elliptic pressure equation we use a finite element formulation, while the velocity is calculated in a flux conserving manner from the pressure. For the parabolic saturation equation we use a two step operator splitting procedure [EE87].

Our computational domain consist of two regions with different physical properties. Domain decomposition is used for adaptive refinement, and to treat the interior boundary. We present an iteration procedure which ensure conservation of mass.

MODEL PROBLEM

For incompressible immiscible displacement of oil by water in a reservoir the following equations yeld

$$\nabla \cdot \mathbf{v} = q_1(\mathbf{x}, t) \tag{1}$$

$$\mathbf{v} = -\mathbf{K}(\mathbf{x}) M(S, \mathbf{x}) \cdot \nabla p \tag{2}$$

$$\phi \frac{\partial S}{\partial t} + \nabla \cdot (f(S)\mathbf{v}) - \epsilon \nabla \cdot (D(S, x)\nabla S) = q_2 . \tag{3}$$

[1] Department of Mathematics, University of Bergen, Johs. Brunsg. 12., 5008 Bergen, Norway, email: resbg@mi.uib.no

[2] Department of Mathematics, University of Bergen, Johs. Brunsg. 12., 5008 Bergen, Norway, email: resme@mi.uib.no

Domain Decomposition Methods in Sciences and Engineering, edited by R. Glowinski *et al.*

We will use Neumann type of boundary conditions.

$\mathbf{v}$ is the total Darcy velocity, which is the sum of the velocity of the oil and water phase. $\mathbf{K}(\mathbf{x})$ is the permeability which depend on the porous medium, $M(S,\mathbf{x})$ is a mobility function that depend on the water saturation S, and p is the total fluid pressure. ϕ is the porosity of the porous media which is considered constant. The fractional flow function $f(S)$ is a nonlinear function of the saturation and is given as

$$f(S) = \frac{\lambda_w(S)}{\lambda_w(S) + \lambda_o(S)} \tag{4}$$

where the mobility of oil and water,

$$\lambda_l, \ l \in \{o, w\}$$

is a given function of S.

The diffusion coeffesient $D(S,x)$ is

$$D(S,x) = K(x)f(S)\lambda_o(S)\frac{\partial P_c}{\partial S}. \tag{5}$$

$K(x)$ is the absolute permeability and P_c is the capillary pressure function It is well known that a nonlinear equation like eq(3) will establish shock like solutions therefore we split the function $f(S)$ into two parts. [EE87] [DR82]

$$f(S) = \bar{f}(S) + b(S)S,$$

The convective part

$$\phi\frac{\partial S}{\partial \tau} \equiv \frac{\partial S}{\partial t} + \bar{f}'(S)\mathbf{v} \cdot \nabla S = 0, \tag{6}$$

is solved by the Modified Method of Characteristics [DR82]. The elliptic part

$$\phi(\mathbf{x})\frac{\partial S}{\partial \tau} + \nabla \cdot (b(S)S\mathbf{v}) - \nabla \cdot (D(S,\mathbf{x}) \cdot \nabla S) = 0 \tag{7}$$

is discretised by bilinear finite elements, with optimal testfunctions [BM84] . The nonlinear terms in the elliptic equation is linearised around the solution of the convective equation.

The pressure equation is solved by an finite element method with bilinear elements and chapau testfunctions. The velocity is derived from the pressure with a second order method.

We use a sequential time-stepping procedure decribed in [DES92].

COMPUTATIONAL DOMAIN

Our computational domain consist of two regions with different physical properties. The first region has permeability K_1 and capillary pressure curve P_{C1}, the second has permeability K_2 and capillary pressure curve P_{C2}.

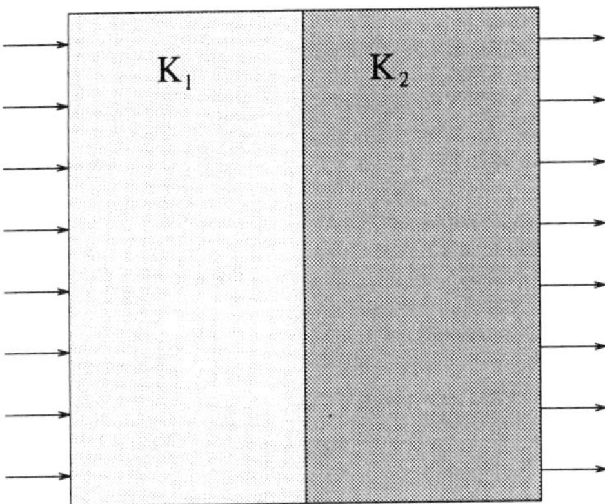

Figure 1 *Computational domain showing two different regions with different physical properties connected trough an interior boundary.*

The capillary pressure function,

$$P_C(S) = 0.09\phi^{-0.9}K^{-0.1}\frac{1-S}{\sqrt{S}} \tag{8}$$

depend on the saturation S and the permeability K [KME92]. Further,

$$\lambda_l = S_l^2, \ l \in \{o, w\}$$

in both regions.

In addition to our outer Neumann boundary condition, the discontinuity in $\mathbf{K}$, leads to a problem with the interior boundary condition. The capillary pressure must be continuous over the interior boundary Γ.

$$P_{C_1}(S_1) = P_{C_2}(S_2) \tag{9}$$

This condition lead to a discontinuous saturation over the boundary.

$$S_1 \neq S_2 \tag{10}$$

In order to conserve the mass over the boundary the flux must be conserved over the interior boundary Γ.

$$(f(S)\mathbf{v} - \epsilon D \cdot \nabla S)^1 \cdot n^\Gamma = (f(S)\mathbf{v} - \epsilon D \cdot \nabla S)^2 \cdot n^\Gamma \tag{11}$$

DOMAIN DECOMPOSITION

In order to solve our equations on the computational domain, we construct a coarse grid Ω_C, where the interior boundary is in the middle of a set of coarce grid elements. In addition we construct a fine grid Ω_F , where the interior boundary is placed in a set of nodes. The fine grid is decomposed in into the same number of overlapping subgrids as there are elements in the coarse grid. We may write

$$\Omega_F = \cup\Omega^i.$$

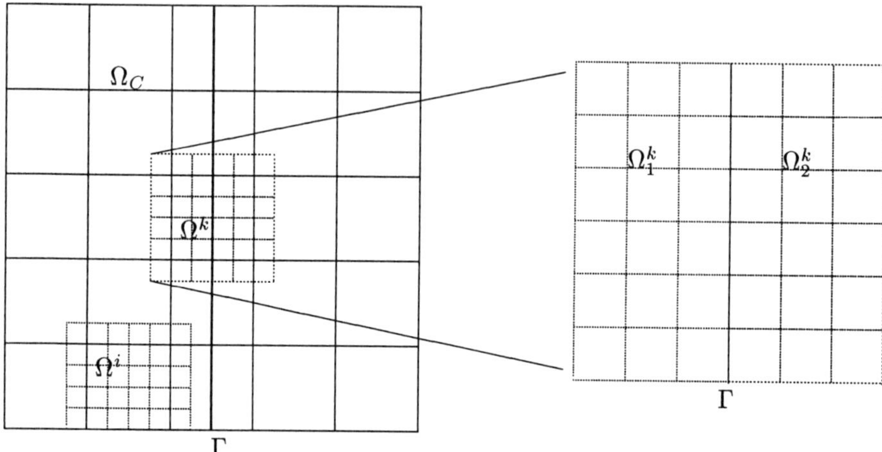

Figure 2   The coarse grid, and fine overlapping grids.

For those coarse elements that contain an interior boundary we need two additional grids. Let Ω^k be a fine grid that contain an interior boundary, then Ω_1^k is the part of Ω^k that has permeability K_1 and Ω_2^k is the part of Ω^k that has permeability K_2 . Ω_1^k and Ω_2^k has equal spacing to Ω^k.

THE SOLUTION PROCEDURE

- Solve the pressure equation, and derive the velocity from the pressure.
- Trace the characteristics with the new velocity.
- Determine the area with large gradients in saturation, and trace the characteristics for the fine grids where the gradients are large.
- if the refined grid has an interior boundary
  MassBalance iteration.
  else
  Solve the elliptic saturation equation.

The three first steps in the procedure is well described and published in [DES92], so we will only describe the new massbalance iteration in the following.
The massbalance iteration ensures that no mass is lost or created at the interior boundary. For the massbalance iteration we use the fine grids which contain an interior

boundary as a control grid. The amount of mass that has been transported in to this
"homogeneous" domain, Ω^k, is the same mass that we have to distribute between the
two subgrids Ω_1^k and Ω_2^k . The massbalance iteration can be outlined like this:

- Trace the characteristics on Ω_1^k.
- Use the jump condition at the boundary, $P_{C1}(S_1) = P_{C2}(S_2)$.
- Trace the charcterisitcs on Ω_2^k, with the new saturation on the boundary.
- Solve the elliptic equation on Ω_1^k and Ω_2^k with Diriclet condition at the
  interior boundary.
- Calculate the massbalance.
- While the mass is not balancing do

  1. Add to the saturation at the left side of the boundary a scaling factor
     times the error in mass.
  2. Use the jump condition at the boundary to determine the saturation
     on the right side of the boundary.
  3. Solve the elliptic equation on both domains with the new Diriclet
     condition at the interior boundary.
  4. Calculate the massbalance.

- end do.

NUMERICAL RESULTS

The numerical results is shown for two regions with different permeabilities. Figure 4
show the effect of the permeability on the diffusion function.

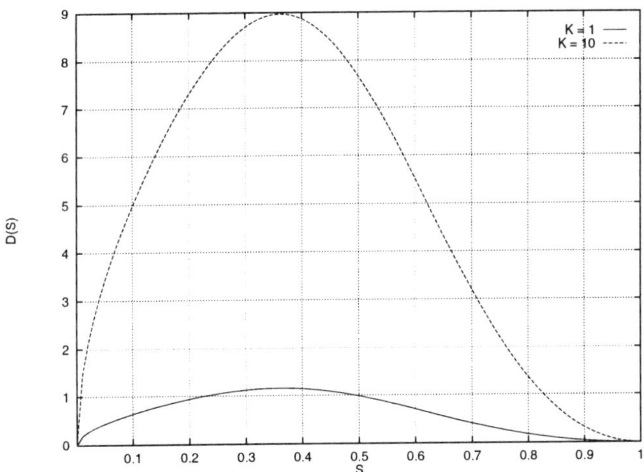

Figure 3   *The diffusion function for $K = 1$ and $K = 10$.*

The following figures shows a domain with an interface at $x = 0.5$. At the
interface we have a jump in the saturation pofile due to the interface-condition

$$P_{C_1}(S_1) = P_{C_2}(S_2).$$

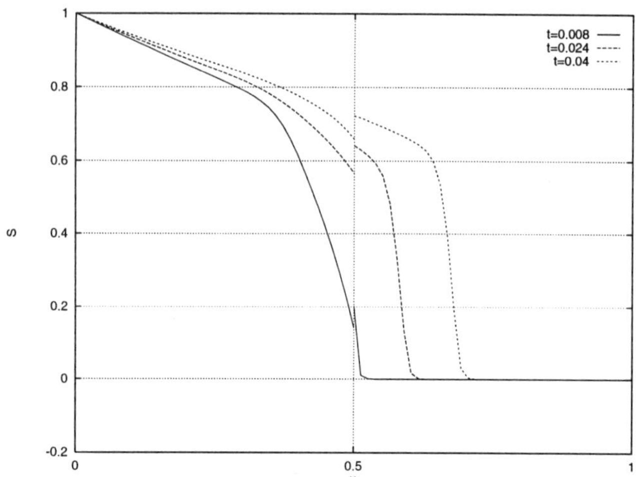

Figure 4   *Saturation profiles when $K_1 = 10$ and $K_2 = 1$. Interior boundary at $x = 0.5$ and $\epsilon = 0.01$.*

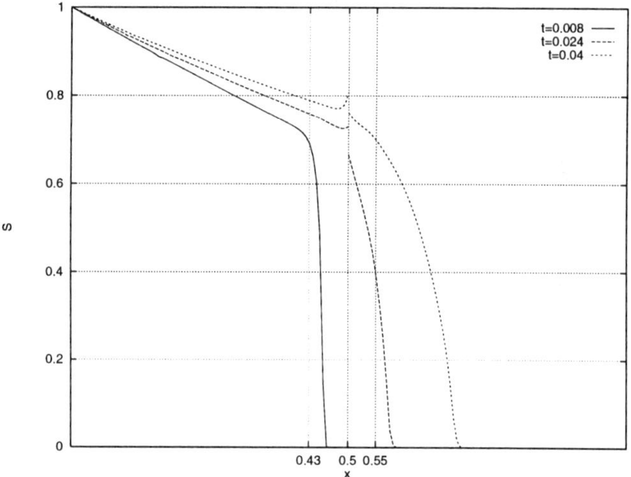

Figure 5   *Saturation profiles when $K_1 = 1$ and $K_2 = 10$. Interior boundary at $x = 0.5$ and $\epsilon = 0.01$.*

In 2D the saturation at the interface is given by the mean of the saturation at the interface, therefore we are not able to see the jump as clearly as in 1D, where we where able to present one point with two different saturation values. In both 1D and 2D the mass is nicely conserved. We can conclude that the massbalance iteration work in both 1D and 2D, and is consistent with the boundary conditions 9 and 11. The maximum number of iterations needed in 2D is eleven.

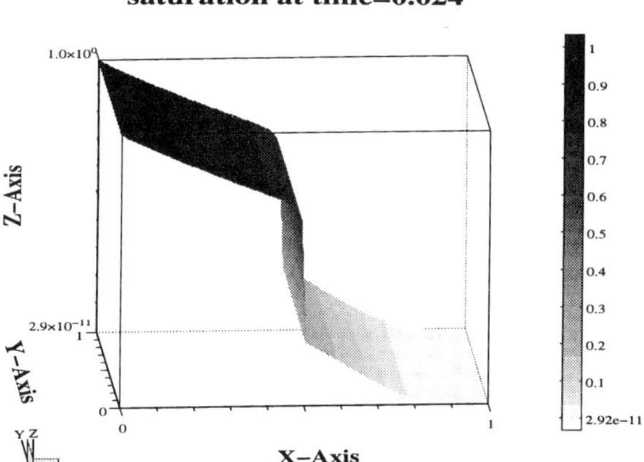

Figure 6 *Saturation profile in 2D when $K_1 = 1$ and $K_2 = 10$. Interior boundary at $x = 0.5$ and $\epsilon = 0.01$.*

ACKNOWLEDGEMENTS

This research was supported by the Norwegian Reasearch counsil.

REFERENCES

[BM84] Barrett J. and Morton K. (1984) Approximate symmetrization and petrov-galerkin methods for diffusion-convection problems. *Computer Methods in Applied Mechanics and Engineering* 45: 97–122.

[DES92] Dahle H., Espedal M., and Sævareid O. (1992) Characteristic, local grid refinement techniques for reservoir flow problems. *International Journal for Numerical Methods in Enginering* 34: 1051–1069.

[DR82] Douglas J. and Russell T. (1982) Numerical methods for convection-dominated diffusion problems based on combining the method of characteristics with finite element or finite difference procedures. *SIAM Journal on Numerical*

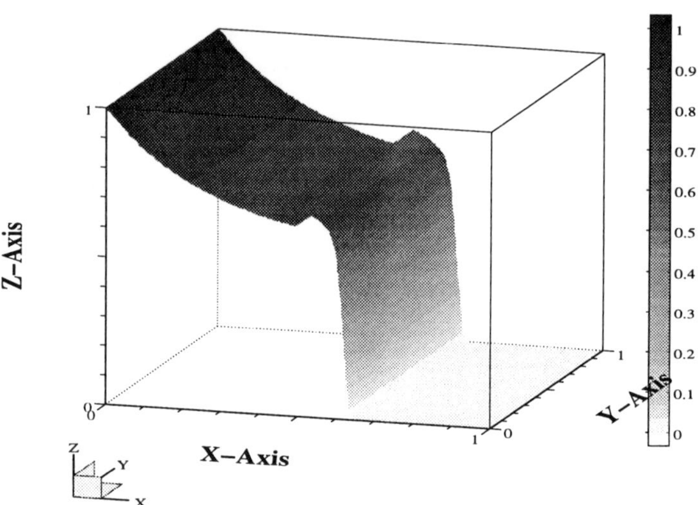

Figure 7   *Saturation profile in 2D when $K_1 = 10$ and $K_2 = 1$. Interior boundary at $x = 0.5$ and $\epsilon = 0.01$.*

Analysis 19: 871–885.

[EE87] Espedal M. and Ewing R. (1987) Characteristic petrov-galerkin subdomain methods for two-phase immiscible flow. *Computer Methods in Applied Mechanics and Engineering.* 64: 113–135.

[KME92] Kristiansen J., Mikkelsen M., and Esbensen K. (1992) A modified leverett approach and pls-regression for integrated formation evaluation. Number 33. Annual Symposium of Society of Professional Well Log Analysis.

A Domain Decomposition Technique for Viscous/Inviscid Coupling

C.-H. LAI[1], A.M. CUFFE[2], & K.A. PERICLEOUS[3]

INTRODUCTION

The numerical simulation of viscous flows past realistic configurations using time-dependent compressible Navier-Stokes for high Reynolds numbers requires a large amount of computing time [cfd94]. Although steady-state solutions of a compressible inviscid flow past over a complete aircraft described by the Euler equation can be obtained in just a few minutes time, modelling 3-D unsteady flows to understand the aerodynamics of a complete aircraft is still a grand challenge. With the limits of computing resources, it is important to develop efficient numerical technqiues which are suitable for implementation on parallel computers and for engineering design purposes.

One such technique relies upon a problem partitioning concept. The concept in relation to viscous flows past over obstacles was originally observed by Prandtl [Pra04], of which the following two observations are involved. When a viscous fluid flow past a solid boundary, tangential and normal components of the velocity must both equal to the components of the velocity of the surface, and fluid sticks to the surface. For an ideal inviscid fluid, only the normal components are equal and there is no restriction on tangential slip of fluid relative to the boundary. The fundamental differences lead

[1] School of Computing & Mathematical Sciences, University of Greenwich, Wellington Street, Woolwich, London SE18 6PF, U.K. email: C.H.Lai@grn.ac.uk
[2] Address same as above. This author is supported by a Bursary of the University.
[3] Address same as above.

Domain Decomposition Methods in Sciences and Engineering, edited by R. Glowinski *et al.*
© 1997 John Wiley & Sons, Ltd.

to the idea of a boundary layer first proposed by Prandtl. It is a thin layer of fluid adjacent to the solid surface within which viscous effects are dominant and that such effects become negligible outside the thin layer. Hence it is valid to divide the flow region into two parts, i.e. the boundary layer and the inviscid flow outside the layer.

Mathematically, one uses either the Navier-Stokes equation or one of its reduced form for flows within the boundary layer and an inviscid model described by either the Euler equation or the potential equation outside the boundary layer. Numerically, such problem partitioning is often referred to as the zonal method [Sch86]. There are two zones namely, inner and outer, with the inner zone being described by the thickness function t. The inner zone is governed by the Navier-Stokes equation and the outer zone is governed by the Euler equation. Due to the amount of computing power, one current industrial practice is to solve a reduced form of the Navier-Stokes equation in the boundary layer and then coupled with a compressible potential model outside the boundary layer. The coupling is obtained by means of an iterative technique similar to the one developed by Schwarz [Sch90] for elliptic problems.

One property of the zonal method is that a prior knowledge is required for the partitioning of the flow field. Hence there is a jump of mathematical models being used in different zones. It would be useful to have a dynamic zonal recognition capability with a smooth transition from one model to another model. It would also be useful to minimize the thickness function t such that accuracy obtained by means of the Navier-Stokes equation for viscous flow simulations can be retained.

In this paper, a summary is given of various truncation techniques, including the scalar truncation method used by Brezzi [BCR89]. An extension is given of the scalar truncation method to the Navier-Stokes equation in the context of a finite volume method in order to build a dynamic zonal recognition capability into an in-house cell-centred finite volume multiphase software UIFS [Cho93].

TRUNCATION TECHNIQUES

Truncation methods have been around for a long time. It has been used, in a different form, for the determination of local hyperbolic and elliptic regions in a plane transonic flows for shock location [MC71] or local adaptive mesh refinement for shock capturing. This Section describes two frequently used truncation techniques related to flows past an obstacle.

Prandtl's truncation

Prandtl introduced the idea of boundary layer by neglecting the effect of viscosity outside the thin layer next to a surface. One assumption to implement the idea in 2-D incompressible viscous flows past an obstacle described by

$$\nabla \cdot \underline{u} = 0 \tag{1}$$

$$\frac{\partial \phi}{\partial t} + \nabla \cdot (\underline{u}\phi - \nu\nabla\phi) = -\frac{1}{\rho}\frac{\partial p}{\partial \underline{n}} \tag{2}$$

where $\phi = u, v$ and $\underline{u} = (u, v)^T$ is $u \gg v$. The assumption leads to the boundary layer equations as described by

$$\nabla \cdot \underline{u} = 0 \tag{3}$$

$$\frac{\partial u}{\partial t} + u \frac{\partial u}{\partial x} + v \frac{\partial u}{\partial y} = -\frac{1}{\rho} \frac{\partial p}{\partial x} + \nu \frac{\partial^2 u}{\partial y^2} \tag{4}$$

$$\frac{1}{\rho} \frac{\partial p}{\partial y} \approx 0 \tag{5}$$

There are other assumptions which reduces the 2-D incompressible viscous flow model to other turbulent boundary layer models [Kno86] or models suitable for regions with separations and wakes [GWB73] [Rau75].

Scalar truncation

Early work involving the concept of a scalar truncation technique can be found in Perkins and Rodrigue paper related to a dynamic zonal recognition process for viscous Burgers' equation [PR89]. The idea is to implement in a finite difference context rather than the manual truncation technique as used by Prandtl. Finite difference representations of the viscous term at various grid points across the whole flow domain are calculated, a viscous region across the flow domain can then be masked out leaving the rest of the flow domain consists of an inviscid region at least numerically. One disadvantage of this method is that the viscous term is calculated across the whole computational domain which is extremely ineffective. Also the method does not provide a smooth transition from one mathematical model to the other, and hence the solution is not necessarily smooth across the boundary of the models. An improved version was introduced by Brezzi [BCR89] such that the transition of the viscous term from the viscous region to the inviscid region is smooth.

The zonal recognition is based on the scalar truncation method,

$$\mathcal{T}(q) = \begin{cases} 0, & |q| \leq \epsilon \\ f(q), & \epsilon < |q| < \sigma \\ q, & |q| \geq \epsilon + \sigma \end{cases} \tag{6}$$

where q is certain physical quantities and ϵ and σ are two threshold parameters which can be adjusted to produce numerical results close enough to the true solution. For the present study q represents the viscous term. Work by Perkins and Rodrigue was to choose $\sigma = 0$ in the above scalar truncation method. Brezzi chose $f(q)$ as a straigt line and Canuto chose $f(q)$ as a cubic spline [AC93]. The main concern here is that a smooth transition of the viscous term should be maintained. Therefore, it is equivalent to the determination of the pair of parameters, ϵ and σ, such that the solution to the Burgers equation is most accurate under certain error norms. Some comparisons of results using various threshold parameters can be found in [AC93][BCR89][PR89]. Since $\mathcal{T}$ is nonlinear which means in general, $\mathcal{T}(\frac{\partial^2 u}{\partial x^2}) \neq \frac{\partial}{\partial x}(\mathcal{T}(\frac{\partial u}{\partial x}))$, and hence it is only suitable in the context of finite difference methods. An extension, given in the following Section, to two-dimensional problems which involves finite volume methods requires different treatment. Similar extensions to finite element methods can be found in [AP93].

A Vector Truncation Method

For two-dimensional incompressible viscous flow, the momentum equation given in (2) is integrated over a control volume Ω using a finite volume technique, i.e.

$$\iint_\Omega \frac{\partial \phi}{\partial t} d\Omega + \iint_\Omega \nabla \cdot (\underline{u}\phi - \nu\nabla\phi) d\Omega = -\frac{1}{\rho} \iint_\Omega \frac{\partial p}{\partial \underline{n}} d\Omega \tag{7}$$

which is re-written as

$$\iint_\Omega \frac{\partial \phi}{\partial t} d\Omega + \int_{\partial\Omega} (\underline{u}\phi - \nu\nabla\phi) \cdot \underline{n} ds = -\frac{1}{\rho} \iint_\Omega \frac{\partial p}{\partial \underline{n}} d\Omega \tag{8}$$

where $\underline{n}$ denotes the unit normal vector. The fluid flow inside a viscous region is certainly a rotational flow, and hence velocity gradients are the major contribution to viscous effect. Therefore it is reasonable to apply truncations to these velocity gradients, i.e.

$$\iint_\Omega \frac{\partial \phi}{\partial t} d\Omega + \int_{\partial\Omega} (\underline{u}\phi - \nu\underline{\mathcal{T}}(\nabla\phi)) \cdot \underline{n} ds = -\frac{1}{\rho} \iint_\Omega \frac{\partial p}{\partial \underline{n}} d\Omega \tag{9}$$

should replaces (8), and $\underline{\mathcal{T}}$ denotes the vector truncation methods,

$$\underline{\mathcal{T}}(\nabla\phi) = (\mathcal{T}(\frac{\partial \phi}{\partial x}) , \mathcal{T}(\frac{\partial \phi}{\partial y}))^{\mathrm{T}} \tag{10}$$

Since $\frac{\partial u}{\partial y}$ and $\frac{\partial v}{\partial x}$ contribute to the rotational effect of the flow in the viscous region, therefore these two velocity gradients are significant compare with $\frac{\partial u}{\partial x}$ and $\frac{\partial v}{\partial y}$. Hence $\frac{\partial u}{\partial x}$ and $\frac{\partial v}{\partial y}$ reduce to a smaller value earlier than $\frac{\partial u}{\partial y}$ and $\frac{\partial v}{\partial x}$ with the application of (10). Thererfore (9) effectively reduces the Navier-Stokes equation to a truncated integral form of Navier-Stokes near the outer zone but remains inside the boundary layer and to the Euler equation further away from the body.

The truncation method is embedded into the finite volume formulation which ensures that the boundary layer is evolved as part of the numerical solution. It allows a smooth transition of mathematical models from the Navier-Stokes to the Euler equation.

NUMERICAL RESULTS

First, numerial tests is provided for a flat plate problem. Instead of using the infinite upper half plane, the finite domain $\{(x,y) : -0.5 \leq x \leq 2 , 0 \leq y \leq 10\}$ is used, where the flat plate is placed in $0 \leq x \leq 1$. There are 50×50 cells along x and y-axis, $1/5$ of the cells along y-axis are located in $0 \leq y \leq 0.2$ and the rest are located in $0.2 < y \leq 10$. The upstream Reynolds number is chosen to be 1000. The vector truncation method is tested with a series of ϵ and σ parameters. The results are compared with the Blasius solution, with the package Phoenics and with the in-house cell-centred finite volume software UIFS [Cho93] and are presented in Figure 1. Figure

2 shows the effect of $||u_{\epsilon,\sigma} - u_{fv}||_\infty$ for different values of ϵ and σ used in the flat plate problem, where $u_{\epsilon,\sigma}$ denotes the numerical solution of u by taking ϵ and σ as the threshold parameters and u_{fv} denotes the numerical solution of u by using the software UIFS.

Second, numerical tests are provided for the NACA0012 aerofoil at zero angle of incidence. Therefore only part of the upper half of the physical domain is considered and a C-type of mesh is used for subsequent computation. The finite physical domain is mapped onto the computational domain $\{(\xi, \eta) : 0 \le \xi \le 2\ ,\ 0 \le \eta \le 10$ where the transformed aerofoil is a straight line located in $0 \le \xi \le 1$. There are 40×100 cells along ξ and η-axis, $1/5$ of the cells along η-axis are located in $0 \le \eta \le 0.2$ and the rest are located in $0.2 \le \eta \le 10$. There are 25 cells along the aerofoil surface. The upstream Reynolds number is chosen to be 1000. Figure 3 shows the velocity profiles against the distance along η-axis, obtained at $\xi = 0.46$ and $\xi = 0.66$. Figure 4 shows the effect of $||u_{\epsilon,\sigma} - u_{fv}||_\infty$ for the aerfoil problem.

In general, the finite volume solution obtained by means of applying the vector truncation technique to the Navier-Stokes equation is close enough, in the sense of engineering purpose, compare to the corresponding finite volume solution obtained by means of the Navier-Stokes equation. From Figures 2 and 4, the solution accuracy of the flat plate problem and the NACA0012 aerofoil probelm is not sensitive to the choice of σ.

CONCLUSIONS

The scalar truncation method is extended to a vector truncation method and is embedded in a cell-centred finite volume method. The effect of the threshold parameter σ is insignificant to the accuracy of the solution. The choice of the threshold parameters namely, σ and ϵ, is effectively equivalent to the determination of a minimal inner region defined by the thickness function t. The present approach shows that the computation of velocity gradients in the finite volume method can be used in both the evaluation of the line integrals as well as for viscous effect comparison.

REFERENCES

[AC93] Arina R. and Canuto C. (1993) A self-adaptive domain decomposition for the viscous/inviscid coupling i: Burgers equations. *J. of Computational Physics* 105: 290–300.

[AP93] Achdou Y. and Pironneau O. (1993) The χ-method for the Navier-Stokes equations. *IMA J. of Numer. Analysis* 13: 537–558.

[BCR89] Brezzi F., Canuto C., and Russo A. (1989) A self-adaptive formulation for the Euler/Navier-Stokes coupling. *Computer Methods in Applied Mechanics and Engineering* 73: 317–330.

[cfd94] (1994) *New Opportunities and Directions in Aeronautical CFD*, number 16 in EASE Programme: CFD Community Club Seminars, Chilton, Didcot, UK.

[Cho93] Chow P. M.-Y. (1993) *Control Volume Unstructured Mesh Procedure for Convection-Diffusion Solidification Process*. PhD dissertation, University of Greenwich.

[GWB73] Green J., Weeks D., and Brooman J. (1973) Prediction of turbulent boundary layers and wakes in incompressible flow by a lag entrainment method. Technical Report and Memo T3792, Royal Aircraft Establishment, Farnborough, UK.

[Kno86] Knott M. (1986) *Numerical Solutions of Two-Dimensional Unsteady Boundary Layers*. PhD dissertation, University of London.

[MC71] Murman E. and Cole J. (1971) Calculation of plane steady transonic flows. *AIAA J.* 9: 114–121.

[PR89] Perkins A. and Rodrigue G. (1989) A domain decomposition method for solving a two-dimensional viscous burgers' equation. *Applied Numerical Mathematics* 6: 329–340.

[Pra04] Prandtl L. (1904) Über Flüssigkeitsbewegung bei sehr kleiner reibung. In *Proceedings of the 3rd International Mathmatics Congress*. Heidelberg.

[Rau75] Raudkivi A.J. & Callander R. (1975) *Advanced Fluid Mechanics*. Edward Arnold (Publisher) Ltd, London.

[Sch90] Schwarz H. (1890) Über einen Grenzübergang durch alternierendes verfahren. *Gesammelte Mathematische Abhandlungen* 2: 133–143.

[Sch86] Schmatz M. (1986) Calculation of strong interactions on airfoils by zonal solutions of the Navier-Stokes equations. In Rues D. and Kordulla W. (eds) *Notes on Numerical Methods in Fluid Mechanics*, number 13, pages 335–342. Vieweg, Braunschweig-Wiesbaden.

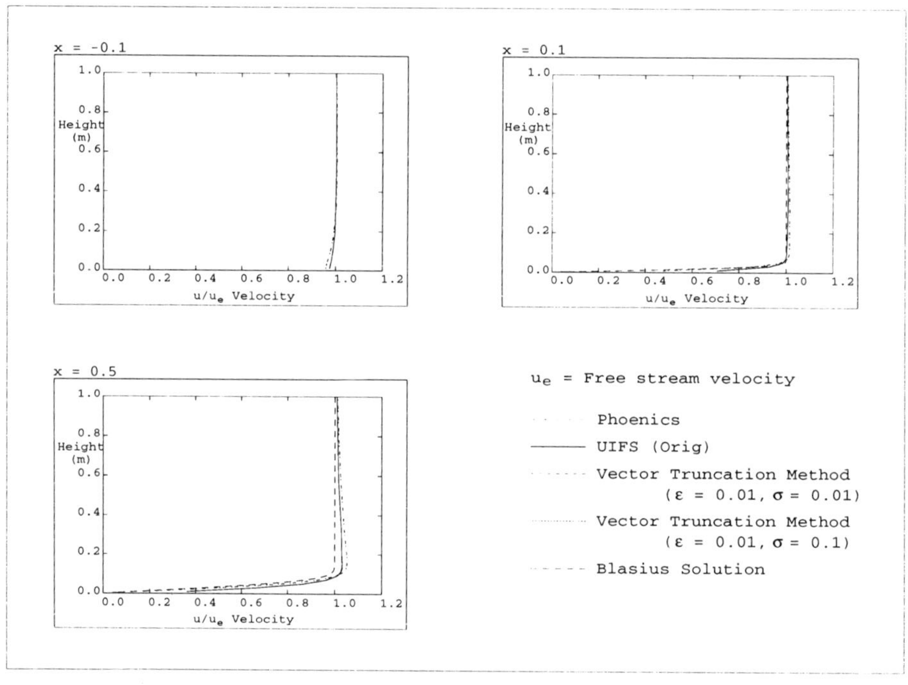

Figure 1 The flat plate problem.

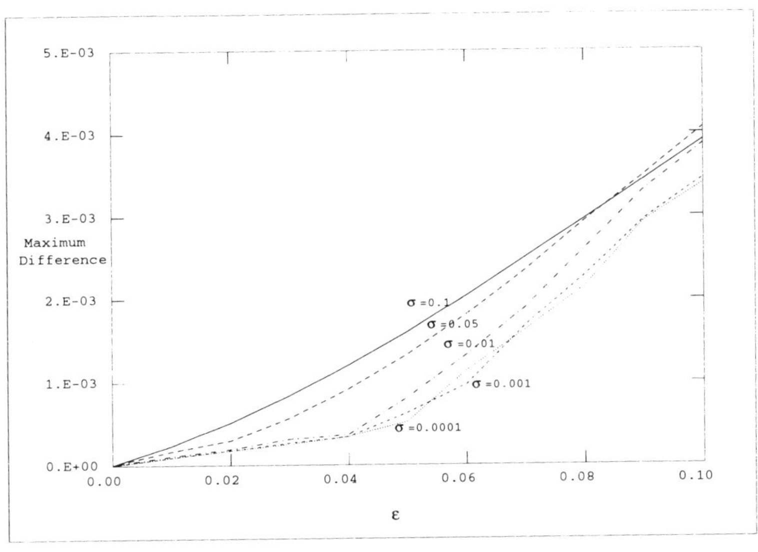

Figure 2 $\|u_{\epsilon,\sigma} - u_{fv}\|_\infty$ for some values of ϵ and σ.

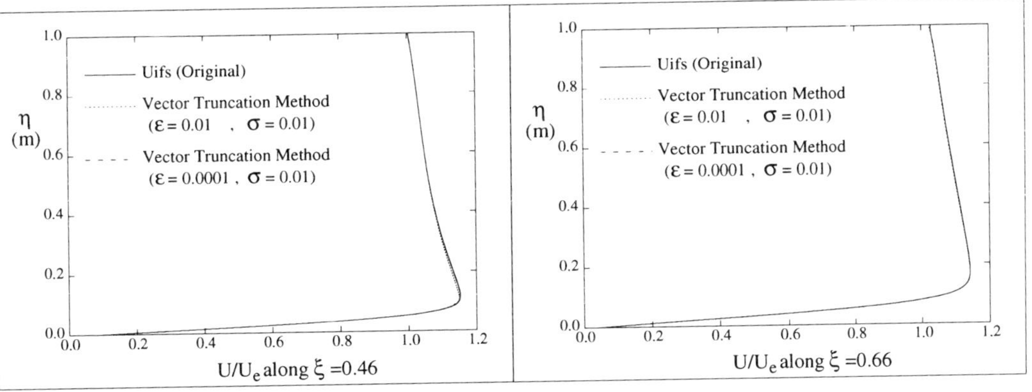

Figure 3 Velocity profile along $\xi = 0.46$ and $\xi = 0.66$.

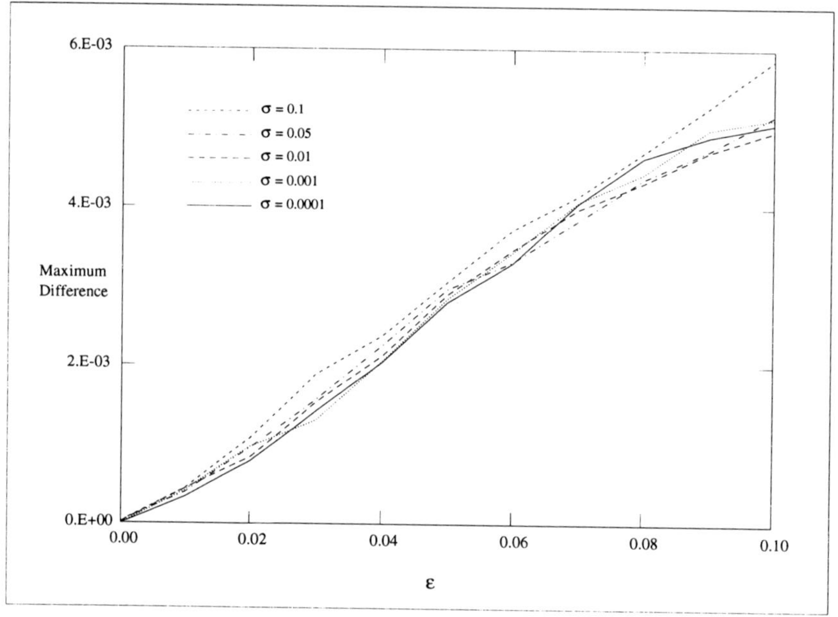

Figure 4 $||u_{\epsilon,\sigma} - u_{fv}||_\infty$ for some values of ϵ and σ.

Convergence Analysis of Parallel Domain Decomposition Algorithm for Navier-Stokes Equations

Kaitai Li and Cuihua Li

1   Introduction

In the paper [1], we developed a Schwarz's domain decomposition algorithm and a convergence analysis for the stationary incompressible Navier-Stokes problem. But this is a serial algorithm. In this paper, we discuss a class of parallel algorithms. We consider the stationary Navier-Stokes equations with the Dirichlet boundary condition:

$$
\begin{cases}
-\nu \Delta u + \sum_{j=1}^{N} u_j \dfrac{\partial u}{\partial x_j} + \mathrm{grad} p = f \\
\qquad\qquad\qquad \mathrm{div} u = 0 \quad \text{in } \Omega \\
\qquad\qquad\qquad u|_{\partial \Omega} = 0,
\end{cases}
\tag{1}
$$

where Ω is a bounded domain of $R^N (N = 2, 3)$ with a Lipschitz-continuous boundary $\partial \Omega$. The vector $u = \{u_i\}_{i=1}^{N}$ is the velocity of the fluid, $\nu > 0$ is its kinematic viscosity (assumed to be constant), p its pressure and the vector $f = \{f_i\}_{i=1}^{N} \in [H^{-1}(\Omega)]^N$ the density of the body forces per unit mass. We introduce the following spaces:

$$
X = [H_0^1(\Omega)]^N, \quad V = \{v|v \in X; \mathrm{div} v = 0, \text{in} \quad \Omega\},
$$

$$
M = L_0^2(\Omega) = \{q|q \in L^2(\Omega); \int_{\Omega} q dx = 0\}.
$$

[0] Project Supported by State Key Basic Research Project & NSFC

[1] College of Science, Xian Jiaotong University, Xian, 710049, P.R.CHINA.
  e-mail: KTLi@XJTU.EDU.CN

Domain Decomposition Methods in Sciences and Engineering, edited by R. Glowinski *et al.*

The weak formulation of (1) is

$$(P) \quad \begin{cases} \text{Find } u \in V \quad \text{such that} \\ a_0(u,v) + a_1(u;u,v) = <f,v>, \forall v \in V, \end{cases}$$

or

$$(Q) \quad \begin{cases} \text{Find } \{u,p\} \in X \times M \quad \text{such that} \\ a_0(u,v) + a_1(u;u,v) - b(p,v) = <f,v>, \forall v \in X \\ b(q,u) = 0, \quad \forall q \in M, \end{cases}$$

where

$$a_0(u,v) = \nu(\mathrm{grad}\,u, \mathrm{grad}\,v), \quad a_1(u;v,w) = \sum_{i,j=1}^{N} \int_\Omega u_j \frac{\partial v_i}{\partial x_j} w_i \, dx,$$

$$b(p,v) = (p, \mathrm{div}\,v), \quad <f,v> = \int_\Omega f \cdot v \, dx, \quad \forall u,v,w \in X.$$

Problem (P) is equivalent to problem (Q) and there exists at least one solution[2,3].

2 The Parallel Algorithm

Assume that Ω is split into m subdomains Ω_j:

$$\Omega = \bigcup_{j=1}^{m} \Omega_j \quad (m \geq 2),$$

satisfying

$$\exists \xi_j \in C_0^{+\infty}(\Omega_j), 0 \leq \xi_j \leq 1, \text{in} \quad \Omega_j; \xi_j = 0, \text{in} \quad \Omega \backslash \Omega_j,$$

$$\text{such that} \quad \sum_{j=1}^{m} \xi_j = 1, \text{in} \quad \Omega. \tag{2}$$

Later on, we denote by $r(x)$ the number of subdomains to which the point x belongs. Let

$$V_j = \{u \in [H_0^1(\Omega_j)]^N; \mathrm{div}\,u = 0, \text{in} \quad \Omega_j\} \quad (j = 1, 2, \cdots, m).$$

and consider V_j as a closed subspace of V by extending its elements to $\Omega \backslash \Omega_j$ by 0. Then we have the following simple result.

Lemma 1 ([1]). *Under the condition (2), we have*

$$V = V_1 + V_2 + \cdots + V_m,$$

and, for all $v \in V$, there exist $v_j \in V_j$ such that

$$v = \sum_{j=1}^{m} v_j \quad and \quad \max_{1 \leq j \leq m} \| v_i \|_1 \leq C_0 \| v \|_1,$$

where C_0 is some positive constant.

The algorithm is designed as follows:

Step 0. Choose an initial value $u^0 \in V$.

Step 1. For $j = 1, 2, \cdots, m$ and $n \geq 0$, solve in parallel the following subproblems:

$$
\begin{cases}
\text{Find } u_j^{n+1} \in u^n + V_j \text{ and } p_j^{n+1} \in L_0^2(\Omega_j) \text{ such that} \\[2mm]
a_0(u_j^{n+1}, v)_{\Omega j} + a_1(u_j^{n+1}; u_j^{n+1}, v)_{\Omega_j} - b(p_j^{n+1}, v)_{\Omega_j} \\[2mm]
\qquad = <f, v>_{\Omega_j}, \quad \forall v \in [H_0^1(\Omega_j)]^N, \\[2mm]
b(q, u_j^{n+1})_{\Omega_j} = 0, \qquad \forall q \in L_0^2(\Omega_j), \\[2mm]
p_j^{n+1} = 0, \qquad \text{in } \Omega \backslash \Omega_j.
\end{cases}
\tag{3}
$$

Step 2. Choose $\theta_j \in (0, 1)$ such that $\sum\limits_{j=1}^{m} \theta_j = 1$ and for $n \geq 0$ set

$$
u^{n+1} = \sum_{j=1}^{m} \theta_j u_j^{n+1}, \quad p^{n+1}(x) = \frac{1}{r(x)} \sum_{j=1}^{m} p_j^{n+1}.
\tag{4}
$$

Set $n = n + 1$, go to Step 1.

Here the notations are given by:

$$
a_0(u, v)_{\Omega_j} = \nu \int_{\Omega_j} \text{grad} u \cdot \text{grad} v dx, \quad a_1(u; v, w)_{\Omega_j} = \sum_{i,j=1}^{N} \int_{\Omega_j} u_j \frac{\partial v_i}{\partial x_j} w_i dx,
$$

$$
b(p, v)_{\Omega_j} = \int_{\Omega_j} p \, \text{div} v dx, \quad <f, v>_{\Omega_j} = \int_{\Omega_j} f \cdot v dx.
$$

3   The Convergence

Let H be a Hilbert space, F a differentiable mapping from H into H' (the dual space of H), $DF(\cdot)$ its derivative, and let $u \in H$ be a solution of the equation $F(u) = 0$. We say that u is a nonsingular solution if there exists a constant $\gamma_0 > 0$ such that

$$
\| DF(u) \cdot v \|_* \geq \gamma_0 \| v \|_H \quad \forall v \in H.
$$

For the Navier-Stokes problem, we define a C^2-mapping $F(\cdot) : V \to V'$ as follows:

$$
< F(u), v > = a_0(u, v) + a_1(u; u, v) - < f, v > \quad \forall u, v \in V.
$$

Clearly, $F(\cdot)$ is infinitely differentiable in V and its derivative $DF(u) \in \mathcal{L}(V; V')$ is given by:

$$
< DF(u)v, w > = a_0(v, w) + a_1(u; v, w) + a_1(v; u, w).
$$

As a consequence, problem (P) can be rewritten as:

$$
\begin{cases}
\text{Find} \quad u \in V \quad \text{such that} \\[2mm]
< F(u), v > = 0, \quad \forall v \in V.
\end{cases}
\tag{5}
$$

In addition, we introduce the abstract Stokes operator: $A \in \mathcal{L}(V; V')$ defined by

$$< Au, v >= a_0(u, v), \quad \forall u, v \in V.$$

Obviously[3] A is symmetric, V-elliptic, and there exists $A^{-1} \in \mathcal{L}(V'; V)$. Furthermore, the mapping $f \in V' \to \| f \|_{V'} =< A^{-1}f, f >^{\frac{1}{2}}$ is a norm on V' that is equivalent to the dual norm[3].

We define the functional

$$J(v) = \frac{1}{2} \| F(v) \|_{V'}^2,$$

Then problem (5) is equivalent to:

$$\begin{cases} \text{Find} \quad u \in V \quad \text{such that} \\ J(u) = \inf_{v \in V} J(v). \end{cases} \tag{6}$$

Lemma 2. *Let u^* be a nonsingular solution of Navier-Stokes problem (1). Then the functional*

$$J(v) = \frac{1}{2} < A^{-1}F(v), F(v) >$$

is strictly convex in a neighborhood of u^* and weakly lower semicontinuous. This means that there exist two constants, $\rho > 0$ and $\alpha > 0$, such that

$$D^2 J(v) \cdot (w, w) \geq \alpha \| w \|_1^2, \quad \forall v \in S(u^*; \rho), \forall w \in V. \tag{7}$$

Here

$$S(u^*; \rho) = \{v \in V; \| v - u^* \|_1 \leq \rho\}.$$

Proof. The strict convexity of $J(v)$ in a neighborhood of a nonsingular solution u^* can be found in [3].

Now, we will prove the weak lower semicontinuity. To do this, let $\{v_i\}$ be a weakly convergent sequence in V. Assume that $v_i \to v_*$ weakly in V. We then have

$$\lim_{i \to \infty} < F(v_i), v >=< F(v_*), v > \quad \forall v \in V.$$

(see chapter 9 in [4]. Let $A^{-1}F(v_*) = g_*, A^{-1}F(v_i) = g_i$, i.e.

$$Ag_* = F(v_*), Ag_i = F(v_i),$$
$$< A(g_* - g_i), v >=< F(v_*) - F(v_i), v >= a_0(g_* - g_i, v).$$

Hence

$$\lim_{i \to \infty} a_0(g_* - g_i, v) = \lim_{i \to \infty} < F(v_*) - F(v_*), v > - < F(v_*) - F(v_*), v >= 0 \quad \forall v \in V.$$

This implies $g_i \to g_*$ weakly in V.

Also, with $Ag = F(v)$

$$J(v) =< A^{-1}F(v), F(v) >=< g, Ag >= a_0(g, g),$$

since

$$a_0(g_i - g_*, g_i - g_*) \geq 0, \quad a_0(g_i, g_i) \geq 2a_0(g_i, g_*) - a_0(g_*, g_*),$$

$$\lim_{i \to \infty} \inf a_0(g_i, g_i) \geq a_0(g_*, g_*).$$

It follows that

$$\lim_{i \to \infty} \inf J(v_i) \geq J(v).$$

$J(v_i)$ is weakly lower semicontinuous. $\qquad \Box$

Let u^0 be an initial value in $S(u^*, \rho)$, and let $R = J(u^0)$. It is clear that $D = \{v \in V; J(v) \leq R\} \cap S(u^*; \rho)$ is not empty because of $u^0 \in S(u^*, \rho)$, and functional $J(v)$ is strictly convex in D. Its derivative $DJ(v)$ is uniformly continuous in $D^{[3]}$. Therefore there exist constants $\alpha, C > 0$ such that (see (7))

$$J(v) - J(u) - < DJ(u), v - u > \geq \alpha \parallel v - u \parallel_1^2, \quad \forall u, v \in D \qquad (8),$$

and

$$\parallel DJ(v) - DJ(u) \parallel_{V'} \leq C \parallel v - u \parallel_1 \quad \forall u, v \in D. \qquad (9)$$

Now, we consider a local minimization problem:

$$\begin{cases} \text{Find} \quad u \in D \quad \text{such that} \\ \\ J(u) = \inf_{v \in D} J(v) \end{cases} \qquad (10)$$

Theorem 1. *There exists a unique solution u^* of (10) which satisfies (6).*

Proof By using lemma 2, $J(v)$ is weakly lower semicontinuous in V and D is a closed subset in $V, J(v) \geq 0$. Then the generalized Weierstrass theorem shows that (10) has a unique solution u^*. $\qquad \Box$

Clearly, the domain decomposition algorithm for (10) consists of

$$(P_n) \begin{cases} \text{Find} \quad u_j^{n+1} \in \{u^n + V_j\} \cap D \quad \text{such that} \\ a_0(u_j^{n+1}, v)_{\Omega_j} + a_1(u_j^{n+1}; u_j^{n+1}, v)_{\Omega_j} = < f, v >_{\Omega_j} \quad \forall v \in V_j. \end{cases}$$

(P_n) is equivalent to

$$\begin{cases} v_j^{n+1} = \underset{\substack{v \in V_j \\ u^n + v \in D}}{\text{arginf}} \quad J(u^n + v) \quad \text{in} \ \Omega_j \\ v_j^{n+1} = 0, & \text{in} \quad \Omega \backslash \Omega_j \\ u_j^{n+1} = u^n + v_j^{n+1} & \text{in} \quad \Omega. \end{cases} \qquad (11)$$

Similarly, (11) has a unique solution sequence $\{u_j^n\}$.

Theorem 2. *Assume that $\{u^*, p^*\}$ is an isolated solution of the N-S Eqs. Then the sequence $\{u^n, p^n\}$, defined by (11) and (4), converges strongly in $X \times M$ to $\{u^*, p^*\}$.*
Proof It follows from (11) that

$$J(u_j^{n+1}) \leq J(u^n), \quad u_j^{n+1} \in D \quad (n \geq 0, j = 1, 2, \cdots, m).$$

Since $J(v)$ is convex in D, we have

$$J(u^{n+1}) = J(\sum_{j=1}^{m}\theta_j u_j^{n+1}) \leq \sum_{j=1}^{m}\theta_j J(u_j^{n+1}) \leq J(u^n), \quad (n \geq 0). \tag{12}$$

Therefore there exists a $q \in R^1$ such that $J(u^n) \to q$, as $n \to +\infty$. Thus, $\{u^n\} \subset D$. Furthermore, (11) yields

$$\begin{cases} < DJ(u_j^{n+1}), v_j >= 0 \quad \forall v_j \in V_j \\ u_j^{n+1} - u^n \in V_j \quad (j = 1, 2, \cdots, m; n \geq 0). \end{cases} \tag{13}$$

Combining (11) with (7), we deduce

$$J(u^n) - J(u_j^{n+1}) \geq \alpha \parallel u^n - u_j^{n+1} \parallel_1^2 .$$

For $\theta_j \in (0, 1)$ satisfying $\sum_{j=1}^{m}\theta_j = 1$, we have

$$\alpha \sum_{j=1}^{m}\theta_j \parallel u^n - u_j^{n+1} \parallel_1^2 \leq J(u^n) - \sum_{j=1}^{m}\theta_j J(u_j^{n+1})$$

$$\leq J(u^n) - J(u^{n+1}) \to 0, \text{as} \quad n \to +\infty.$$

Therefore,

$$\parallel u^n - u_j^{n+1} \parallel_1 \to 0 \quad (j = 1, 2, \cdots, m), \quad \text{as} \quad n \to +\infty.$$

By using Lemma 1, for each $v \in V$, there exist $v_j \in V_j$ such that

$$v = \sum_{j=1}^{m} v_j \quad \text{and} \quad \max_{1 \leq j \leq m} \parallel v_j \parallel_1 \leq C_0 \parallel v \parallel_1 .$$

Combining this with (7) and (8), we deduce

$$| < DJ(u^n), v > | = |\sum_{j=1}^{m} < DJ(u^n), v_j > | = |\sum_{j=1}^{m} < DJ(u^n) - DJ(u_j^{n+1}), v_j > |$$

$$\leq C_0 \sum_{j=1}^{m} \parallel DJ(u^n) - DJ(u_j^{n+1}) \parallel_{V'} \parallel v \parallel_1 \leq C_0 C \sum_{j=1}^{m} \parallel u^n - u_j^{n+1} \parallel_1 \parallel v \parallel_1 .$$

Hence

$$\parallel DJ(u^n) \parallel_{V'} \leq C_0 C \sum_{j=1}^{m} \parallel u^n - u_j^{n+1} \parallel_1 \to 0, \quad \text{as} \quad n \to +\infty.$$

Since u^* is a solution of problem (6), we find

$$J(u^*) \leq J(u^n) \quad (n \geq 0).$$

Using again (7), we get

$$\alpha \parallel u^n - u^* \parallel_1^2 \leq J(u^*) - J(u^n) + < DJ(u^n), u^n - u^* > .$$

$$\leq\ <DJ(u^n), u^n - u^* >\ \leq\ \|DJ(u^n)\|_{V'}\|u^n - u^*\|_1 .$$

Therefore

$$\|u^n - u^*\|_1 \leq \frac{1}{\alpha}\|DJ(u^n)\|_{V'} \to 0, \quad \text{as} \quad n \to +\infty.$$

Thus, the sequence $\{u^n\}$ converges strongly in V to the isolated solution u^* of problem (6).

Next, we consider the convergence of the pressure sequence $\{p^n\}$.

Let the pair $\{u^*, p^*\}$ be a solution of Problem (Q). We know that

$$\|u^n - u^*\|_1 \to 0, \|u_j^{n+1} - u^*\|_1 \to 0, \quad \text{as} \quad n \to +\infty (j = 1, 2, \cdots, m).$$

Moreover, using (Q), we have for any v_j in $[H_0^1(\Omega_j)]^N$ with $v_j = 0$ in $\Omega\backslash\Omega_j$:

$$b(p_j^{n+1} - p, v_j) = a_0(u_j^{n+1} - u^*, v_j) + a_1(u_j^{n+1}; u_j^{n+1}, v_j)$$

and

$$-a_1(u^*; u^*, v_j) \to 0^{[3]}, \quad \text{as} \quad n \to +\infty.$$

Similarly, if $\Omega_i \cap \Omega_j \neq \Phi$, then we have for any v_{ij} in $[C_0^\infty(\Omega_i \cap \Omega_j)]^N$ with $v_{ij} = 0$ in $\Omega\backslash(\Omega_i \cap \Omega_j)$

$$b(p_i^{n+1} - p_j^{n+1}, v_{ij}) \to 0, \quad \text{as} \quad n \to +\infty.$$

Since $C_0^{+\infty}(\Omega_i \cap \Omega_j)$ is dense in $L^2(\Omega_i \cap \Omega_j)$, we get

$$b(p_i^{n+1} - p_j^{n+1}, v)_{\Omega_j \cap \Omega_i} \to 0, \forall v \in [H^1(\Omega_i \cap \Omega_j)]^N, \text{as} \quad n \to +\infty.$$

Hence, we get for all $v_j \in [H_0^1(\Omega_j)]^N$, with $v_j = 0$ in $\Omega\backslash\Omega_j$:

$$b(p^{n+1} - p^*, v_j)_{\Omega_j} = b(p_j^{n+1} - p^*; v_j)_{\Omega_j} + b(\frac{1}{r(x)}\sum_{k=1}^M p_k^{n+1} - p_j^{n+1}, v_j)_{\Omega_j}$$

$$= b(p_j^{n+1} - p^*, v_j)_{\Omega_j} + \frac{1}{r(x)}\sum_{k=1}^m b(p_k^{n+1} - p_j^{n+1}, v_j)_{\Omega_j \cap \Omega_k} \to 0, \quad \text{as} \quad n \to +\infty.$$

Set $v_j = \xi_j v$, for all $v \in [H_0^1(\Omega)]^N$. We assert that

$$b(p^{n+1} - p^*, v) = \sum_{j=1}^m b(p^{n+1} - p^*, v_j)_{\Omega_j} \to 0, \quad \text{as} \quad n \to +\infty,$$

i.e. $\forall v \in [H_0^1(\Omega)]^N$, $(p^{n+1} - p^*, div v) \to 0$ as $n \to +\infty$. Since div maps $[H_0^1(\Omega)]^N$ onto $L_0^2(\Omega)([3])$, we obtain

$$p^n \to p^* \quad \text{weakly in } M \quad \text{as } n \to +\infty.$$

The inf-sup condition [3] yields

$$p^n \to p^* \quad \text{strongly in} \quad M \quad \text{as } n \to +\infty.$$

Hence, Theorem 2 is valid. $\qquad\qquad\qquad\qquad\qquad\qquad\qquad\qquad\qquad\qquad\qquad\qquad\qquad\square$

REFERENCES

[1] Li Cuihua & Li Kaitai (1995) *Domain-decomposition Method for Navier-Stokes Equations* (in Chinese), J. Xian Jiaotong Univ., 6(29).

[2] Temam R. (1984) *Navier-Stokes Equations, Theory and Numerical Analysis.* North-Holland, Amsterdam, New York.

[3] Girault V. and Raviart P. A. (1985) *Finite Element Method of the Navier-Stokes Equations*, Springer-Verlag.

[4] Li Kaitai, Ma Yichen (1992) *Hilbert Space Method for Mathematical Physical Equations* (in Chinese), Xian Jiaotong University Press.

Domain Decomposition Finite Volume Method for Three-dimensional Inviscid Flow Calculations

Tiejun Nie and Jianhu Feng

1 Introduction

For a fighter aircraft, the flow quality at the inlet can have a large impact on the performance of the propulsion system. It is obvious that the flow quality depends not only on the shape of the inlet, but also on the shape of the forebody and the inlet mounting position. So the integrated solution of internal and external flows is required.

Solutions of the Euler equations can give a more physical representation of inviscid subsonic, transonic and supersonic flowfields than potential flow methods. Some important phenomena such as the vortex in a S-shaped inlet, separation induced by shock, can be nicely modeled with the use of Euler equations. Furthermore, solutions of Euler equations are considered as a stepping stone before the solution of time averaged Navier-Stokes equations can be attempted.

However, there are many difficulties to solve the 3-D Euler equations for complex geometries, such as grid generation, computer memory size, speed limitation and so on. This paper presents an Euler code designed for integrated combinations. The code is based on a finite volume Runge-Kutta method [1, 2]. In order to relieve the computer memory requirement, a domain decomposition technique is employed. This also makes the mesh generation much easier. Iterations are performed for each zone in turn. A newly-designed model of a fighter forebody-inlet combination and a missile model have been used to validate the code. Satisfactory results have been obtained and the solution cost is affordable.

---

[1] Department of Applied Mathematics Northwestern Polytechnical University, Xi'an, Shaanxi, 710072, P.R.China

Domain Decomposition Methods in Sciences and Engineering, edited by R. Glowinski *et al.*

2   Numerical Algorithm

The finite volume scheme is derived from the integral form of the Euler equations

$$\frac{\partial}{\partial t} \int_{\Omega} \vec{W} \, d\Omega + \oint_{\partial \Omega} \vec{F} \cdot \vec{n} \, ds = 0 \tag{1}$$

where Ω denotes a fixed region with boundary $\partial\Omega$ and outward normal $\vec{n}$, $\vec{W}$ represents the vector of conserved quantities, and $\vec{F}$ is the corresponding flux tensor,

$$\vec{W} = \begin{bmatrix} \rho \\ \rho u \\ \rho v \\ \rho w \\ e \end{bmatrix}, \quad \vec{F} = \begin{bmatrix} \rho u \vec{i} + \rho v \vec{j} + \rho w \vec{k} \\ (\rho u u + p)\vec{i} + \rho u v \vec{j} + \rho u w \vec{k} \\ \rho v u \vec{i} + (\rho v v + p)\vec{j} + \rho v w \vec{k} \\ \rho w u \vec{i} + \rho w v \vec{j} + (\rho w w + p)\vec{k} \\ (e + p)u\vec{i} + (e + p)v\vec{j} + (e + p)w\vec{k} \end{bmatrix} \tag{2}$$

where ρ and p are the fluid density and pressure, u, v, w are the Cartesian velocity components, $\vec{i}, \vec{j}, \vec{k}$ denote the unit vectors of the Cartesian coordinate system, and e, the total energy linked with other variables by the equation of state

$$p = (\gamma - 1)(e - \frac{1}{2}\rho(u^2 + v^2 + w^2)) \tag{3}$$

γ is the ratio of specific heats.

The computational domain Ω is divided into hexahedral cells. Values of the dependent variable $\vec{W}$ are defined at cell centers. A discrete approximation to the spatial terms yields

$$\frac{d}{dt}(V_{i,j,k} \cdot \vec{W}_{i,j,k}) + \vec{Q}_{i,j,k} = 0 \tag{4}$$

where $V_{i,j,k}$ is the (i, j, k) cell volume. This equation represents the discrete flux balance. The discrete flux term $\vec{Q}_{i,j,k}$ represents the net flux out of the cell,

$$\vec{Q}_{i,j,k} = \sum_{l=1}^{6} (\vec{F} \cdot \vec{S})_l \tag{5}$$

where $\vec{S}_l(l = 1, 2, \cdots, 6)$ denotes the lth surface vector of the (i, j, k) cell, and the flux $\vec{F}_l$ are evaluated at the lth surface of the (i, j, k) cell. It is usually calculated using the averages of the quantities $\vec{W}_{i,j,k}$ from adjacent cell centers.

The present finite volume spatial discretization reduces to a central-difference scheme which is formally of second-order accuracy for smooth grids. In order to suppress its well-known tendency for odd-even points decoupling, to capture shocks and to minimize pre- and post-shock oscillations, an adaptive dissipation term $\vec{D}_{i,j,k}$ is added to the system

$$\frac{d}{dt}(V_{i,j,k} \cdot \vec{W}_{i,j,k}) + \vec{Q}_{i,j,k} - \vec{D}_{i,j,k} = 0' \tag{6}$$

This dissipation formulation uses blended second and forth differences in each of the three parametric directions [1].

To integrate the system (6) of ordinary differential equations, the present code uses Runge-Kutta time stepping schemes [1, 2]. Several convergence acceleration techniques like local time stepping, enthalpy damping and implicit residual smoothing are adopted [2].

3 Grid System and Flow Solution Coupling

In the present approach, several computational zones are used to model integrated flowfield. For example, zone (I) is the subdomain forward of the inlet highlight; zone (II) is the subdomain of the internal inlet and diffuser; zone (III), the external region aft of the highlight. The computational grid system of each zone is generated independently by means of TTM [3].

The flow solutions are execute on each zone in turn. Separate solutions are coupled by interpolating the fluxes across the interface between adjacent zones during the in two steps. First, for each zonal boundary cell, a ghost point is introduced in the neighboring zone. The ghost point is located in the plane formed by the neighboring zone boundary cell centers but has the same y, z values as the present zonal boundary cell centers. The flow values at ghost points are determined by interpolation using the values of neighboring zone. The second step evaluates the fluxes across the interface by interpolation using values of the point of the present cell and its ghost point.

The treatment of other boundary conditions is the same as that in the literature [4].

4 Flow Solution Algorithm

The flow computations are initiated by assuming uniform in coming flow quantities. If this coming flow is supersonic, then the flow field inside the inlet is initiated with a subsonic uniform flow assuming there is a normal shock located right at inlet entrance. After that several iterations have been performed for one subdomain, the flow values are updated and the flow solution is stored in a data file. Then the computation shifts to another subdomain. A sequence of iterative updates for all subdomains is referred to as a 'cycle'. Usually hundreds of cycles are required to achieve a satisfactory, converged solution.

5 Numerical Results

The first example is a forebody-inlet combination model. The flight conditions are: $M_\infty = 2.047$, angle of attack $\alpha = 0^0$, and the engine mass ratio $\bar{m} = 0.94$. The S-shaped inlet is a rectangle at the entrance and changes into circle at the exit. Even though the grid is very coarse ($13 \times 17 \times 15$ for zone (I), $9 \times 23 \times 15$ for zone (II), and $43 \times 19 \times 10$ for zone (III)), the result agrees with the design requirement.

The second example is a total missile-inlet combination model. The computational domain is divided into eight subdomains. The total grid number is 38495. The following

flight conditions have been modeled: 0 $M_\infty = 2.0, \alpha = 0^0, \bar{m} = 0.98$
$M_\infty = 2.0, \alpha = 1^0, \bar{m} = 0.98$
$M_\infty = 2.0, \alpha = 2^0, \bar{m} = 0.98$
This result is also very satisfactory.

REFERENCES

[1] Jameson A. (1985) Numerical Solution of the Euler Equations for Compressible Inviscid Fluids. in*Numerical Methods for the Euler Equations of fluid Dynamics*, SIAM, 199-245.

[2] Turkel E. (1984) Acceleration to a Steady State for the Euler Equations, ICASE Report No. 84-32.

[3] Thompson J. F., Thames F. C. and Mastin C. S. (1974) Boundary-fitted Curvilinear Coordinate Systems for Solution of Partial Differential Equations Containing any Number of Arbitrary Two Dimensional Bodies, *J. Comp. Phys.*, Vol. 15: 299-319.

[4] Feng Jianhu (1994) *On Numerical Simulation of Three-Dimensional Aircraft Inlet Systems*, Ph. D. Dissertation, Northwestern Polytechnical University.

Index

Index compiled by Geoffrey C. Jones